High-Field EPR Spectroscopy on Proteins and their Model Systems
Characterization of Transient Paramagnetic States

High-Field EPR Spectroscopy on Proteins and their Model Systems
Characterization of Transient Paramagnetic States

Klaus Möbius and Anton Savitsky
Department of Physics, Free University Berlin, Germany

RSCPublishing

ISBN: 978-0-85404-368-2

A catalogue record for this book is available from the British Library

© 2009 Klaus Möbius and Anton Savitsky

All rights reserved

Apart from fair dealing for the purposes of research for non-commercial purposes or for private study, criticism or review, as permitted under the Copyright, Designs and Patents Act 1988 and the Copyright and Related Rights Regulations 2003, this publication may not be reproduced, stored or transmitted, in any form or by any means, without the prior permission in writing of The Royal Society of Chemistry or the copyright owner, or in the case of reproduction in accordance with the terms of licences issued by the Copyright Licensing Agency in the UK, or in accordance with the terms of the licences issued by the appropriate Reproduction Rights Organization outside the UK. Enquiries concerning reproduction outside the terms stated here should be sent to The Royal Society of Chemistry at the address printed on this page.

Published by The Royal Society of Chemistry,
Thomas Graham House, Science Park, Milton Road,
Cambridge CB4 0WF, UK

Registered Charity Number 207890

For further information see our web site at www.rsc.org

Preface

This book is devoted to high-field/high-frequency EPR spectroscopy in biology and (bio)chemistry. In the last 15 years, this segment of magnetic resonance has grown rapidly and has led, together with advanced pulse EPR, to the currently observed renaissance of EPR spectroscopy. Application of higher and higher magnetic fields and microwave frequencies results in spectacular improvements of spectral and time resolution. Thereby, more complex spin systems can be studied by EPR, very much in analogy to what has happened in modern NMR spectroscopy more than a decade ago. The reasons for this time lag are obvious: EPR has to cope with resonance frequencies three orders of magnitude larger than in NMR and correspondingly shorter electron-spin relaxation times. The resulting requirements in adequate microwave technology were challenging and needed their time to be solved. It was only after the end of the Cold War at the early 1990s when low-noise mm microwave sources, fast switches and detectors became available for unclassified research. These devices deliver and process microwave power in the range of several tens of milliwatts necessary for fast pulsed high-field EPR at frequencies up to several 100 GHz. Nowadays, we witness multifrequency and multidimensional EPR benchmark experiments at the technical limits of quasioptical sub-mm microwave bridges and sweepable wide-bore cryomagnets.

The unique potential of high-field/high-frequency EPR spectroscopy for studying complex spin systems was recognized by a few groups already more than 30 years ago, but Yakov S. Lebedev (Moscow) was the first to start a dedicated high-field/high-frequency research program. In 1976, he completed a versatile *continuous-wave* (cw) high-field EPR spectrometer employing 2 mm microwaves and a 5 T superconducting magnet. The history of *pulsed* high-field EPR starts with Jan Schmidt (Leiden) who built the first 3 mm pulse spectrometer in 1989. Since the early days of high-field EPR, a relatively small number of laboratories in physics and physical chemistry have succeeded in building

---

High-Field EPR Spectroscopy on Proteins and their Model Systems: Characterization of Transient Paramagnetic States
By Klaus Möbius and Anton Savitsky
© 2009 Klaus Möbius and Anton Savitsky
Published by the Royal Society of Chemistry, www.rsc.org

their own cw or pulsed high-field EPR spectrometers at different microwave frequencies, ranging from 95 to about 400 GHz, and constructed powerful instrumentation to perform ENDOR (electron–nuclear double resonance), ESEEM (electron spin-echo envelope modulation) or PELDOR (pulsed electron–electron double resonance) experiments. In the early 1990s a broader application of high-field EPR methodology to fundamental and applied research projects became feasible for many other research groups with the introduction of commercial high-field EPR spectrometers or microwave components thereof. Here, BRUKER Analytik (Karlsruhe) took the lead with the 94 GHz ELEXSYS E600/680 spectrometers. In this book, in addition to the technical details of advanced EPR, also a chronological account is given to trace the important steps of the development of high-field EPR.

No exhaustive coverage of the rapidly expanding field of high-field/high-frequency EPR spectroscopy has been attempted in the book. We have rather decided to include, in addition to examples from our own laboratory, a representative cross section of EPR activities of dedicated laboratories around the world. The chosen examples profit from distinct advantages of high-field/high-frequency EPR in relation to conventional X-band EPR, as there are: enhanced spectral resolution and orientational selectivity even for disordered samples of molecules with small g-tensor anisotropy; enhanced separation of radicals with only slightly different g-values; enhanced time resolution for tracing transient radical intermediates; enhanced detection sensitivity for small samples with low spin concentration; increased sensitivity towards molecular motion and concomitant relaxation effects by transforming motionally narrowed EPR spectra into the slow-motion regime; increased selectivity for reaction-induced conformational changes of disordered radicals with small g-tensor anisotropy by ENDOR or PELDOR at high Zeeman fields.

The book combines recent instrumental developments of high-field EPR and its extensions with typical applications in molecular biology and photochemistry. Illustrative examples of organic radicals are chosen as well as topical proteins and their model systems. The book also includes state-of-the-art quantum-mechanical approaches to calculate the g-, hyperfine-, and quadrupole tensor components with an accuracy that is adequate to the precision of the measured parameters of the spin Hamiltonian. We hope that the balanced mixture of introductory and specialized sections of the book will be interesting for both experts and newcomers in this fascinating field of active research and instrumental development. Such a mixture is anticipated to help to reach a wider audience of undergraduate and graduate students as well as postdoctoral and senior scientists in their work in the laboratory.

We want to express our gratitude to the Royal Society of Chemistry and the editorial administration staff for their support to get this book published in due time. We are especially thankful for their tolerance when postponing several deadlines set for submitting the manuscript but that we were unable to meet.

That said, we should like to thank also our wives, Uta and Tanya, for their "tea and sympathy" contributions, understanding our frequent bad mood and occasional despair in view of the still unwritten chapters. Also, for encouraging us to finish "the book" and to get back to a situation with a better balance of the many important things to do.

Klaus Möbius and Anton Savitsky

Contents

Summary   xiii

Acknowledgements   xv

Chapter 1   Introduction

    1.1   Why EPR at High Magnetic Fields?   3
    1.2   NMR *versus* EPR   7
    1.3   From Basic to Advanced Multifrequency EPR, a Chronological Account   9
    References   17

Chapter 2   Principles and Illustrative Examples of High-Field/High-Frequency EPR

    2.1   Spin Hamiltonians and EPR Experiments at High Magnetic Fields   23
         2.1.1   Organic Radicals and Low-Spin Transition-Metal Ions ($S = 1/2$)   23
         2.1.2   Triplet States and High-Spin Transition-Metal Ions ($S > 1/2$)   33
    2.2   High-Field EPR, ENDOR, TRIPLE, ESEEM, PELDOR and RIDME   35
         2.2.1   ENDOR and TRIPLE Hyperfine Spectroscopy   35
              2.2.1.1   Liquid-Solution Steady-State ENDOR and TRIPLE   37
              2.2.1.2   Liquid-Solution Transient EPR and ENDOR in Photochemistry   53

High-Field EPR Spectroscopy on Proteins and their Model Systems: Characterization of Transient Paramagnetic States
By Klaus Möbius and Anton Savitsky
© 2009 Klaus Möbius and Anton Savitsky
Published by the Royal Society of Chemistry, www.rsc.org

		2.2.1.3	Solid-State Pulse ENDOR and TRIPLE	73
		2.2.1.4	ESEEM Hyperfine Spectroscopy	79
	2.2.2	Electron–Electron Dipolar Spectroscopy		93
		2.2.2.1	PELDOR	97
		2.2.2.2	RIDME	100
		2.2.2.3	High-Field RIDME and PELDOR on Nitroxide Radical Pairs	102
References				113

Chapter 3 Instrumentation

3.1	Experimental Techniques			125
	3.1.1	Continuous-Wave EPR (cw EPR)		125
	3.1.2	Time-Resolved EPR (TREPR)		126
	3.1.3	Pulse EPR		129
3.2	Historical Overview of High-Field/High-Frequency EPR Spectrometers			131
	3.2.1	First Generation		133
	3.2.2	Second Generation		134
3.3	Technical Aspects of High-Field/High-Frequency EPR			137
	3.3.1	Sensitivity Considerations		137
	3.3.2	Detection Schemes		138
	3.3.3	Microwave Sources		143
	3.3.4	Resonators		145
		3.3.4.1	Single-Mode Cavities	146
		3.3.4.2	Fabry–Perot Resonators	148
		3.3.4.3	Loop-Gap Resonators	152
		3.3.4.4	Dielectric Resonators	153
		3.3.4.5	Nonresonant Systems	154
	3.3.5	Microwave Transmission Lines		156
	3.3.6	Magnet Systems		158
3.4	High-Field Multipurpose Spectrometers Built at FU Berlin			160
	3.4.1	The 95-GHz Spectrometer		160
		3.4.1.1	Microwave Bridge Design	162
		3.4.1.2	Magnet and Cryostat	163
		3.4.1.3	Probeheads	164
		3.4.1.4	EPR and ENDOR Performance	169
		3.4.1.5	Field-Jump PELDOR	170
		3.4.1.6	Dual-Frequency PELDOR	171
	3.4.2	The 360-GHz Spectrometer		172
		3.4.2.1	Quasioptical Microwave Propagation	172
		3.4.2.2	Microwave-Bridge Design	174

Contents   xi

		3.4.2.3	Quasioptical Components	176
		3.4.2.4	Induction-Mode Operation	176
		3.4.2.5	Magnet and Cryostat	178
		3.4.2.6	Probeheads	178
		3.4.2.7	ENDOR	179
		3.4.2.8	Transient EPR Bridge with Reference Arm	180
		3.4.2.9	Pulsed Orotron Source	182
	References			186

Chapter 4 Computational Methods for Data Interpretation

References   203

Chapter 5 Applications of High-Field EPR on Selected Proteins and their Model Systems

5.1	Introduction			206
5.2	Nonoxygenic Photosynthesis			213
	5.2.1	Multifrequency EPR on Bacterial Photosynthetic Reaction Centers (RCs)		217
		5.2.1.1	X-Band EPR and ENDOR Experiments	219
		5.2.1.2	95-GHz EPR on Primary Donor Cations $P^{\bullet+}$ in Single-Crystal RCs	223
		5.2.1.3	360-GHz EPR on Primary Donor Cations $P^{\bullet+}$ in Mutant RCs	228
		5.2.1.4	Results of g-tensor Computations of $P^{\bullet+}$	232
		5.2.1.5	95-GHz EPR and ENDOR on the Acceptors $Q_A^{\bullet-}$ and $Q_B^{\bullet-}$	234
		5.2.1.6	95-GHz ESE-Detected EPR on the Spin-Correlated Radical Pair $P^{\bullet+}Q_A^{\bullet-}$	248
		5.2.1.7	95-GHz RIDME and PELDOR on the Spin-Correlated Radical Pair $P^{\bullet+}Q_A^{\bullet-}$	250
		5.2.1.8	Multifrequency EPR on Primary Donor Triplet States in RCs	268
	5.2.2	Multifrequency EPR on Bacteriorhodopsin (BR)		272
		5.2.2.1	Site-Directed Nitroxide Spin Labelling	274
		5.2.2.2	Hydrophobic Barrier of the BR Proton-Transfer Channel	275

		5.2.2.3	Modelling of Solute–Solvent Interactions	278
		5.2.2.4	Conformational changes during the BR photocycle	281
	5.3	Oxygenic Photosynthesis		283
		5.3.1	Multifrequency EPR on Doublet States in Photosystem I (PS I)	284
		5.3.2	Multifrequency EPR on Doublet States in Photosystem II (PS II)	288
	5.4	Photoinduced Electron Transfer in Biomimetic Donor–Acceptor Model Systems		290
		5.4.1	Introduction	290
		5.4.2	Covalently Linked Porphyrin–Quinone Dyad and Triad Model Systems	294
		5.4.3	Base-Paired Porphyrin–Quinone and Porphyrin–Dinitrobenzene Complexes	308
	5.5	DNA Repair Photolyases		314
		5.5.1	Introduction	314
		5.5.2	High-Field EPR and ENDOR Experiments	319
	5.6	Colicin A Bacterial Toxin		323
		5.6.1	Introduction	323
		5.6.2	Models of Transmembrane Ion-Channel Formation	326
		5.6.3	95-GHz EPR Studies of Membrane Insertion	327
	References			330

Chapter 6 Conclusions and Perspectives

    References   364

Subject Index   366

Summary

A hot topic in molecular biology, particularly in the emerging field of proteomics is the attempt to understand, on the basis of structural and dynamics data, the dominant factors that control the specificity and efficiency of electron- and ion-transfer processes in proteins. To this end, a multitude of tasks remains to be accomplished before definite solutions will be found and, hence, many experimental techniques are being used to offer method-specific answers to inherently difficult questions. Among the spectroscopic techniques, modern EPR, operated at high magnetic fields and microwave frequencies and extended by multiple-resonance capabilities, is playing an important role in this endeavor, particularly in view of the fact that for high spectral resolution no single-crystal protein preparations are needed to obtain detailed structural information.

During the last decade, the combined efforts of biologists, chemists and physicists in developing high-field EPR techniques and applying them to functional proteins demonstrated that this type of spectroscopy is particularly powerful for characterizing the structure and dynamics of transient states of proteins in action on biologically relevant time scales from nanoseconds to hours.

The book describes how high-field EPR methodology, in conjunction with mutation strategies for site-specific isotope or spin labelling and with the support of modern quantum-chemical computation methods for data interpretation, is capable of providing new insights into biological processes. This is achieved by characterizing structure and dynamics of the reaction partners at the molecular level and by revealing their electronic properties in great detail. Specifically, we discuss the theoretical and instrumental background of continuous-wave and pulse high-field EPR and its multiple-resonance extensions ENDOR, TRIPLE and PELDOR as well as high-field RIDME and ESEEM. Some emphasis is put on a balanced description of both historical spadework and present-day performance of advanced EPR at 95 and 360 GHz. This culminates in a coherent treatment of state-of-the-art research of high-field EPR in terms of both instrumentation development and application to representative protein complexes and their biomimetic model systems. Specific aspects of structure–dynamics–function relations are highlighted that have been revealed by combining high-field EPR with genetic engineering techniques to study site-specific mutants, for example, nitroxide spin-labelled membrane proteins. The spectrally resolved tensor components of the nitroxide Zeeman, hyperfine and quadrupole interactions allow the polarity and proticity properties of binding sites of the protein matrix, which are relevant for tuning the biological function, to be probed. The importance of DFT-based quantum-chemical interpretation of the experimental data (g-, zero-field-, hyperfine-, and quadrupole-tensors) is emphasized. The information obtained is complementary to that of protein crystallography, solid-state NMR and laser spectroscopy.

The book focuses on reviewing recent 95- and 360-GHz high-field studies (EPR, ENDOR, TRIPLE, ESEEM, RIDME, PELDOR) of the following protein systems: (*i*) Light-induced electron-transfer intermediates in wild-type and mutant reaction centers from photosynthetic organisms as well as in biomimetic donor–acceptor model systems. (*ii*) Light-induced proton-transfer intermediates of site-specifically nitroxide spin-labelled mutants of bacteriorhodopsin. (*iii*) Light-generated flavin radical intermediates in DNA photolyase enzymes, which utilize long-wavelength light to repair UV-induced DNA damages. (*iv*) Refolding intermediates of site-specifically nitroxide spin-labelled mutants of the colicin A bacterial toxin, which initiate the formation of lethal transmembrane ion channels. The book concludes with a summary of specific advantages of high-field EPR pertinent to biomolecular research. In an outlook, some recent accomplishments and future challenges of this type of magnetic resonance spectroscopy are sketched to indicate where modern EPR and NMR meet again after fifty years of alienation.

Acknowledgements

First, we want to acknowledge the merits of those colleagues and friends who, for many years, had a strong impact on our high-field EPR work at FU Berlin: Yakov Lebedev (1935–1996, *Moscow*) and Arnold Hoff (1939–2002, *Leiden*), Pier Luigi Nordio (1936–1998, *Padova*) and Mel Klein (1921–2000, *Berkeley*), Arthur Schweiger (1946–2006, *Zurich*) and Dietmar Stehlik (1939–2007, *Berlin*), Alexander Dubinskii (*Moscow*), Kev Salikhov (*Kazan*) and Yuri Grishin (*Novosibirsk*), Jack Freed (*Cornell*), George Feher (*San Diego*) and Giovanni Giacometti (*Padova*), Harry Kurreck (*Berlin*) and Haim Levanon (*Jerusalem*), Daniella Goldfarb (*Rehovot*) and Edgar Groenen (*Leiden*), Peter Dinse (*Darmstadt*) and Jan Schmidt (*Leiden*), Heinz-Jürgen Steinhoff (*Osnabrück*) and Seigo Yamauchi (*Sendai*), Martin Plato (*Berlin*) and Wolfgang Lubitz (*Mülheim/Ruhr*). To all of them we want to express our gratitude, and our thoughts turn often also to those who sadly are no longer with us.

Most of our high-field EPR applications to biological systems have been, or still are being, performed in cooperation with numerous groups from biology, chemistry and physics around the world. In addition to the colleagues mentioned above, we want to thank (in alphabetical order) Adelbert Bacher (*Munich*), Robert Bittl (*Berlin*), Sergei Dzuba (*Novosibirsk*), Martin Engelhard (*Dortmund*), Jack Fajer (*Brookhaven*), Marco Flores and Maurice van Gastel (*Mülheim/Ruhr*), Martina Huber (*Leiden*), Alexey Osintsev (*Kemerovo*), Thomas Prisner (*Frankfurt/M*), Günther Rist (*Basel*), Renad Sagdeev (*Novosibirsk*), Hugo Scheer (*Munich*), Alexey Semenov (*Moscow*), Jonathan Sessler (*Austin*), Alexander Tikhonov (*Moscow*), Yuri Tsvetkov (*Novosibirsk*), Giovanni Venturoli (*Bologna*), Emanuel Vogel (*Cologne*) and Herbert Zimmermann (*Heidelberg*) for their contributions. Their individual share in the work becomes evident by the references cited throughout the book.

Over the years, many coworkers in the Möbius group – diploma and PhD students, postdocs and visiting senior scientists – have contributed to the high-field EPR work reviewed in this book. In the course of these activities it was the enthusiasm and tenacity of students and postdocs alike that allowed the numerous technical problems in developing high-field EPR, ENDOR, ESEEM and PELDOR instrumentation operating at 95 GHz and 360 GHz to be solved, and these techniques to be applied to novel molecular systems from biology and biochemistry. In this respect, we mention especially the former students (in chronological order) Olaf Burghaus, Anna Toth-Kischkat, Thomas Götzinger, Robert Klette, Martin Rohrer, Moritz Knüpling, Jens Törring, Gordon Elger, Michael Fuhs, Ingo Köhne, Andreas Bloeß, Martin Fuchs, Marcus Gallander, Alexander Schnegg, Evgenia Kirilina, Asako Okafuji; and the former postdocs (in alphabetical order), Edmund Haindl, Martina Huber, Chris Kay, Friedhelm Lendzian, Wolfgang Lubitz, Thomas Prisner, Anton Savitsky and Stefan Weber. K. M. wants to thank them all for their eminent contributions to performing exciting experiments in high-field EPR spectroscopy.

And finally, our sincere appreciation to Martin Plato (*Berlin*), Alexander Dubinskii (*Moscow*) and Andrzej Szyczewski (*Poznan*) for their generous assistance in the search activities necessary for writing this book: The thorough discussions with M. P. and A. D. on the theoretical background of high-field ESEEM and PELDOR measuring strategies, and on state-of-the-art quantum-chemical calculations for analyzing the measured parameters of the spin Hamiltonian in terms of geometric and electronic structure, were always clarifying the point and essential for forming a basis for the book. As was the help of A. S. for keeping track of the exploding amount of literature in molecular biology and biochemistry for which a really large number of publications has to be considered to be potentially relevant for EPR spectroscopy in general and high-field EPR in particular. Cordial thanks to the three of you!

Over the years, our high-field EPR activities have received sustaining support by the Deutsche Forschungsgemeinschaft (DFG) in the framework of the priority programs SFB 337, SFB 498, III P 5 – MO 132/14-1, SPP 1051 and the project MO 132/19-2, by the Volkswagenstiftung priority programs I/70 382, I/73 145, I/73 146 and by the HCM (CHRX-CT9-30 328), TMR (FMRX-CT98-0214), INTAS (01-483) and COST (P15) network programs of the European Union, which is gratefully acknowledged.

CHAPTER 1
Introduction

During the last decade, the chemistry, biology and physics communities have apparently witnessed a boost of new EPR (electron paramagnetic resonance) applications. This is largely due to technological breakthroughs in the development of pulsed microwave sources and components, sweepable cryomagnet design and fast data-acquisition instrumentation. They enable the EPR spectroscopists to introduce multiple-pulse microwave irradiation schemes, very much in analogy to what is common practice in modern NMR (nuclear magnetic resonance), and to apply advanced multifrequency high-field EPR techniques as powerful spectroscopic tools with unique potential for the elucidation of structure and dynamics of complex systems, for example membrane proteins in biological action.

This assessment is corroborated by the substantial increase of publications related to high-field/high-frequency EPR since the last 15 years (see Figure 1.1). The growing appreciation is mirrored also by the rising number of research groups in Europe, the US and Japan dedicated to the development and/or application of high-field EPR spectroscopy. This was made possible by increased financial support from national and international funding agencies. The European Union, for example, supported the Human Capital and Mobility (HCM) project "High-Field EPR: Technology and Applications" (coordinator J. Schmidt, Leiden, 1993–1996) and the EU network project "SENTINEL" ("Service Enhancement through Infrastructure Networking for Electron Paramagnetic Resonance Spectroscopy with Large Fields", coordinator M. Martinelli, Pisa, 2001–2005). Exceptionally strong support was granted by the DFG (Deutsche Forschungsgemeinschaft) through the Priority Program "High-Field EPR in Biology, Chemistry and Physics" (coordinator K. Möbius, Berlin, 1998–2004). These initiatives acted like seeding programs for the rapid development of high-field EPR spectroscopy in Europe, including Israel and

---

High-Field EPR Spectroscopy on Proteins and their Model Systems:
Characterization of Transient Paramagnetic States
By Klaus Möbius and Anton Savitsky
© 2009 Klaus Möbius and Anton Savitsky
Published by the Royal Society of Chemistry, www.rsc.org

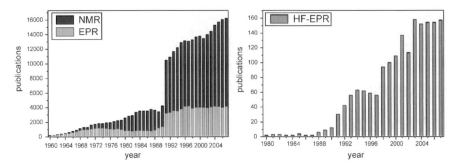

Figure 1.1 Annual distributions of NMR and EPR publications (statistics made March 2008 using proper search profiles on *ISI Web of Knowledge©, The Thomson Corporation*). The diagrams show phases of stagnation followed by phases of strong growth reflecting active research periods. In NMR the drastic increase of the number of publications on structure determinations since 1990 clearly correlates with technological breakthroughs (three-dimensional NMR spectroscopy in combination with selective isotope labelling). In EPR, at about the same years, also technological breakthroughs (pulse EPR techniques, high-field EPR) gave rise to a strong increase of EPR publications. This is particularly true for the subset of high-field (HF) EPR. We attribute this observation to the availability of commercial W-band EPR spectrometers and the increased funding of research activities on biological systems by a growing number of dedicated high-field EPR laboratories.

Russia. In Figure 1.2 the present distribution of high-field EPR groups in Europe is shown. There is a noticeable congestion of such groups in Germany, apparently as a benefit from the sustaining support by the DFG.

In the US, there is a strong representation of high-field EPR spectroscopy with about ten research groups throughout the country. Particularly renowned high-field EPR groups are concentrated in dedicated national facilities located in Ithaca, NY, at Cornell University (ACERT, the "National Biomedical Center for Advanced ESR Technology", headed by J.H. Freed,[1]) in Tallahassee, FL, (National High Magnetic Field Laboratory, the EPR group headed until recently by L.-C. Brunel), in Milwaukee, WI, at the Medical College of Wisconsin (National Biomedical EPR Center, headed by J.S. Hyde) and in Cambridge, MA, at MIT (Francis Bitter Magnet Laboratory, headed by R.G. Griffin).

In Japan, high-field EPR research is traditionally devoted to physics with the focus on novel magnetic materials. There exist about ten mm and sub-mm high-field EPR facilities in Japanese universities and national institutes working in the field of physical sciences.[2] They cover broad field ranges up to 150 T using superconducting and hybrid magnets, either in constant-field or repetitive pulse-field mode of operation. But also in chemical and biological sciences a growing interest is noticeable in Japan, both in terms of instrument developments and scientific applications. For example, in Sendai at the Tohoku University, S. Yamauchi and coworkers have demonstrated the advantages of W-band

Introduction

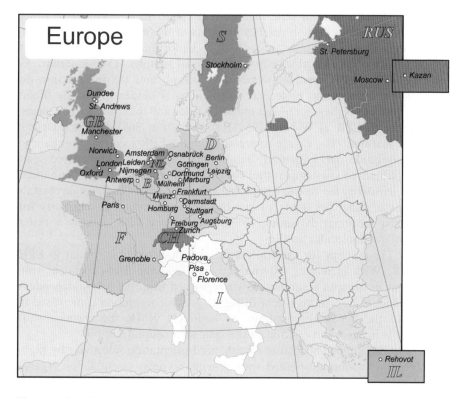

Figure 1.2 High-field EPR groups in Europe (2008).

high-field EPR for their studies of organic excited multiplet states in fluid solution over recent years.[3,4]

1.1 Why EPR at High Magnetic Fields?

Induced chemical reactions in condensed phases often proceed via the formation of radical pairs as reaction intermediates. Radical pairs are formed, for example, after appropriate excitation of donor molecules, by one-electron transfer to acceptor molecules leading to ionic radical pairs. Such electron-transfer reactions are found in many fields of chemistry and biology, for example in tribochemistry, *i.e.* during chemical reactions initiated by mechanical activation of solid mixtures of compounds by pressure, in photochemistry, *i.e.* during chemical reactions initiated by the absorption of infrared, visible and ultraviolet light, and in photosynthesis, *i.e.* during photoinduced chemical reactions in green plants, algae and certain bacteria by which carbon dioxide and water are converted to carbohydrates and oxygen.

In principle, EPR appears to be a promising spectroscopic technique to study both stable and transient radical-pair intermediates. In practice, however, for large spin systems in solid-state reactions, standard EPR – similar to other types of spectroscopy – soon reaches its limits of useful information content, unless single-crystal samples are available. Unfortunately, large molecular complexes are often available only as disordered samples. Their standard X-band (9.5 GHz) EPR spectra are poorly resolved, and the information on magnetic parameters and molecular orientations is hidden under the broad lines. By going to higher and higher magnetic fields and microwave frequencies, for example to EPR at W-band (95 GHz) or even at 360 GHz, at least five important features, *(i)–(v)*, are emerging from the EPR spectra: *(i)* enhanced spectral resolution; *(ii)* enhanced orientational selectivity in disordered samples; *(iii)* enhanced low-temperature electron-spin polarization; *(iv)* enhanced detection sensitivity for restricted-volume samples such as small single crystals of proteins or fullerenes, and *(v)* last but not least enhanced sensitivity for probing fast motional dynamics, *i.e.* high-frequency EPR acts as a faster "snapshot" for molecular motion.

Ad (i): The strategy for spectral resolution enhancement is similar in EPR and NMR: With increasing external Zeeman field the field-dependent spin interactions in the spin Hamiltonian are separated from the field-independent ones (see Figure 1.3). In high-field EPR, the *g*-factor resolution is increased in relation to the hyperfine couplings, in high-field NMR the chemical-shift resolution is increased in relation to the spin–spin couplings.

Ad (ii): The important feature of enhanced orientation selectivity by high-field EPR on randomly oriented spin systems becomes essential for organic radicals with only small *g*-anisotropy (see Figure 1.4) Well below room temperature, the overall rotation of, for example, a protein complex becomes so slow that powder-type EPR spectra are obtained. If the anisotropy of the leading interaction in the spin Hamiltonian is larger than the inhomogeneous linewidth, even from disordered powder-type EPR spectra the canonical orientations of the dominating interaction tensor can be resolved. As a consequence, single-crystal-like information on the hyperfine interactions can be extracted by performing orientation-selective ENDOR at the field values of resolved spectral features. In the case of transition-metal complexes the hyperfine anisotropy of the metal ion may provide this orientation selectivity from the entire orientational distribution of the molecules. Often, their *g*-anisotropy is large enough to allow for distinct orientational selectivity already in X-band EPR allowing for single-crystal-like ENDOR.[5–7] The best approach for elucidating molecular structure and orientation in detail is, of course, to study single-crystal samples. Unfortunately, to prepare them for large biological complexes like membrane proteins is often difficult or even impossible.

Ad (iii): The enhanced low-temperature electron-spin polarization at high Zeeman fields allows to extract the absolute sign of the zero-field splitting parameter, D, of a two-spin system like a biradical or triplet state. At high fields, considerable thermal spin polarization can be achieved already well above helium temperature, provided that the sample temperature becomes comparable with the Zeeman temperature, $T_Z = g \cdot \mu_B \cdot B_0 / k_B$ (*g*: electron

Introduction

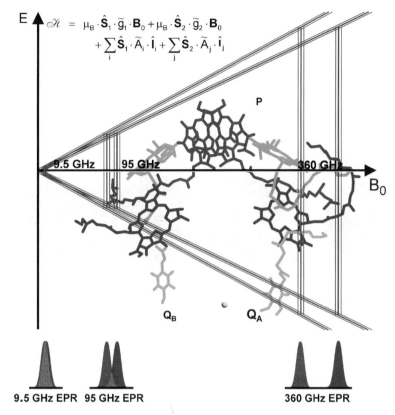

Figure 1.3 Enhanced spectral resolution by high-field EPR, taking the cofactor radical ions in bacterial photosynthetic reaction centers as example, with P the primary donor, Q_A and Q_B the quinine acceptors. The spin Hamiltonian in the inset describes two radicals in an external Zeeman field B_0 and contains the Zeeman interactions of the two electron spins S_1, S_2 and their hyperfine interactions with the nuclear spins I_i, I_j in the radicals. For details, see Chapter 2.

g-factor, μ_B: Bohr magneton, B_0: Zeeman field; k_B: Boltzmann constant). At $T \gg T_Z$, the characteristic triplet powder EPR spectrum (see Figure 1.5) is symmetric at its low- and high-field sides and, hence, contains no information of the sign of D. At $T < T_Z$, the Boltzmann distribution leads to increased populations of the low-energy levels, resulting in asymmetric lineshapes from which the absolute sign of D can be directly read off. Thermal spin polarization as a means to determine the absolute sign of D in high-spin systems has been used at a variety of EPR frequencies, for example at 9.5 GHz ($T_Z \approx 0.4$ K),[8] at 95 GHz ($T_Z \approx 4$ K),[9] 140 GHz ($T_Z \approx 6.5$ K),[10] 360 GHz ($T_Z \approx 15.5$ K).[11]

Ad (iv): With respect to detection sensitivity and its enhancement with increasing microwave frequencies, one has to distinguish between the *absolute* and *relative* (concentration) sensitivities. The absolute sensitivity is defined by

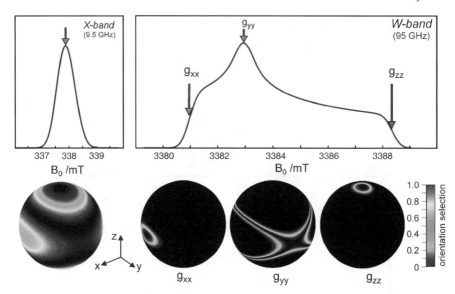

Figure 1.4 Enhanced orientational selectivity by high-field EPR, taking the anion radical of the ubiquinone acceptor cofactor in frozen-solution bacterial photosynthetic reaction centers as example.

the minimum detectable number of spins in the sample, N_{min}, the relative (or concentration) sensitivity is given by N_{min}/V_S, *i.e.* is scaled by sample volume, V_S. This is limited by the amount of sample that can be introduced into the cavity of high-field EPR spectrometers that, of course, is usually significantly smaller than standard X-band cavities. Consequently, if the amount of sample available is limited, as in single crystals of proteins, the sensitivity of high-field EPR can be superior by orders of magnitude because the absolute sensitivity grows with increasing frequencies much more strongly than the relative sensitivity does. Under certain experimental conditions, for constant incident microwave power and unsaturated EPR lines, one obtains theoretical expressions for the absolute sensitivity, $N_{min} \propto \omega_0^{-9/2}$, and for the relative sensitivity, $N_{min}/V_S \propto \omega_0^{-3/2}$ (see the detailed sensitivity discussions in Chapters 2 and 3, which are based on refs. [12,13]).

Ad (v): The faster "snapshot" capability for complex motional dynamics with increasing EPR frequency can be used in a multifrequency continuous-wave (cw) EPR approach at the same temperature to probe fast internal modes of motion and to discriminate them from the slow restricted motion of a macromolecule in solution. In high-frequency cw EPR spectra, slow motions appear to be frozen out, whereas fast motions dominate the observed spectral lineshape.[14–17]

In many fields of chemistry the precise nature of primary reactions with an involvement of paramagnetic intermediates is not well understood. This situation motivated us, when writing this book, to put particular emphasis on

Introduction

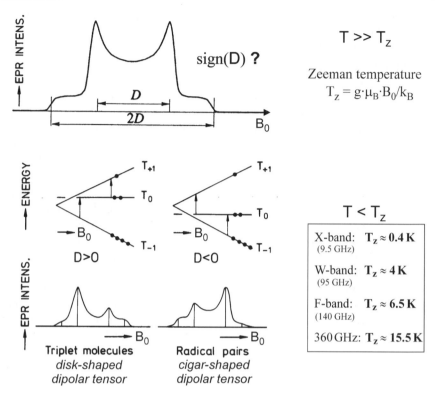

Figure 1.5 Enhanced thermal spin polarization by high-field EPR, taking mechanically generated radical pairs in a donor–acceptor mixture as example. For details, see ref. [9].

high-field/high-frequency EPR techniques used in (bio)chemistry as a means to elucidate the structural and dynamical details of short-lived paramagnetic intermediates involved in primary reactions.

1.2 NMR *versus* EPR

Up to 15 years ago, when asking scientists from outside the magnetic resonance community the question "what is magnetic resonance?" most of them, no matter which specific discipline they were working in, probably would have responded positively, but only with reference to NMR. When asking biologists, chemists and physicists the specific question "what is EPR?", probably only a small minority would have responded positively, and this only with reference to conventional cw X–band EPR spectroscopy (operating at a microwave frequency of around 9 GHz). Admittedly, such an imbalance of appreciation of the two magnetic resonance sisters was totally justified, because over the last decades NMR, in contrast to EPR, had grown up much faster and had become well established in the material sciences, in the life sciences and even in medical diagnostics as a unique tool for obtaining details of information inaccessible so far on molecular structure, chemical kinetics and medical imaging. Not surprisingly, therefore, that as many as four Nobel prizes for NMR methodology and applications have been awarded within the last 15 years (to R.R. Ernst in Chemistry in 1991, to K. Wüthrich in Chemistry in 2002, to P.C. Lauterbur and to P. Mansfield in Physiology and Medicine in 2003). But none, so far, for EPR!

EPR seemed to be hopelessly lagging behind its famous (though younger) sister NMR, and it is only during the last decade that the chemistry, biology and physics communities have started to appreciate the dramatic catching up of EPR in modern molecular spectroscopy. The reasons for EPR's broad jump ahead are to be found in the remarkable technological breakthroughs in pulsed microwave technology, sweepable cryomagnet fabrication and fast data-acquisition and -handling instrumentation. Indeed, modern EPR is apparently booming now, rather similar to what had happened with NMR 15–20 years earlier. And when nowadays the question "what is magnetic resonance?" is posed again, many chemists, biologists and physicists would probably respond differently from what they had said 20 years ago: They would agree that EPR spectroscopy matured to an attractive sister of NMR spectroscopy, both exhibiting unique and complementary capabilities in elucidating structure and dynamics of complex (bio)chemical systems in the fluid, glassy or solid state.

Why is there such a discrepancy between the technical requirements for pulsed NMR and pulsed EPR? The answer is related to the different time scales of the NMR and EPR phenomena that, in turn, are a consequence of the vastly different magnetic moments of nuclei and electrons (for example for ^{1}H the magnetic moment ratio is 1.5×10^{-3}, for ^{14}N it is 1.1×10^{-4}). Thus, the time scales are determined by the nuclear and electron resonance frequencies (in the radiofrequency (rf) and microwave (mw) domains, respectively), and the characteristic frequency separations in the respective spectra (Hz *versus* MHz) and the relaxation times T_1, T_2 (ms *versus* ns) are vastly different. Because of the long nuclear T_1 and T_2 times in diamagnetic molecules, NMR pulses need not be shorter than 10 μs, which do not pose technical problems to generate and detect coherently. The electronic transverse relaxation times (T_2), however, are typically in the 100 ns range and, consequently, in EPR the mw pulses have to be as short as a few ns. To generate and detect them coherently poses great technical problems even today. This holds, for example, for the mw sources with adequate output power, for fast mw switches and mixers as well as for fast electronic semiconductor components and computers for controlling the pulse trains, likewise for detecting and handling the transient signals in the ns time scale.

Nowadays, pulse NMR has completely replaced cw NMR, culminating in multidimensional spectroscopy. High-field steady-state cryomagnets and nuclear resonance frequencies up to 1 GHz (for protons) have dramatically improved the detection sensitivity and chemical-shift separations. And even formerly exotic nuclei have now become routinely observable. Despite the spectacular breakthroughs in mm and sub-mm microwave technologies in the last decade, in EPR the cw *versus* pulse situation is still very different from that

Introduction 9

in NMR. In fact, the general prognosis is that coexistence between cw and pulse EPR will continue to persist. Which option to choose will be determined entirely by the sample under study. The specific sample properties and relaxation times ultimately dictate the preference for either a cw or pulse experiment to be performed.

Regarding spectral resolution and detection sensitivity, modern EPR and NMR spectroscopies both follow similar strategies: to apply higher and higher static Zeeman fields to separate field-dependent from field-independent spin interactions in the molecular system. By this strategy not only can otherwise overlapping lines be disentangled, but also the population difference and quantum energy of the driven transitions between electron and nuclear spin energy levels will be increased, allowing for the detection of fewer and fewer spins.

1.3 From Basic to Advanced Multifrequency EPR, a Chronological Account

EPR and NMR phenomena were originally observed in radiofrequency spectroscopy experiments employing cw electromagnetic fields, EPR in 1944 by E.K. Zavoisky at Kazan University, NMR in 1946 by E.M. Purcell, H.G. Torrey and R.V. Pound at Harvard and, independently, by F. Bloch, W. Hansen and M.E. Packard at Stanford. These classic NMR experiments were honoured as early as 1952 by the Nobel Prize in Physics to Bloch and Purcell. Zavoisky's discovery of EPR, on the other hand, was only inadequately recognized on the western side of the Iron Curtain – in contrast to the eastern side: In 1957, Zavoisky was awarded the Lenin Prize, the highest sign of recognition in the former USSR for his discovery of electron paramagnetic resonance; and in 1970 the State Committee on Inventions and Discoveries enlisted "The Electron Paramagnetic Resonance Phenomenon" into the State Register of the USSR. It was as late as 1977, when finally Zavoisky was honoured also internationally. Not by a Nobel Prize, though, but at least by the prestigious ISMAR Award of the International Society of Magnetic Resonance, presented at the ISMAR Conference in Banff, Canada, on May 21, 1977 – alas posthumously[18] as he had died October 9, 1976, in Moscow just after having been informed about the decision of the ISMAR Prize Committee (see Figure 1.6).

Up to the 1960s, both NMR and EPR remained as cw methods, *i.e.* the samples placed in a static magnetic field were irradiated by continuous radiofrequency (rf) and microwave (mw) fields to drive NMR and EPR transitions, respectively. But it was as early as 1949 when E.L. Hahn at Urbana applied rf pulses and invented the nuclear spin-echo detection. This, with the introduction of powerful computers for fast Fourier-transform (FT) NMR in the late 1970s, opened the arena for pulse NMR spectroscopy and with all its potential for recording multidimensional spectra of complex biosystems in the liquid and solid state and, of course, for applications in medical imaging. It took almost a

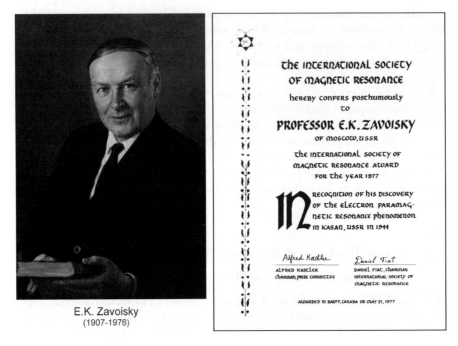

Figure 1.6 Evgeny K. Zavoisky and the certificate of the 1977 ISMAR Award of the International Society of Magnetic Resonance.

decade before R.J. Blume at Columbia[19] observed, for the first time, electron-spin echoes analogous to Hahn's nuclear spin echoes. It took many more years before electron spin-echo-detected EPR (ESE) methodologies gained sufficient experimental and theoretical backing to revolutionize EPR spectroscopy in a way rather similar to what had happened in NMR spectroscopy decades earlier.

Pulse EPR spectroscopy has many founders, both from the theoretical and the experimental side. In the early 1970s, K.M. Salikhov at Novosibirsk[20] laid the theoretical foundation for several advanced pulse EPR methods. For example, he developed the theory of electron-spin phase relaxation by stochastic modulation of the dipole–dipole interaction between paramagnetic centers and its effect on the ESE decay.[21] He suggested the first pulse ELDOR (PELDOR) protocol to observe the modulation of the ESE signal due to the electron–electron dipolar interaction of weakly coupled biradical systems in disordered solids.[22] Some years later, Salikhov theoretically predicted new spin phenomena, such as quantum beats of the EPR line intensity[23] and out-of-phase spin echoes of correlated radical pairs.[24] These spin phenomena should be observable in time-resolved EPR spectra of transient spin-polarized radical-pair intermediates. And, indeed, such new spin phenomena were observed in several laboratories shortly after they had been predicted. A prominent example was the photosynthetic reaction center with its donor excited to the

Introduction

singlet state by short laser flashes leading to spin-correlated donor–acceptor radical pairs (for a review of these early experiments, see ref. [25]. A rich variety of mw pulse sequences and sophisticated experiments is found in more recent reviews and text books, for example that of A. Schweiger and G. Jeschke.[26] Pioneering work was done, among others, by W.B. Mims at Bell Telephone Laboratories, Yu.D. Tsvetkov in Novosibirsk, J.H. Freed at Cornell and A. Schweiger at ETH Zurich.

It was a milestone in the history of magnetic resonance spectroscopy when, in 1956, G. Feher at Bell Labs,[27] invented a reconciliatory "EPR meets NMR" experiment for which he coined the name ENDOR (electron–nuclear double resonance), having been inspired by the Old Testament (King Saul did the unthinkable asking a medium, the witch of Endor, to make visible the invisible [*1st Samuel 28:7–20*]). His ingenious concept was to apply simultaneously two electromagnetic fields, one in the mw, the other in the rf range, to drive EPR and NMR transitions having an energy level in common. Thereby, the advantages of EPR (high detection sensitivity) are combined with those of NMR (high resolution capability). Feher's first cw ENDOR experiment was technically feasible only because the sample – phosphorus-doped silicon – was kept at low temperature, where all the relaxation times are sufficiently long to easily saturate both EPR and NMR transitions, a necessary condition for cw ENDOR. The cw ENDOR technique was later extended to ESE-detected pulse versions for solid-state samples by W.B. Mims (1965) at Bell Labs[28] and E.R. Davies (1974) at the Clarendon.[29] W.B. Mims is acknowledged to be the driving force in pulse EPR in general, and in pulse ENDOR in particular. It is certainly justified to call him the *Spiritus Rector* of pulse EPR.[26]

In contrast to solid-state ENDOR at low temperatures, for radicals in liquid solution, the electronic and nuclear relaxation times are much shorter – of the order of 10^{-5} to 10^{-7} s – and, consequently, ENDOR-in-solution experiments are technically much more demanding since much larger saturating mw and rf fields have to be applied. This is why liquid-state ENDOR experiments required many more years than solid-state ENDOR experiments before they were successful. The pioneering work was performed by A.L. Cederquist at Washington University[30] in 1963, who studied metal–ammonia solutions, and by J.S. Hyde and A.H. Maki at Varian Associates and Harvard, respectively, in 1964,[31] who investigated a stable organic radical dissolved in n-heptane solution. The further development of ENDOR-in-solution spectroscopy was highly stimulated by J.H. Freed at Cornell, whose general theory of saturation and double-resonance proved to be very appropriate for describing amplitude, width, and shape of ENDOR lines in great detail (see, for example, ref. [32]).

Admittedly, ENDOR also observes the rule that there are no advantages without disadvantages. There are apparent weaknesses of ENDOR in comparison to EPR concerning sensitivity (typically one order of magnitude lower) and relative line intensities (they no longer reflect the number of contributing nuclei). These weaknesses can be overcome by extending ENDOR to electron–nuclear–nuclear triple resonance. For the special case of only one set of hyperfine-coupled nuclei, such a triple resonance extension was proposed early on by G. Feher[33] and J.H. Freed.[34] Its experimental realization, however, had to wait until 1974 when K.P. Dinse in the Möbius group at FU Berlin accomplished "Special TRIPLE" on radicals in liquid solution.[35] In cw Special TRIPLE the two frequency-swept rf fields are applied at frequencies always symmetrically placed around the Larmor frequency of the respective nucleus. This variant of triple resonance enhances the signal intensity and allows relative line intensities to be related to the number of responsible nuclei. Thereby, the assignment of ENDOR lines to molecular positions is made possible, which is a vital task, but notoriously difficult in ENDOR spectroscopy.[36–38]

About a year later, it was demonstrated by R. Biehl in the same group at FU Berlin,[39] that additional information about relative signs of hyperfine couplings of radicals in solution – and thereby about their assignment – can be obtained by generalizing the triple resonance experiment to include NMR transitions of different nuclei in the radical ("General TRIPLE"). In cw General TRIPLE two rf fields with independently variable frequencies are applied, one pumping a selected NMR transition while the other is swept through the ENDOR spectrum. From the resulting characteristic intensity changes in the ENDOR spectrum the relative signs of the hyperfine couplings can be directly read off. The analogue of this experiment for solid-state samples at low temperature (77 K) was performed earlier by R.J. Cook and D.H. Whiffen (1964) at Teddington National Physical Laboratory.[40] They called it "double ENDOR", and applied it to X-irradiated organic crystals to determine relative signs of hyperfine couplings. The advantages of TRIPLE over ENDOR – enhanced sensitivity and resolution, information about multiplicity and relative signs of hyperfine couplings from line intensity variations – often justify the extra experimental efforts inherent in the triple resonance spectroscopy. This technique was shown later to be extremely powerful in elucidating the electronic structures not only of organic radicals in solution[36] but also of transient cofactor radical-ion intermediates in primary photosynthesis.[41,42]

To measure the electron–nuclear hyperfine and nuclear quadrupole interactions by EPR techniques, the nuclear transitions can be driven either directly by rf fields as in ENDOR or, as a more indirect alternative, by ESEEM (electron spin-echo modulation). This single-resonance technique was introduced by W.B. Mims[43] in 1972. He applied a mw spin-echo pulse train with varying interpulse separation and observed, on top of the exponential echo-decay trace, echo-amplitude modulations from hyperfine and quadrupole interactions. To obtain detectable echo modulations efficient mixing of the nuclear and electron-spin eigenfunctions by the dipolar hyperfine interaction is mandatory. Consequently, the strength of the external magnetic field has to be properly chosen to approximately balance the Zeeman splitting of the nuclear sublevels and the respective hyperfine splittings ("cancellation condition").

Concerning the detection of short-lived transient states or reaction intermediates, besides pulsed EPR also specific cw EPR strategies can be used to obtain time-resolved signals. Employing field modulation at a frequency as high as 1 MHz the time resolution could be extended to the μs range. A decisive step forward to drastically higher time resolution was achieved for fast photoreactions by abandoning field modulation and generating the time-dependent EPR signal *via* wavelength-selective pulsed laser excitation. Subsequent direct, *i.e.* broadband detection of the transient EPR signal at a fixed Zeeman-field value is accomplished by employing sufficiently fast data-acquisition systems. The Zeeman field is stepped through the spectrum establishing a time-resolved transient (TR) technique known as "TREPR". The pioneering experiment was done by S.I. Weissman and coworkers[44] at Washington University in 1979. In TREPR the inherent loss of sensitivity for broadband detection of transient paramagnetic states can often be compensated by accumulation of the spectra after each light flash. Moreover, an orders-of-magnitude signal enhancement *via* electron-spin-polarization effects can be utilized that occur in many photoreactions. They appear in reactions with instantaneously generated excited states and subsequent fast detection of the transient reaction intermediates, *e.g.*, triplets, radicals, radical pairs, before spin-lattice relaxation can thermalize them (for overviews of electron-spin-polarization effects, see for example refs. [25,45,46]). By now, the time resolution of TREPR has been pushed to the 10 ns range, and TREPR has proved to be extremely powerful in a broad range of mw frequencies from S-band (3 GHz), X-band (9 GHz), K-band (24 GHz), Q-band (35 GHz) up to the high-field EPR frequencies 95 GHz, 120 GHz and 240 GHz.[25,47–49] For many applications in photochemistry a multifrequency approach of TREPR experiments with a wide range of Zeeman fields turned out to be essential for the detailed analysis of spin-polarized spectra in the case of competing polarization mechanisms.

Long-range distance measurements on chemical and biological systems with the scale of a few nm are an important area of application for pulse EPR spectroscopy.[50] They are based on selective measurements of the electron-electron dipolar coupling between two spin-carrying domains, which is a function of their interspin distance and relative orientation. The main advantage of pulse *versus* cw EPR techniques in this endeavor is the ability to separate the electron-electron coupling from other interactions, such as electron-nuclear hyperfine interactions, and to reduce inhomogeneous line broadening. Thereby, the distance range that can be probed is extended to 7–8 nm. Yu.D. Tsvetkov and his coworkers at Novosibirsk[22,51] established the 3-pulse electron-electron double resonance technique (PELDOR or DEER) in 1981. Later it was extended to a 4-pulse sequence for deadtime-free detection.[26,52] Other powerful pulse sequences for measuring electron-electron frequencies and, thereby, distances have been invented, for example the single-frequency techniques DQC (double-quantum-coherence) by J.H. Freed and coworkers at Cornell[53] and RIDME (relaxation-induced dipolar modulation enhancement) by Yu.D. Tsvetkov and coworkers in Novosibirsk.[54]

Eventually, we are arriving in this chronological digression at high-field/high-frequency EPR, the actual subject of this book. It is certainly adequate to emphasize that with the development of high-field EPR another jump ahead in the capacity of EPR spectroscopy was accomplished. The first publications on this subject appeared around 1970.[55–58] It is, nevertheless, fair to state that Ya.S. Lebedev at the Institute of Chemical Physics in Moscow was the first to start a dedicated research program on high-field/high-frequency EPR in physical chemistry. Together with his coworkers he ultimately did the pioneering work on molecular systems from organic chemistry, starting in the 1970s[59] and being inspired by the earlier preliminary experiments of D.J.E. Ingram at the University of Keele.[55,56] Yakovs Lebedev's contributions to high-field EPR certainly set quality benchmarks and widened horizons in EPR spectroscopy. His early death in 1996 at the age of only 61 was a tragic loss for the EPR community. This book is an additional opportunity to trace his footsteps in high-field EPR, coming into the picture even in most recent accomplishments: memory is at the beginning of the new. As an aside, it is intriguing to notice a remarkable coincidence of names and location: Pjotr N. Lebedev in Moscow generated already in 1895 the first mm microwaves ($\lambda = 6$ mm) using a spark-gap transmitter with miniaturized electrodes as the radiation source.[60]

Naturally, the first laboratory-built high-field EPR spectrometers were operating in the cw mode of the microwave bridge. A few laboratories have developed the instrumentation for millimeter and submillimeter high-field EPR spectrometers with the goal in mind to take advantage of the benefits of high-field/high-frequency EPR, such as improved spectral resolution of complex spin systems, orientational selectivity, increased detection sensitivity for limited-volume samples or faster "snap-shot" capabilities for motional dynamics. No doubt that they were ploughing the ground for a promising new field of molecular spectroscopy of complex systems. Until the second half of the 1990s, laboratory-built high-field EPR spectrometers were described for cw microwave irradiation at 95 GHz ($\lambda \approx 3$ mm),[61–64] around 150 GHz ($\lambda \approx 2$ mm),[65–67] at 250 GHz ($\lambda \approx 1$ mm),[68–70] and in the sub-mm region,[71–73] even reaching 360 GHz/14 T EPR.[74] Only a few additional EPR spectrometers operating at frequencies above 200 GHz were described up to the end of the 1990s.[75–78] High-field EPR spectrometers with pulsed microwave irradiation were also described during this time, operating at wavelengths of 3 mm [47,79,80] and 2 mm.[67,81,82] The extension to high-field ENDOR experiments has been realized at the 3 mm [47,61,83,84] and 1 mm [85] microwave frequency bands. Whereas W-band high-field cw ENDOR was accomplished first in Berlin,[83] high-field pulse ENDOR was accomplished first in Leiden.[80,86] After these early years of high-field EPR spectroscopy, several groups continued and other groups started to build their dedicated high-field/high-frequency instrumentation. For instance, the Berlin group developed a cw 360-GHz EPR spectrometer with quasioptical microwave configuration, employing a Fabry–Perot resonator probehead,[74] and the Leiden group constructed a cw and pulse EPR spectrometer for 275 GHz with a quasioptical microwave-bridge design, coupling the bridge to a single-mode cavity.[87] And, to give two more examples: The Cornell group developed a 170/240-GHz EPR spectrometer based on quasioptical technology, for which high sensitivity was achieved even for aqueous biological samples using specially designed Fabry–Perot resonators,[88] and the Frankfurt group built a pulsed 180-GHz spectrometer with a quasioptical circulator and single-mode cavity; the instrument is designed for optional pulsed ENDOR and

PELDOR experiments.[89] Also commercially available EPR and ENDOR spectrometers, mostly operating at W-band frequencies, were introduced at the end of the 1990s, the major manufacturer was, and still is Bruker Biospin, Germany.[90,91] Appropriate references to the laboratories that completed the construction of mm- and sub-mm high-field EPR spectrometers since the end of the 1990s, are included in recent overview articles, for instance see refs. [15,92–103]. A discussion of the spectrometer performance specifications will be presented in Chapter 3.

Time-resolved EPR spectroscopy is an important issue in general, and under high-field conditions in particular. The arena of time-resolved high-field EPR was opened in 1989 by the first pulsed W-band EPR spectrometer built in the Schmidt group in Leiden.[64,86] Soon after, T.F. Prisner and M. Rohrer in the Möbius group in Berlin completed a versatile pulsed EPR spectrometer at W-band, which served also for the first high-field echo-detected TREPR experiment on pulsed laser-generated transient radicals in photosynthetic reaction centers.[47] However, the introduction of time-resolution capability of the sub-mm high-field EPR at 360 GHz with a quasioptical microwave bridge had to wait until 2004. The design of this spectrometer in Berlin was planned to start with the cw mode of operation[74] using a solid-state 120-GHz source with subsequent tripling. The 360-GHz output power at the Gaussian horn antenna is not larger than 1 mW, limiting the B_1 field at the sample in the Fabry–Perot resonator to values too small for fast pulsing. As an unconventional novel approach to time-resolved sub-mm high-field EPR, a dedicated 360-GHz pulsed Orotron ("Oro" refers to Russian abbreviations for "open resonator plus reflecting grid" in the high-voltage vacuum-tube generator) was constructed and integrated in the quasioptical microwave bridge of the heterodyne induction-mode EPR spectrometer in Berlin.[104] Orotron microwave radiation is generated by the interaction of a nonrelativistic electron beam with a diffraction grating. The resulting weak stimulated Smith–Purcell radiation is amplified by feedback with the electromagnetic field in an open Fabry–Perot resonator construction of high Q-value. In the pulsed Orotron version, an additional gate electrode in combination with a high-voltage pulsing unit controls the electron-beam current. The generated pulses at 360 GHz have pulse lengths from 100 ns to 10 µs and a pulse power of about 25 mW. Within a 10-µs time period, incoherent pulse trains of arbitrary duration can be generated. First EPR free-induction-decay (FID) measurements at 360 GHz and a magnetic field of 12.9 T could be done, using polycrystalline perylenyl ions as a test sample. The preliminary results are very encouraging and give rise to all sorts of speculations about future applications of multifrequency Orotron-EPR spectroscopy.[104]

As mentioned before, under certain conditions ESEEM can be a competitive alternative to ENDOR for measuring hyperfine and quadrupole interactions.[43] The first successful W-band high-field ESEEM measurements of nitrogen hyperfine and quadrupole interactions in disordered powder samples were performed in Berlin in 1998.[105] Most recently, this work was largely extended by elaborate W-band ESEEM studies on nitroxide spin-label molecules to explore the sensitivity of the *g*-, hyperfine- and quadrupole-tensors for probing polarity and proticity effects of the solvent matrix.[106]

In 2006/2007 pulsed high-field electron dipolar spectroscopy, specifically PELDOR or DEER, was introduced for resolving the relative orientation of weakly coupled biradical partners in a frozen-solution sample, in addition to measuring their distance. This extension of electron dipolar spectroscopy, established at X-band frequencies for determining large interradical distances (see recent overviews,[107–109]) to high frequencies and fields was accomplished independently by the group of T.F. Prisner (Frankfurt)[101,110,111] at 180 GHz, the group of G. Jeschke (Konstanz)[112] at 95 GHz and the group of K. Möbius (Berlin) at 95 GHz.[113]

Figure 1.7 shows the microwave and radiofrequency irradiation schemes of a variety of cw and pulse high-field EPR techniques that will be dealt with in this book.

Some more statements remain to be added: From inspection of the literature we feel that a comprehensive coverage of high-field EPR studies of biological systems will go beyond the scope of this book. Hence, instead of attempting to compose an encyclopedic overview on this subject, we will select aspects of high-field EPR that touch on, or even reflect the main topics of our own

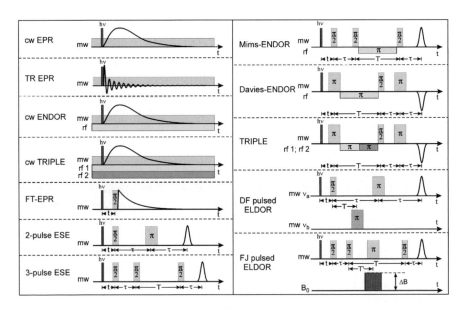

Figure 1.7 Microwave (mw) and radio-frequency (rf) cw and pulse-irradiation schemes of various time-resolved EPR techniques. The initial laser excitation pulse *hν* starts the photoreaction with paramagnetic intermediates. In stable paramagnetic systems the laser pulse is omitted. Abbreviations: cw (continuous wave), TR (transient), FT (Fourier transform), ESE (electron-spin echo), DF (dual frequency), FJ (field jump). For details, see Chapters 2 and 3.

research. In essence, we will focus on high-field EPR studies on biochemical systems that were performed in our laboratory at FU Berlin in teamwork with undergraduate and graduate students, postdocs and senior scientists from around the world. An exception to this restriction will be oxygenic photosynthesis (Section 5.3), for which predominantly results from other laboratories will be reviewed. Hence, right from the outset we have to ask our esteemed colleagues for their understanding when their achievements could not adequately be considered within the framework set for this book. To partly compensate for what had to be excluded from the presentation, the reader is referred to the overview articles cited abundantly throughout the book. They cover the wide subject of high-field EPR at more length and provide a multitude of relevant references.

Concluding this introductory chapter we point out that a major challenge of molecular biology, biochemistry and biophysics, especially in the growing field of proteomics, is to understand the function and reaction mechanism of highly specialized proteins on the level of their molecular and electronic structure. We take the view that the arsenal of modern EPR techniques in general, and of high-field EPR in particular, provides powerful and versatile tools that are highly needed for the endeavor of elucidating structure and dynamics of complex biosystems. These tools can provide details of molecular information complementary to what can be learned from other biophysical techniques already established in the field.

References

1. J. H. Freed, *EPR Newsletter*, 2007, **16**, 10.
2. H. Ohta, S. Okubo, K. Kawakami, D. Fukuoka, Y. Inagaki, T. Kunimoto and Z. Hiroi, *J. Phys. Soc. Jpn.*, 2003, **72 Suppl. B**, 26.
3. S. Yamauchi, *Bull. Chem. Soc. Jpn.*, 2004, **77**, 1255.
4. V. F. Tarasov, I. S. M. Saiful, Y. Iwasaki, Y. Ohba, A. Savitsky, K. Möbius and S. Yamauchi, *Appl. Magn. Reson.*, 2006, **30**, 619.
5. G. Rist and J. S. Hyde, *J. Chem. Phys.*, 1968, **49**, 2449.
6. G. Rist and J. S. Hyde, *J. Chem. Phys.*, 1969, **50**, 4532.
7. G. Rist and J. S. Hyde, *J. Chem. Phys.*, 1970, **52**, 4633.
8. A. W. Hornig and J. S. Hyde, *Mol. Phys.*, 1963, **6**, 33.
9. S. D. Chemerisov, O. Y. Grinberg, D. S. Tipikin, Y. S. Lebedev, H. Kurreck and K. Möbius, *Chem. Phys. Lett.*, 1994, **218**, 353.
10. D. S. Tipikin, G. G. Lazarev and Y. S. Lebedev, *Zh. Fiz. Khim.*, 1993, **67**, 176.
11. M. Fuchs, A. Schnegg, K. Möbius, 2004, unpublished results.
12. C. P. Poole, *Electron Spin Resonance*, J. Wiley, New York, 1983.
13. G. Feher, *Bell System Tech. J.*, 1957, **36**, 449.
14. Z. C. Liang and J. H. Freed, *J. Phys. Chem. B*, 1999, **103**, 6384.
15. J. H. Freed, *Annu. Rev. Phys. Chem.*, 2000, **51**, 655.

16. P. P. Borbat, A. J. Costa-Filho, K. A. Earle, J. K. Moscicki and J. H. Freed, *Science*, 2001, **291**, 266.
17. J. H. Freed, In: *Very High Frequency (VHF) ESR/EPR, Biological Magnetic Resonance,* Vol. **22,** O. Grinberg, L.J. Berliner, Eds., Kluwer/ Plenum Publishers, New York, 2004, pp. 19.
18. K. Hausser, *J. Magn. Reson.*, 1978, **29**, 179.
19. R. J. Blume, *Phys. Rev.*, 1958, **109**, 1867.
20. K. M. Salikhov, A. G. Semenov and Y. D. Tsvetkov, *Electron-spin Echo and its Applications*, Nauka, Moscow, 1976.
21. A. D. Milov, K. M. Salikhov and Y. D. Tsvetkov, *Zh. Eksper. Teor. Fiz.*, 1972, **63**, 2329.
22. A. D. Milov, K. M. Salikhov and M. D. Schirov, *Fiz. Tverd. Tel.*, 1981, **23**, 957.
23. K. M. Salikhov, C. H. Bock and D. Stehlik, *Appl. Magn. Reson.*, 1990, **1**, 195.
24. K. M. Salikhov, Y. Kandrashkin and A. K. Salikhov, *Appl. Magn. Reson.*, 1992, **3**, 199.
25. D. Stehlik and K. Möbius, *Annu. Rev. Phys. Chem.*, 1997, **48**, 745.
26. A. Schweiger and G. Jeschke, *Principles of Pulse Electron Paramagnetic Resonance*, Oxford University Press, Oxford, 2001.
27. G. Feher, *Phys. Rev.*, 1956, **103**, 834.
28. W. B. Mims, *Proc. R. Soc. London, Ser. A*, 1965, **283**, 452.
29. E. R. Davies, *Phys. Lett. A*, 1974, **A 47**, 1.
30. A. Cederquist, PhD Thesis, Washington University, 1963.
31. J. S. Hyde and A. H. Maki, *J. Chem. Phys.*, 1964, **40**, 3117.
32. J. H. Freed, In: *Multiple Electron Resonance Spectroscopy*, M. M. Dorio, J.H. Freed, Eds., Plenum Press, New York, 1979, pp. 73.
33. G. Feher, *Physica*, 1958, **24**, S80.
34. J. H. Freed, *J. Chem. Phys.*, 1969, **50**, 2271.
35. K. P. Dinse, R. Biehl and K. Möbius, *J. Chem. Phys.*, 1974, **61**, 4335.
36. H. Kurreck, B. Kirste and W. Lubitz, *Electron Nuclear Double Resonance Spectroscopy of Radicals in Solution*, Wiley, New York, 1988.
37. F. Gerson and W. Huber, *Electron Spin Resonance Spectroscopy of Organic Radicals*, Wiley-VCH, Weinheim, 2003.
38. F. Gerson and G. Gescheidt, In: *Biological Magnetic Resonance,* Vol. **24a**, S. S. Eaton, G. R. Eaton, L. J. Berliner, Eds., Springer, New York, 2005, pp. 145.
39. R. Biehl, M. Plato and K. Möbius, *J. Chem. Phys.*, 1975, **63**, 3515.
40. R. J. Cook and D. H. Whiffen, *Proc. R. Soc. London*, 1964, **84**, 845.
41. H. Levanon and K. Möbius, *Annu. Rev. Biophys. Biomol. Struct.*, 1997, **26**, 495.
42. W. Lubitz and F. Lendzian, In: *Biophysical Techniques in Photosynthesis, Advances in Photosynthesis,* Vol. **3**, J. Amesz, A. J. Hoff, Eds., Kluwer, Dordrecht, 1996.
43. W. B. Mims, *Phys. Rev. B*, 1972, **5**, 2409.
44. S. S. Kim and S. I. Weissman, *J. Am. Chem. Soc.*, 1979, **101**, 5863.

45. *Advanced EPR, Applications in Biology and Biochemistry*; A. J. Hoff, Ed., Elsevier, Amsterdam, 1989.
46. K. A. McLauchlan and M. T. Yeung, In: *Electron Spin Resonance*, Vol. **14**, N. M. Atherton, E. R. Davies, B. C. Gilbert, Eds., Royal Society of Chemistry, Cambridge, 1994, pp. 32.
47. T. F. Prisner, M. Rohrer and K. Möbius, *Appl. Magn. Reson.*, 1994, **7**, 167.
48. A. Savitsky, M. Kühn, D. Duche, K. Möbius and H. J. Steinhoff, *J. Phys. Chem. B*, 2004, **108**, 9541.
49. J. van Tol, L. C. Brunel and R. J. Wylde, *Rev. Sci. Instrum.*, 2005, **76**, 074101.
50. *Distance Measurements in Biological Systems by EPR*; Biological Magnetic Resonance, Vol. **19**; L. J. Berliner, S. S. Eaton, G. R. Eaton, Eds.; Kluwer/Plenum Publishers, New York, 2000.
51. A. D. Milov, A. B. Ponomarev and Y. D. Tsvetkov, *Chem. Phys. Lett.*, 1984, **110**, 67.
52. M. Pannier, S. Veit, A. Godt, G. Jeschke and H. W. Spiess, *J. Magn. Reson.*, 2000, **142**, 331.
53. S. Saxena and J. H. Freed, *J. Phys. Chem. A*, 1997, **101**, 7998.
54. L. V. Kulik, S. A. Dzuba, I. A. Grigoryev and Y. D. Tsvetkov, *Chem. Phys. Lett.*, 2001, **343**, 315.
55. D. J. E. Ingram, *Biological and Biochemical Applications of Electron Spin Resonance*, Adam Hilger, London, 1969.
56. E. F. Slade and D. J. E. Ingram, *Proc. R. Soc. London, Ser. A*, 1969, **312**, 85.
57. I. Amity, *Rev. Sci. Instrum.*, 1970, **41**, 1492.
58. Y. Alpert, Y. Couder, J. Tuchendl and H. Thome, *Biochim. Biophys. Acta*, 1973, **322**, 34.
59. O. Grinberg and A. A. Dubinskii, In: *Very High Frequency (VHF) ESR/ EPR, Biological Magnetic Resonance*, Vol. **22**, O. Grinberg, L. J. Berliner, Eds., Kluwer/Plenum Publishers, New York, 2004, pp. 1.
60. P. Lebedew, *Ann. Phys. Chem.*, 1895, **56**, 1.
61. O. Burghaus, M. Rohrer, T. Götzinger, M. Plato and K. Möbius, *Meas. Sci. Technol.*, 1992, **3**, 765.
62. E. Haindl, K. Möbius and H. Oloff, *Z. Naturforsch. A*, 1985, **40**, 169.
63. W. Wang, R. L. Belford, R. B. Clarkson, P. H. Davis, J. Forrer, M. J. Nilges, M. D. Timken, T. Walczak, M. C. Thurnauer, J. R. Norris, A. L. Morris and Y. Zhang, *Appl. Magn. Reson.*, 1994, **6**, 195.
64. R. T. Weber, J. A. J. M. Disselhorst, L. J. Prevo, J. Schmidt and W. T. Wenckebach, *J. Magn. Reson.*, 1989, **81**, 129.
65. O. J. Grinberg, A. A. Dubinskii, V. F. Shuvalov, L. G. Oranskii, V. I. Kurochkin and J. S. Lebedev, *Dokl. Akad. Nauk SSSR*, 1976, **230**, 884.
66. T. Tatsukawa, T. Maeda, H. Sasai, T. Idehara, I. Mekata, T. Saito and T. Kanemaki, *Int. J. Infrared Milli.*, 1995, **16**, 293.
67. L. R. Becerra, G. J. Gerfen, B. F. Bellew, J. A. Bryant, D. A. Hall, S. J. Inati, R. T. Weber, S. Un, T. F. Prisner, A. E. McDermott,

K. W. Fishbein, K. E. Kreischer, R. J. Temkin, D. J. Singel and R. G. Griffin, *J. Magn. Reson. A*, 1995, **117**, 28.
68. W. B. Lynch, K. A. Earle and J. H. Freed, *Rev. Sci. Instrum.*, 1988, **59**, 1345.
69. K. A. Earle, D. S. Tipikin and J. H. Freed, *Rev. Sci. Instrum.*, 1996, **67**, 2502.
70. K. A. Earle and J. H. Freed, *Appl. Magn. Reson.*, 1999, **16**, 247.
71. A. L. Barra, L. C. Brunel and J. B. Robert, *Chem. Phys. Lett.*, 1990, **165**, 107.
72. F. Muller, M. A. Hopkins, N. Coron, M. Grynberg, L. C. Brunel and G. Martinez, *Rev. Sci. Instrum.*, 1989, **60**, 3681.
73. V. F. Tarasov and G. S. Shakurov, *Appl. Magn. Reson.*, 1991, **2**, 571.
74. M. R. Fuchs, T. F. Prisner and K. Möbius, *Rev. Sci. Instrum.*, 1999, **70**, 3681.
75. E. J. Reijerse, P. J. van Dam, A. A. K. Klaassen, W. R. Hagen, P. J. M. van Bentum and G. M. Smith, *Appl. Magn. Reson.*, 1998, **14**, 153.
76. H. P. Moll, C. Kutter, J. van Tol, H. Zuckerman and P. Wyder, *J. Magn. Reson.*, 1999, **137**, 46.
77. G. M. Smith, J. C. G. Lesurf, R. H. Mitchell and P. C. Riedi, *Rev. Sci. Instrum.*, 1998, **69**, 3924.
78. M. Rohrer, J. Krzystek, V. Williams and L. C. Brunel, *Meas. Sci. Technol.*, 1999, **10**, 275.
79. T. F. Prisner, In: *Advances in Magnetic and Optical Resonance*, Vol. 20, W. Warren, Ed., Academic, New York, 1997, pp. 245.
80. J. Allgeier, J. A. J. M. Disselhorst, R. T. Weber, W. T. Wenckebach, J. Schmidt, In: *Modern Pulsed and Continuous-Wave Electron Spin Resonance*, L. Kevan, M. K. Bowman, Eds., Wiley, New York, 1990, pp. 267.
81. A. Y. Bresgunov, A. A. Dubinskii, V. N. Krimov, Y. G. Petrov, O. G. Poluektov and Y. S. Lebedev, *Appl. Magn. Reson.*, 1991, **2**, 715.
82. T. F. Prisner, S. Un and R. G. Griffin, *Israel J. Chem.*, 1992, **32**, 357.
83. O. Burghaus, A. Toth-Kischkat, R. Klette and K. Möbius, *J. Magn. Reson.*, 1988, **80**, 383.
84. J. W. A. Coremans, M. van Gastel, O. G. Poluektov, E. J. J. Groenen, T. den Blaauwen, G. van Pouderoyen, G. W. Canters, H. Nar, C. Hammann and A. Messerschmidt, *Chem. Phys. Lett.*, 1995, **235**, 202.
85. L. Paschedag, J. van Tol and P. Wyder, *Rev. Sci. Instrum.*, 1995, **66**, 5098.
86. J. A. J. M. Disselhorst, H. van der Meer, O. G. Poluektov and J. Schmidt, *J. Magn. Reson. A*, 1995, **115**, 183.
87. H. Blok, J. A. J. M. Disselhorst, S. B. Orlinskii and J. Schmidt, *J. Magn. Reson.*, 2004, **166**, 92.
88. K. A. Earle, B. Dzikovski, W. Hofbauer, J. K. Moscicki and J. H. Freed, *Magn. Reson. Chem.*, 2005, **43**, S256.
89. M. M. Hertel, V. P. Denysenkov, M. Bennati and T. F. Prisner, *Magn. Reson. Chem.*, 2005, **43**, S248.

90. D. Schmalbein, G. G. Maresch, A. Kamlowski and P. Höfer, *Appl. Magn. Reson.*, 1999, **16**, 185.
91. P. Höfer, A. Kamlowski, G. G. Maresch, D. Schmalbein and R. T. Weber, In: *Very High Frequency (VHF) ESR/EPR, Biological Magnetic Resonance*, Vol. **22**, O. Grinberg, L.J. Berliner, Eds., Kluwer/Plenum Publishers, New York, 2004, pp. 401.
92. K. Möbius, *Chem. Soc. Rev.*, 2000, **29**, 129.
93. G. M. Smith and P. C. Riedi, In: *Electron Paramagnetic Resonance*, Vol. **17**, N. M. Atherton, M. J. Davies, B. C. Gilbert, Eds., Royal Society of Chemistry, Cambridge, 2000, pp. 164.
94. T. Prisner, M. Rohrer and F. MacMillan, *Annu. Rev. Phys. Chem.*, 2001, **52**, 279.
95. P. C. Riedi and G. M. Smith, In: *Electron Paramagnetic Resonance*, Vol. **18**, B. C. Gilbert, M. J. Davies, D. M. Murphy, Eds., Royal Society of Chemistry, Cambridge, 2002, pp. 254.
96. J. H. Freed, In: *Very High Frequency (VHF) ESR/EPR, Biological Magnetic Resonance*, Vol. **22**, O. Grinberg, L. J. Berliner, Eds., Kluwer/Plenum Publishers, New York, 2004, pp. 21.
97. K. A. Earle and A. I. Smirnov, In: *Very High Frequency (VHF) ESR/EPR, Biological Magnetic Resonance*, Vol. **22**, O. Grinberg, L. J. Berliner, Eds., Kluwer/Plenum Publishers, New York, 2004, pp. 469.
98. T. F. Prisner, In: *Very High Frequency (VHF) ESR/EPR, Biological Magnetic Resonance*, Vol. **22**, O. Grinberg, L. J. Berliner, Eds., Kluwer/Plenum Publishers, New York, 2004, pp. 249.
99. P. C. Riedi and G. M. Smith, In: *Electron Paramagnetic Resonance*, Vol. **19**, B. C. Gilbert, M. J. Davies, D. M. Murphy, Eds., Royal Society of Chemistry, Cambridge, 2004, pp. 338.
100. K. Möbius, A. Savitsky, A. Schnegg, M. Plato and M. Fuchs, *Phys. Chem. Chem. Phys.*, 2005, **7**, 19.
101. M. Bennati and T. F. Prisner, *Rep. Prog. Phys.*, 2005, **68**, 411.
102. P. C. Riedi, In: *Electron Paramagnetic Resonance*, Vol. **20**, B. C. Gilbert, M. J. Davies, D. M. Murphy, Eds., Royal Society of Chemistry, Cambridge, 2006, pp. 245.
103. K. Möbius and D. Goldfarb, In: *Biophysical Techniques in Photosynthesis*, Vol. **II**, T. J. Aartsma, J. Matysik, Eds., Springer, Dordrecht, 2008, pp. 267.
104. Y. A. Grishin, M. R. Fuchs, A. Schnegg, A. A. Dubinskii, B. S. Dumesh, F. S. Rusin, V. L. Bratman and K. Möbius, *Rev. Sci. Instrum.*, 2004, **75**, 2926.
105. A. Bloeß, K. Möbius and T. F. Prisner, *J. Magn. Reson.*, 1998, **134**, 30.
106. A. Savitsky, A. A. Dubinskii, M. Plato, Y. A. Grishin, H. Zimmermann, K. Möbius, *J. Phys. Chem. B*, 2008, **112**, 9079.
107. G. Jeschke, *EPR Newsletter*, 2005, **14**, 14.
108. O. Schiemann and T. F. Prisner, *Q. Rev. Biophys.*, 2007, **40**, 1.
109. P. P. Borbat and J. H. Freed, *EPR Newsletter*, 2007, **17**, 21.

110. V. P. Denysenkov, T. F. Prisner, J. Stubbe and M. Bennati, *Appl. Magn. Reson.*, 2005, **29**, 375.
111. V. P. Denysenkov, T. F. Prisner, J. Stubbe and M. Bennati, *Proc. Natl. Acad. Sci. USA*, 2006, **103**, 13386.
112. Y. Polyhach, A. Godt, C. Bauer and G. Jeschke, *J. Magn. Reson.*, 2007, **185**, 118.
113. A. Savitsky, A. A. Dubinskii, M. Flores, W. Lubitz and K. Möbius, *J. Phys. Chem. B*, 2007, **111**, 6245.

CHAPTER 2
Principles and Illustrative Examples of High-Field/High-Frequency EPR

In this chapter we will present a rather phenomenological description of the theoretical principles together with illustrative examples of those EPR techniques that stood the test for high-field applications in biophysics and biochemistry. For a more indepth theoretical treatment of advanced EPR spectroscopy we refer to renowned textbooks, for example.[1–4] We also refer to rather recent overview articles on high-field/high-frequency EPR in the biosciences,[5–14] and to the special issues of periodicals or anthologies dedicated to this subject.[15–17]

2.1 Spin Hamiltonians and EPR Experiments at High Magnetic Fields

We first describe the basic spin Hamiltonians with interaction terms that determine the energy levels and EPR transition frequencies, thereby determining the characteristics of the EPR spectrum.

2.1.1 Organic Radicals and Low-Spin Transition-Metal Ions ($S = 1/2$)

In molecular radicals or transition-metal ion complexes with unpaired electron spins $S = 1/2$, the electron and nuclear spins will align with respect to the total magnetic field they experience. This is the vector sum of the external Zeeman field, B_0, the local field originating from the residual orbital angular momentum of the unpaired electron and spin-orbit coupling that leads to effective g-tensor components shifted from the free-electron value, and additional local hyperfine and quadrupole fields from nearby magnetic nuclei, for example ^{14}N nuclei with $I=1$ or protons with nuclear spin $I=1/2$. They may be located within the radical molecule or in the solvent microenvironment of the unpaired electron ("matrix" nuclei).

For such $S=1/2$ systems, the static spin Hamiltonian, $\hat{H}_0$, that describes the time-independent spin interaction energies, consists of the terms

$$\hat{H}_0/h = \frac{\mu_B}{h} \cdot \boldsymbol{B}_0 \cdot \tilde{g} \cdot \hat{S} - \sum_i \frac{g_{ni} \cdot \mu_K}{h} \cdot \boldsymbol{B}_0 \cdot \hat{I}_i + \sum_i \hat{S} \cdot \tilde{A}_i \cdot \hat{I}_i \\ + \sum_i \hat{I}_i \cdot \tilde{P}_i \cdot \hat{I}_i \qquad (2.1)$$

i.e. $\hat{H}_0$ contains the field-dependent electron and nuclear Zeeman interactions as well as the field–independent electron–nuclear hyperfine and nuclear quadrupole interactions with magnetic nuclei (for the quadrupole interaction to exist, nuclei with $I>1/2$ in an asymmetric electronic environment are required). Here, $\boldsymbol{B}_0$ is the external magnetic field (a vector), $\tilde{g}$, $\tilde{A}$ and $\tilde{P}$, are the corresponding interaction tensors (matrices) of the electron Zeeman, hyperfine and quadrupole interactions (h: Planck constant; μ_B, μ_K: Bohr and nuclear magnetons; g_n: nuclear g-factors; $\hat{S}$, $\hat{I}$, electron and nuclear spin vector operators; the summation is over all nuclei).

The quadrupole interaction between a nuclear electric quadrupole moment, Q, and an electric field gradient at the position of the nucleus is described by the Hamiltonian $\hat{I} \cdot \tilde{P} \cdot \hat{I}$. In its principal axes system the quadrupole tensor, $\tilde{P}$, is traceless, and the quadrupole Hamiltonian can be written as

$$\hat{H}_Q/h = P_{xx} \cdot \hat{I}_x^2 + P_{yy} \cdot \hat{I}_y^2 + P_{zz} \cdot \hat{I}_z^2 \\ = \frac{e^2 \cdot q \cdot Q}{4I \cdot (2I-1) \cdot h} \cdot \left[3 \cdot \hat{I}_z^2 + 2 \cdot I \cdot (I+1) + \eta \cdot (\hat{I}_x^2 - \hat{I}_y^2) \right] \qquad (2.2)$$

where $e \cdot q$ is the electric field gradient of the electron plus nuclear charge distribution along the z-direction, and $\eta = (P_{xx} - P_{yy})/P_{zz}$ is the asymmetry parameter of the charge distribution with $|P_{zz}| \geq |P_{yy}| \geq |P_{xx}|$ and $0 \leq \eta \leq 1$.[3,4] The largest principal value of the quadrupole tensor is given by $P_{zz} = e^2 \cdot q \cdot Q/[2I \cdot (2I-1) \cdot h]$. Commonly, the quadrupole tensor components for any given value $I \geq 1$ are characterized by the two quantities $e^2 \cdot q \cdot Q/h$ (in linear frequency units) and η; for quantifying the magnitude of the interaction the factor $2I \cdot (2I-1)$ has to be included.

The hyperfine interaction between an electron and a nuclear spin, $\hat{S} \cdot \tilde{A} \cdot \hat{I}$ in eqn (2.1), can be written as the sum of the isotropic (or Fermi contact) interaction, $\hat{H}_{Fermi}/h = A_{iso} \hat{S} \cdot \hat{I}$; with (in SI units)

$$A_{iso} = \frac{2\mu_0}{3h} \cdot g_e \cdot \mu_B \cdot g_n \cdot \mu_K \cdot |\Psi(0)|^2 \qquad (2.3)$$

and the anisotropic dipole–dipole interaction between the magnetic moments of the electron and nuclear spins

$$\widehat{H}_{\text{SI}}/h = \frac{\mu_0}{4\pi \cdot h} \cdot g_e \cdot \mu_B \cdot g_n \cdot \mu_K \cdot \left[\frac{3 \cdot (\hat{S} \cdot \mathbf{r}) \cdot (\mathbf{r} \cdot \hat{I})}{r^5} - \frac{\hat{S} \cdot \hat{I}}{r^3} \right] \quad (2.4)$$

Here, $\mu_0 = 4\pi \times 10^{-7}\,\text{N} \cdot \text{A}^{-2}$ is the vacuum permeability, $|\Psi(0)|^2$ the electron density at the nucleus, r is the distance between the electron and nuclear spins, g_e is the free-electron g-factor.

For nucleus *i* the Hamiltonian of the anisotropic part of the electron–nuclear hyperfine interaction, the electron–nuclear dipolar (END) term, $\widehat{H}_{\text{SI}}$, can be written as[18]

$$\widehat{H}_{\text{SI}}/h = \frac{\mu_0}{4\pi \cdot h} \cdot \frac{g_e \cdot \mu_B \cdot g_n \cdot \mu_K}{r^3} \cdot (A + B + C + D + E + F) \quad (2.5)$$

The six terms $A, B, \ldots, F$ represent products of electron and nuclear spin operators and angular functions in spherical coordinates with the polar angles θ and φ describing the orientation between the dipolar axis and the external Zeeman field $\boldsymbol{B_0}$:

$$\left. \begin{aligned}
A &= \hat{S}_z \cdot \hat{I}_z \cdot (1 - 3 \cdot \cos^2 \theta) \\
B &= -\tfrac{1}{4} \cdot (\hat{S}_+ \cdot \hat{I}_- + \hat{S}_- \cdot \hat{I}_+) \cdot (1 - 3 \cdot \cos^2 \theta) \\
C &= -\tfrac{2}{3} \cdot (\hat{S}_+ \cdot \hat{I}_z + \hat{S}_z \cdot \hat{I}_+) \cdot \sin\theta \cdot \cos\theta \cdot e^{-i\varphi} \\
D &= -\tfrac{2}{3} \cdot (\hat{S}_- \cdot \hat{I}_z + \hat{S}_z \cdot \hat{I}_-) \cdot \sin\theta \cdot \cos\theta \cdot e^{i\varphi} \\
E &= -\tfrac{4}{3} \cdot \hat{S}_+ \cdot \hat{I}_+ \cdot e^{-2i\varphi} \\
F &= -\tfrac{4}{3} \cdot \hat{S}_- \cdot \hat{I}_- \cdot e^{2i\varphi}
\end{aligned} \right\} \quad (2.6)$$

Here, the x- and y-components of the spin operators are expressed in terms of the raising and lowering shift operators. Depending on the specific way these spin operators are acting on a state of the unperturbed Hamiltonian (the electron Zeeman interaction, being the leading term in X-band EPR and even more so in high-field EPR) the $A, B, \ldots, F$ terms are classified as secular (A-term), pseudosecular (B-term) and nonsecular (C-, D-, E-, F-terms). Only the secular and pseudosecular parts in the END term contribute in first order to the dipolar splitting of the unperturbed energy levels. In EPR (or NMR or ENDOR) the coherent microwave (or radio frequency) fields are normally applied in a direction perpendicular to the static external Zeeman field so that the selection rules $\Delta m_S = \pm 1$, $\Delta m_I = \pm 1$ hold. As a consequence, the nonsecular terms are unimportant for determining the line positions of the spectra to first-order perturbation theory. Higher-order satellites due to small state admixtures will have vanishingly small intensities as long the Zeeman terms are much larger than the *C*-, *D*-, *E*-, *F*-terms. This does not mean, however, that the nonsecular END parts, when becoming time dependent, are unimportant for the electron and nuclear relaxation pathways. They are induced by the pertaining randomly fluctuating local fields, for instance owing to Brownian motion of the radicals in liquid-state samples. Then, the time-dependent nonsecular interactions induce relaxation transitions of the electron spins (W_e) and nuclear spins (W_n) as well as coupled "flop-flop" crossrelaxation transitions (W_{x2}), see Section 2.2.1.1.

The tensors $\tilde{g}$, $\tilde{A}$ and $\tilde{P}$ of the electron Zeeman interaction, the hyperfine interaction and the quadrupole interaction probe the electronic structure (*i.e.* the electron wavefunction and energy) of the molecule either globally (*g*-tensor) or locally (hyperfine- and quadrupole-tensors). The *g*- and *A*-tensors contain isotropic and anisotropic contributions, whereas the *P*-tensor is traceless, *i.e.* contains only anisotropic contributions. In isotropic fluid solution at temperatures high enough for fast molecular tumbling, the anisotropic interaction components are averaged out so that only the isotropic values, $1/3 \cdot \text{Tr}(\tilde{g})$ and $1/3 \cdot \text{Tr}(\tilde{A}_i)$, contribute to the observed line positions. In frozen solutions, powders or single crystals, anisotropic tensor contributions also become observable, providing that the necessary spectral resolution conditions prevail, *i.e.* the separations of the lines are larger than their linewidths. For this situation, the information content of the EPR spectra is, of course, considerably enhanced by spatial information and, for example, molecular orientations with respect to $\boldsymbol{B}_0$ or electron–nuclear distances can be extracted.

In the strong-field approximation, the energy eigenvalues of eqn (2.1) are classified by the magnetic spin quantum numbers, m_S and m_I, and are given (without quadrupole contribution), to first order (in frequency units) by

$$E(m_S, m_{I_i})/h = \frac{g \cdot \mu_B}{h} \cdot B_0 \cdot m_S - \sum_i \frac{g_{ni} \cdot \mu_K}{h} \cdot B_0 \cdot m_{I_i} + \sum_i A_i \cdot m_S \cdot m_{I_i} \quad (2.7)$$

where the scalar interaction parameters *g*, *A* are the square root-values of the squared tensors and contain the desired information about magnitude and orientation of the interaction tensors.[3] To keep the energy expression simple, the contributions due to the quadrupole interaction, *P*, are omitted for the moment. In thermal equilibrium between the spin system and the lattice, the energy levels are populated according to the Boltzmann distribution. When irradiating the sample with microwaves at a fixed frequency and sweeping the external field through the resonance region, EPR transitions occur according to the first-order selection rules $\Delta m_S = \pm 1$, $\Delta m_{I_i} = 0$. This leads to an EPR spectrum of absorption lines that is characteristic for the electron-spin interactions of the sample molecules. The intensity distribution of the hyperfine lines is determined by the number of symmetry-equivalent nuclei. Hence, the intensity ratio of the hyperfine lines is a valuable aid for assigning the lines to specific nuclear positions in the molecule. From eqn (2.7) it follows that the purely nuclear-spin interactions, *i.e.* the nuclear Zeeman and quadrupole interactions, do not show up in the EPR spectrum as long as the first-order approximation for the selection rules holds. Second-order contributions from the nuclear

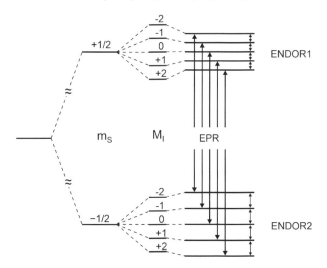

Figure 2.1 High-field spin energy levels of a radical ($S=1/2$) with a group of four equivalent protons ($I=1/2$) dissolved in fluid solution. $M_I = \sum m_{I_i}$ is the total nuclear quantum number of the group. The allowed EPR and ENDOR transitions are marked by arrows. The energy splittings are drawn according to eqn (2.7) with $A>0$.

Zeeman and quadrupole energy terms can often be neglected. However, this is only justified as long as the nuclear spin terms are considerably smaller than the other two terms. This might no longer be the case at high Zeeman fields or for substantial quadrupole couplings.

As a simple example, a doublet radical ($S=1/2$) is considered, containing four symmetry-equivalent protons (each with $I=1/2$) in a strong B_0 field. Figure 2.1 shows the energy-level scheme according to eqn (2.7) for the case that the radical is dissolved in isotropic fluid solution. In this situation, only scalar interactions prevail provided that fast molecular tumbling occurs. As a consequence of the strong-field EPR selection rules, five EPR lines are observed with binominal intensity distribution owing to the first-order transition-frequency degeneracies of equivalent nuclei. Their isotropic hyperfine coupling constant (hfc) $A_{\text{iso}} = 1/3 \cdot \text{Tr}(\tilde{A}_i)$ can be directly read off from the separation of the hyperfine lines in the EPR spectrum.

For single-crystal samples of the same radical, the complete g- and hyperfine-tensor information, *i.e.* both the isotropic and anisotropic contributions, can be extracted from the angular dependence of the EPR lines ("rotation patterns") when the crystal is rotated about three orthogonal axes. Note that there exists no quadrupole interaction with protons or other nuclei with $I=1/2$. If the tensors $\tilde{g}$ and $\tilde{A}_i$ are collinear, *i.e.* have the same principal axes system (α, β, γ), their rotation patterns will have the same angular dependence. When the crystal is mounted on a proper wedge in such a way that the crystal rotation axes are parallel to the molecular axes system (x, y, z) and remain under rotation perpendicular to the B_0 field direction, the actual rotation occurs in a molecular plane defining an angle θ between the rotation axis and the field direction. Iterated measurements of the angular dependence of the apparent g- and A-values at different orientations give the elements of the tensors $\tilde{g}^2$ and $\tilde{A}^2$, which can be transformed to principal axes. One could begin by rotating the field in the xy plane, followed by rotations in the yz and zx planes. Then, the dependence of the apparent g-value on the rotation angle θ_{ij} in the ij plane takes the form:[3,19]

$$g(\theta_{xy}) = \left(g_{xx}^2 \cdot \sin^2\theta_{xy} + g_{yy}^2 \cdot \cos^2\theta_{xy} + g_{xy}^2 \cdot \sin 2\theta_{xy}\right)^{1/2} \quad (2.8a)$$

(notice that g_{ij}^2 denotes the components of the squared $\tilde{g}$-tensor). An analogous expression holds for the hyperfine coupling value A_i of nucleus i in the limit of small g-anisotropy, as is typical for bio-organic systems as long as they do not contain paramagnetic transition-metal ions. For the other two molecular planes, yz and zx, the corresponding tensor components are found by cyclic replacement $y \to z \to x \to y$, i.e.,

$$g(\theta_{yz}) = \left(g_{yy}^2 \cdot \sin^2\theta_{yz} + g_{zz}^2 \cdot \cos^2\theta_{yz} + g_{yz}^2 \cdot \sin 2\theta_{yz}\right)^{1/2} \quad (2.8b)$$

$$g(\theta_{zx}) = \left(g_{zz}^2 \cdot \sin^2\theta_{zx} + g_{xx}^2 \cdot \cos^2\theta_{zx} + g_{zx}^2 \cdot \sin 2\theta_{zx}\right)^{1/2} \quad (2.8c)$$

with the rotation angles θ_{ij} defined as the respective angles between the i- and j-axis and the field direction.

For many bio-organic samples, in particular in frozen solution, conventional X-band EPR runs into problems with spectral resolution. This is because several radical species or different magnetic sites with rather similar g-values may be present or because a small g-tensor anisotropy does not allow canonical orientations of the powder spectrum to be resolved. From the spin Hamiltonian eqn (2.1) one sees that some interactions are magnetic-field-dependent (the Zeeman interactions), while others are not (the hyperfine and quadrupole interactions). Consequently, if paramagnetic species with different g-factors or with anisotropic g-tensor components are present, the difference in resonance field positions ΔB_0 is proportional to the frequency of the electromagnetic irradiation:

$$\Delta B_0 = \frac{h \cdot \nu}{\mu_B} \cdot \left(\frac{1}{g_1} - \frac{1}{g_2}\right) \quad (2.9)$$

This shows, in analogy to modern NMR spectroscopy, that higher mw EPR frequencies, ν, and corresponding resonance fields, B_0, should lead to enhanced spectral resolution, at least as long as the linewidths do not increase with field. Except for transition-metal complexes, for many bio-organic systems $g \cong 2$, and relative g-value variations $\Delta g/g$ rarely exceed 10^{-4}–10^{-3}. At X-band frequencies, therefore, the line separation due to g-value differences is only

$\Delta B_0 = 0.03$–0.3 mT, which can easily be masked in disordered solid-state samples with typical linewidths around 1 mT. An increase of the mw frequency by, for instance, a factor of 10 (W-band, 95 GHz) improves the spectral resolution accordingly. This is a consequence of the increasing importance of the g-tensor components in the electronic Zeeman interaction as the magnetic field is increased. Moreover, the spectral analysis is generally simpler in high-field EPR because the first-order approximation for the energies often applies.

The question arises, what defines "high" in high-field EPR, *i.e.* how large should the Zeeman field be for a particular sample? For all cases of delocalized spin systems, in which unresolved hyperfine interactions dominate the inhomogeneous EPR linewidth, a "true" high-field experiment must fulfil the condition

$$\frac{\Delta g}{g_{\mathrm{iso}}} \cdot B_0 > \Delta B_{1/2}^{\mathrm{hf}} \qquad (2.10)$$

which relates the Zeeman field B_0 of the spectrometer with properties of the sample, *i.e.* the anisotropic electron Zeeman interaction must exceed the inhomogeneous linewidth, $\Delta B_{1/2}^{\mathrm{hf}}$. Apparently, there are two options to fulfil this condition, either to make the Zeeman interaction large enough or to reduce the linewidth sufficiently. Hence, instead of fixing a minimum field/frequency value to meet the "high-field EPR" benchmark, one should ensure that the chosen B_0 value renders the Zeeman splitting larger than the inhomogeneous linewidth. Nevertheless, it has become common practice in the EPR community to use the term "high-field EPR" for microwave frequencies in the W-band or higher and Zeeman fields produced by superconducting magnets. For example, for *deuterated* samples, Q-band EPR might already fulfil the high-field condition eqn (2.7) in the case of semiquinone radicals with a rather large g-anisotropy,[20–22] whereas for *protonated* samples with inherently larger linewidths, it does not. On the other hand, in the case of chlorophyll ion radicals, due to their very small g-anisotropy, even W-band EPR might not meet the high-field condition for protonated samples. Then, deuteration of the sample will be necessary or, as an alternative, a further increase of the mw frequency and B_0 field is required, for instance by resorting to 360-GHz EPR (see Section 5.2.1.3). Fortunately, for many protein systems with no paramagnetic transition-metal ion sites or, at least, no substantial spin density at such a site, the increase of line separation ΔB_0 with increasing Zeeman field directly translates into an increase of spectral resolution, because often no noticeable line broadening due to "g-strain" effects occurs with increasing B_0. For the primary donor cation radical in reaction centers (RCs) from *Rb. sphaeroides*, for example, up to 24 T were applied,[23] and g-strain broadening was found to be negligible.

The term "g-strain" is used to describe a spread of principal g-factors caused by heterogeneities of the local environment of the spins and leading to additional line broadening that increases linearly, or approximately linearly with the Zeeman field. This phenomenon is well known in the case of paramagnetic transition-metal complexes such as metallo-proteins[24,25] where large spin-orbit coupling can produce dramatic variations in the g-values with varying crystal fields. Such effects are, however, expected to be small in the case of organic radicals in proteins for which small g-anisotropies are the rule. Apparently, the primary donor cation radical in RCs from *Rb. sphaeroides* is not an exception of this rule.

In addition to the improved g-resolution mentioned above, high-field EPR can also improve the detection sensitivity. In this respect one has to distinguish between the *absolute* and *relative* (or *concentration*) sensitivities because the amount of sample that can be introduced into the cavity of high-field EPR spectrometers is usually significantly smaller than in standard X-band spectrometers. Consequently, the amount of sample available has to be taken into account and, when this is limited, like in single crystals of proteins, the sensitivity of high-field EPR can be superior by orders of magnitude.

This statement has to be taken with a grain of salt: In principle, the sensitivity should increase more than linearly with increasing microwave resonance frequency (or Zeeman field). In practice, however, most existing high-field EPR spectrometers fall short of the theoretical sensitivity. For cw mode of operation, the minimum detectable number of spins, N_{min}, depends on a number of parameters such as the Zeeman frequency, ν_e, the temperature, the sample volume, the Q-value and microwave coupling of the resonator, its filling factor, the EPR linewidth and optimum modulation amplitude, the noise figures and insertion losses of the microwave source, detector and components, the microwave power, the noise figure of the preamplifier and the effective bandwidth of the lock-in amplifier.[26] Only if one assumes the unrealistic case that the noise characteristics of the microwave sources, components, detectors and amplification stages are frequency independent up to the mm and sub-mm range, can simple expressions for the frequency dependence of the detection sensitivity be derived.[26,27] For constant incident microwave power and unsaturated EPR lines, one obtains *(i)* an *absolute* value of $N_{min} \propto \nu_e^{-9/2}$ and *(ii)* a *relative* value $N_{min}/V_S \propto \nu_e^{-3/2}$, where V_S is the sample volume. Case *(i)* applies to the situation in which the sample size cannot be varied, *i.e.* when V_S is constant and the filling factor is proportional to V_S/V_C, the cavity volume V_C scaling with ν_e^{-3}. This corresponds to small-size single crystals. Case *(ii)* applies to the situation in which there is enough sample material available to put into the resonator without lowering its Q-value significantly. In real practice, the noise characteristics of source, components and detectors at frequencies up to about 100 GHz (W-band) have improved in recent years so drastically that the detection sensitivity of dedicated W-band EPR spectrometers clearly exceeds that of standard X-band spectrometers (see Section 3.3.1). The situation is still very unfavorable in the sub-mm microwave regime (far-IR) where the noise characteristics of sources, components and detectors perpetuate to limit the actual sensitivity of ultrahigh-field EPR to values at best comparable to those of X-band spectrometers (see Section 3.3.2).

These sensitivity considerations are also appropriate when tackling the problem of accurately measuring, by EPR methods, the concentration of paramagnetic molecules in a given sample. This is a very relevant field of EPR application but, admittedly, somewhat neglected in the general academic discussion, unless specialized meetings are attended. We notice, however, that there is a growing need for quantitative EPR not only in industrial and academic research, but also in public-health institutions to provide protocols for accurate quantitative results from EPR measurements in different laboratories that can be compared in a meaningful way. Both the relative intensity quantification of EPR samples with respect to reliable EPR concentration standards and the absolute spin concentration of samples are often of interest.

Typical application areas of EPR methodologies for measuring the concentration of primary or induced radicals are tooth-enamel dosimetry of ionizing radiation after radioactive fallout from nuclear test explosions or accidents in nuclear power or fuel-reprocessing plants,[28–33] food control to trace γ-irradiated fruits and spices[34,35] and determination of intra- and extracellular oxygen concentrations by means of the EPR spectra of specially designed nitroxide radicals.[36,37]

Just to give a few important examples of nuclear accidents since the mid-1950s with severe radioactive fallout and the necessity to carry out large-scale dosimetry of the potential victims of nuclear pollution: *(i)* On September 29, 1957, the Soviet nuclear fuel-reprocessing plant near Chelyabinsk in the Ural mountains had a severe explosion accident with subsequent heavy radioactive contamination of the environment. *(ii)* Only 11 days later, on October 10, 1957, the Sellafield nuclear fuel-reprocessing facility on the Cumbrian cost, UK, had the "Windscale fire", the most severe nuclear accident on British soil. The fire destroyed one of the "Windscale Piles" and released a substantial amount of radioactive material to the environment. *(iii)* On March 28, 1979, the Three Mile Island nuclear power plant located near Harrisburg, PA, USA, had the most significant accident with a partial core meltdown in the history of the American commercial power generating industry. It resulted in the release of a significant amount of radioactivity to the environment. *(iv)* On April 26, 1986, reactor number four of the Chernobyl nuclear power plant located near Pripyat, Ukraine, in the Soviet Union exploded. Further explosions and the resulting fire sent a plume of highly radioactive fallout into the atmosphere. Nearly thirty to forty times more fallout was released than by the nuclear bomb on Hiroshima. The radioactive plume drifted over parts of the western Soviet Union, eastern, western and northern Europe and North America. According to official post-Soviet data, an extensive geographical area was badly contaminated, resulting in the evacuation and resettlement of over 336 000 people. *(v)* On September 30, 1999, Japan's worst nuclear radiation accident took place at a uranium fuel-processing facility in Tokaimura, northeast of Tokyo. The direct cause of the criticality accident was attributed to workers putting uranyl nitrate solution containing about 16.6 kg of uranium, which exceeded the critical mass, into a precipitation tank, thereby setting off a nuclear chain reaction. Several workers were exposed to neutron- and γ-radiation doses in excess of tolerable limits and died. Hundreds of other plant and emergency workers as well as nearby residents received lesser doses of radiation but had to be hospitalized. Many thousands of other people were forced to remain indoors for a few days to reduce radiation exposure.

Immediately after such large-scale nuclear accidents, public health authorities ought to obtain reliable data of radiation doses to which people in the vicinity of the accident and further away had been exposed. For example, EPR laboratories in the USSR got involved in screening measurements of many individuals exposed to radiation fallout from the Chernobyl accident, and concentrations of radiation-induced radicals have been determined by EPR from the enamel of teeth extracted from the suspected victims.[28] Moreover, very recently scenarios have been described in the scientific literature to use, in cases of nuclear fallout from accidents, terrorism or war, *in-vivo* EPR for dosimetry of populations that have been exposed to unknown doses of ionizing radiation.[33] In such situations of emergency, there is an urgent need to determine, for a large number of potential victims, the magnitude of the radiation exposure to each individual. For those who received high doses of radiation with significant health risk, appropriate clinical treatments have to be initiated immediately, while those with low doses of exposure and corresponding low probability of acute health effects can be considered for later treatment by the medical emergency institutions. *In-vivo* EPR measurements of radiation-induced changes in the enamel of teeth is a promising method, perhaps the only such method, that can differentiate among radiation doses with sufficient accuracy and speed for classifying individuals into categories of urgency for medical treatment. In its current state, the *in-vivo* EPR dosimeter for repeated measurements of a tooth in the mouth of the subjects can provide estimates of absorbed dose with an error of approximately ± 25 cGy over the range of interest for acute biological effects of radiation. The time required for data acquisition, the lower detection limit, and the precision of the determined dose values are expected to improve with ongoing improvements of the in-mouth resonator design, the magnet system and the algorithm for acquiring the data and calculating the dose.

During recent years, it is interesting to observe that specialized workshops are increasingly being offered to the EPR community that discuss the various aspects of quantitative EPR in detail. Examples of those aspects are related to proper standard samples, adequate instrument hardware and suitable data-handling software requirements that should be considered for obtaining meaningful results. Specific topics to be discussed include: choosing a reference standard that can be used (and reproduced) in any EPR laboratory, optimum resonator configuration (Q-value, microwave field B_1, modulation frequency and amplitude), mw power-saturation characteristics, sample positioning and homogeneity of static and oscillating fields. All these factors have to be put together to provide a calculation model for obtaining an accurate spin concentration of the sample.

2.1.2 Triplet States and High-Spin Transition-Metal Ions ($S > 1/2$)

Also for high-spin systems ($S > 1/2$), such as enzymatic proteins with one or several transition-metal cofactors, EPR spectroscopy at high magnetic fields might be advantageous. For such systems "fine-structure" and exchange terms have to be added to the spin Hamiltonian, *i.e.* eqn (2.1) has to be extended by

$$\hat{H}_{\text{ZFS}}/h + \hat{H}_{\text{exchange}}/h = \hat{S} \cdot \tilde{D} \cdot \hat{S} - J \cdot (\hat{S}^2 - 1) \quad (2.11)$$

with the total spin $S = (S_1 + S_2 + \ldots)$. Here, $\tilde{D}$ is the traceless zero-field splitting (ZFS) tensor, and J is the isotropic exchange interaction parameter (in standard convention, $J < 0$ corresponds to antiferromagnetic coupling, $J > 0$ to ferromagnetic coupling). In the principal axes system of $\tilde{D}$, the anisotropic part in eqn (2.11) is normally rewritten in terms of the zero-field parameters D, E:

$$\hat{H}_{\text{ZFS}}/h = \hat{S} \cdot \tilde{D} \cdot \hat{S} = D \cdot \left(\hat{S}_z^2 - \frac{1}{3} \cdot S \cdot (S+1)\right) + E \cdot \left(\hat{S}_x^2 - \hat{S}_y^2\right) \quad (2.12)$$

with $D = 3/2 \cdot D_{zz}$ and $E = (D_{xx} - D_{yy})/2$, where D_{xx}, D_{yy} and D_{zz} are the principal values of the ZFS tensor. For a triplet state, they are related to the zero-field energy eigenvalues (energy levels) of the triplet-spin eigenfunctions $|T_x\rangle, |T_y\rangle, |T_z\rangle$, as follows:[19]

$$E_x/h = 1/3 \cdot D - E, \quad E_y/h = 1/3 \cdot D + E, \quad E_z/h = -2/3 \cdot D \quad (2.13)$$

Obviously, it might happen that EPR transitions of high-spin systems with large zero-field splittings cannot be observed at all at standard X-band frequencies because the energy of the mw quantum is too small. For such cases, the higher quantum energy of high-frequency microwaves can drive the transitions.[38,39] An example for such a biological high-spin system is metmyoglobin with $S = 5/2$ ferric heme, for which EPR transitions at 130 GHz became observable that had been undetectable at X-band due to the large zero-field splitting.[40]

Although in the spin Hamiltonian the fine-structure term is not field dependent it leads, in combination with the electronic Zeeman term, to a field-dependent mixing of the electron-spin eigenfunctions. At zero field, the triplet-spin eigenfunctions, $|T_x\rangle, |T_y\rangle, |T_z\rangle$, are quantized along the molecular axes system (x, y, z). At high field, the magnetic spin quantum number, $m_S = +1, 0, -1$, is a good quantum number, and the spin eigenfunctions become $|T_{+1}\rangle$, $|T_0\rangle, |T_{-1}\rangle$. If the external field values B_0 are such that the electron Zeeman and the fine-structure splittings are comparable in magnitude, the spin functions become mixed functions of both bases, the degree of mixing depending on B_0 and the relative orientation of the molecule with respect to the field. As a consequence, the triplet energy eigenvalues of the different electron wavefunctions are not linearly related to the strength of B_0. This intermediate region requires more complicated calculations to analyze the EPR spectrum. Hence, another reason to perform high-field EPR on high-spin systems ($S > 1/2$) is to simplify the spectrum analysis.

There is one more benefit: the EPR lines of high-spin systems usually get narrower at higher magnetic fields than in X-band EPR spectra, again because of second-order effects. If we take Mn^{2+} centers ($S = 5/2$, $I = 5/2$) in disordered protein samples as an example, the EPR transitions are strongly broadened by contributions from the zero-field tensor. Their linewidth, $\Delta B_{1/2}$, is determined by second-order contributions from the zero-field coupling D, $\Delta B_{1/2} \propto D^2/B_0$. No wonder, therefore, that "needle sharp" manganese hyperfine lines are normally observed in the 95-GHz high-field EPR spectra of Mn^{2+} containing protein complexes, such as photosystem II of oxygenic photosynthesis, even in disordered frozen-solution samples.

This behavior is exploited by many high-field EPR spectroscopists who use Mn^{2+} ions doped into MgO powder as a reference sample for magnetic field calibration and precise g-factor measurements.[41] Up to second order, the EPR resonance fields of the six Mn^{2+} hyperfine components, $m_I = -5/2, \ldots +5/2$, are given by

$$B_{m_I} = B_0 - A_{iso} \cdot \left[m_I - \frac{A_{iso}}{2 \cdot B_0} \cdot \left(I \cdot (I-1) - m_I^2 \right) \right] \quad (2.14)$$

where A_{iso} is the isotropic Mn^{2+} hyperfine coupling in field units. The electronic g-factor is contained in the Zeeman field $B_0 = F \cdot v/g$ (v microwave frequency, $F = h/\mu_B = 71.447751 \text{ mT GHz}^{-1}$). The high-precision reference data are: $g(Mn^{2+}) = 2.00101 \pm 0.00005$ and $A_{iso}(Mn^{2+}) = -(8.710 \pm 0.003)$ mT.[41]

To say it again: High-field EPR is particularly useful for half-integer high-spin systems ($S = (2 \cdot n + 1)/2$, $n = 1, 2, \ldots$), such as Mn(II) and Fe(III) with $S = 5/2$, and Gd(III) with $S = 7/2$. For such systems the inhomogeneous linewidth of orientationally disordered samples, $\Delta B_{1/2}$, of the central $|-1/2> \rightarrow |1/2>$ EPR transition is determined by second-order contributions from the zero-field coupling. Hence, when $g \cdot \mu_B \cdot B_0/h \gg D$ the broadening becomes negligible and narrow signals are obtained also in orientationally disordered samples. This leads to increased sensitivity and resolution. In this case the first-order approximation is valid and the resonance fields are given by:

$$B(m_S \rightarrow m_S + 1) = B_0 - \frac{h}{g \cdot \mu_B} \cdot A_{iso} \cdot m_I + \frac{h}{g \cdot \mu_B}(2m_S + 1) \cdot \nu_D$$
$$\nu_D = \frac{D}{2} \cdot \left[(3\cos^2\theta - 1) + \eta \cdot \sin^2\theta \cdot \cos 2\varphi \right] \quad (2.15)$$

where A_{iso} is the isotropic hyperfine coupling of ^{55}Mn (in frequency units), m_I is the corresponding nuclear spin projection and $\eta = 3 \cdot E/D$. The angles θ and φ describe the orientation of the magnetic field relative to the principal axes system of the ZFS tensor.

Another important aspect of high-field EPR on high-spin systems, for example organic triplet states and radical pairs with $S=1$, is the possibility to determine the *absolute* sign of the zero-field parameter D by sufficient thermal polarization of the triplet levels already at moderately low temperatures.[42,43] If the temperature is low enough to fulfil the condition $k_B T < g \cdot \mu_B \cdot B_0$ the lowest spin level, corresponding to $m_S = -1/2$, is predominantly populated, and the EPR spectra of disordered samples become asymmetric. The "Zeeman temperature" $T_Z = g \cdot \mu_B \cdot B_0 / k_B$ is defined accordingly as the temperature around which this asymmetry becomes pronounced, *i.e.* a higher line intensity is observed either on the high-field side ($D > 0$) or on the low-field side ($D < 0$) of the spectrum. Approximate values of the Zeeman temperature at different EPR frequencies are: $T_Z = 0.4$ K (at 9.5 GHz), 4 K (at 95 GHz), 6.5 K (at 140 GHz) and 15.5 K (at 360 GHz). The sign of D is indicative of the shape of the dipolar tensor. Organic triplet states generally have disk-shaped dipolar tensors ($D > 0$), whereas weakly coupled radical pairs have cigar-shaped ones ($D < 0$).

2.2 High-Field EPR, ENDOR, TRIPLE, ESEEM, PELDOR and RIDME

We now turn to several extensions of high-field/high-frequency EPR, such as ENDOR, TRIPLE, ESEEM, PELDOR and RIDME, in some more detail to elaborate on what can be additionally learned about (bio)chemical and biological systems when going beyond conventional X-band techniques.

2.2.1 ENDOR and TRIPLE Hyperfine Spectroscopy

Thorough accounts of high-field/high-frequency ENDOR and TRIPLE spectroscopy have recently been published, and we suggest them for further reading.[14,44,45]

For large low-symmetry radicals with the unpaired electron delocalized over many spin-carrying nuclei, for example the cofactor ion radicals occurring in photosynthetic electron transfer, with each set of inequivalent nuclei the number of EPR lines increases in a *multiplicative* way, according to the EPR selection rules $\Delta m_S = \pm 1$, $\Delta m_{Ii} = 0$. This results in strongly inhomogeneously broadened EPR spectra because individual hyperfine lines can no longer be resolved in the available spectral range that, for $g = 2$ systems, is restricted to *ca.* 3 mT due to the normalization condition for the unpaired electron spin density. For such cases, by resorting to ENDOR techniques the spectral resolution can be greatly improved because ENDOR is inherently a variant of NMR on paramagnetic systems, the unpaired electron serving as a highly sensitive detector for the NMR transitions $\Delta m_{Ii} = \pm 1$. Each group of equivalent nuclei – no matter how many nuclei are involved and of what value their individual spin is – contributes only two ENDOR lines because, within an m_S manifold, the hyperfine levels are equidistant to first order (see Figure 2.1).

Hence, in ENDOR with each set of inequivalent nuclei the number of resonance lines increases merely in an *additive* way. Double-resonance excitation thus offers the advantage of NMR in terms of high resolution *via* reduced number of redundant hyperfine lines in conjunction with the advantage of EPR in terms of detecting low-intensity rf transitions *via* high-intensity mw transitions, *i.e.* by means of quantum transformation.

In cw ENDOR, the sample is irradiated simultaneously by two electromagnetic fields, a mw field (to drive EPR transitions $\Delta m_S = \pm 1$) and an rf field (to drive NMR transitions $\Delta m_{Ii} = \pm 1$). Under appropriate experimental conditions, which are more stringent for cw than for pulse-irradiation schemes.[1,4,46] ENDOR signals are observed by monitoring the changes of EPR line intensities when sweeping the rf field through the nuclear resonance frequencies. In cw Special TRIPLE resonance[47] the two rf fields are applied at frequencies symmetrically placed around the nuclear Larmor frequency to enhance the signal intensity. In cw General TRIPLE resonance[48] two rf fields with independently variable frequencies are applied, one pumping a selected ENDOR transition while the other is swept through the ENDOR spectrum. From the resulting characteristic intensity changes in the ENDOR spectrum the relative signs of the hyperfine couplings can be directly read off.

In the pulsed version of triple resonance,[49] the first rf pulse (pump pulse) with fixed frequency pumps a specific nuclear transition, while the frequency of the second rf pulse is swept to cover the resonance region. The time separation between the two rf pulses can be varied. Again, assignment of hyperfine couplings and determination of their relative signs are the main goals. When the frequencies of the two rf pulses are varied independently, a two-dimensional triple-resonance spectrum is obtained.[50]

Apparently, the gain in resolution of ENDOR *versus* EPR, becomes very pronounced for low-symmetry molecules with increasing number of groups of symmetry-related nuclei. The resolution enhancement becomes particularly drastic when nuclei with different magnetic moments are involved. Their ENDOR lines appear in different frequency ranges and, providing their Larmor frequencies are separated at the chosen Zeeman field value B_0, the different nuclei can be immediately identified. In the case of an accidental overlap of ENDOR lines from the different nuclei at X-band (9.5 GHz, 0.34 T) the lines can be separated when working at higher Zeeman fields and mw frequencies, for instance at 3.4 T, 95 GHz[51] or even at 12.9 T, 360 GHz.[52] This disentangling of ENDOR lines by different field/frequency settings for the EPR condition is depicted in Figure 2.2. In biological molecules with several nonproton magnetic nuclei this separation of accidentally overlapping ENDOR lines is of great help for analyzing complex spin systems by means of their nuclear Zeeman and hyperfine interactions.

For a doublet radical with electron–nuclear hyperfine interaction, but without nuclear quadrupole interactions, according to its spin Hamiltonian only two ENDOR lines of a particular group of equivalent nuclei with $I = 1/2$, appear, to first order, at

$$\nu^{\pm}_{i\text{ENDOR}} = |\nu_n \pm A_i/2| \qquad (2.16)$$

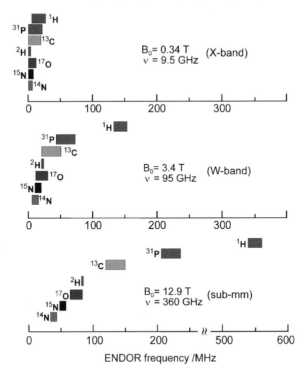

Figure 2.2 Gain in ENDOR resolution for doublet-state systems ($S=1/2$, $g=2$) with increasing Zeeman field B_0 and microwave frequency v. Spectral lines of typical nuclei in organic biomolecules, largely overlapping at traditional X-band ENDOR ($B_0 = 0.34$ T, $v = 9.5$ GHz), become completely separated at 360 GHz/12.9 T.

Here, the nuclear Larmor frequency is given by $v_n = g_n \cdot \mu_K \cdot B_0/h$, and the hyperfine coupling parameter A contains isotropic and anisotropic contributions and is defined analogous to eqn (2.8). Obviously, the two ENDOR lines are symmetrically displayed about v_n or $A/2$, whichever is larger. In an isotropic solution, the hyperfine couplings (hfcs) are given by $A_{iso} = 1/3 \, \text{Tr}(\tilde{A})$.

2.2.1.1 Liquid-Solution Steady-State ENDOR and TRIPLE

In this section we present a phenomenological explanation of liquid-phase cw steady-state ENDOR signal intensities and their alteration by applying a second rf field in TRIPLE resonance experiments. In an isotropic liquid solution, at sufficiently elevated temperatures, the anisotropic contributions of the g-tensor are effectively averaged out. Nevertheless, EPR and ENDOR studies at higher frequency/field settings than for standard X-band EPR can be of great advantage: If in the sample several radical species with only slightly different isotropic g-values are present, for example as intermediate or final reaction products with

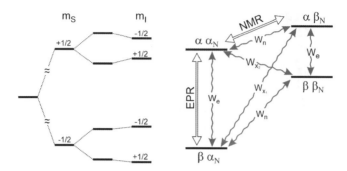

Figure 2.3 Four-level diagram of a system with $S = 1/2$, $I = 1/2$ with relaxation rates $W_{\alpha\beta}$ (a positive sign of the isotropic hyperfine coupling constant is assumed). Only one EPR and one NMR induced transition are shown. The Greek letters $\alpha\alpha_N$, $\alpha\beta_N$, ... denote the values $+1/2, +1/2$; $+1/2$, $-1/2$; ... of the electron and nuclear magnetic spin quantum numbers m_S and m_I, respectively. The wavy lines indicate the relaxation transitions with their rates. For details, see text.

g_1 and g_2, they can be spectroscopically resolved and identified at sufficiently high Zeeman fields according to eqn (2.9). Running high-field ENDOR on the respective g_1 and g_2 positions in the EPR spectrum would provide access to the individual hyperfine coupling constants of the two radicals. They carry detailed information about the identity of the radicals and their electronic structure. Another good reason for applying multifrequency EPR and ENDOR on liquid-state samples would hold for time-resolved measurements of spin-polarized transient reaction intermediates to elucidate the prevailing polarization mechanism via its magnetic field dependence (see Section 2.2.1.2).

Figure 2.3 shows the energy-level diagram and transition scheme of induced mw and rf and relaxation driven EPR and NMR transitions for the simplest case $S = 1/2$, $I = 1/2$. For doublet-state radicals in isotropic solution, each group of equivalent nuclei contributes, to first order, only two ENDOR lines to the spectrum:

$$\nu_{\text{ENDOR}}^{\pm} = |\nu_n \pm A_{\text{iso}}/2| \qquad (2.17)$$

with the nuclear Larmor frequency $\nu_n = g_n \cdot \mu_K \cdot B_0/h$ and the isotropic hyperfine coupling constant A_{iso}.

The isotropic hyperfine interaction, termed the Fermi contact interaction, represents the energy of the nuclear moment in the local magnetic field produced at the nucleus by electric currents that are associated with the spinning electron.[19] It has the form (see eqn (2.1)):

$$\hat{H}_{\text{Fermi}}/h = A_{\text{iso}} \cdot \hat{S} \cdot \hat{I} = A_{\text{iso}} \cdot (\hat{S}_x \cdot \hat{I}_x + \hat{S}_y \cdot \hat{I}_y + \hat{S}_z \cdot \hat{I}_z) \qquad (2.18)$$

The coupling constant A_{iso} is proportional to the squared amplitude of the electronic wavefunction at the nucleus, see eqn (2.3).

Hence, the contact interaction can only occur when the unpaired electron has a finite probability density ("spin density") in its s-orbitals at the nucleus. In other words: Since the p-, d-, f- or higher orbitals have nodes at the nucleus, for an isotropic hyperfine interaction to occur the molecular orbital of the unpaired electron must have some s-character of its atomic orbital components at a magnetic nucleus ($I \neq 0$).

The gain in resolution by ENDOR becomes particularly pronounced when nuclei with different magnetic moments are involved and their ENDOR lines appear in different Larmor frequency ranges. If this is not the case at X-band ENDOR, the lines can be disentangled by performing ENDOR at higher Zeeman fields and correspondingly higher mw frequencies.[52] This strategy will be highlighted in Sections 3.4.1.4 for 95-GHz ENDOR and 3.4.2.7 for 360-GHz ENDOR.

At this point, a brief interruption of our digression on experimental techniques for measuring hyperfine coupling constants is appropriate to delineate a few theoretical approaches that were attempted in the history of EPR to relate hyperfine couplings to the electronic (and geometric) structure of the molecule. To this end, isotropic hyperfine coupling constants were analyzed by quantum-chemistry methods of various levels of approximation to translate the spectroscopic data into electronic structure, the actual goal in molecular science.

Since the beginning of EPR applications to chemistry in general, and to organic π-radicals in particular, the hyperfine interaction of the unpaired electron spin with magnetic nuclei in conjugated molecules has been among the most studied phenomena in physical chemistry. Probably, these studies have played an important role in initiating the rapid development of quantum chemistry, starting from the basic Hückel π-electron approach and soon stretching to a variety of powerful semiempirical and, recently, to topical all-electron *ab-initio* and DFT computational methods (see Chapter 4). Over the last forty-five years, neutral and charged doublet-state radicals in liquid solution have been the subject of countless EPR investigations, and a stupendous number of isotropic hyperfine couplings has been accumulated in the literature, ready for comparison with theoretical predictions. It is now generally accepted that the crucial link between the quantum-mechanical description of the electronic structure and the measured hyperfine couplings is rather well understood.[53,54] Such a link between theory and experiment has been revealed at various degrees of approximation in the theoretical approach. Due to the nodal plane of the $2p_z$ atomic orbital of the unpaired π-electron in the singly occupied molecular orbital (SOMO), all the isotropic hyperfine-coupling constants are proportional to the net s-spin density at the nuclei and result exclusively from core-polarization mechanisms. These s-spin densities may be positive or negative and can be directly computed only by using spin-unrestricted procedures. Such calculations led to polarization expressions that relate the isotropic hyperfine coupling of a nucleus with the unpaired π-electron, *i.e.* with the spin density ρ_C^π at a specific atom i of the conjugated molecular skeleton. In the endeavor to find proper descriptions of the spin-polarization mechanisms, it started with the famous McConnell equation.[55] This is a proportionality relation between the hyperfine coupling of an α-proton, which is σ-bonded to a carbon atom in the molecular skeleton, and the π-spin density ρ_C^π in the 2p$_z$-orbital of this carbon. For ion radicals, this relation was extended to account for charge corrections of the σ-π-polarization according to either Colpa–Bolton[56] or Giacometti–Nordio–Pavan.[57] Nowadays, this link between theory and experiment is often provided by large-scale modern quantum-chemical computations, as is discussed in Chapter 4.

One might argue that π-electron theory has served its time. Certainly, the rapid development of DFT and other sophisticated methods in quantum chemistry, coupled with an incredible growth in computer power in recent years, and the availability of corresponding efficient program packages have made it possible to carry out large-scale calculations including all valence electrons, even for quite large biomolecules and even at affordable computation costs. Nonetheless, we feel that in relying too much on the output data of spin densities at all nuclei in the molecule, one may lose an understanding of the physical meaning of the data in terms of electronic structure. Such a feel for the electronic properties of the molecule, and for the structural details that might have an effect on the distribution of the unpaired electron, one can obtain using π-electron theory and McConnell-type relations for the hyperfine couplings. We share this assessment with N.M. Atherton, Sheffield, who – in his renowned monograph *Principles of Electron Spin Resonance*[3] – strikes a blow for keeping a place for π-electron calculations in combination with McConnell-type relations when analyzing hyperfine interactions in organic π-radicals. When Hückel theory is really too crude to reproduce the spin-density distribution, then established extensions of the simple theory should be tried. There exist various methods for semiempirical π-electron calculations that stood the test over the years, among them particularly the McLachlan method[58] and the RHF-INDO/SP method[59,60] should be recommended. The latter method turned out to be very powerful for MO calculations on molecules as large as the chlorophyll radical ions in photosynthesis.[61]

The σ-π-polarization factors, which provide the link between the π-spin densities and the isotropic hyperfine coupling constants, have different form and magnitude for the various nuclei in the molecule. The widely used McConnell equation

$$a_H = Q_{CH}^H \cdot \rho_C^\pi \text{ with the σ-π-polarization factor } Q_{CH}^H \approx -70\,\text{MHz} \quad (2.19)$$

holds for α-protons, and rather precise polarization factors have been determined empirically for many specific classes of organic molecules.

Naturally, spin polarization of the C–H bond, which produces an unpaired 1s-spin density at the hydrogen nucleus, also produces 1s- and 2s-spin densities at the nucleus of an sp^2-hybridized carbon atom. Also, the polarization effects from the neighboring π-spin densities are of great importance regarding the ^{13}C nucleus; they may even lead to considerable magnitudes of negative hyperfine coupling constants. The theoretical analysis of the polarization mechanisms led to the well-known Karplus–Fraenkel relation[62] between an isotropic ^{13}C hyperfine coupling and π-spin densities both at this C-atom and at neighboring conjugated atoms X_i:

$$a_C = \left(S_C + \sum_i Q^C_{CX_i}\right) \cdot \rho^\pi_C + \sum_i Q^C_{X_iC} \cdot \rho^\pi_{X_i} \qquad (2.20)$$

in which the polarization contribution of the inner-shell 1s electrons at the carbon by its ρ^π_C value is described by S_C, and the contributions of the valence-shell 2s electrons are described by the various Qs. Here, Q^A_{BC} is the σ-π-polarization factor for the nucleus of atom A resulting from the interaction between the bond BC and the π-spin density on atom B. A corresponding expression holds for the general case that atom A is being bonded to atom B conjugated in the molecular skeleton. For a planar C–CH–C fragment model, the following σ-π-polarization factors have been calculated:[3,62]

$$S_C = -35.6 \text{ MHz}, Q^C_{CH} = +54.6 \text{ MHz},$$
$$Q^C_{CC'} = +40.3 \text{ MHz}, \qquad (2.21)$$
$$Q^C_{C'C} = -38.9 \text{ MHz}$$

Note that the σ-π-spin-polarization parameters relating the π-spin densities at the carbon atom in question and at the neighboring conjugated atoms with the ^{13}C hyperfine coupling are roughly of equal magnitudes but opposite signs. This leads to a partial compensation of the various polarization effects and, thus, turns the ^{13}C couplings into very sensitive probes for details of the molecular and electronic structure. The experimental challenge, of course, is the fact that the natural abundance of ^{13}C is only about 1%, and isotope enrichment or site-specific isotope labelling has to be done for measuring high-quality ^{13}C hyperfine data.

In addition to the α-protons in C–H bonds of the aromatic skeleton, also the β-protons of alkyl-substituted π-radicals exhibit characteristic hyperfine splittings in the EPR spectra. A β-proton is bonded to a carbon adjacent to the π-skeleton, for instance in a methyl group. Then, the β-proton might get its 1s-spin density both by direct orbital overlap with the π-orbitals ("hyperconjugation") and by core-polarization from ρ^π_C on the carbons in the conjugated π-skeleton. It is found experimentally and theoretically that the link between the β-proton hyperfine splitting and the π-spin density takes the form[3,63,64]

$$a_\beta = (B_0 + B_2 \cdot \cos^2 \theta) \cdot \rho^\pi_C \qquad (2.22)$$

where θ is the angle of twist between the α-carbon $2p_z$ orbital and the plane containing the β-proton C–H bond. The coefficient B_2 reflects the s-spin density arising from hyperconjugation, and B_0 accounts for the orientation-independent spin-polarization mechanism. If there is fast rotation of the β-protons about the C–H bond, e.g., in the case of freely rotating methyl groups, then an averaged value $\langle a_\beta \rangle$ will be observed for each proton: $\langle a_\beta \rangle = (B_0 + 1/2 \cdot B_2)\rho^\pi_C$. Empirical values for the coefficients B_0 and B_2 are found in the literature.[3,65]

A Liquid-Phase ENDOR Intensities and Lineshapes

A detailed theoretical study of *steady-state* multiresonance experiments in the liquid phase has been carried out by J.H. Freed and coworkers in a series of papers[66–68] using the density matrix formalism and Redfield's approximate treatment of relaxation. In one of these papers,[68] subtle lineshape effects – broadenings and splittings – were described that are due to the coherent nature of the applied strong rf and mw fields. A specific coherence effect is particularly interesting because it can be exploited to assign ENDOR lines to molecular positions, *i.e.* when applying ENDOR as an analytical tool. It requires nuclear spins $I > 1/2$ or a set of at least two equivalent nuclei of $I = 1/2$. The magnitude of the coherence splitting is dependent on the hyperfine transitions being mw saturated and on the rf field strength.

This coherence effect was optimized by K.P. Dinse *et al.*[69,70] to assign hyperfine splittings in ENDOR-in-solution spectra of various low-symmetry radicals by counting the number of protons contributing to a specific ENDOR line. A cylindrical ENDOR cavity (TE_{011} mode) was constructed to achieve cw rf fields up to 3 mT (rotating frame). The internal NMR coil is part of the power stage of a 1-kW cw rf transmitter station. To secure thermal stability of the cavity frequency, effective water cooling was employed both for the cavity body and the two-loop NMR coil.[70]

On the basis of Freed's relaxation theory for radicals in fluid solution, M. Plato *et al.*[71] carried out a systematic investigation of the cw ENDOR sensitivity of various heteronuclei, *i.e.* nuclei other than protons, in organic radicals. Optimum ENDOR conditions, such as temperature and viscosity of the solvent, mw and rf field strengths, were formulated as a function of a few nuclear and molecular properties. They include relaxation from fluctuating spin-rotation interaction, electron–nuclear dipolar and nuclear quadrupolar couplings and Heisenberg spin exchange. The theoretical results were found to be in good agreement with experimental observations on ^{2}H, ^{13}C, $^{14/15}$N, ^{19}F, ^{31}P and alkali nuclei in different molecular systems, thus allowing predictions to be made on the ENDOR detectability of other chemically interesting nuclei, such as $^{10/11}$B, ^{17}O, ^{27}Al, ^{29}Si, ^{33}S and $^{35/37}$Cl. In the meantime, most of these nuclei have indeed been detected by cw ENDOR in solution.[65,72,73]

In the following, we will summarize the main aspects of liquid-phase cw ENDOR intensities and lineshapes[74] to give a feel of how to choose the experimental conditions for optimum resolution and sensitivity. We believe that in doing so we address a substantial fraction of the EPR community. Let's face it: Despite the spectacular achievements of advanced multifrequency pulse EPR techniques, there are many scientists in academic and industrial research laboratories who are still applying cw EPR and ENDOR or TRIPLE on radicals in solution because they want to measure hyperfine coupling constants to characterize their specific molecules.

The magnitude of the cw ENDOR effect, *i.e.* the ENDOR signal intensity, is largely determined by the relaxation behavior of the radical in solution. Conceptually, this strong dependence of the signal intensity on the various relaxation rates can be best understood for the simplest case $S = 1/2$, $I = 1/2$, *i.e.* in a four-level energy diagram (see Figure 2.3). For a clear presentation of the various induced and relaxation transitions, the energy-level diagram on the right side has been stretched apart. The relaxation transition rates are indicated by wavy lies: W_e describes the relaxation rate of the electron spins and W_n that of the nuclear spins. W_{x1} and W_{x2} are crossrelaxation rates that describe the flip-flop and flop-flop processes of the coupled electron and nuclear spins. In a phenomenological description, the ENDOR experiment, in which an ESR and a NMR transition are saturated simultaneously, can be visualized as the creation of an alternative relaxation path for the pumped electron spins ($1 \leftrightarrow 2$). This path is activated by driving the NMR transition ($1 \leftrightarrow 3$) and completed via W_e ($3 \leftrightarrow 4$) and W_n ($4 \leftrightarrow 2$) or, even better, via W_{x1} ($3 \leftrightarrow 2$). The extent to which this relaxation bypass can compete with the direct W_e route ($1 \leftrightarrow 2$) determines the degree of desaturation of the ESR line and thereby the ENDOR intensity. In the absence of crossrelaxation, the optimum steady-state ENDOR effect is obtained when $W_n = W_e$.[75,66–68]

Apparently, the intensity pattern of ENDOR lines will generally not reflect the number of contributing nuclei but, rather, is governed by a delicate interplay between the various induced and relaxation transition rates. This is different from EPR and NMR and causes problems in assigning ENDOR lines arising from the same nuclear type to specific positions in the radical. For obtaining good signal-to-noise ratios this interplay of the various transition rates has to be optimized by varying radical concentration, solvent viscosity, temperature, and strength of the microwave and rf fields. Although, in the best case, only about 10% of the ESR sensitivity can be achieved, cw ENDOR is still a very sensitive variant of NMR.

In actual practice, the experimentalist needs theoretical assistance in the multiparameter ENDOR signal-optimization procedure, easier to apply than the full ENDOR relaxation theory. The simplest theoretical approach to predicting the magnitude of the steady-state ENDOR effect as a function of the various spin-lattice relaxation rates is based on rate equations for the level populations. This leads to an electric circuit analogue of the kinetic scheme.[46,76–78] Although such an approach is very helpful in providing a first understanding of the influence of the most critical parameters on the ENDOR intensity, it is not suited to explain the ENDOR spectrum in all details, such as linewidths, saturation behavior, and subtle coherence effects caused by the two strong radiation fields.[70,79,80]

As pointed out above, the most detailed theoretical treatment of steady-state ENDOR in solution has been performed by J.H. Freed and coworkers[75,66–68] using a density-matrix method. This rather complex approach has proved to be extremely powerful in explaining intensity and shape of ENDOR spectra in great detail. Based on their concept, we have introduced as many simplifying assumptions as possible in order to obtain a manageable, though sufficiently realistic, method in which the number of critical parameters is reduced to a minimum.[46,71] What are the critical parameters that govern the ENDOR intensity? To answer this question the complete spin Hamiltonian of the radical solution has to be inspected:

$$\widehat{H}(t) = \widehat{H}_0 + \widehat{H}_1(t) + \widehat{\varepsilon}(t) \qquad (2.23)$$

where $\hat{H}_0$ is the time-independent Hamiltonian (see eqn (2.1)) of the isotropic interactions of radicals in solution:

$$\hat{H}_0/h = \frac{g \cdot \mu_B}{h} \cdot B_0 \cdot \hat{S} - \sum_i \frac{g_{ni} \cdot \mu_K}{h} \cdot B_0 \cdot \hat{I}_i + \sum_i A_{iso}^i \cdot \hat{S} \cdot \hat{I}_i \qquad (2.24)$$

giving the first-order energy levels and transition frequencies (g and A_{iso} are the isotropic parts of the g- and hyperfine-tensors). $\hat{H}_1(t)$ contains interaction energies that are randomly time modulated by the Brownian tumbling motion of the molecule in solution (internal molecular motions and exchange processes are neglected here), and $\hat{\varepsilon}(t)$ represents the sum of interactions of the spins with the coherent radiation fields in the multiresonance experiment. The relaxation term $\hat{H}_1(t)$ is of primary interest since it determines the magnitude of all spin-lattice relaxation rates $W_{\alpha\beta}$ and linewidth parameters $(1/T_2)_{\alpha\beta}$ between any two levels α, β of the spin system. The Brownian rotational motion creates relaxation via modulation of the spin-rotational interaction resulting in W_e, $(1/T_2)_e$, the electron Zeeman interaction resulting in W_e, $(1/T_2)_e$, the electron–nuclear dipole interaction resulting in W_e, W_n, W_{x1}, W_{x2}, $(1/T_2)_e$, $(1/T_2)_n$ and the nuclear quadrupole interaction resulting in W_n, $(1/T_2)_n$.

For most organic radicals, dominant relaxation contributions stem from spin–rotational (SR) and electron–nuclear dipole (END) interactions. SR and END produce the following relaxation effects:[81]

$$\text{SR}: \quad W_e = (2T_{2e})^{-1} = B \cdot \tau_R^{-1} \qquad (2.25)$$

$$\text{END}: \quad W_{\alpha\beta}, (1/T_2)_{\alpha\beta} \propto j^{(A)}(\omega_{\alpha\beta}) \qquad (2.26)$$

where the SR parameter B can be estimated[71] from the deviation of the g-tensor elements from the free-electron value

$$B = 1/2 \, \text{Tr}(\tilde{g} - g_e \cdot \tilde{1})^2 \qquad (2.27)$$

The rotational correlation time, τ_R, can be estimated via the Einstein–Debye relation, $\tau_R = V_{\text{eff}} \cdot \eta/kT$, from the effective molecular volume and viscosity and temperature of the solvent. The various spin-lattice relaxation and linewidth contributions of the END interaction are related to each other by spectral density functions

$$j^{(A)}(\omega) = \frac{\pi^2}{5} \cdot \text{Tr}(\tilde{A}^2) \cdot \frac{\tau_R}{1 + \omega^2 \cdot \tau_R^2} \qquad (2.28)$$

where $\text{Tr}(\tilde{A}^2)$ is the trace of the squared dipolar hyperfine-tensor $\tilde{A}$, it measures the anisotropy of the END interaction.

This strongly reduces the number of independent fundamental parameters that determine the ENDOR intensity and linewidth to A_{iso}, B, $\text{Tr}(\tilde{A}^2)$, V_{eff}, $\eta(T)$, and the irradiation fields, i.e. the mw field B_1 and the rf field B_2. These

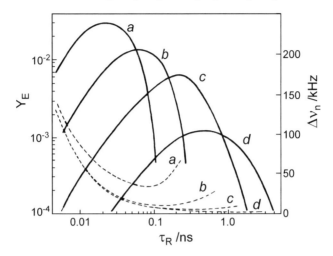

Figure 2.4 ENDOR signal amplitude Y_E and saturated ENDOR linewidth $\Delta\nu_n$ (in kHz, broken lines) as functions of the rotational correlation time τ_R (in ns). The hyperfine anisotropy $\mathrm{Tr}(\tilde{A}^2)$ is used as a parameter: $\mathrm{Tr}(\tilde{A}^2)$ = 1000 MHz² (curves a), 100 MHz² (b), 10 MHz² (c), and 1 MHz² (d). The calculations were performed for one nucleus with $I = 1/2$, spin-rotational parameter $B = 10^{-6}$, saturation degrees $\sigma_e = 3$, $\sigma_n = 3$, no exchange. For details, see ref. [71].

parameters are required as input parameters for the computational procedure for solving the equation of motion of the density matrix.[71] As an example, Figure 2.4 shows the calculated ENDOR amplitude and linewidth as a function of the rotational correlation time, τ_R, for a particular B value. $\mathrm{Tr}(\tilde{A}^2)$ is chosen as a parameter with values ranging from 1 to 1000 MHz². This range covers the majority of nuclei in free radicals. The curves have well-pronounced maxima at optimum values of τ_R. The steep rise and fall of the ENDOR amplitude on either side of τ_R^{opt} demonstrate how critically the various relaxation rates have to be balanced by proper choice of solvent and temperature. For $\tau_R < \tau_R^{\mathrm{opt}}$ one obtains $W_n < W_e$ (W_n bottleneck), for $\tau_R > \tau_R^{\mathrm{opt}}$ one obtains $W_e < W_n$ (W_e bottleneck).

Evaluation of the computer results for wide ranges of B and $\mathrm{Tr}(\tilde{A}^2)$ yields approximate expressions for optimum values of the correlation time, and mw and rf field strengths, B_1 and B_2, respectively:[71]

$$\tau_R^{\mathrm{opt}} \cong 200 \cdot \sqrt{B/\mathrm{Tr}(\tilde{A}^2)} \qquad (2.29)$$

$$\frac{\nu_{\mathrm{ENDOR}}}{\nu_H} \cdot B_2^{\mathrm{opt}} \cong 100 \cdot \sqrt{B \cdot \mathrm{Tr}(\tilde{A}^2)} \qquad (2.30)$$

$$B_1^{\mathrm{opt}} \propto \sqrt{B \cdot \mathrm{Tr}(\tilde{A}^2)} \qquad (2.31)$$

where $\text{Tr}(\tilde{A}^2)$ is given in MHz^2, τ_R in ns, and the field strengths are given in 0.1 mT in the rotating frame.

The factor in front of B_2^{opt} in eqn (2.30), being the ratio of the actual ENDOR transition frequency of the observed nucleus and the free proton frequency, takes account of the different nuclear magnetic moments and the hyperfine enhancement[82] of the applied rf field by the isotropic hyperfine interaction. In a simple physical picture, this enhancement originates in the hyperfine field that has a projection along the rf-field direction oscillating at the frequency of the rf field. For nuclei other than protons and deuterons, the hyperfine enhancement often becomes so large that even nuclei with small magnetic moments can be detected without needing excessively large rf power levels.

In general, however, it is the requirement of sufficiently strong microwave and rf fields which causes the experimental difficulties in ENDOR-in-solution experiments. For broadband operation, a practical upper limit for the rf field strength is given by currently available rf amplifiers. Extremely important for successful ENDOR in solution is the microwave cavity containing the ENDOR coil. It is the central part of broadband electron–nuclear multiple resonance spectrometers. Sophisticated constructions of cavity/coil arrangements have been described in the literature over the last 20 years, and specific references can be found in several overview articles, for example refs. [46, 83–85], but also in a recent research article[86] on an improved TM$_{110}$ cavity for cw X-band ENDOR.

B TRIPLE Resonance as an Extension of ENDOR

From many applications in chemistry, biology and physics it became clear that steady-state cw ENDOR in solution, though extremely powerful in resolving complex hyperfine structures of low-symmetry radicals, is suffering from sensitivity problems: Only less than 10% of the EPR intensity is normally observable as an ENDOR effect that has to be maximized by carefully controlling temperature and viscosity of the solvents, thereby optimizing the delicate interplay between electron and nuclear relaxation rates. Cw ENDOR also suffers from problems of assigning the measured hyperfine couplings to molecular positions: The ENDOR line intensities are determined primarily by electron and nuclear relaxation and not by the multiplicity of the NMR transitions. These drawbacks were the motivation to extend liquid-solution ENDOR to electron–nuclear–nuclear triple resonance[47,48] in which two high-power rf sources are connected to the NMR coil inside the EPR cavity.

According to the different irradiation schemes involving only one nucleus or two inequivalent nuclei, we distinguish between Special TRIPLE and General TRIPLE resonance:

Special TRIPLE Resonance

There are mainly two drawbacks of the ENDOR method: *(i)* In the frequently occurring case of a W_n bottleneck, *i.e.* when $W_n \ll W_e$, the ENDOR effect becomes very weak if crossrelaxation is absent. *(ii)* The intensity pattern of ENDOR lines generally does not reflect the number of nuclei involved in the various transitions. Both drawbacks can, at least in part, be overcome by applying two NMR rf fields at a frequency separation of exactly the hfc of a particular set of equivalent nuclei. This Special TRIPLE resonance experiment can be understood with the aid of the four-level scheme of Figure 2.3. In addition to the first NMR rf field at, for instance, frequency v^- (1 ↔ 3), a second NMR rf-field at frequency v^+ short circuits the W_n bottleneck (2 ↔ 4). Thereby, the efficiency of the NMR-induced relaxation bypass is enhanced. Provided that both rf fields are applied at saturating levels, a considerable increase in the signal intensity can be achieved and, additionally, the line intensities become rather independent of W_n. For $W_n \ll W_e$, this results in an intensity pattern that, similar to NMR, is dominated by the number of nuclei involved in a particular transition. This facilitates the assignment of hfcs to specific molecular positions. Furthermore, Special TRIPLE has the advantage of narrower lines, *i.e.* at a given power level the effective NMR saturation, which determines the observed linewidth, is smaller in TRIPLE than in ENDOR. Linewidth reductions of typically 30–50% are observed, in agreement with model calculations.[83]

In Figure 2.5, the ENDOR and special TRIPLE spectra are shown for the 3-pyridylphenyl ketone anion radical as an example.[46] The spectra were recorded at a rather high temperature so that the rings are rotating rapidly on the EPR time scale. At this temperature, however, $W_n \ll W_e$ and, as a result, the ENDOR signal-to-noise ratio is poor. Since the NMR transitions are already saturated at the applied rf field amplitude, all the ENDOR lines show equal intensity within the noise limits. For the Special TRIPLE spectrum, the same total rf power was distributed into two side bands and, hence, the rf-field amplitudes per NMR transition are reduced by a factor $\sqrt{2}$. In qualitative agreement with theoretical predictions, a reduction of the linewidth in the TRIPLE spectrum occurs, the signal-to-noise is increased, and the approximate intensity ratios 2:1:2:1:1:1:1 of the TRIPLE lines reflect the number of protons involved in the NMR transitions (the *meta* and *ortho* protons of the phenyl ring are equivalent).

General TRIPLE Resonance
In the special triple resonance version, both rf fields are applied at a separation of the hfc A_{iso} (for which we will sometimes also use the short-hand notation "a") of the same nucleus. Triple resonance can, however, be generalized to several inequivalent nuclei with the aim to obtain hfcs together with their relative signs.[48] If we consider, for example, two inequivalent protons, the first-order energy levels in the basis $|m_S m_{I_1} m_{I_2}\rangle$ can be arranged to form the eight corners of a cube, Figure 2.6. In such a three-dimensional representation, the various desaturation bypasses for a pumped electron transition, involving the NMR transitions of the two nuclei, can be visualized more clearly than in the conventional two-dimensional transition schemes. In Figure 2.6, this energy-level arrangement is depicted for the two different cases, $a_1, a_2 > 0$ and $a_1 > 0, a_2 < 0$. Every two corners are connected by the various relaxation and

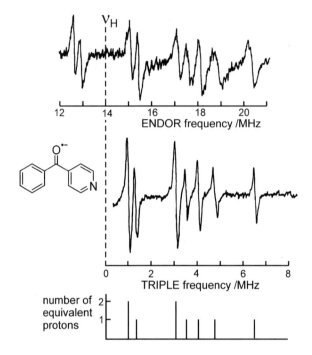

Figure 2.5 ENDOR and Special TRIPLE resonance spectra of the 3-pyridylphenyl ketone anion radical (solvent, DME; counterion, Na^+; $T = 240$ K). The intensity ratios of the stick diagram reflect the number of equivalent protons involved in the respective NMR transitions. The Special TRIPLE spectrum was recorded at the same total NMR power level as the ENDOR spectrum.[46]

induced transitions, the EPR transitions occurring vertically, the NMR transitions horizontally. All transitions are doubly degenerate to first order. If we now consider the level *populations*, the different multiple-resonance experiments can be represented by different geometrical figures (see left part of Figure 2.6). These figures are derived from the cubes by contracting those corners that are connected by induced NMR transitions. They represent the limiting case of highly saturated transitions where the populations of the connected levels are equalized. In this representation, an ENDOR experiment forms a prism, and a Special TRIPLE experiment forms a square. If we consider the case that both rf fields drive the low-frequency transitions (v_1^- and v_2^-), in General TRIPLE two different cases have to be distinguished depending on the relative signs of the hfcs. If they have the same sign, all the NMR transitions are saturated in the same plane resulting in a pyramid. If the hfcs have opposite signs, the NMR transitions for the two nuclei are saturated in the opposite planes and a tetrahedron is formed. The sign of the hyperfine coupling is an important parameter in molecular spectroscopy. For example, in the case of the isotropic hyperfine constant, A_{iso}, the sign provides additional insight into the electronic

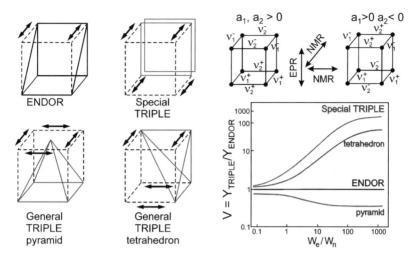

Figure 2.6 Topology of ENDOR and TRIPLE resonance experiments for the three-spin system $S=1/2$, $I_1=1/2$, $I_2=1/2$. The NMR transitions with their frequencies are given for the two cases of equal and opposite signs of the hyperfine couplings. For simplification, the EPR transitions are not distinguished. The graph shows the TRIPLE amplification factor V as a function of the ratio W_e/W_n, obtained by analyzing the electric-circuit analogue of the various relaxation networks. The curves shown are valid for induced NMR rates 100 times larger than W_n; crossrelaxation was neglected. For details, see ref. [48].

structure when finding an answer to the question: how is an unpaired electron density produced at the nucleus? Is it by a spin-polarization mechanism ($A_{iso} < 0$, for example α-protons) or by conjugation or hyperconjugation mechanisms ($A_{iso} > 0$, for example β-protons)?

The relevance of these topological games is shown in Figure 2.6 giving the result of a theoretical analysis of the relaxation networks of the various geometrical figures for a wide range of W_e/W_n values.[48] This analysis was performed within the approximation of the electric-circuit analogue of rate equations (mentioned above) applying Kirchhoff's laws for branching networks. When a TRIPLE amplification factor, V, is defined as the ratio of TRIPLE and ENDOR line amplitudes, always $V > 1$ for a "tetrahedron" experiment and $V < 1$ for a "pyramid" experiment. The difference between pyramid and tetrahedron becomes particularly pronounced in those cases where W_n is much smaller than W_e. Such cases are typical for many ENDOR-in-solution experiments. In the extreme situation $W_e/W_n \gg 1$, Special TRIPLE can even reach 100% EPR sensitivity!

Obviously, relative signs of hfcs can easily be determined from intensity changes in General TRIPLE spectra. As a representative example, Figure 2.7 shows for the fluorenone $\cdot^-/\text{Na}^+$ ion pair how the intensity patterns of proton and sodium lines change in a characteristic way when extending ENDOR to

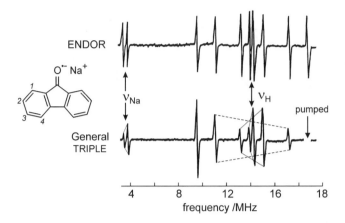

Figure 2.7 ENDOR and General TRIPLE spectra of the fluorenone anion radical (solvent: tetrahydrofurane, counter ion: Na$^+$, $T = 226$ K). For details, see ref. [87].

General TRIPLE. From such intensity patterns the signs of the various hfcs are revealed relative to the sign of the hfc belonging to the pumped transition.[87] Different relative signs of the hfcs of different types of protons and of the sodium counter ion are reflected by inversion of the amplitude ratios.

In the following subsections, selected examples of liquid-solution ENDOR and TRIPLE experiments are given that were performed at FU Berlin together cooperation partners from chemistry and biology departments around the globe with the aim to elucidate the electronic structure of complex molecular systems:

C ENDOR in Liquid Crystals

Evidently, magnetic resonance spectroscopy in liquids excels by narrow lines, but sacrifices information on anisotropic interactions as long as isotropic solvents are used. This is because the anisotropic parts of tensor interactions are averaged out by rapid Brownian tumbling. However, by using liquid crystals as anisotropic solvents, valuable information about anisotropic interactions can be retrieved from line positions while retaining narrow hyperfine lines typical for liquid-solution spectra.[88] In the nematic mesophase of a liquid crystal, solute molecules can be partially aligned in the external Zeeman field of an EPR spectrometer. This results, for axial symmetry of either the interaction or ordering tensor, in a shift of the measured interaction parameter relative to its isotropic value, $F - F_{\text{iso}} = O_{33} \cdot F'_{33}$. Here, O_{33} is the temperature-dependent ordering parameter, and F'_{33} is the principal component of the traceless interaction tensor that refers to the axis of highest symmetry of the solute molecule. F stands for any second-rank interaction tensor, for example the g-, hyperfine- or quadrupole-tensors.

The most striking aspect of ENDOR in liquid crystals is the possibility to directly determine, for nuclei with $I > 1/2$, components of the quadrupole interaction tensor of radicals in fluid solution from their ENDOR line positions. EPR in liquid crystals is not suitable in this respect because, to first order, the quadrupole interaction shifts all EPR-connected levels equally. The first determination of ^{14}N quadrupole couplings in an organic radical was achieved by ENDOR in liquid crystals by K.P. Dinse in the Möbius group at FU Berlin.[89] When cooling the liquid crystal from its isotropic to its nematic phase one observes shifts or even splittings of the ENDOR lines of the quadrupole nucleus (e.g., $I=1$), depending on which EPR line ($m_I = +1, 0, -1$) is saturated. The quadrupole splitting is given by

$$\delta\nu_Q = 3/2 \cdot O_{zz} \cdot e^2 \cdot q_{zz} \cdot Q/h \qquad (2.32)$$

from which $e^2 q_{zz} Q$ can be deduced when the ordering parameter O_{zz} is known. Even the small deuterium quadrupole coupling along the C–D bond of the aromatic radical perinaphthenyl (PNT), $e^2 \cdot q_{CD} \cdot Q/h = +188$ kHz, could be measured with this technique by R. Biehl in the same group.[90] For small quadrupole couplings of radicals in an anisotropic matrix, ENDOR is probably the only method of choice. As an illustrative example, the ^{2}H ENDOR spectra of the partially deuterated PNT radical in isotropic and nematic solution[90] are presented in Figure 2.8.

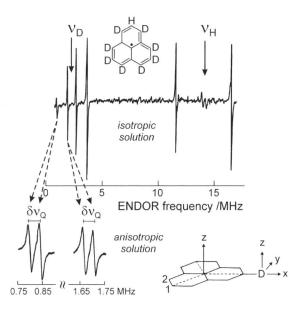

Figure 2.8 Proton and deuterium ENDOR lines of a partially deuterated perinaphthenyl radical in isotropic and anisotropic phases of a liquid crystal. In the nematic mesophase, quadrupole splittings of the deuterons, $\delta\nu_Q$, are resolved. For both lines $\delta\nu_Q = 42.2$ kHz at 293 K. From $\delta\nu_Q = (3/2) \cdot O_{zz} \cdot e^2 \cdot q_{zz} \cdot Q/h$ with $O_{zz} = -0.3$ at 293 K the value $e^2 \cdot q_{zz} \cdot Q/h = -94$ kHz ($e^2 \cdot q_{CD} \cdot Q/h = +188$ kHz) was determined. For details, see ref. [90].

D Porphyrinoid and Chlorophyll Ions

Since the mid-1970s a growing interest in chlorophylls and structural variants of porphyrins has been evident, one reason is their potential to model photosynthetic chromophores. As examples of such ionic porphyrinoid systems, porphycene radical anions[91] and bacteriopurpurin radical cations[92] have been studied by liquid-phase EPR, ENDOR and TRIPLE (see Figure 2.9). The determination of the spatial and electronic structures of porphyrinoid model systems in general and, in particular, of chlorophyll ion radicals *in vitro*, i.e. the isolated chromophores in nonaqueous solvents, is considered as an assistance for understanding their role in the photoinduced electron-transfer chain of *in-vivo* systems, for example in the photosynthetic chromophore-protein complexes. ENDOR-in-solution techniques in conjunction with elaborate MO methods turned out to be extremely powerful for resolving and analyzing the complex hyperfine structures of the chromophores.[54,61,65,73,93–97]

In this respect, the bacteriochlorophyll cation radical, $BChl^{\cdot+}$, plays a prominent role as a research target for ENDOR spectroscopy because it is involved as an electron-transfer partner in bacterial photosynthetic reaction centers

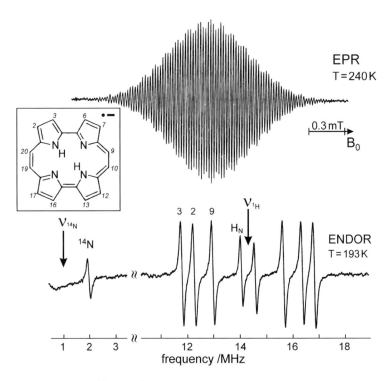

Figure 2.9 EPR and ^{14}N and ^1H ENDOR spectra from free-base porphycene radical anion in tetrahydrofurane. The signs of the hfcs are from General TRIPLE, the assignment to molecular positions is based on MO calculations. For details, see ref. [91].

(RCs) from *Rb. sphaeroides* or *Rps. viridis* (see Section 5.2). BChl$^{\cdot+}$ contains many inequivalent sets of protons and nitrogens and, consequently, the EPR spectrum is inhomogeneously broadened with only an indication of hyperfine-structure resolution. This gets even worse for the dimer of bacteriochlorophylls that form the primary-donor cation in RC protein complexes from purple bacteria, for which only a Gaussian envelope EPR line can be observed, with no indication of hyperfine structure at all. In terms of hyperfine-coupling resolution, liquid-solution ENDOR and TRIPLE at elevated temperatures are superior to frozen-solution ENDOR/TRIPLE because, owing to the motional averaging of anisotropic hyperfine interactions in fluids, the linewidths are smaller so that several *isotropic* hyperfine coupling constants (hfcs) from the magnetic nuclei in the radicals can be obtained.[61,98–100] Particularly powerful for elucidating the hyperfine structure and, thereby, the spin-density distribution over the primary-donor dimer cation radical was the room-temperature study of single-crystal RCs by ENDOR and TRIPLE.[101] For details of these X-band ENDOR/TRIPLE studies on the primary donor and acceptor radical ions in bacterial photosynthetic RCs, see Section 5.2.1.1.

2.2.1.2 Liquid-Solution Transient EPR and ENDOR in Photochemistry

Realistic sensitivity estimates for successful *steady-state* ENDOR/TRIPLE in solution show that a minimum of 10^{13} radicals in the cavity during the detection period is needed. As a consequence, *steady-state* ENDOR on short-lived radicals in solution is restricted to lifetimes longer than ms, even when using fast-flow systems to supply fresh sample to the cavity. To extend the applicability to transient radical intermediates with lifetimes as short as sub-μs, a goal we found appealing for elucidating complex reaction mechanisms, we took advantage of chemically or photolytically induced electron-spin-polarization effects as signal boosters. These effects are subsumed under the acronym CIDEP (chemically induced dynamic electron polarization) and show up by EPR lines in emission or enhanced absorption in time-resolved and pulse EPR and ENDOR spectra. Multifrequency/multifield EPR and ENDOR techniques with fast time-resolved detection of spin-polarized molecular systems are being heavily used now in studies of photoinduced transient radicals, radical pairs of donor–acceptor complexes and excited states in organic photochemistry[102] photosynthetic reaction centers and their biomimetic model systems[95,103] as well as of DNA photolyases.[104] As a rule, the electron-spin system of reactive radicals exhibits CIDEP effects, *i.e.* the momentary populations of the spin states deviate strongly from thermal equilibrium leading to lines in emission or enhanced absorption. This phenomenon can arise from a variety of mechanisms, and electron-spin polarizations can be created either when the radicals are suddenly initiated or during their bimolecular termination reactions from preferred spin states. A large amount of information on these spin polarizations has been collected during the past 35 years, and the CIDEP phenomenon has

A CIDEP Mechanisms in Photochemical Reactions

Many reviews concerning CIDEP effects in transient paramagnetic species have been published in recent years. Therefore, in the following only a short overview of polarization mechanisms will be given that are relevant for the photochemical systems described in this book.

A.1 Triplet Mechanism (TM)

Photochemical radical generation most often occurs by reaction of singlet excited molecules. Then, the Zeeman energy levels α and β become equally populated, and the radical system has zero spin polarization immediately after creation. Alternatively, singlet excited molecules may first undergo fast selective singlet–triplet intersystem crossing (ISC) to specific sublevels of their excited triplet state and produce radicals from there by fast quenching reactions. It is well known that the ISC process can lead to differently populated triplet sublevels dictated by the symmetry of the molecule.[105] The resulting initial spin polarization of the triplet state can then be transferred to the reaction products, a pair of radicals, provided the decay into radicals occurs fast enough to compete with spin relaxation in the triplet molecule. The TM CIDEP is independent of m_I, the nuclear magnetic spin quantum number, *i.e.* the TM CIDEP creates a *"net effect"* polarization of the same phase for both radicals Figure 2.10.

The zero-field sublevel spin states of a triplet molecule, $|T_x\rangle$, $|T_y\rangle$ and $|T_z\rangle$, are nondegenerate due to the dipolar or zero-field coupling between the electrons (x, y and z denote the principal axes of the zero-field tensor). ISC often occurs, under molecular selection rules, predominantly into one of these states, thus producing triplet molecules with pronounced spin polarization in the

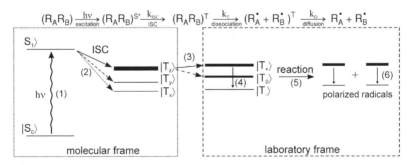

Figure 2.10 Reaction scheme and schematic representation of polarization development by the triplet mechanism. The sequence of events and the approximate time scales (for carbonyl chromophores): (1) Photo excitation, $<10^{-14}$ s; (2) Singlet-triplet ISC, 10^{-12} to 10^{-9} s; (3) Zeeman and ZFS interactions, 10^{-11} to 10^{-10} ; (4) Triplet spin relaxation, 10^{-10} to 10^{-8} s; (5) Reaction, 10^{-6} to 10^{-9} s; (6) Radical spin relaxation, 10^{-7} to 10^{-5} s.

molecular frame. This is a commonly observed phenomenon in the solid state, see for example, ref. [106]. As soon as a triplet is formed, its spin magnetic moment will start to interact with the external magnetic field B_0 of the EPR spectrometer. The interaction can transform a substantial part of the molecular-frame spin polarization into a Zeeman-level spin polarization with respect to B_0 (*i.e.* in the laboratory frame).

The transformation may be envisioned by an adiabatic model in which, after triplet formation, the magnetic field is slowly turned on in such a direction that the highest-energy molecular spin state will evolve into the highest Zeeman spin state $|T_+\rangle$. However, the actual transformation process is more complicated because of essentially two factors. Firstly, the exposure to an external field is not an adiabatic process, as the magnetic field is already present when the triplets are formed. Secondly, in liquids the orientations of the triplet molecules with respect to the magnetic field are subject to stochastic changes because of rapid rotational diffusion. In fact, the rigorous quantum-mechanical treatment shows[107–111] that the amount of spin polarization that is transferred from the molecular to the laboratory frame depends on the ratio of zero-field splitting (ZFS) and Zeeman interaction, v_{ZFS}/v_e, as well as the rotational correlation time τ_R. Spin polarization vanishes if either $|v_{ZFS}/v_e| \ll 1$ or $v_e \cdot \tau_R \ll 1$. In both cases, the molecular-frame polarization will be destroyed faster than it is transformed, either by Zeeman precession or molecular tumbling. Fortunately, neither of these cases is met for a wide variety of triplet molecules in solutions of low viscosity.

The Zeeman polarization will be destroyed by spin-lattice relaxation unless the triplet reacts rapidly to yield a pair of radicals. Spin-lattice relaxation of triplet molecules in solution is very fast because of the strong anisotropic magnetic dipole–dipole interaction between the two triplet spins combined with the rapid change of this interaction as the molecule tumbles in solution. The time scale of this process is of the order of 10^{-8} to 10^{-10} s.[112] Consequently, the triplet molecule must react on a comparable time scale in order to transfer a substantial part of the triplet polarization to the radical products. It should be mentioned that even if the Zeeman polarization of the triplets has completely relaxed to the equilibrium one, p_{eq}, an excess absorption of $4/3 \cdot p_{eq}$ will be carried to the radicals due to twice the Zeeman splitting $g \cdot \mu_B \cdot B_0$ between $|T_+\rangle$ and $|T_-\rangle$. This small enhanced absorption, however, is often negligible in comparison with the population differences in the electron spin levels produced initially in the radical system by the TM effect.[113]

A.2 Radical-Pair Mechanism (RPM)

The radical-pair mechanism (RPM) of CIDEP in liquid solution is more general than the TM in that the radicals do not have to originate from a triplet state. The polarization arises during the radical lifetime as a result of magnetic interactions between radicals when forming spin-correlated radical pairs. In the RPM CIDEP, the spin polarization originates in spin-correlated radical-pair formation by encounters, diffusive separations and re-encounters of the radicals. While being in close contact, ST_0 singlet–triplet mixing occurs in the radical pair as a result of the different local fields due to different g-factors (Δg) and/or hyperfine couplings (ΔA_{iso}) of the radical-pair partners. When the Δg term in the ST_0 mixing coefficient dominates, a net effect polarization of opposite phase for the two radicals is observed. Dominating hyperfine terms (ΔA_{iso}), on the other hand, lead to *"multiplet effect"* polarization, which is m_I dependent with half the hyperfine lines in emission (E), the other half in enhanced absorption (A). Radicals derived from a triplet radical-pair precursor show an E/A line pattern (low field/high field) (see Figure 2.11), whereas singlet-derived radicals exhibit an A/E pattern (low field/high field).

Apparently, RPM CIDEP has three elements:

(i) The reaction mechanism involves a pair of radicals either created together via dissociation or some other reaction of an excited molecule ("geminate" or "G-pair"), or formed by a random encounter of separately generated radicals ("free" or "F-pair").
(ii) The reactivity of the radical pair is determined by its electron-spin state that can be singlet, triplet or a mixture of singlet and triplet states. The radicals in the singlet state will attract each other to form a recombination (or some other) product, whereas the triplet state is repulsive and does not lead to product formation.
(iii) The radicals comprising the pair may separate to a point where the short-range valence or exchange forces are negligible. Then, magnetic interactions within the radicals (electron–nuclear hyperfine interactions and electron interactions with the external magnetic field) can mix these singlet and triplet states, converting a reactive singlet pair into a triplet pair, and *vice versa*.

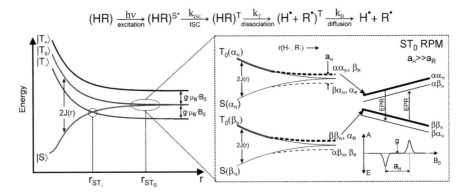

Figure 2.11 *Right:* Energy levels of the singlet and triplet spin states of a radical pair in an external magnetic field as a function of separation r. The regions of level crossing are accentuated. $J(r)$ is the exchange energy. *Left:* Quasiadiabatic model of the radical-pair polarization by ST_0 mixing in a radical pair. The dashed curves indicate adiabatic separation, the solid curves fast (diabatic) separations. The radicals are generated in the triplet state according to the reaction given on the top. For details, see text.

The detailed theory of the RPM version of CIDEP requires application of the stochastic Liouville formalism to describe the time evolution of the radical-pair spin state under the action of the spin Hamiltonian. The resulting models of the process, including a simple vector model that also embodies the effect of spin exchange,[114,115] have been described in the literature.[116–119] However, a simple qualitative picture of the RPM polarization development as well as the polarization rules can be given using the quasiadiabatic model of Adrian.[113,120]

The main assumption of this model is that the separation and re-encounter of the radical pair causes the spin state to evolve partially as it would have done if the initial separation of the radicals were very slow (adiabatic). The actually observed polarization is much less than that achievable in the limiting case of adiabatic radical separation. According to theory,[121–125] the polarization is approximately given by

$$p(ST_0) \approx \frac{\pi}{2 \cdot \delta \cdot d} \sqrt{\frac{Q \cdot d^2}{D}} \tag{2.33}$$

where $p(ST_0)$ is the difference of the probabilities for a radical pair being initially in the $T_0(\alpha_N)$ state to separate as $\alpha\alpha_N,\beta_R$ and $\beta\alpha_N,\alpha_R$, with analogous definitions holding for $T_0(\beta_N)$, $S(\alpha_N)$, and $S(\beta_N)$. Here, (α, β) and (α_N, β_N) denote the orientation of electron and nuclear spins of the observer radical with respect to the external magnetic field. The hyperfine interaction of the counter radical is considered to be negligible, therefore only electron-spin projections (α_R, β_R) are considered. D is the relative diffusion coefficient of the two radicals forming the pair, Q is half the difference of their Larmor frequencies in angular frequency units, and d is the distance of closest approach. The definition of $1/\delta$, the characteristic range of the exchange interaction $J(r)$, follows from its exponential dependence on the radical separation r,

$$J(r) = J_0 \cdot \exp[-\delta \cdot (r - d)] \tag{2.34}$$

The quantity d^2/D is the time required for the radical pair to diffuse to a distance of closest approach, typically 10^{-12} to 10^{-10} s. As for realistic systems $\delta \cdot d \gg 1$ holds, eqn (2.33) yields typical RPM polarizations of 5×10^{-3} to 5×10^{-2} for $\delta \cdot d \approx 4$ and hyperfine splittings of about 2 mT, i.e. $Q = 3.5 \times 10^8$ rad s^{-1}.

For the initial, i.e. geminate triplet radical pair the EPR spectrum of the observer radical appears E/A polarized (low-field EPR lines in emission, high-field lines in enhanced absorption). For the initial singlet pair the polarization pattern is reversed (A/E). F-pairs of radicals form in an initial singlet or triplet state. Usually, these pairs polarize like triplet-state geminate pairs, because the singlet F-pairs are removed by reaction upon the first encounter.

Electron-spin polarization can also be generated at the ST_- level-crossing region. In this region, the singlet–triplet splitting due to the exchange interaction is balanced by the triplet Zeeman splitting. The adiabatic terms avoid the

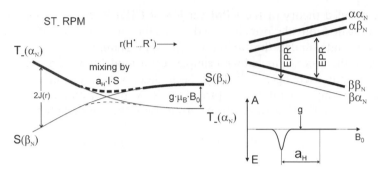

Figure 2.12 Quasiadiabatic model of the radical pair polarization by ST_- mixing. For details, see text.

level crossing because of singlet–triplet mixing by the hyperfine interaction, thus causing the initial $T_-(\alpha_N)$ state to be converted to $S(\beta_N)$, and *vice versa*, see Figure 2.12. In contrast to the ST_0 mixing, this process produces a net magnetic polarization connected with a change in electron-spin orientation and accompanied by an opposite change in nuclear-spin orientation, as required by the conservation rule of angular momentum.

Starting from an initial triplet state, this process would effectively underpopulate the $\beta\alpha_N$ level and the β_R electron spin level of the counter radical R·, because conversion of $T_-(\alpha_N)$ to $S(\beta_N)$ removes $\beta\alpha_N$ as well as β_R states and generates additional $\alpha\beta_N$, $\beta\beta_N$ as well as α_R states. In this case, the low-field line is emissively polarized and the high-field line has zero polarization. However, this is the case only for the radical whose hyperfine interaction causes the singlet–triplet mixing. All hyperfine levels of the counter radical R· are equally polarized. Finally, for an initial singlet radical pair the polarization pattern will be reversed, *i.e.* the high-field line will show an enhanced absorption.

The ST polarization is generally less important in liquid solutions than ST_0 polarization, because the ST_- crossing region in high fields is distance-wise very small compared to the ST_0 crossing region, thus restricting the amount of the ST_- mixing. Nonetheless, ST polarization can be significant in two cases: *(i)* one radical of the pair has a large hyperfine splitting (for instance phosphorus-centered radicals);[126] *(ii)* diffusion is slow and the system spends a relatively long time in the crossing region,[127] alternatively the external magnetic field is low enough.

A.3 Spin-Correlated Radical Pair Mechanism (CCRP)

Here, we summarize the theoretical description of the correlated coupled radical-pair (CCRP) mechanism taking charge-separated donor–acceptor systems as an example.[128–131] After light-induced electron transfer, one unpaired spin is localized on the donor site, the other unpaired spin on the acceptor site. The spin dynamics of this radical pair are described by the stochastic Liouville equation, which includes charge separation, charge recombination and spin relaxation. The spin Hamiltonian of the system for a specific orientation of the molecule with respect to the external magnetic field B_0 is given by

$$\hat{H} = \hat{H}_0(\hat{S}_1) + \hat{H}_0(\hat{S}_2) + \hat{H}_{\text{exchange}}(\hat{S}_1, \hat{S}_2) + \hat{H}_{SS}(\hat{S}_1, \hat{S}_2). \quad (2.35)$$

The static spin Hamiltonian $\hat{H}_0$ of each electron spin is given by eqn (2.1). The electron–electron spin interaction terms are given by

$$\hat{H}_{\text{exchange}}/h = -J \cdot \hat{S}_1 \cdot \hat{S}_2 \quad (2.36)$$

$$\hat{H}_{SS}/h = \frac{\mu_0}{4\pi \cdot h} \cdot g_1 \cdot g_2 \cdot \mu_B^2 \cdot \left[\frac{\hat{S}_1 \cdot \hat{S}_2}{r^3} - \frac{3 \cdot (\hat{S}_1 \cdot \mathbf{r}) \cdot (\mathbf{r} \cdot \hat{S}_2)}{r^5} \right] \quad (2.37)$$

where g_1 and g_2 are the g-values of two electron spins orientation-selected by the external magnetic field. Eqn (2.37) can be represented in the form:

$$\hat{H}_{SS}/h = d \cdot \left[\hat{S}_{1z} \cdot \hat{S}_{2z} - \frac{1}{2} \cdot (\hat{S}_{1x} \cdot \hat{S}_{2x} + \hat{S}_{1y} \cdot \hat{S}_{2y}) \right] \quad (2.38)$$

where

$$d(\theta) = \frac{\mu_0}{4\pi \cdot h} \cdot \frac{g_1 \cdot g_2 \cdot \mu_B^2}{r^3} \cdot (1 - 3 \cdot \cos^2 \theta) \quad (2.39)$$

is the dipolar coupling frequency, which depends on the angle θ between the dipolar axis $\mathbf{r}$ and the external magnetic field $\mathbf{B}_0$.

Depending on the relative strengths of the various magnetic interactions, some assumptions can be made concerning the proper eigenfunctions. In the high-field limit, when the dipolar and exchange interactions are small compared to the Zeeman interaction, the eigenstates of the 4-level system are approximated by the unperturbed triplet states $|T_+\rangle$ and $|T_-\rangle$ and the mixtures $|2\rangle$, $|3\rangle$ of $|S\rangle$ and $|T_0\rangle$. The amount of state mixing (see Figure 2.13) is expressed in terms of the mixing angle, φ_{mix}:

$$\tan \varphi_{\text{mix}} = \frac{\Delta \nu}{J + d/2 + F}; \quad F = \sqrt{(J + d/2)^2 + \Delta \nu^2} \quad (2.40)$$

Here, the half-difference of the Larmor frequencies of the interacting spins is given by

$$\Delta \nu = \frac{\mu_B}{2h} \cdot (g_1 - g_2) \cdot B_0 \quad (2.41)$$

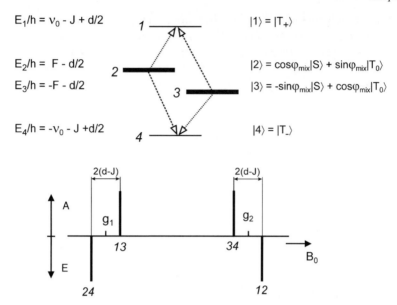

Figure 2.13 Energy levels of the eigenstates of an initial singlet-state, weakly coupled radical pair; EPR stick spectrum for the transitions between the numbered levels for the case $(J-d(\theta)) > 0$. For details, see refs. [130,132].

if the hyperfine, quadrupole and nuclear Zeeman terms in eqn (2.1) are neglected.

According to the CCRP model, all four possible one-quantum transitions of light-induced radical pairs have the same intensity because differences in transition moments are compensated by the respective population differences.[130,131] These population differences are due to spin polarization that results from singlet electron transfer as well as from triplet electron transfer, *i.e.* the populations of the four states differ significantly from the Boltzmann equilibrium. Thus, for each orientation of the molecule with respect to the external magnetic field, two EPR lines in absorption and two lines in emission should be detected, all equal in amplitude. The transient EPR spectra of frozen-solution samples reflect the powder average over all possible orientations. Because of the anisotropy in the magnetic interactions, TREPR spectra contain information about the molecular structure, in addition to the information about the prevailing polarization mechanisms.

For strongly coupled spins, *i.e.* $|J| \gg |\Delta \nu|$, singlet and triplet are approximately eigenstates, *i.e.* $|2\rangle \approx |S\rangle$ and $|3\rangle \approx |T_0\rangle$, while for weakly coupled spins the eigenstates are given by the respective eigenfunctions of each radical. It will be demonstrated later in this book that by means of multifrequency time-resolved EPR on covalently linked porphyrin-quinone systems the different cases from weakly to strongly coupled radical pairs can be distinguished (see Section 5.4).

B CIDEP-Enhanced ENDOR

As was shown at the beginning of this section, steady-state ENDOR on short-lived radicals in solution is restricted to lifetimes longer than 1 ms. To extend the applicability to transient radical intermediates with lifetimes as short as sub-µs, we took advantage of chemically or photolytically induced electron-spin-polarization effects and invented "CIDEP-enhanced ENDOR". This experiment was developed in the Möbius group at FU Berlin in 1983–85, first together with R.Z. Sagdeev as visiting scientist from Novosibirsk in the cw detection mode,[133] somewhat later it was extended in the same group by F. Lendzian and P. Jaegermann to the time-resolved direct-detection mode, TREPR.[134] Depending on whether dominant net-effect or multiplet-effect polarizations are exploited, different EPR saturation conditions have to be fulfilled.[134] The critical point of this experiment is the available rf-field strength necessary to saturate NMR transitions of the transient radicals during their (often short) lifetime.[135]

Many important chemical reactions in photochemistry and radiolysis, involve transient radicals of unknown structure with lifetimes much shorter than 1 ms. When aiming at structure information via hyperfine couplings, which are typically of the order of some MHz, intermediates with lifetimes not shorter than µs become particularly interesting. To study these intermediates by ENDOR spectroscopy with its inherent resolution potential, advantage has to be taken of the large non-Boltzmann population differences created by dynamic polarization effects such as CIDEP in all its variations (see subsection A). There is a broad coverage of dynamic polarization effects in the literature, as an introduction, we refer in particular to ref. [136].

The large deviations from Boltzmann equilibrium populations (spectra in emission and/or in enhanced absorption are observed) are due to singlet–triplet (ST) mixing in the intermediate species. Two established mechanisms are responsible for spin-polarization effects in solution: (i) the triplet mechanism, where the polarization originates in selective singlet–triplet intersystem crossing in the excited molecule, which reacts chemically to produce polarized radicals, and (ii) the radical-pair mechanisms, where the polarization originates in radical encounters, diffusive separations and re-encounters. If the chemical reactions proceed in times shorter than the spin-lattice relaxation times of the radical intermediates, polarization can be transferred to more stable products. In CIDEP experiments deviations from the Boltzmann distribution may show up in the EPR spectrum of paramagnetic intermediate or final reaction products.

Mechanism (i) results in net-effect polarization, i.e. all lines in the spectrum are either in emission or in enhanced absorption. Mechanism (ii) is more complex: In high magnetic fields the ST_0 mixing coefficient, Δv, which is the difference in electron-spin precession frequencies of radicals 1 and 2 in their local magnetic fields, contains contributions from the Zeeman and hyperfine interactions:

$$\Delta v = \frac{\mu_B}{2h} \cdot (g_1 - g_2) \cdot B_0 + \frac{1}{2} \cdot \left(\sum_j {}^{(1)}a_j \cdot m_{I_j} - \sum_k {}^{(2)}a_k \cdot m_{I_k} \right) \qquad (2.42)$$

where the a_i are short-hand notations for A_{iso}^i. Consequently, when the Δg contribution to Δv dominates, there will be no m_I dependence, and a net polarization effect results, whereas for dominating hyperfine contribution the polarization pattern becomes dependent on m_I, with some lines in emission and other lines in enhanced absorption (the multiplet effect). Often, a superposition of net- and multiplet-effect polarizations is observed.

To explain the CIDEP-enhanced ENDOR detection strategy, the simplest model case of a radical ($S = 1/2$) containing one proton ($I = 1/2$) is considered. We assume that spin-lattice relaxation can be ignored, *i.e.* signals are detected after photolytic generation in a time shorter than the spin-lattice relaxation time. For optimizing the detection strategy, the net-effect CIDEP has to be distinguished from the multiplet-effect CIDEP.

In Figure 2.14 case (a) represents a pure net-effect polarization, whereas case (b) represents a pure multiplet-effect polarization. It is easy to see that in case (a) a population difference between the nuclear-spin levels has to be established first, for instance by a saturating microwave field. Driving the NMR transition further transfers population from the $\beta\alpha_N$ level to the empty $\beta\beta_N$ level so that the population difference between the levels connected by the EPR transition is increased, resulting in an enhanced EPR signal (an enhanced emission signal in this case). For the second NMR transition an enhancement is also obtained.

In the case of a multiplet-effect polarization, see Figure 2.14(b), there is already an initial population difference between the nuclear-spin levels, and for the observed EPR transition saturation must be avoided in order not to reduce this polarization. Driving the NMR transition decreases the population difference between the levels connected by the EPR transition, which results in a decreased EPR signal. Net-effect polarization can, therefore, easily be discriminated from multiplet-effect polarization, since in the former case EPR and ENDOR signals will have the same phase, whereas in the latter case they have opposite phases.

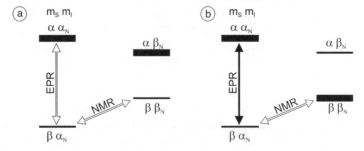

Figure 2.14 Spin-energy and population diagram for an electron in a magnetic field interacting with an $I = 1/2$ nucleus and showing a net-effect polatization (a) and a multiplet-effect polarization (b). The saturated EPR and NMR transitions are indicated by open arrows.

The critical condition for this experiment is the rf-field strength, B_2, necessary to saturate NMR transitions of transient radicals during their lifetime, τ. The NMR saturation condition for a transient radical is given by

$$\left(\frac{g_n \cdot \mu_K}{\hbar}\right)^2 \cdot B_2^2 \cdot T_{1n}^* \cdot T_{2n}^* \approx 1 \qquad (2.43)$$

with $\hbar = h/2\pi$, $1/T_{1,2n}^* = 1/T_{1,2n} + 1/\tau + 1/\tau_D$ (τ_D is the delay time between the photolysing laser pulse and the detection gate). For the p-benzosemiquinone radical (PBQH$\cdot$) in ethylene glycol solution at room temperature ($\tau = 1$ ms, $T_{1n} \approx T_{2n} \approx \tau_D = 4$ μs) the saturating rf-field strength is estimated to be 2 mT (rotating frame). To obtain this field strength with reasonable rf power is not an easy task, typically 1 kW is required in our TM_{110} cavity/coil configuration.[134,135]

As a test system for net-effect polarization,[134] UV-irradiated p-benzoquinone in ethylene glycol solution was chosen, because earlier EPR studies have shown[137] that the observed CIDEP is due to a dominant triplet mechanism with only very weak contributions of the radical-pair mechanism. The photoreaction proceeds via singlet-excited benzoquinone, intersystem crossing to triplet-excited benzoquinone and abstraction of a hydrogen atom from the solvent under formation of the polarized PBQH$\cdot$ radical. Figure 2.15(a) depicts the EPR and ENDOR spectra of PBQH$\cdot$ detected in the time-resolved mode. All EPR lines appear in emission, as do the ENDOR lines; this reflects an increase of the EPR amplitudes upon rf irradiation. This demonstrates that the net effect dominates the polarization mechanism. The ENDOR intensity amounts to *ca.* 10% of the EPR intensity. It could be increased by a factor of 2 when, in a Special TRIPLE experiment, the two NMR transitions belonging to one particular hyperfine coupling were driven simultaneously. Furthermore, relative signs of the hyperfine couplings could be determined from the General TRIPLE version of the experiment.

As a test system for the multiplet-effect polarization,[134] UV-irradiated dibenzylketone in propanediol solution was chosen. Earlier EPR experiments[138–141] have revealed that in this photoreaction benzyl radicals with almost pure multiplet CIDEP are produced. In the radical-pair polarization scheme of this system two identical benzyl radicals form the radical pair. They are produced via photoinduced bond cleavage and decarbonylation reactions. Consequently, in eqn (2.42) the Δg term vanishes, and the population difference of the two energy levels connected by an NMR transition, and thereby the ENDOR effect, becomes proportional to the hyperfine coupling a_j. Large ENDOR effects are, therefore, expected for lines corresponding to large a_j values, whereas lines corresponding to small a_j should be weaker. Figure 2.15(b) shows the strong multiplet-effect polarization in the time-resolved mode of detection after pulsed laser excitation: the low-field half of the EPR spectrum is in emission, the high-field half in enhanced absorption. Depending on the magnetic spin quantum number M_I of the EPR line and its polarity being selected for the ENDOR experiment, the

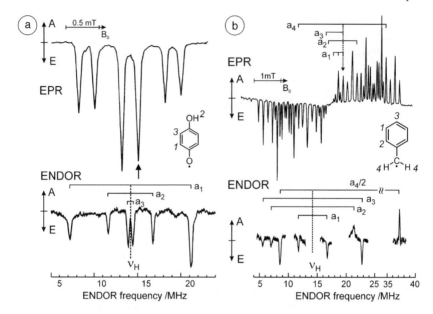

Figure 2.15 (a) Time-resolved EPR and ENDOR spectra of p-benzosemiquinone radicals in ethylene glycol solution, detected 5 µs after creation by 10-ns laser pulses at 308 nm at $T = 300$ K. For ENDOR the EPR line indicated by an arrow was chosen for detection applying saturating microwave and rf fields. The measured hyperfine constants are $a_1 = -14.00 =$ MHz; $a_2 = -5.13$ MHz; $a_3 = 0.53$ MHz. (b) Time-resolved EPR and ENDOR spectra of benzyl radicals in propanediol solution detected 4 µs after creation by a 10-ns laser pulse at 308 nm at $T = 298$ K. For ENDOR the EPR line indicated by an arrow was chosen for detection applying a nonsaturating microwave and a saturating rf field. The measured hyperfine constants are $a_1 = 4.95$ MHz, $a_2 = -14.39$ MHz, $a_3 = -17.35$ MHz and $a_4 = -45.7$ MHz. For details, see ref. [134].

ENDOR effect shows up either in a decreased absorption or a decreased emission in accordance with a dominating multiplet effect. Almost 100% changes in the EPR intensity could be observed for the largest hyperfine coupling.[134]

From these first results it is concluded that CIDEP-enhanced ENDOR and TRIPLE of transient reaction intermediates have a promising future potential for the characterization of photochemically generated radicals and photoexcited triplet states exhibiting pronounced CIDEP effects in their EPR spectra.

C Transient Intermediates in Light-Induced Reactions

C.1 Introduction

Since the early 1970s, EPR methods were extensively applied to characterize the transient free radicals occurring in liquid-solution reactions. Time-resolved

EPR techniques have been applied to a wide variety of photochemical systems, including investigations of radical reaction kinetics, identification and characterization of reaction intermediates, elucidation of reaction mechanisms with regard to the spin states involved. One broad area of EPR studies is to investigate the chemical reactivity of radicals, especially concerning the radical–radical recombination, as well as addition reactions of the radicals to other chemical compounds. The addition reactions are of crucial importance for industrial coating processes as well as in medical treatments. The second field of EPR applications is to study spin-polarization (CIDEP) effects and spin-relaxation dynamics in free-radical systems.

C.2 Photochemical Reactions in Liquid Solution, Illustrative Examples
In the following we demonstrate, by means of two photochemical systems, the power of multifrequency EPR in reaction and spin chemistry of free radicals in liquid solution.[103]

C2.1 Photoexcited 2,2-Dimethoxy-1,2-Diphenylethan-1-one
The compound 2,2-dimethoxy-1,2-diphenylethan-1-one (DMPA) is an industrial agent for efficient photocuring (IRAGACURE® 651, Ciba Speciality Chemicals) used mainly for acrylic and unsaturated polyester/styrene resins. Due to its technical relevance, the DMPA photochemistry has been the subject of many investigations, including EPR experiments.[142–144]

The primary steps of photolysis of DMPA consist of singlet excitation and fast intersystem crossing (ISC) to its triplet state, immediately ($\tau_T < 100$ ps) followed by Norrish type-I photocleavage of the triplet, yielding benzoyl and dimethoxy benzyl radicals, R_1 and R_2, respectively (see Figure 2.16(a)). The quantum efficiency, Φ, of the triplet cleavage is $\Phi > 0.7$ (dependent on solvent). The radicals R_1 and R_2 appear highly electron-spin polarized. This polarization originates from the triplet (TM) and radical-pair (RPM) mechanisms, see subsections A.1 and A.2 in Section 2.2.1.2.

The W-band high-field EPR spectra obtained after photolysis of 65 mM DMPA in oxygen-free benzene solution are shown in Figure 2.16(b) for several time delays after the laser flash. For comparison, the X- and Q-band spectra[144] are also shown. At W-band, the TREPR spectra of the benzoyl, R_1, and dimethoxibenzyl radicals, R_2, appear completely resolved owing to the significant difference of their g-values, $g_1 = 2.00690$ and $g_2 = 2.00295$. Directly after the laser flash the spectra of both radicals appear in net emission at all microwave frequencies and Zeeman fields. This indicates that TM CIDEP dominates. However, the amount of initial emissive polarization of R_1, compared to that of R_2, decreases with increasing Zeeman field. This is not unexpected: On the one hand, the polarization due to Δg-RPM CIDEP in the initial radical pair increases with increasing field, giving rise to enhanced absorption of radical R_1 and emission of R_2. On the other hand, the net emissive polarization, which the radicals gain from the triplet precursor, gets smaller at higher fields (see below), thereby enhancing the contribution of Δg-RPM. The increasing Δg-RPM polarization in F-pairs of R_1 and

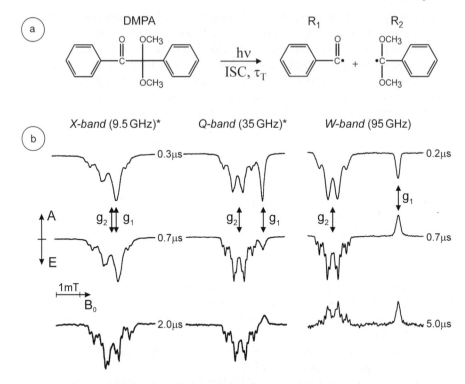

Figure 2.16 (a) Reaction scheme for the primary steps of the photolysis of DMPA in liquid solution. (b) X-, Q- and W-band time-resolved EPR spectra taken at delay times indicated after flash photolysis (352 nm) of 65 mM (W-band) DMPA in benzene at room temperature. The X- and Q-band spectra (marked by *) are adapted from ref. [144].

R_2 has been proposed to be responsible for the phase inversion in Q-band (35 GHz) EPR (emission to absorption) of the benzoyl radical signals at later times. This argument should also hold for W-band EPR. However, this is not the only possible explanation: The W-band EPR spectra, recorded 5 μs after laser irradiation, show that both signals appear in absorption. This strongly suggests that in this situation the radicals are observed in their thermal (Boltzmann) equilibrium. The equilibrium polarization $P_{eq} = \frac{1}{2} \cdot g \cdot \mu_B \cdot p_{eq} \cdot N_A \cdot x_i/x$ (N_A = Avogadro's number, x_i/x = weight factor of the EPR transition under study) is proportional to the population difference ($N_\alpha - N_\beta$) between the two Zeeman levels α ($m_S = +1/2$) and β ($m_S = -1/2$) in Boltzmann equilibrium, i.e.

$$p_{eq} = \frac{N_\alpha^0 - N_\beta^0}{N_\alpha^0 + N_\beta^0} \approx \frac{h \cdot \nu}{2 \cdot kT}, \quad (2.44)$$

taking into account that $h\nu \ll kT$. At room temperature, p_{eq} corresponds to about 7.8×10^{-4} at $\nu = 9.5$ GHz (X-band), and to 7.8×10^{-3} at 95 GHz

(W-band). Thus, in X-band EPR the same initial population difference decays to a 10-times smaller magnitude than in W-band EPR. This means that, if the radical system starts with a net polarization corresponding to a population difference of the two Zeeman levels of, say, $(N_\alpha - N_\beta)/(N_\alpha + N_\beta) = 0.1$ both at X- and W-band, after relaxation and in the absence of chemical reactions the observed signal contains only about 0.8% of the initial polarization in X-band, but 8% in W-band. Thus, the observed phase inversion is caused both by Δg-RPM in F-pairs and by increased equilibrium polarization at high magnetic fields.

The results of this multifrequency investigation of DMPA photolysis led to the following conclusions: *(i)* For a full understanding of the polarization mechanisms in the radical system it is necessary to perform EPR experiments at different microwave frequencies and Zeeman fields; *(ii)* at high magnetic fields it is possible to observe the radicals in their thermally equilibrated states, *i.e.* Boltzmann polarized. This chance to observe, by TREPR at high fields, transient radicals in thermal equilibrium opens new perspectives for investigations of CIDEP phenomena and radical kinetics. First, there is the possibility now for direct determination of the initial polarization, since the initial signal can be scaled to the thermally equilibrated one. Secondly, analysis of the time dependence of the EPR signals can provide directly the kinetic information about the radical reactions.

C.2.2 Photoexcited 2,4,6-Trimethylbenzoyl-Diphenyl-Phosphineoxide

The investigation of the photolysis of a polymerization photoinitiator of the acrylphosphine oxide type, the compound 2,4,6-trimethylbenzoyl-diphenyl-phosphineoxide (TMDPO) demonstrates the full power of multi-frequency EPR in spin-chemistry studies as well as in investigations of radical reactions. The photochemistry of TMDPO is well established.[145–147] According to the reaction scheme in Figure 2.17, photolysis of TMDPO yields 2,4,6-trimethylbenzoyl, R_1, and diphenylphosphinoyl, R_2, radicals via Norrish type-I cleavage ($\Phi \approx 0.6$) from the short-lived ($\tau_T < 1$ ns) triplet excited state that is formed from the photoexcited singlet state of TMDPO via ISC.

S-, X- and W-band transient EPR spectra after laser-flash photolysis of TMDPO in benzene solution are shown in Figure 2.18. The spectra consist of three lines: the inner line stems from the carbon-centered radical R_1 ($g_1 = 2.00055$), and the outer two lines are due to the phosphorus-centered radical R_2 ($g_2 = 2.00412$, $a_P = 36.25$ mT). At short delay times after the laser flash, the W-band TREPR spectra of R_1 and R_2 exhibit net absorptive polarization, see Figure 2.18. For the radical R_2, an additional multiplet polarization of E/A type (low-field line in emission/high-field line in absorption) is revealed, leading to a reduced absorption-line intensity at the low-field side as compared to the high-field side. The net absorption observed for both radicals reflects the TM polarization transferred from excited TMDPO. The second contribution to the net polarization of the radicals comes from the Δg–RPM in the geminate triplet radical pairs (RPs) of R_1 and R_2. For the radical R_1, the Δg-RPM contribution

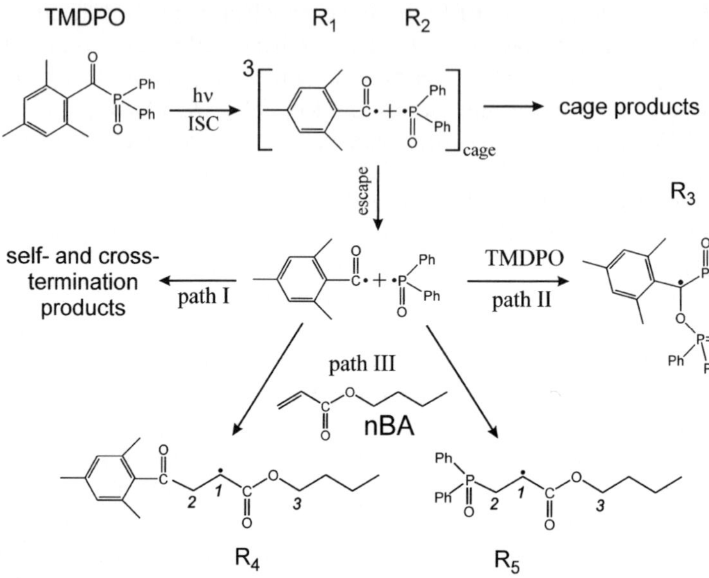

Figure 2.17 Reaction scheme for the primary and secondary reaction steps of the photolysis of TMDPO. For details, see ref. [103].

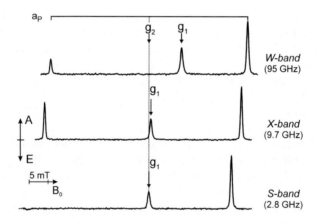

Figure 2.18 Time-resolved EPR spectra of R_1 and R_2 recorded at different microwave frequencies after laser-flash photolysis (355 nm) of TMDPO in benzene solution at room temperature with a time delay of 200 ns. For details, see ref. [149].

is positive (absorption), and for the radical R_2 it is negative (emission) since $g_1 < g_2$. The E/A multiplet polarization contribution in the spectrum of R_2 is generated by ST_0 mixing of the radical-pair mechanism (ST_0-RPM) in the geminate RPs of R_1 and R_2. The initial polarization pattern of the radicals is different in all EPR frequency bands (S-, X-, W-band), which is clearly seen in Figure 2.18. This difference is explained in terms of a field dependence of the CIDEP-generating mechanisms. For example, the initial multiplet ST_0–RPM polarization is, to a good approximation, magnetic-field independent,[123–124,148] whereas the polarization due to Δg–RPM and TM changes when going from S- to W-band EPR.

At long times after the laser flash the difference between spectra taken at different microwave frequencies becomes more dramatic. At X-band the enhanced absorption observed for R_2 decays within about 1 μs after the laser flash, leaving the spectrum purely multiplet polarized (E/A). However, at W-band the intensities of both high- and low-field lines become equal after the same delay. Both lines appear in absorption that slowly decays to zero. To explain this phenomenon, the EPR time profiles of the high- and low-field lines of R_2, as observed in X- and W-band, were compared.[142] The time profiles were analyzed in terms of net and multiplet contributions, which were obtained by adding and subtracting the intensities of the low- and high-field lines of R_2, respectively. At X-band, the observed picture is well described by the CIDEP mechanisms and the reaction scheme of Figure 2.17. At W-band, the observation can only be explained by assuming equilibrium polarization, because the initial net polarization does not decay to zero, but to the thermally equilibrated polarization.

The time profile of the net-polarization contribution at W-band can be analyzed using the analytical solution of modified Bloch equations that account for an initial net polarization and a first-order chemical reaction. The perpendicular magnetization in the rotating frame, to which the EPR signal is proportional, is given by:

$$v(t) = P_{eq} \cdot [R]_0 \cdot \frac{g_e \cdot \mu_B \cdot B_1}{\hbar} \cdot T_2 \cdot \left[e^{-t/\tau} + (P_n - 1) \cdot e^{-t/T_1} \right] \quad (2.45)$$

if the conditions $T_2 \ll T_1$, $(g_e \cdot \mu_B \cdot B_1/\hbar)^2 \ll T_1^{-1} \cdot T_2^{-1}$ and $T_1/\tau \ll 1$ are fulfilled. Here, B_1 is the microwave field amplitude, P_{eq} the radical equilibrium polarization, $[R]_0$ the initial radical concentration, and τ the first-order lifetime constant of reaction path II with excess TMDPO (see Figure 2.17), P_n is the net polarization contribution in units of P_{eq}. The conditions for validity of eqn (2.45) were found to be fulfilled for R_2. The best agreement with experiment was obtained for $P_n = 8.5$, $T_1 = 600$ ns and $1/\tau = k_1 = 1.7 \times 10^5 \, s^{-1}$.

The possibility to read out the chemical lifetime of R_2 from W-band EPR time profiles allows the study of the addition reaction of diphenylphosphinoyl radicals to functional monomers. This option was used to investigate the addition reaction of R_2 to n-butylacrylate (nBA),[103] path III in the reaction scheme of Figure 2.17. Figure 2.19 depicts the cw and TREPR spectra taken without and in the presence of 0.5 M nBA in the toluene stock solution of 48 mM TMDPO. Without nBA, Figure 2.19(a), the TREPR spectra show the same behavior as described above. In the cw EPR spectra, the signals of long-lived radicals are observed. One of the signals around $g_3 = 2.00288$ is assigned to a carbon-centered radical R_3 ($a(P1) = 2.87$ mT, $a(P2) = 1.79$ mT and $a(H) = 0.13$ mT)

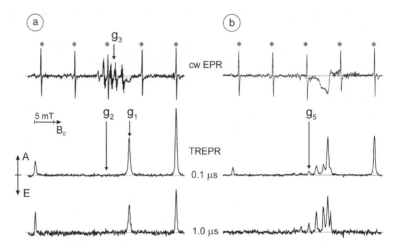

Figure 2.19 W-band cw and time-resolved EPR spectra observed after photolysis of (a) 48 mM TMDPO; (b) 48 mM TMDPO and 0.5 M *n*-butylacrylate in toluene solution at $T = 298$ K. The cw and TREPR spectra were taken simultaneously with a magnetic-field-modulation amplitude of 5 μT. The asterisks mark the signals of the Mn^{2+} field standard that was positioned in the microwave cavity. For details, see ref. [103].

formed according to path II in the reaction scheme. An addition rate constant of the reaction "R_2 adds to TMDPO" of $3.5 \times 10^6 \text{ s}^{-1} \text{ M}^{-1}$ is estimated from k_1 and the TMDPO concentration. Radical R_3 is not observed in the sample containing nBA, Figure 2.19(b). The unresolved signal around $g = 2.005$ observed in the cw EPR spectrum is probably due to a mixture of radicals from the polymerization chain reaction. At 0.1 μs after the laser flash, the TREPR spectrum shows the signals of R_1 and R_2; however, the ratio of the EPR intensities changes in favor of the benzoyl radical R_1. Furthermore, the signal of the adduct radical, R_5, appears at $g_5 = 2.0033$ ($a(P) = 5.8$ mT, $a(1H,2H) = 1.93$ mT and $a(3H) = 0.2$ mT). At longer times, 1 μs in Figure 2.19(b), the R_2 signal disappears, but this is accompanied by a growing-in of the R_5 signal. The signal due to R_1 decays approximately with the same rate constant as in the nBA-free sample.

The net signals of R_2 at low nBA concentrations show biexponential decay character according to eqn (2.45). At nBA concentrations above 100 mM the signals decay monoexponentially, see Figure 2.20(a). The simultaneous fit of the net decay profiles to eqn (2.45), varying only the chemical lifetime constant, provided the τ values shown in Figure 2.20(b). Analysis of the data in terms of

$$1/\tau = k_1 + k_{\text{add}} \cdot [\text{nBA}] \qquad (2.46)$$

yields a second-order addition rate constant of "R_2 to nBA" of $k_{\text{add}} = 3.2 \times 10^7 \text{ s}^{-1} \text{ M}^{-1}$ for low adduct concentrations. The constant k_{add} is in good agreement with the values $2.8 \times 10^7 \text{ s}^{-1} \text{ M}^{-1}$ in acetonitrile[150] and 3.3×10^7 $\text{s}^{-1} \text{ M}^{-1}$ for addition to methyl methacrylate in hexane that were obtained by

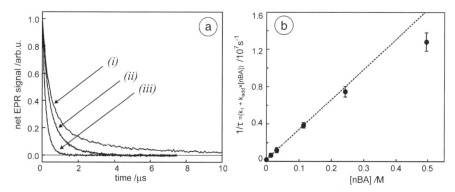

Figure 2.20 (a) Net-polarization EPR time profiles of radical R_2 observed without nBA (*i*) and with 32 mM (*ii*) and 114 mM (*iii*) nBA in the toluene solution of 48 mM TMDPO, respectively. (b) Plot of the inverse chemical lifetime $1/\tau$ of R_2 vs. nBA concentration, as determined from the fits of net-polarization EPR time profiles. The dashed line gives a linear fit for nBA concentrations below 250 mM. For details, see ref. [103].

laser flash photolysis.[146] At high nBA concentrations (>250 mM) the measured reciprocal chemical lifetime values significantly deviate from a linear dependence of [nBA], see Figure 2.20(b). The recent investigation of the addition reaction of differently substituted benzoyl radicals to nBA established two domains of radical reactivity in dependence on the concentration of nBA.[151] At alkane concentrations above 1 M, the addition reaction is slowed down by about one order of magnitude. This decrease coincides with a dramatic increase of the macroscopic viscosity of the reaction medium.

We conclude: The example given above demonstrates the power of high-field EPR for obtaining detailed information on radical reactions in solution. The possibility to observe the radicals in their thermally relaxed state allows their chemical kinetics to be directly followed. Moreover, the high time resolution of high-field EPR allows measurement of the short radical chemical lifetimes directly from the EPR signal decays. For example, the EPR signal of R_2 in the solution containing 0.5 M nBA decays with a time constant of about 70 ns, Figure 2.20(b). This value is already comparable with the time resolution of conventional X-band EPR spectrometers. It is, however, about a factor of 10 longer than the time resolution of our W-band spectrometer, see Section 3.4.1.

The analysis of the net EPR signals yields, besides the chemical lifetime of the radicals, also the initial net polarization, P_n, in absolute units. For the determination of the initial multiplet polarization, P_m, the ratio of the net and multiplet EPR intensities, at short times after the laser pulse, can used. A ratio $P_n/P_m = 1.6 \pm 0.1$ was found for the initial W-band net and multiplet CIDEP magnitude of the diphenylphosphinoyl radical system after photolysis of TMDPO. The radicals are produced with a net polarization $P_n = 8.5 \pm 1$, which corresponds to an initial population difference $P_n \cdot p_{eq}$(W-band) = 0.066. This net polarization is superimposed by multiplet polarization (E/A) of

$P_m = 5.5 \pm 0.8$, which is equivalent to a population difference of 0.043. Because only triplet and radical-pair mechanisms are responsible for the production of the initial multiplet and net polarizations in R_2, the values obtained at W-band can be used to determine the absolute polarizations at lower frequencies. The experiments at S-, X- and Q-bands provided the field dependence of the ratio P_n/P_m of R_2 under the same experimental conditions.[149] Analysis of these ratios in terms of the TM and the RPM by ST_0, ST_- mixing, which becomes extremely important at low fields, yielded the absolute values of triplet polarization that R_2 gains from the polarized TMDPO triplet, see Figure 2.21. The observed values of TM polarization are extremely high. They exceed the equilibrium polarization by 410 times at X-band and nearly 1000 times at S-band.

The magnetic field B_0 (or corresponding microwave frequency) is an important parameter for TM polarization that determines, in conjunction with other parameters, the efficiency of spin-polarization transfer from the molecular frame of the excited triplet molecule to the laboratory frame and then to the radicals. In contrast to TM polarization, RPM polarization is practically field independent for organic molecules. The reason is that the field-dependent Δg-PRM contribution, see eqn (2.42), is normally small in organic radicals in relation to the field-independent Δa term in the ST_0 mixing. Qualitatively, the effect of microwave frequency on the magnitude of the TM polarization is well understood. Moreover, it is also now possible to test the theory quantitatively. The field

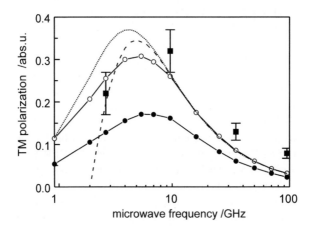

Figure 2.21 TM polarization of the phosphinoyl radical R_2 (■) after photolysis of TMDPO in comparison with the theoretical calculations (●) by numerical solution of the stochastic Liouville equation (SLE) and using the triplet-state parameters from ref. [152] and triplet rotational correlation time from a spherical approximation of the molecule; (○) best fit of the numerical solution of SLE for the lifetime of the triplet state and its initial population, as given in the text. The dashed and dotted lines show the polarization calculated by using analytical expressions of Atkins–Evans[110] and Pedersen–Freed[111] with the parameters of (○). The figure is adopted from ref. [149].

dependence of TM polarization was examined by comparison of the experimental data with the numerical solution of the stochastic Liouville equation (SLE) that describes the evolution of the spin system.[149] Figure 2.21 presents the experimental dependence of the TM on microwave frequency (squares) and the TM frequency dependence as calculated from the SLE (filled circles), using the TMDPO triplet parameters from the literature[152] and a triplet rotational correlation time of 48 ps. This value was estimated from the spherical approximation of the molecule and using the Debye–Stokes equation. It is obvious that the TM polarization is underestimated by the SLE calculation. Its field dependence, however, is reproduced correctly. In a next step, the SLE numerical solution was fitted to the experimental data with the reorientational correlation time and initial population of the T_z state as variable parameters, see Figure 2.21 (open circles). The optimum fit requires a slightly higher population of the T_z level (1.0 instead of 0.8) as well as an increase of the correlation time by a factor of 2 (98 ps, instead of 48 ps). Such a discrepancy is not surprising because the simplified model for the correlation time was used. Therefore, the rotational correlation time should be considered as an effective fitting parameter.

The numerical solution of SLE explains the observed experimental TM polarization quite well. To check the validity of approximate analytical formulas derived by Atkins and Evans[110] and Pedersen and Freed,[111] TM polarization was calculated with the triplet parameters obtained above, see in Figure 2.21 the dashed and dotted lines, respectively. It was found that both approximations are valid only at high microwave frequencies (beyond X-band), although the Atkins–Evans approach remains qualitatively correct also at low mw frequencies. Investigations of the TM polarization of two other acrylphosphine oxide photoinitiator systems confirm the experimental results of the TMDPO system and their theoretical analysis.[149] Moreover, the parameters of the excited triplet state of phenylbis(2,4,6-trimethylbenzoyl)phospine oxide could be extracted from the dependence of its TM polarization on the microwave frequency.

2.2.1.3 Solid-State Pulse ENDOR and TRIPLE

At first, in subsection A, we treat the ENDOR and TRIPLE phenomena without taking nuclear quadrupole interaction explicitly into account. This will be done in the subsequent subsection B.

A Systems Without Quadrupole Interaction
ENDOR measurements can also be carried out in pulse mode with electron spin-echo detection as long as the relaxation times are long enough to allow multiple pulse trains to prepare and to detect the desired spin states. This is normally not the case in liquid-phase samples at elevated temperatures and, hence, echo-detected ENDOR experiments are a typical domain for solid-state samples. Excellent textbooks and specialized review articles cover this subject in detail, for example refs. [4,9,153–155]. Here, we base our discourse mainly on a chapter in ref. [14], which was published recently.

Although the Mims-type[156] and Davies-type[157] ENDOR pulse sequences (shown in Figure 1.7) were introduced already several decades ago, they are still the most common versions of pulse ENDOR spectroscopy. In both versions, the ENDOR spectrum is obtained by recording the echo intensity as a function of the frequency of the rf pulse. A change in the echo amplitude occurs when the rf is on-resonance with an NMR transition in the coupled electronic–nuclear spin system, thereby generating an ENDOR signal. The Davies-ENDOR experiment is based on the selective mw excitation of only one of the EPR multiplet transitions and, therefore, the first mw π pulse has to be selective with respect to the hyperfine splitting. Consequently, this sequence is most suited for systems with medium to large hyperfine couplings ($A > 2$ MHz). Mims-ENDOR, on the other hand, does not require selective pulses but suffers from "blind spots". This is because the ENDOR signal efficiency is scaled by $1/2 \cdot \sin^2(\pi \cdot A \cdot \tau)$, where A is the hyperfine splitting measured in Hz. Hence, maximum ENDOR efficiency is obtained for $\tau = (2n+1)/2A$, whereas blind spots (zero ENDOR efficiency) occur at $\tau = n/A$ with $n = 0, 1, 2, \ldots$ (see ref. [4]). As a consequence, a maximum hyperfine coupling, A_{max}, exits for which the Mims-ENDOR spectrum remains undistorted by blind spots. It is given by $A_{max} = 1/2 \cdot \tau$. Correspondingly, the first blind spot in the ENDOR spectrum defines the minimum hyperfine coupling, A_{min}, that can be observed by short enough τ values: $A_{min} = 1/\tau$. Accordingly, Mims-ENDOR is usually applied when the hyperfine splitting is small, $A < 2$ MHz, where typical τ values of 0.15–0.25 µs place the blind spots well outside the ENDOR spectral range of interest. If lower values of τ are required and cannot be used due to spectrometer "deadtime", it is possible to apply the remote-detection Mims (ReMims)-ENDOR sequence.[158] In general, ^{1}H spectra are usually recorded by the Davies-ENDOR sequence, whereas Mims-ENDOR is preferred for ^{2}H measurements.

In general, the intensity of the ENDOR signal, *i.e.* the ENDOR effect, is defined as:

$$F_{ENDOR} = \frac{I_{rf}(\text{off}) - I_{rf}(\text{on})}{I_{rf}(\text{off})} \tag{2.47}$$

where, in the case of pulse ENDOR, $I_{rf}(\text{on})$ and $I_{rf}(\text{off})$ correspond to the spin-echo intensity with the rf pulse switched on and off, respectively. Usually, the ENDOR effect is the same for the α ($m_S = +1/2$) and β ($m_S = -1/2$) electron spin manifolds. The maximum ENDOR effect is obtained with rf pulses for 180° rotation of the nuclear spin (π pulse).

When the hyperfine coupling becomes large, the nuclear nutation frequency, ν_2, also depends on the magnitude of the hyperfine coupling. For the simple case of an isotropic interaction, it is given by ref. [4]

$$\nu_2 = \frac{g_n \cdot \mu_K}{h} \cdot E \cdot B_2; \quad E = \left(1 + \frac{m_S \cdot A_{iso}}{\nu_I}\right) \tag{2.48}$$

where E is the so-called hyperfine-enhancement factor and B_2 is the rf-field amplitude. This shows that when the hyperfine coupling is large compared to the nuclear Larmor frequency, v_I, E becomes large, and a lower rf power is required for generating an rf π pulse. For low-g_n nuclei, that exhibit large enough hyperfine couplings, like ^{14}N and ^{55}Mn, the hyperfine-enhancement factor becomes very useful as it shortens the required rf pulse time for a 180° rotation.

Also, in cw ENDOR the hyperfine-enhancement factor plays an important role for the ENDOR intensity since it amplifies the rf amplitude seen by the nucleus, see Section 2.2.1.1.

The majority of pulse ENDOR experiments are carried out at low temperatures because long enough T_1 values are required to allow insertion of rf pulses, typically 10–40 μs, during the period T without a considerable loss of the electron polarization generated by the first pulses. For this reason, pulse ENDOR experiments are usually not applicable to liquid-solution samples because of their short T_1s. For nitroxide and other organic radicals, liquid-nitrogen cooling is sufficient, whereas for paramagnetic transition-metal ions measurements in the range of 1.2–15 K are usually necessary and require helium cooling. Caution should be exercised for high-field EPR at high mw frequencies, such as 95 and 140 GHz, the cavity tuning and the phase of the signal are highly sensitive to small changes in the temperature because of fluctuations of the He gas flow in the cryostat. This is particularly problematic for broad peaks and/or very weak ENDOR effects. There are solutions described in the literature to overcome this problem.[159] In solids, the advantages of pulse over cw ENDOR are often dramatic in terms of distortion-free lineshapes the pulse techniques produce. Moreover, they have the prospect of introducing a variety of one- and two-dimensional experiments for resolution, assignment and correlation purposes, quite in the spirit of established NMR techniques.

As mentioned above when describing cw multiple resonance experiments, one can get over certain weaknesses of ENDOR concerning sensitivity and relative line intensities by extending it to electron–nuclear–nuclear triple resonance techniques, such as Special and General TRIPLE.[83] Both the Special and General TRIPLE experiments have their analogous pulse sequences,[49,159] see Figure 1.7. The Special TRIPLE technique requires two rf pulses, one exciting v_α and the other the corresponding v_β. Hence, the relation between the two ENDOR frequencies should be known *a priori*. This condition is often satisfied at high field where the two ENDOR frequencies are symmetrical with respect to the nuclear Larmor frequency, as occurs in liquid samples. For solids, the frequency doublet is symmetrical only when the first-order approximation applies, as is often encountered at a high Zeeman field. Using Special TRIPLE, a maximum of a twofold increase in the ENDOR effect can be achieved.[159] In the General TRIPLE version, one rf frequency is set to a particular ENDOR transition while the other is scanned.[49] As in the cw experiment, the resulting spectrum shows which ENDOR signals belong to the same m_S manifold as that excited by the first rf pulse, thus yielding the *relative* sign of the hyperfine coupling. When a number of paramagnetic centers contribute to the ENDOR spectrum, the General ENDOR sequence correlates signals belonging to the same center and the same m_S manifold. Scanning the frequency of both rf pulses produces a two-dimensional spectrum with crosspeaks between ENDOR lines belonging to the same m_S manifold.[50] In orientationally disordered samples the shape of the crosspeaks can provide the relative orientation between the principal directions of the hyperfine interactions involved.[44]

The sign of the hyperfine coupling is an important parameter. For example, in the case of the isotropic hyperfine constant, A_{iso}, the sign provides additional insight into the electronic mechanism producing unpaired electron density at the nucleus, either by a spin-polarization mechanism ($A_{iso} < 0$, e.g., for α-protons) or by conjugation/hyperconjugation mechanisms ($A_{iso} > 0$, e.g., for β-protons). The sign is an important reference when quantum-chemical calculations are carried out for interpreting the hyperfine interaction in terms of structure. The experimental determination of the *absolute* sign of the hyperfine coupling requires large thermal polarization at very low temperatures. This is usually not achieved at X-band frequencies but is rather easily obtained at high Zeeman fields and moderately low temperatures. This strategy is analogous to a common procedure for high-spin systems ($S > 1/2$) to determine the sign of the zero-field splitting parameter by substantial thermal polarization in high Zeeman fields already at moderately low temperatures.[43]

Extension to high-field ENDOR or TRIPLE has the additional advantage of providing single-crystal-like hyperfine information in the reference frame of the g-tensor, even from disordered samples with very small g-anisotropy. This will be treated to some detail later in the application sections, for example the Zeeman magnetoselection of the canonically oriented radical anions of ubiquinone-10 in disordered frozen solution by W-band pulse EPR and ENDOR.[160] In the case of organic radicals with small hyperfine interactions one has to resort to the anisotropy of the Zeeman interaction. This can become large enough at high B_0 fields to provide the desired orientation selectivity for ENDOR experiments on disordered systems. In the case of transition-metal complexes, the large hyperfine anisotropy may provide this orientation selectivity for the random distribution of the molecules in the sample. The best approach for elucidating molecular structure and orientation in detail is, of course, to study single-crystal samples. Unfortunately, to prepare them is often difficult or even impossible for large biological complexes. Nonetheless, a few such EPR and ENDOR studies have been carried out at high fields to take advantage of the high sensitivity for size-limited samples, the high resolution and the simplification of the spectra.[161–163]

B Systems with Quadrupole Interaction

We now turn to spin systems containing nuclei ($I > 1/2$) with quadrupole interactions. In biological systems, examples of common nuclei with $I > 1/2$ are the nitrogen ^{14}N and deuterium ^{2}H isotopes with $I = 1$, the copper ^{63}Cu and ^{65}Cu isotopes with $I = 3/2$, the oxygen ^{17}O and manganese ^{55}Mn isotopes with

$I = 5/2$. If the molecule under study contains nuclei with $I > 1/2$ in an electronic environment with a nonvanishing electric-field gradient, q, the ENDOR line pattern given by eqn (2.16) is extended by a quadrupolar contribution from the traceless nuclear quadrupole tensor, *i.e.* in the first-order approximation:[164]

$$\nu^{\pm}_{i\text{ENDOR}}(m_I) = \left| \nu_n \pm A_i/2 + \frac{3}{2} \cdot P_i \cdot (2m_I - 1) \right|, -I + 1 \leq m_I \leq I \quad (2.49)$$

P_i is a function of the quadrupole coupling constant $e^2 \cdot q \cdot Q/h$, the asymmetry parameter η and the relative orientation of the quadrupole tensor with respect to the Zeeman field. The quadrupole coupling splits or shifts the ENDOR lines ($m_I = I, I-1, \geq 0$) provided the nucleus is in an anisotropic electric environment.[21,90,165,166] Deuterium quadrupole couplings in paramagnetic molecular states are sensitive probes for local-structure information on H-bond patterns. Empirical relations between $e^2 \cdot q \cdot Q$ values and O–H distances (a $1/R^3_{O \cdots H}$ correlation) were established long ago,[167] but have been verified also by semiempirical MO[165] and DFT calculations.[168] It is noteworthy that the DFT calculations also reproduce a $1/R^2_{O \cdots H}$ correlation for the g_{xx} tensor component.

Another virtue of high-field ENDOR is the possibility to detect weakly coupled, low-g_n nuclei such as ^{2}H. Their ENDOR signals appear around the corresponding Larmor frequency, which at X-band is very low (≈ 2 MHz) and usually hard to detect. Therefore, X-band ^{2}H ENDOR is not often reported, however the situation is already different at Q-band (35 GHz), see ref. [22] and references therein. At high fields, like in W-band ENDOR, these signals appear around 20 MHz (see Figure 2.2) and are easily detected.[169] It should be noted that ^{2}H is most useful as a probe for exchangeable protons.[22] It can be used to derive additional structural information if the ^{2}H quadrupolar splittings can be resolved. Such resolution can be obtained if the ENDOR signals are recorded at the resolved principal components of the g-tensor in the W-band EPR spectrum allowing for single-crystal-like ENDOR spectra.[22,170]

Finally, at high fields the ENDOR frequencies can often be described using first-order approximations, thereby simplifying the spectral analysis. This is important in the case of quadrupolar nuclei with half-integer spins, for instance ^{17}O, which have many potential applications for local-structure determination of biological systems. In this case, the ENDOR frequencies are given by eqn (2.49) including the quadrupolar splitting term. For nuclei with half-integer nuclear spin, such as $I = 3/2$ and $5/2$, this term becomes zero for the ENDOR transitions corresponding to $m_I = 1/2 \rightarrow -1/2$ that are thus independent of the nuclear quadrupole interaction to first order. To second order, they exhibit an anisotropy that is proportional to $1/\nu_n [3e^2 \cdot q \cdot Q/4h \cdot I \cdot (2I-1)]^2$. Consequently, the higher the magnetic field is the simpler is the ENDOR spectrum, and a better resolution is observed for orientationally disordered systems. Similarly, simpler ^{55}Mn ENDOR spectra are expected at high fields,[171] which may be useful for applications to the oxygen-evolving complex in photosystem II of oxygenic photosynthesis.

Below, we give some representative examples of applications of high-field EPR and ENDOR on proteins in general, which demonstrate the power of the methods for analyzing structure and dynamics beyond X-ray crystallography:

The degree of dynamic and structural information on Mn^{2+} sites ($S = 5/2$, $I = 5/2$) in proteins that can be derived from single-crystal high-field EPR and ENDOR was demonstrated on the protein concanavalin A.[163,171,172] This protein is a member of the plant lectin family, a ubiquitous group of proteins that bind saccharides. They frequently appear on the surface of cells for specific molecular recognition of individual sugars. Concanavalin A has been extensively studied in the past, and its 3D X-ray structure was determined to an exceptionally high resolution of 0.94 Å.[173] The W-band EPR/ENDOR investigation emphasized the complementarity of high-field EPR and ENDOR information with respect to X-ray crystallography data by focusing on local motions and the location of protons in the protein. The protein contains two metal binding sites, one occupied by Mn^{2+} and the other by Ca^{2+}. The Mn^{2+} site has a slightly distorted octahedral geometry in which the Mn^{2+} is coordinated to three carboxyl groups, one imidazole and two water molecules. The role of the transition-metal ion in this protein is structural, serving as a stabilizer for the loops that constitute the saccharide binding site.[174] The ZFS tensor of the Mn^{2+} was determined from W-band EPR single-crystal rotation patterns acquired at room temperature and at 4.5 K.[172] The highly resolved spectra allowed to distinguish two types of Mn^{2+} sites, Mn_A^{2+} and Mn_B^{2+}, with different D and E values. This is in contrast to room-temperature measurements, which revealed only one type of Mn^{2+} site, as expected from the X-ray structure. The temperature-dependent EPR data showed that, as the temperature increases, the two well-resolved sextets of Mn_A^{2+} and Mn_B^{2+} lines shift and gradually coalesce into a single sextet at room temperature. These spectral changes were analyzed in terms of a two-site exchange process, with rates in the range of 10^7–10^8 s^{-1} for the temperature range of 200–266 K and an activation energy of 23.8 kJ/mol. This dynamic process was attributed to a conformational equilibrium within the Mn^{2+} binding site that freezes into two conformations at low temperatures.[172]

W-band single-crystal ENDOR measurements, which determined the position of the water protons in the Mn^{2+} sites, were also carried out.[163] The 3D structure shows that the histidine and water protons are the closest to the Mn^{2+} ions and, therefore, they dominate the ^{1}H ENDOR spectrum. ^{1}H ENDOR rotation patterns were collected in two crystallographic planes. The signals of the water and the imidazole ring protons were assigned by comparing the ^{1}H ENDOR spectra of crystals grown in H_2O and D_2O, where the ^{1}H water signals are absent from the spectra of D_2O crystals. The signs of the hyperfine splittings were determined through assignment of the ENDOR signals to their respective $m_S = \pm 1/2$ manifolds. This was achieved by recording the ENDOR spectra at magnetic-field positions where the contributions of the $|-1/2\rangle \to |1/2\rangle$ transitions are negligible and those of the $|-3/2\rangle \to |-1/2\rangle$ and $|-5/2\rangle \to |-3/2\rangle$ are substantial.[175] Under such conditions the contribution of the $m_S = 1/2$ ENDOR line is considerably diminished.

Simulations of the rotation patterns[163] provided the principal components of the hyperfine-tensors and their direction relative to the crystallographic axes. The ENDOR spectra recorded at magnetic-field positions corresponding to either Mn_A^{2+} and Mn_B^{2+} were the same. This shows that the Mn–H distances and orientations are the same for the two sites. Therefore, the water and the imidazole ligands are not involved in the local motion described above, but rather one (or more) of the carboxylate ligands. Using the point-dipole approximation, the coordinates of the water and imidazole protons as well as the O–H distances were calculated and added to the X-ray determined 3D structure. The resulting coordinates of the protons reveal that they do not participate in hydrogen bonds. Apparently, besides acting as ligands, the water molecules also play a role in stabilizing the structure of the protein.

Another indication for the source of difference between Mn_A^{2+} and Mn_B^{2+} found at low temperature can be obtained from ^{55}Mn ENDOR, which provides the ^{55}Mn hyperfine and nuclear quadrupole interactions. While the ^{55}Mn hyperfine interaction has already been determined through EPR measurements[172] and was found to be insensitive to the difference between the two sites, the quadrupole interaction turned out to be a more sensitive probe. ENDOR measurements were carried out at a crystal orientation ($a \parallel B_0$) where the signals of Mn_A^{2+} and Mn_B^{2+} are well resolved.[171] The ENDOR spectra were measured with the magnetic field set to the lowest-field ^{55}Mn hyperfine component, $m_I = 5/2$, corresponding either to Mn_A^{2+} or to Mn_B^{2+}. The spectra reveal significant shifts, indicating that at this orientation the quadrupole interactions of Mn_A^{2+} and Mn_B^{2+} are different, $P = 0.57$ and 0.26 MHz, respectively. Here, the orientation-dependent quadrupole-interaction parameter P is defined as $P = [3 \cdot e^2 \cdot q \cdot Q]/[8I \cdot (2I-1) \cdot h] \cdot [(3 \cdot \cos^2 \theta' - 1) + \eta \cdot \sin^2 \theta' \cdot \cos(2\phi')]$ and $\eta = (P_{xx} - P_{yy})/P_{zz}$, where the angles θ' and ϕ' describe the orientation of the principal axes system of the quadrupole-tensor with respect to the field B_0. This shows that if the ^{55}Mn quadrupole interaction can be resolved and fully determined, it can provide further important information regarding the spatial and electronic structure of the protein site.

Other instructive examples of high-field EPR and ENDOR applications were published recently on flavin radicals in DNA photolyase from *Escherichia (E.) coli* in frozen-solution samples.[104,176–178] They are described to some detail in Section 5.5.

Quite an impressive number of instructive applications of multifrequency high-field EPR and ENDOR to biological systems have been published recently, for example frozen-solution samples of flavin radicals in different DNA photolyases; for an overview, see ref. [178]. A few of the high-field/high-frequency EPR and ENDOR experiments on flavin radicals in different DNA photolyases, e.g., from *E. coli* and *X. laevis*, are described to some detail in Section 5.5.

2.2.1.4 ESEEM Hyperfine Spectroscopy

When applying a single-frequency microwave spin-echo pulse train with varying interpulse separation to a molecular system of coupled electronic and nuclear spins, one may observe amplitude modulations of the exponentially decaying echo signal. This phenomenon, known as electron spin-echo envelope modulation (ESEEM),[179] requires certain conditions to be fulfilled by the spin system, as will be shown below. The modulated echo-decay time trace is normally analyzed by means of fast Fourier transformation into the frequency domain to extract hyperfine and quadrupole couplings in a way that is complementary to ENDOR.

In the following, we summarize a recent high-field EPR and ESEEM investigation of the ^{14}N quadrupole interaction of nitroxide spin labels in disordered solids.[180] This investigation was done to explore the experimental and theoretical background for using the quadrupole-tensor components of nitroxide spin labels for probing polarity and proticity effects of their microenvironment. It is an attempt to obtain an additional means for separating polarity from proticity effects, which is expected to complement the matrix information obtained from using the g- and nitrogen hyperfine-tensor components of the nitroxide spin probe. We consider the high-field ESEEM experiments on nitroxide spin labels in organic frozen solutions as a step towards differentiation between polarity and proticity matrix effects on the biological function of proteins.

For protein systems adopting solely diamagnetic states of their reactants, *e.g.*, the intermediate states of the light-driven proton pump bacteriorhodopsin, EPR techniques can still serve for probing environmental effects on the process efficiency. This is possible by resorting to site-directed spin-labelling (SDSL) mutagenesis techniques, using specific nitroxide spin label side chains as reporter groups, for example the MTS [(1-oxyl-2,2,5,5-tetramethylpyrroline-3-methyl) methanethiosulfonate] spin label.[181,182] SDSL has matured to an extremely important branch of bio-EPR spectroscopy, and nitroxide side chains can be introduced at almost any desired site in a protein (see Section 5.2.2.1). It has been shown (for reviews, see refs. [183,184] that the isotropic and anisotropic components of the g- and hyperfine-tensors of the nitroxide spin label can be used to reveal the polarity and proticity properties of the immediate environment of the reporter group. Since single crystals of membrane proteins are often difficult to prepare, if at all, frozen-solution protein preparations, lacking long-range order, are commonly used for EPR studies of matrix effects.

For nitroxide spin-labelled molecules with rather small anisotropies of their spin interactions, the application of high-field EPR techniques with correspondingly high microwave frequencies is preferable in comparison to standard X-band EPR techniques to ensure high orientational selectivity as well as high spectral and time resolution.[13,185] Such a magnetoselection can be further exploited by using double-resonance techniques, *e.g.*, ENDOR or pulsed ELDOR (PELDOR) or the ESEEM technique. They are capable of providing single-crystal-like information from orientationally selected fractions of molecules in the disordered sample. By the combination of high-field EPR and SDSL techniques subtle changes of the polarity and proticity profiles can be measured, for example, along proton-transfer pathways in proteins embedded in natural and artificial membranes.[185] This information is obtained by resolving the g_{xx} and A_{zz} components of the nitroxide interaction tensors of a series of molecules with the spin label attached to specific molecular sites. The theoretically predicted[186–190] and experimentally established[191,192] linear correlations g_{xx} vs A_{zz} allowed differentiation between polar and nonpolar protein and membrane regions. Moreover, different slopes occasionally observed in the g_{xx} vs. A_{zz} plots were assigned to either polarity or proticity effects on the magnetic parameters of the spin label from its local protein or membrane environment. So far, however, the scatter of the data has prevented the different slopes becoming an established method to differentiate between polarity and proticity effects of the matrix.

In contrast to the g- and nitrogen hyperfine-tensors, the ^{14}N ($I=1$) quadrupole interaction tensor of the nitroxide spin label has not been exploited in EPR for probing effects of the microenvironment of functional protein sites. Precise knowledge of the ^{14}N quadrupole coupling constant $e^2 \cdot q \cdot Q$ and the asymmetry parameter η of the electric field gradient at the ^{14}N nucleus in the nitroxide would enlarge the arsenal of sensitive probes for environmental effects on specific sites of the molecule, both in terms of polarity[193] and hydrogen-bond effects.[194]

To measure directly the nuclear quadrupole interaction by EPR techniques, the nuclear transitions are normally driven directly by rf fields as in ENDOR. A more indirect alternative is offered by ESEEM, *i.e.* by applying solely a mw pulse train with varying pulse separation and observing spin-echo modulations from hyperfine and quadrupole interactions. To obtain detectable modulations of the echo decays it is mandatory that "forbidden" transitions, flipping the electron and the nuclear spins simultaneously, become partially allowed.[179] This requires an efficient mixing of the nuclear and electron-spin eigenfunctions by the dipolar hyperfine interaction. Consequently, the strength of the external magnetic field has to be properly chosen to approximately balance the Zeeman splitting of the nuclear sublevels and the respective hyperfine splittings ("cancellation condition").[195] This means that an optimum Zeeman field value exists for each nucleus and hyperfine coupling, as has been demonstrated by multi-frequency ESEEM experiments on $S=1/2$, $I=1/2$ as well as on $S=1/2$, $I=1$ systems.[195–198] Pulsed ENDOR and ESEEM techniques are complementary to each other[4,199] concerning their ability to reveal large or small nitrogen dipolar hyperfine interactions ("geminate" or "distal" nitrogens), respectively. Only a few high-field nitrogen ESEEM experiments on single-crystalline[200] and disordered samples[198] have been reported in the last decade.

It was the aim of the reviewed work[180] to investigate, by means of high-field EPR and ESEEM experiments in conjunction with DFT calculations, the problems *(i)* whether nitrogen quadrupole-tensor components can be determined with high accuracy from frozen-solution samples and *(ii)* what kind of information on the polarity and proticity properties of the nitroxide spin-label environment can be extracted from the interaction between the electric-field gradient (tensor with elements $e \cdot q_{ij}$) at the site of the ^{14}N nitrogen nucleus and its electric quadrupole moment ($e \cdot Q$). Specifically, the question was addressed

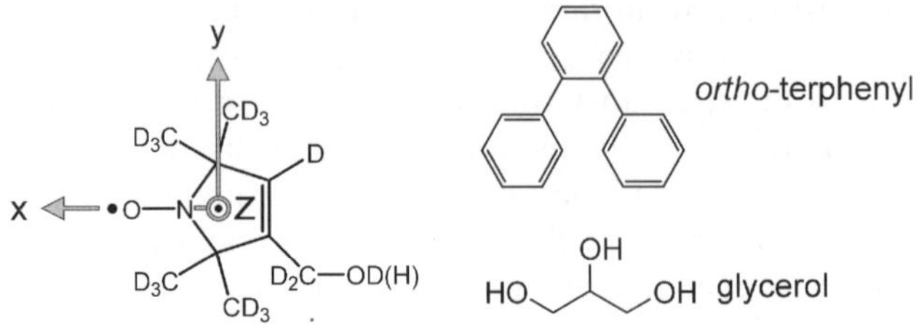

Figure 2.22 Molecular structure of the perdeuterated nitroxide radical R1; the conventional principal axes of the g-tensor are indicated; *ortho*-terphenyl and glycerol hosts for the diluted nitroxide glassy solutions.

whether the ^{14}N quadrupole information on matrix effects is similar or complementary to that obtained from the established spin-probe parameters, g_{xx} and A_{zz}.

We focused on the spectroscopic and quantum-chemical aspects of measuring and calculating the quadrupole interaction parameters of the perdeuterated nitroxide radical R1 dissolved in frozen solutions of either nonpolar, aprotic *ortho*-terphenyl or polar, protic glycerol, see Figure 2.22.

Figure 2.23 shows W-band cw EPR spectra of R1-^{14}N and R1-^{15}N nitroxide radicals in frozen ortho-terphenyl solution at 180 K. The spectra exhibit the typical powder-pattern lineshape expected for a dilute distribution of nitroxides. The spectra are clearly resolved into three separate regions corresponding to the principal values of the g-tensor, g_{xx}, g_{yy} and g_{zz}. Moreover, due to the reduction of the inhomogeneous linewidth by perdeuteration of the radicals, the nitrogen hyperfine splitting (doublets for R1-^{15}N, $I=1/2$, triplets for R1-^{14}N, $I=1$) is observed in all g-regions. The spectra were analyzed by numerical solution of the spin Hamiltonian given in eqn (2.1). At first, the spectrum of R1-^{15}N was analyzed. The best-fit spectrum is shown in Figure 2.23(a). Perfect agreement with the experimental spectrum is achieved by using the set of magnetic parameters, *i.e.* g-tensor, nitrogen hyperfine-tensor, homogeneous EPR linewidth, orientation-dependent EPR inhomogeneous linewidths, as listed in Table 2.1. In the next step, the spectrum of R1-^{14}N was calculated using the parameters obtained from R1-^{15}N, *i.e.* tentatively omitting quadrupole contributions, but rescaling the γ_N and A values by the factor $(1.4)^{-1}$ according to the ratio of the nitrogen nuclear g-values $g_n(^{15}N)/g_n(^{14}N) = 1.4$. Although there is good agreement of the *positions* of the EPR lines in all spectral regions of the g-tensor components (see Figure 2.23(a)), the EPR signal *intensities* agree only in the g_{zz} spectral region. This disagreement of intensities in the g_{xx} and g_{yy} regions cannot be improved by varying the corresponding linewidths. However, switching on the ^{14}N quadrupole interaction term in the spin Hamiltonian allows reproduction of the intensities of the experimental EPR spectrum, Figure 2.23(b). From the simulation, the values $P_{xx} = +1.1$ MHz and $P_{yy} = +0.6$ MHz were obtained. Thus, the quadrupole effects in the cw W-band EPR spectrum of perdeuterated ^{14}N nitroxide radicals are directly observed. This is only possible

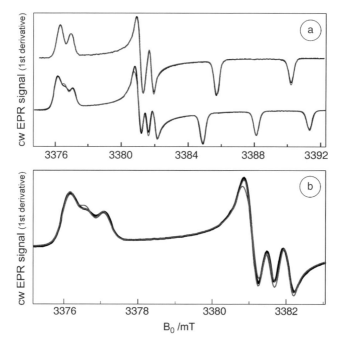

Figure 2.23 (a) Experimental W-band cw EPR spectra of 1 mM R1-^{15}N (upper spectrum) and R1-^{14}N (lower spectrum) nitroxide radicals in frozen solution of *ortho*-terphenyl taken at 180 K. They are overlaid with the corresponding best-fit spectra (red lines) obtained without taking the quadrupole interaction into account. For the derived magnetic parameters, see text. (b) Expanded view of the g_{xx}, g_{yy} regions of the R1-^{14}N spectrum. The best-fit spectra calculated without quadrupole contributions and with the quadrupole couplings $P_{xx} = 1.1$ MHz; $P_{yy} = 0.6$ MHz are shown by red and green lines, respectively. For details, see ref. [180].

Table 2.1 The magnetic parameters of the R1-^{14}N nitroxide in *ortho*-terphenyl and glycerol obtained from the analysis of W-band cw EPR spectra and nitrogen nuclear ESEEM patterns at 180 K.

	otho-terphenyl			gycerol		
	xx	yy	zz	xx	yy	zz
g_{ii}	2.00905(2)	2.00617(2)	2.00227(2)	2.00841(2)	2.00604(2)	2.00223(2)
$A_{ii}{}^a$/MHz	13.60(15)	13.50(15)	93.2(15)	14.7(15)	14.7(15)	101.4(15)
$\Delta B_{1/2}{}^b$/MHz	10.6	6.2	7.8	21.6	9.2	10.5
P_{ii}/MHz	+1.26(3)	+0.53(3)	−1.79(4)	+1.23(3)	+0.36(3)	−1.59(4)

^{a}The measured ^{15}N hyperfine values are a factor of 1.4 larger, which corresponds to the ratio of magnetic moments of ^{15}N to ^{14}N isotopes.
^{b}The inhomogeneous (FWHM) linewidth due to unresolved internal ^{2}H and external ^{1}H hyperfine couplings (Gaussian distribution). The Lorentzian homogeneous linewidth of 3.9 MHz is orientation independent.

because of a favorable disposition of the ^{14}N nuclear Zeeman and hyperfine interaction energies in the g_{xx} and g_{yy} regions of the R1-^{14}N spectrum at W-band EPR. And, of course, because of the narrow EPR linewidth (below 10 MHz) of the perdeuterated spin label. This allows the nitrogen hyperfine structure in the g_{xx} and g_{yy} regions and to detect their rather fine spectral details to be detected.

Although, the ^{14}N quadrupole interaction is observed in the cw W-band EPR spectra, the P_{xx} and P_{yy} values can be evaluated only with quite a large error. This is because they are obtained by a multiparameter fit of fine details of the experimental cw EPR spectra. Hence, we decided to resort to ESEEM experiments from which the quadrupole interactions can be determined in a more straightforward way (see Figures 2.24 and 2.25).

The stimulated ESE decays show distinct modulations for both the R1-^{15}N and the R1-^{14}N nitroxides. In the case of R1-^{15}N, the echo modulation is

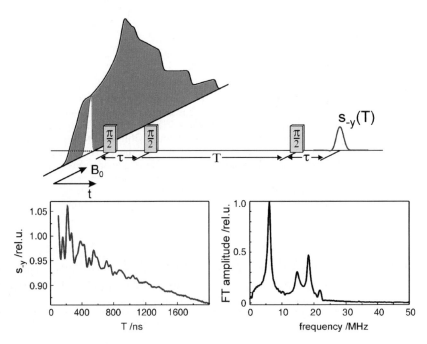

Figure 2.24 W-band microwave pulse sequence for the stimulated high-field ESEEM experiment on the nitroxide radical R1. Top: The echo-detected EPR spectrum of the R1-^{14}N radical as well as the microwave excitation bandwidth for typical microwave pulse-length settings are shown. Typical settings are $\tau = 40$ ns, $\pi/2$-pulse length $t_p = 30$ ns. The time T is stepped from $T_0 = 100$ ns in 5 ns steps. Bottom left: A representative example of a nuclear modulation echo-decay trace at the indicated magnetic-field position is shown. Bottom right: The Fourier-transformed spectrum of the ESEEM decay example is given, the cutoff frequency of the high-pass filter was set to 1.5 MHz. For details, see ref. [180].

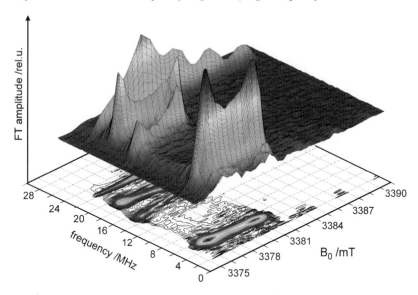

Figure 2.25 3D representation of the experimental ESEEM surface. In the frequency-field plane the contour plot of the ESEEM intensities is shown, which is used in the following figures.

dominated by a frequency of about 22.1 MHz in the whole spectral region of nitroxide EPR absorption. This modulation is assigned to the nitroxide methyl deuterons, see Figure 2.22. Additional ESEEM frequencies are detected in the g_{xx}, g_{yy} region. Figure 2.26 shows in a contour-plot representation the R1-^{15}N ESEEM spectrum, as derived by Fourier transformation of the primary ESEEM recording from the time to the frequency domain. Beyond the ridge parallel to the magnetic-field axis at 22.1 MHz, that was mentioned above, two additional ridges are clearly distinguishable in the spectrum. They are spread apart on the field axis in the g_{xx}, g_{yy} region, and terminated, *i.e.* confined, by the peaks at 5.05 MHz and 24.2 MHz (low-field terminals in the g_{xx} region) and at 5.15 MHz and 24.1 MHz (high-field terminals in the g_{yy} region). These features are assigned to the ^{15}N nuclear modulations.[180] Indeed, when the direction of the external magnetic field B_0 is chosen to be parallel to the principal x- or y-axis of the g-tensor (these orientations are selected for ESEEM measurements performed at the x- or y regions), the nuclear transition frequencies approach the values:

$$\left\{ \begin{matrix} \nu^+_{x(y)} \\ \nu^-_{x(y)} \end{matrix} \right\} = 1.4 \cdot \left\{ \begin{matrix} \nu_N + \frac{1}{2} \cdot A_{xx(yy)} \\ \nu_N - \frac{1}{2} \cdot A_{xx(yy)} \end{matrix} \right\} \quad (2.50)$$

Here, the factor 1.4 is introduced to rescale for the ^{15}N nucleus the A_{ii} and ν_N values that were previously determined for ^{14}N. The observed terminal peaks in the ESEEM contour plots are positioned symmetrically about 14.63 MHz,

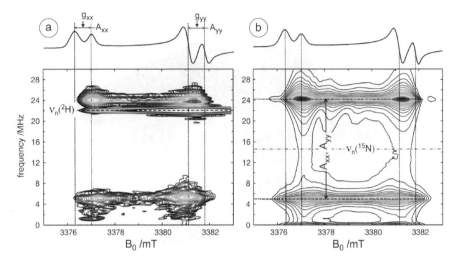

Figure 2.26 (a) Frequency-field dependence of the experimental three-pulse stimulated ESEEM of 1 mM R1-^{15}N/*ortho*-terphenyl glass at 180 K in the g_{xx},g_{yy} regions. Beyond the g_{yy} region no nuclear modulations due to ^{15}N are observed. The contour lines are shown as isohypses from 0.05 to 0.4 of the maximum FT ESEEM intensity. (b) Contour plot of nitrogen W-band ESEEM intensities calculated for disordered R1-^{15}N nitroxides using the parameters read out from (a) and cw EPR. The ESEEM contributions from the deuterons in R1 ($v_n(^2H)$ in (a)) were omitted in the calculations. The contour lines are shown as isohypses from 0.05 to 1 of the maximum calculated FT ESEEM intensity. On top, the corresponding experimental (a) and calculated (b) cw EPR spectra are shown. For details, see text.

which is close to the ^{15}N Larmor frequency, $1.4 \cdot v_N(^{14}N) = 14.6$ MHz, at a Zeeman field of $B_0 = 3.38$ T. The frequency differences $v^+_{x(y)} - v^-_{x(y)}$ of 19.15 MHz (18.95 MHz) are in a good agreement with the hyperfine coupling constants $1.4 \cdot A_{xx}(^{14}N) = 19.2 \pm 0.2$ and $1.4 \cdot A_{yy}(^{14}N) = 19.0 \pm 0.2$ MHz, as determined from the analysis of the cw EPR spectrum of R1-^{14}N.

One observation deserves extra attention: The appearance of considerable ESEEM spectral densities close to the transition frequencies of principal hyperfine components is, at first sight, somewhat surprising in view of the following facts: For $S=1/2$, $I=1/2$ systems with an axial A-tensor ($A_{zz}=A_\parallel$, $A_{xx}=A_{yy}=A_\perp$), the ESEEM modulation-depth factor, $k(\theta)$, is given by:[201]

$$k(\theta) = \left(\frac{B(\theta) \cdot v_N}{v_\alpha(\theta) \cdot v_\beta(\theta)} \right)^2 \qquad (2.51)$$

where

$$v_{\alpha,\beta}(\theta) = [(v_N \pm A(\theta)/2)^2 + B(\theta)^2/4]^{1/2} \qquad (2.52)$$

with

$$A(\theta) = A_\| \cdot \cos^2\theta + A_\perp \cdot \sin^2\theta, \quad B(\theta) = (A_\| - A_\perp) \cdot \sin\theta \cdot \cos\theta \qquad (2.53)$$

Here, θ is the angle between the B_0 field direction and the A-tensor z-axis. Evidently, the nuclear modulation should vanish when orientations of the nitroxides approach the field direction parallel to any principal axis of the A-tensor. The maximum modulation-depth factor is expected at θ angles near the perpendicular orientation, where the cancellation condition, $v_N = A(\theta)/2$, is met. For the R1 radical, this angle is approximately 82°.

To understand this apparent contradiction to the observed ESEEM spectral densities, we first performed a simplified simulation of the R1-^{15}N ESEEM spectrum in a direct way, i.e. without calculating the time-domain responses and their further transformation to the frequency domain.[180] In this simulation, resonance EPR fields and nuclear transition frequencies were calculated for a particular orientation of the radicals by solving the Hamiltonian equation (2.1). Partial density contributions were evaluated that are proportional to the statistical weight of the chosen orientation, and scaled by the ESEEM factor $k(\theta)$ given in eqn (2.51). Such contributions were then accumulated from all orientations. The resulting pattern was convoluted along the field axis with an EPR inhomogeneity function, as was done in the successful cw-EPR simulations above. However, this pattern does not give the sharp peak singularities observed in the experimental spectra of Figure 2.26(a). Thus, the "direct" calculation of the ESEEM spectrum contradicts the experimental result.

In order to get a more exact description the ESEEM spectra that accounts for the specific action of semiselective pulses in the echo experiments and deadtime effects of the spectrometer, numerical simulations were performed via calculation of the time-domain ESEEM recordings and their subsequent Fourier transformation to frequency-domain spectra. Figure 2.26(b) shows the contour-plot representation of the calculated ESEEM spectrum of R1–^{15}N (^{2}H contribution not included). This simulation is in perfect agreement with the experimental spectrum in Figure 2.26(a). The terminal peaks in the calculated spectrum appear not only at exactly the same frequencies, but also at the same magnetic-field positions as in the experiment. Although the observed ESEEM signals can be correctly reproduced by these numerical calculations, we admit that the physical background of observing ESEEM at the canonical orientations of the nitrogen hyperfine-tensor of randomly oriented frozen-solution samples remains somewhat blurred.[180]

The experimental ESEEM spectrum of R1-^{14}N at 180 K is shown as a contour plot in Figure 2.27(a). The 22.1-MHz ^{2}H ridge is observed with the same intensity as in the R1-^{15}N spectrum. In contrast to the pattern of R1-^{15}N (Figure 2.26(a)), this spectrum is dominated by the contribution from the ^{14}N nuclear modulations. The observed nitrogen ESEEM pattern can be interpreted as follows: The two ESEEM ridges at upper frequencies correspond to the upper ridge in Figure 2.26, which is shifted on the frequency axis by the down-scaling factor 1.4, accounting for the reduced magnetic moment of ^{14}N,

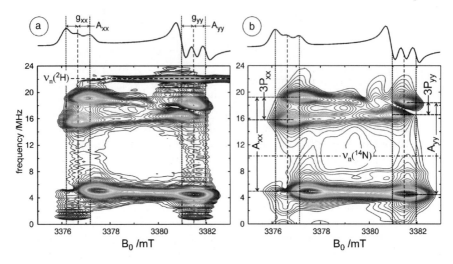

Figure 2.27 (a) Frequency–field dependence of the experimental three-pulse stimulated ESEEM of 1 mM R1-^{14}N/*ortho*-terphenyl glass at 180 K in the g_{xx}, g_{yy} regions. Beyond the g_{yy} region no nuclear modulations due to ^{14}N are observed. The contour lines are shown as isohypses from 0.05 to 1 of the maximum FT ESEEM intensity. (b) Contour plot of nitrogen W-band ESEEM intensities calculated for disordered R1-^{14}N nitroxides using the parameters read out from (a) and cw EPR. The ESEEM contributions from the deuterons in R1 ($\nu_n(^2H)$ in (a)) were omitted in the calculations. On top, the corresponding experimental (a) and calculated (b) cw EPR spectra are shown. For details, see text.

and is split by the quadrupole interaction. The frequencies of the split terminal peaks are given by:

$$\begin{Bmatrix} \nu_{x(y)}^{+,+} \\ \nu_{x(y)}^{+,-} \end{Bmatrix} = \begin{Bmatrix} \nu_N + \frac{1}{2} \cdot A_{xx(yy)} + \frac{3}{2} \cdot P_{xx(yy)} \\ \nu_N + \frac{1}{2} \cdot A_{xx(yy)} - \frac{3}{2} \cdot P_{xx(yy)} \end{Bmatrix} \quad (2.54)$$

The lower-frequency ridge in Figure 2.27(a) represents only one of the two quadrupole-split components that is shifted upwards; the component shifted downwards is close to zero frequency, where the modulations are strongly damped.

The A_{xx}, and A_{yy} values of 13.6 MHz and 13.5 MHz, as read off from the ESEEM spectrum, are in excellent agreement with the 13.60 ± 0.15 MHz and 13.50 ± 0.15 MHz values determined from cw EPR. The absolute values of the quadrupole couplings $[P_{xx}; P_{yy}] = [1.26; 0.53]$ MHz are obtained from the positions of the high-frequency ESEEM peaks around 17 MHz. The P_{zz} value of -1.79 MHz is calculated from the traceless property of the quadrupole tensor. Figure 2.27(b) shows the calculated ESEEM spectrum of R1-^{14}N using experimentally determined g-, A- and P-values. Four trial calculations were performed with different signs of the P_{xx}, P_{yy} values and assuming $A_{xx,yy} > 0$. The remarkably different ESEEM spectra allowed to determine the proper signs of the P_{xx}, P_{yy} values (for details, see ref. [180]). The spectrum shown in Figure 2.27(b), as calculated for $[P_{xx}; P_{yy}; P_{yy}] = [+1.26; +0.53; -1.79]$ MHz, perfectly fits the experimental pattern in Figure 2.27(a). Thus, W-band ESEEM spectroscopy of ^{14}N nitroxides allows for the accurate evaluation of the principal quadrupole-tensor values. The accuracy of the quadrupole frequencies is estimated to be ±30 kHz. This accuracy enables the spectroscopist to use the quadrupole interaction as a sensitive sensor for environmental properties of proteins.

However, when probing matrix properties by quadrupole splittings, it is appropriate to first clarify their temperature dependence. For this purpose, stimulated ESEEM experiments on R1-^{14}N were performed not only at 180 K, but also at 90 K. Evaluation of the ESEEM spectrum yielded the quadrupole couplings [+1.28; +0.51; −1.79] MHz and the hyperfine A_{xx} and A_{yy} values of 12.1 MHz and 12.2 MHz, respectively (13.60 and 13.50 MHz at 180 K). From the cw EPR spectrum an A_{zz} value of 93.50 MHz (93.20 MHz at 180 K) is obtained. Apparently, for R1-^{14}N the quadrupole values are largely temperature independent, whereas the hyperfine splittings show distinct variations with temperature. The hyperfine couplings have to be decomposed into isotropic (A_{iso}) and dipolar (D) contributions. When going from 180 K to 90 K, the increase of D from 26.5 MHz to 27.1 MHz is accompanied by a decrease of A_{iso} from 40.1 MHz to 39.2 MHz.

The temperature dependence of the dipolar hyperfine coupling can be explained by the model of fast librations of the nitroxide molecule in the host matrix.[202,203] The large change of the isotropic hyperfine coupling with temperature, however, is not due to the overall motion of the nitroxide molecule in the matrix. It probably rather originates in an intramolecular motion.[204,205] The observed positive temperature coefficient of A_{iso} was explained by out-of-plane vibration of the oxygen with respect to the planar nitrogen-containing ring structure. Moreover, *ab-initio* and DFT calculations show that the angle between the NO bond and the plane of the attached five-membered ring is a very soft geometrical parameter, *i.e.* large changes in this angle lead only to small changes in the total energy.[186,206,207] Our DFT calculations[180] show a strong dependence of A_{iso} on this angle. An increased average squared angular deviation from the planar equilibrium geometry at the nitrogen atom with increasing temperature would explain the observed temperature behavior of A_{iso}. It is important to emphasize that this distortion does not, however, influence the quadrupole values.

The results of the elaborate DFT calculations of the spin-interaction parameters (the *g*-, *A*-, and *P*-tensors) and their dependence on temperature as well as on polarity and proticity effects of the microenvironment of the nitroxide spin label can be summarized as follows:[180] The P_{yy} value is especially sensitive to the proticity and polarity of the nitroxide environment in H-bonding and nonbonding situations. The quadrupole tensor is shown to be rather insensitive to structural variations of the nitroxide label itself. When using P_{yy} as a testing probe of the microenvironment, its ruggedness towards temperature changes represents an important advantage over the g_{xx} and A_{zz} parameters, which are usually employed for probing matrix effects on the spin-labelled molecular site. Thus, the DFT calculations predict that, beyond measurements of g_{xx} and A_{zz} of spin labelled protein sites in disordered solids, W-band high-field ESEEM studies of ^{14}N quadrupole interactions should allow for reliably probing subtle environmental effects on the electronic structure of biomolecules.

In order to validate the theoretical predictions we have examined the R1-^{14}N nitroxide also in glassy glycerol solution, which emulates a polar/protic protein environment, and compared the results with those from the nonpolar *ortho*-terphenyl solution.[180] The measured interaction parameters are collected in Table 2.1.

Figure 2.28(a) shows the cw EPR spectra of R1-^{14}N in frozen solutions of *ortho*-terphenyl and glycerol, recorded at 180 K. As expected, the change from the nonpolar to a polar/protic environment produces pronounced shift of both g_{xx} and A_{zz}. The A_{zz} value increases from 93.2 MHz to 101.4 MHz and the g_{xx} value decreases from 2.00905 to 2.00841, which is in a good agreement with previous results.[186] The analysis of the glycerol-solution spectrum shows slightly increased EPR linewidths in the y and z spectral regions, but an almost doubled linewidth of the g_{xx} component as compared to the *ortho*-terphenyl

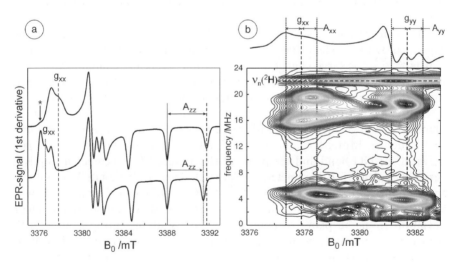

Figure 2.28 (a) Comparison of the experimental W-band cw EPR spectra of 1 mM R1-^{14}N nitroxide radicals in frozen solutions of *ortho*-terphenyl (lower trace) and glycerol (upper trace), taken at 180 K. The asterisk marks the nonpolar component in glycerol solution. (b) Frequency-field dependence of the experimental three-pulse stimulated ESEEM of 1 mM R1-^{14}N/glycerol glass at 180 K in the g_{xx}, g_{yy} regions. Beyond the g_{yy} region no nuclear modulations due to ^{14}N are observed. The contour lines are shown as isohypses from 0.05 to 1 of the maximum FT ESEEM intensity. For details, see ref. [180].

solution. The increase of the g_{xx} linewidth reflects a heterogeneity of the local polarity/proticity properties in glassy glycerol, which leads to a static distribution of the g_{xx} values, producing so-called g-strain. Moreover, a second g_{xx} component with an overall weight of about 15% is present in glycerol solution, see the asterisk-marked region in Figure 2.28(a). The relative weight as well as the lineshape of this component was found to be independent of the freezing protocol of the sample, *i.e.* slow cooling or shock freezing. The g_{xx} value of this component is roughly equal to that in *ortho*-terphenyl indicating that the nitroxide radicals are partially exposed to a nonpolar local environment in glycerol glass. Figure 2.28(b) shows the experimental ESEEM spectrum of R1-^{14}N in glycerol at 180 K. Similar to the *ortho*-terphenyl solution, see Figure 2.27(a), the ESEEM spectrum is dominated by the contribution from ^{14}N nuclear modulations. However, the ratio of ^{2}H to ^{14}N lines is increased in favor of the ^{2}H modulations due to the increased of EPR linewidth, especially in the x spectral region. The A_{xx} and A_{yy} values of 14.7 MHz and 14.7 MHz are read off from the positions of the terminal peaks in the ESEEM spectrum. These values are larger than for the *ortho*-terphenyl solution, which is in agreement with the DFT prediction. The absolute values of the quadrupole couplings, $[P_{xx}; P_{yy}] = [1.23; 0.36]$ MHz, are obtained from the positions of the ESEEM peaks around 18 MHz. Interestingly, the P_{xx} value in polar/protic environment does not reveal a significant change compared to the nonpolar case. The P_{yy} value, however, decreases by 0.17 MHz, in agreement with the predictions from the DFT calculations (for details, see ref. [180]). Moreover, the analysis of the ESEEM spectrum recorded at 90 K yields the quadrupole couplings $[P_{xx}; P_{yy}] = [1.24; 0.37]$ MHz, thus revealing that the quadrupole parameters also in the polar/protic case are temperature independent – again in accordance with the predictions from the DFT calculations.

At 90 K, the ^{14}N hyperfine couplings $[A_{xx}; A_{yy}; A_{zz}] = [13.4; 13.5; 101.7]$ MHz were obtained from the combined analysis of cw EPR and ESEEM spectra. The isotropic hyperfine coupling, $A_{\text{iso}} = 1/3 \cdot \text{Tr}\left(\bar{A}\right)$ decreases from 43.6 MHz to 42.9 MHz when lowering the temperature from 180 K to 90 K. The similar A_{iso} variation with temperature in glycerol (0.7 MHz) and *ortho*-terphenyl (0.8 MHz) strongly supports the proposed out-of-plane vibration mechanism of the NO bond with respect to the five-membered ring plane as responsible for the observed changes of the hyperfine coupling parameters.

In concluding this digression on high-field ESEEM on nitroxide radicals as spin probes for matrix effects we emphasize that our theoretical and experimental results show that P_{yy} has the strongest dependence on different environmental situations compared with P_{xx} and P_{zz}. This is essentially different from the behavior of the g- and A-tensors, where changes dominate along the other two axes, x and z, respectively. The dependence of P_{yy} on environmental effects is at least as strong as for A_{zz}, as is quantified in ref. [180]. Some more interesting aspects emerge when using quadrupole interactions of nitroxide spin labels for probing environmental effects:

Quantum-chemical methods at different levels including DFT generally predict a significant dependence of the ^{14}N isotropic hyperfine coupling constant A_{iso} on the out-of-plane (or tilt) angle α of the NO bond. In NO spin labels of the pyrroline type, the reference plane is the adjacent five-membered ring. This tilt angle α is a very "soft" structural parameter, requiring only very weak external disturbing forces to produce large angles of up to 15°. The consequences for A_{zz}, which is given by sum of the Fermi contact contribution A_{iso} and classical dipolar contributions, can be quite dramatic. In the presently studied spin label R1, DFT methods predict a change of 4 MHz for α = 15°.

Since α can be affected by thermally excited vibrations, spatial crowding effects and hydrogen bonding to the oxygen atom (*e.g.*, inside a locally rigid hydrogen-bond network), A_{zz} can become temperature dependent as well as dependent on environmental effects in an unpredictable way. In unfavorable situations difficult to foresee, such structural ambiguities can cast doubt on the reliability of A_{iso} or A_{zz} values of the NO spin label as a polarity probe. The same holds true for the value of g_{xx}, although there the dependence on the tilt angle α is not quite as pronounced as for A_{zz}.

In contrast to A_{zz} and g_{xx}, the value of P_{yy} is predicted and experimentally verified, to be practically independent of the tilt angle α, varying by at most 0.04 MHz between α = 0° and α = 15°. This variation is close to the measuring error of 30 kHz. The other two quadrupole components, P_{xx} and P_{zz}, show a similarly weak dependence on α.[180] It should be pointed out that even for relatively large values of α (≤15°) the *A*- and *g*-tensors remain collinear since their principal axes follow the direction of the NO bond. The same holds true for the *P*-tensor.

Thus, measurements of P_{yy} can already, in itself, provide important contributions to the study of environmental effects or, additionally, serve as a consistency check for polarity and/or proticity results obtained from the often employed g_{xx} *vs.* A_{zz} correlation.

Differentiation between Polarity and Proticity Effects

Polarity effects from intermolecular electric fields have been widely investigated by observing shifts of A_{zz} and/or g_{xx} on NO spin labels in various environments. Whereas A_{zz} reacts to polarity changes in nonbonding as well as H-bonding situations predominantly through changes in the spin-density distribution of the NO bond (as a consequence of charge displacements between N and O), g_{xx} is also significantly affected by additional perturbations of the n–π energy gap of the O-atom in H-bonding situations. Thus, the observation of g_{xx} shifts may be desirable for the detection of H-bond formation (proticity), but can also lead to ambiguous results in trying to separate proticity from pure polarity effects quantitatively. If measurement of A_{zz} alone does not safely yield the desired information on polarity changes, because of the reasons given above, measurement of P_{yy} is therefore the appropriate choice. Qualitatively, P_{yy} has the same probing properties as A_{zz} in detecting polarity and proticity effects. However, apart from its relative insensitivity to structural and temperature variations, this parameter is exclusively dependent on the electronic charge distribution of the NO bond and, therefore, particularly qualified for polarity

Table 2.2 Theoretical results for the principal components of the P- and A- and g-tensors of a nitroxide radical in microenvironment of different polarity and proticity properties.

Case	P_{xx}	P_{yy}	P_{zz}	A_{xx}	A_{yy}	A_{zz}	A_{iso}	g_{xx}	g_{yy}	g_{zz}	g_{iso}
1	1.46	0.62	−2.08	2.42	1.93	83.62	29.32	2.008640	2.006220	2.002230	2.005697
2	1.49	0.29	−1.78	3.87	3.90	95.42	34.40	2.008170	2.006000	2.002240	2.005470
3	1.49	0.47	−1.96	3.29	3.08	89.94	32.10	2.008090	2.005920	2.002210	2.005407
4	1.47	0.43	−1.90	3.32	3.16	91.29	32.59	2.008360	2.006090	2.002230	2.005560
5	1.50	0.34	−1.84	3.90	3.92	95.27	34.64	2.007900	2.005830	2.002210	2.005313
6	1.50	0.33	−1.83	3.94	3.97	95.58	34.50	2.007900	2.005830	2.002210	2.005313

The principal A and P values are given in MHz.
Case: 1: no H-bond, Vacuum, $E_x=0$
       2: no H-bond, Vacuum, $E_x=0.02$ a.u.
       3: H-bond,   Vacuum, $E_x=0$
       4: no H-bond, H$_2$O, $E_x=0$
       5: H-bond,   H$_2$O, $E_x=0$
       6: H-bond,   H$_2$O, $E_x=0.02$ a.u.

studies in nonbonding as well as H-bonding situations. A theoretical manifestation of this behavior can be deduced from Table 2.2 by plotting P_{yy} versus A_{zz} for the cases 1 (*in vacuo*), 2 (with electric field) and 3 (with H-bond). This plot yields a straight line in spite of the inclusion of the H-bond. This linear relation between P_{yy} and A_{zz} is a consequence of the linear relation (to first order) between charge and spin densities and of the fact that direct energetic contributions from changes in the n–π excitation energy are absent.[180]

Finally, collecting all the above arguments, it appears justified to state that in NO spin-label studies of the microenvironment of reactants in proteins the measurement of P_{yy} can provide important additional information on polarity effects. This information is complimentary to that obtained by measurements of g_{xx} and A_{zz}. This statement is supported by our preliminary results on R1-^{14}N nitroxide in different glassy solutions. The polarity and proticity sensitivity of the quadrupole parameters is not only in good agreement with theoretical predictions described here, but certainly will allow us to distinguish and characterize a multiple heterogeneous local environment of a nitroxide radical in glassy alcohol solutions.[192,208] Moreover, we believe that, in addition to W-band high-field EPR measurements of g_{xx} and A_{zz} of spin-labelled protein sites, W-band ESEEM studies of ^{14}N quadrupole interactions open a new avenue to reliably probe subtle environmental effects on the electronic structure of cofactors in their protein-binding pockets. This is a significant step forward on the way to differentiating between effects from matrix polarity and hydrogen-bond formation on the protein function.

2.2.2 Electron–Electron Dipolar Spectroscopy

The determination of distance and orientation of protein domains and their changes in the course of biological action are of primary concern in proteomics in an attempt to elucidate the relation between structure, dynamics and function. Hence, a variety of biophysical techniques is currently developed and applied to measure distances (and orientations, if possible) in large biosystems. Often, they are available only as disordered samples such as frozen solutions, so X-ray crystallography is not applicable. For disordered systems, EPR spectroscopy offers a powerful approach to structural information over wide distance ranges depending on which anisotropic spin interaction is measured. By this approach, established techniques like FRET (fluorescence energy transfer) or solid-state NMR are complemented. For anisotropic electron–nuclear hyperfine interactions, preferentially measured by ENDOR, the accessible distance range stays well below 1 nm (10 Å). For the dipolar electron–electron interaction, however, measured by specialized microwave pulse sequences as, for instance, in pulsed electron–electron double resonance (PELDOR), the distance range can be extended dramatically, in ideal cases to about 8 nm (80 Å),[209] in proteins realistically to about 5 nm (50 Å), *i.e.* even interspin distances across the photosynthetic membrane can be measured.

For large distances between well-localized electron spins A and B in a radical pair, for which the point-dipole approximation holds and the exchange coupling, J, can be neglected, the electron–electron dipolar Hamiltonian, $\hat{H}_{SS}$, is commonly written as

$$\hat{H}_{SS}/h = \frac{\mu_0}{4\pi \cdot h} \cdot \frac{g_A(\theta_A, \varphi_A) \cdot g_B(\theta_B, \varphi_B) \cdot \mu_B^2}{r_{AB}^3} \cdot (A + B + C + D + E + F) \quad (2.55)$$

where $g_A(\theta_A, \varphi_A)$, $g_B(\theta_B, \varphi_B)$ are the orientation-dependent g-values of radicals A and B selected by the external magnetic field B_0, μ_B is the Bohr magneton, μ_0 the vacuum permeability, and h the Planck constant. The six terms $A, B, \ldots, F$ represent products of spin-component operators and angular expressions in a spherical coordinate system (r_{AB}, θ, φ), in which r describes the radial distance of spin B from spin A located at the origin (or *vice versa*), and the zenith and azimuth polar angles θ and φ describe the orientation of the radical-pair axis with respect to the external Zeeman field B_0:[210]

$$\left.\begin{array}{l} A = \hat{S}_z^A \cdot \hat{S}_z^B \cdot (1 - 3 \cdot \cos^2 \theta) \\ B = -\frac{1}{4} \cdot (\hat{S}_+^A \cdot \hat{S}_-^B + \hat{S}_-^A \cdot \hat{S}_+^B) \cdot (1 - 3 \cdot \cos^2 \theta) \\ C = -\frac{2}{3} \cdot (\hat{S}_z^A \cdot \hat{S}_+^B + \hat{S}_+^A \cdot \hat{S}_z^B) \cdot \sin\theta \cdot \cos\theta \cdot e^{-i\varphi} \\ D = -\frac{2}{3} \cdot (\hat{S}_-^A \cdot \hat{S}_z^B + \hat{S}_z^A \cdot \hat{S}_-^B) \cdot \sin\theta \cdot \cos\theta \cdot e^{i\varphi} \\ E = -\frac{4}{3} \cdot \hat{S}_+^A \cdot \hat{S}_+^B \cdot e^{-2i\varphi} \\ F = -\frac{4}{3} \cdot \hat{S}_-^A \cdot \hat{S}_-^B \cdot e^{2i\varphi} \end{array}\right\} \quad (2.56)$$

The $\hat{S}_x^{A(B)}$ and $\hat{S}_y^{A(B)}$ spin operators are expressed in terms of the raising and lowering shift operators $\hat{S}_+^{A(B)}$ and $\hat{S}_-^{A(B)}$. Note the formal analogy of the electron–electron dipolar interactions with the electron–nuclear dipolar (END) interaction discussed in Section 2.1.1. Following the same arguments as used for the END interaction, in high magnetic fields, when the dipolar coupling of the two unlike electron spins is small compared to the difference of their

Zeeman interactions, the dipolar splitting of the EPR transitions is predominantly determined by the "secular" $\hat{S}_z^A \cdot \hat{S}_z^B$ term A in eqn (2.55), whereas the "pseudosecular" $\hat{S}_{+(-)}^A \cdot \hat{S}_{-(+)}^B$ term B can be neglected to first order. The "nonsecular" terms C, D, E, and F are not important at all for the dipolar energy splitting.[18]

The pseudosecular term becomes important in the case of like spins, *i.e.* when the dipolar coupling is large compared to the difference of their Zeeman interactions. This leads to a scaling of the dipolar coupling frequency by a factor of 3/2 as compared to the case of unlike spins.[211] The intermediate case with comparable magnitudes of the dipolar coupling and Zeeman splitting is more complex, and adequate data analysis requires simulation of the spectra on the basis of the full spin Hamiltonian of the molecular system, including both secular and pseudosecular terms of the dipolar interaction. When working at a particular microwave frequency, the validity of the unlike-spin limit is guaranteed only for a large enough distance of the spins, *e.g.*, $r_{AB} \geq 2$ nm (20 Å) at X-band.[211] In the case of a pair of two nitroxide radicals, which are frequently used for distance measurements on doubly site-specifically labelled proteins, the two nitroxide spins usually have their Zeeman (and hyperfine) frequencies substantially different and, hence, fulfill the condition of the unlike-spin limit. The difference in resonance frequencies of the two radicals arises from their different orientations with respect to the Zeeman field B_0, so that the effective g- and hyperfine-values of their g- and hyperfine-tensors are different.

Neglecting the exchange coupling, in the unlike-spin limit the dipolar coupling frequency, ν_{AB}, is given by

$$\nu_{AB}(\theta) = \nu_d \cdot (1 - 3 \cdot \cos^2\theta) \qquad (2.57)$$

with the dipole–dipole coupling parameter

$$\nu_d = \frac{\mu_0}{4\pi \cdot h} \cdot \frac{\mu_B^2 \cdot g_A \cdot g_B}{r_{AB}^3} \qquad (2.58)$$

Here, the g_A and g_B values of the radical partners A and B are weighted principal components of their g-tensors, *i.e.* weighted according to their orientations in the pair, r_{AB} their distance and θ the angle between the Zeeman field B_0 and the interspin distance vector r_{AB}. For an isotropic distribution of angles θ, *i.e.* isotropic frozen sample without orientational selectivity of the experiment, a characteristic powder-type spectrum, the "Pake pattern", is obtained from which r_{AB} can be deduced, provided g_A and g_B are known from independent EPR experiments. At sufficiently good signal-to-noise ratio, the dipole–dipole coupling parameter ν_d can be directly read off from the singularities of the Pake pattern. The methodological challenge is to devise an EPR strategy that separates the electron–electron coupling from other interactions, such as electron–nuclear hyperfine interactions and inhomogeneous line broadening. Common to most electron–electron dipolar EPR methods for weakly coupled radical-pair systems, in which the dipolar coupling is smaller than the cw EPR linewidth, is the electron-spin-echo (ESE)-detected mode of operation. It is an inherent strength of the ESE method that it is "blind" towards static energy contributions, thus eliminating the masking effects of any source of inhomogeneous EPR line broadening.[212] Only when the interactions become time dependent and, thus, contribute to T_2 relaxation of the spin system, do they determine the spin-echo formation. In cases of anisotropic spin interactions, such as electron–nuclear dipolar or quadrupolar interactions or electron–electron dipole interaction, the echo amplitude may become modulated (ESEEM, spin-echo-envelope modulation) when the time between the pulses in the echo sequence is varied. In a biradical with interspin distances less than 80 Å the modulation pattern is often determined by the electron–electron dipole coupling frequency. The inhomogeneous broadening of nitroxide lineshapes from unresolved intramolecular proton or deuteron hyperfine interactions must be considered when extracting relaxation data from cw spectra, whereas the ESE measurements are independent of any source of inhomogeneous broadening. The pulse electron–electron dipolar spectroscopy was originally introduced in Novosibirsk as a 3-pulse method employing two microwave frequencies[213–218] to measure weak electron–electron dipolar couplings from ESE decays, and soon received the acronym PELDOR (pulsed electron–electron double resonance). About a decade later, the first publication on pulse electron–electron dipolar spectroscopy appeared from outside Novosibirsk,[219] the authors discussed orientational selection and created the acronym DEER (double electron–electron resonance). In the 4-pulse DEER method[220] an additional refocusing mw pulse is introduced that provides deadtime-free data collection. Since the evolution of electron double-quantum-coherence (DQC) also depends on the electron dipole–dipole interaction, a powerful 6-pulse DQC method was introduced by the Freed group at Cornell.[221,222] Commonly used pulse EPR schemes as well as dedicated cw EPR experiments for distance measurements are described in many review articles and books, for example in the anthology *Distance Measurements in Biological Systems by EPR*, edited by L.J. Berliner, G.R. Eaton and S.S. Eaton.[223] In this book, experts in the field report in detail about the pros and cons of distance measurements by cw and pulse EPR methods when applied to a broad variety of biological two-spin systems ranging from organic radicals and transition-metal ions, dipolarly coupled nitroxide spin labels to donor–acceptor cofactor radical ions in photosynthetic reaction centers. Another important book covering in depth the subject of distance measurements by electron dipolar spectroscopy is the monograph *Principles of Pulse Electron Paramagnetic Resonance* by A. Schweiger and G. Jeschke.[4] Several excellent reviews have appeared in the last few years covering both methodologies and applications of pulse dipolar EPR spectroscopy, primarily on structural biology of spin-labelled systems which have attracted rapidly growing attention, for example refs. [211, 224–231].

The high-field extension of 3-pulse PELDOR, developed with the aim to resolve the relative orientation of the radical-pair partners, is very powerful for large protein systems.[232] Also, the conceptually related, but single-frequency dipolar spectroscopy method RIDME (relaxation induced dipolar modulation enhancement)[233] has been extended to high magnetic fields and microwave frequencies, and interesting applications have been reported already, for example at 130 GHz on nitroxide biradicals[234] or at 95 GHz on donor–acceptor radical pairs in photosynthetic reaction centers.[232]

Here, we will briefly describe only a few aspects of PELDOR and RIDME at high Zeeman fields pertinent for measuring both distance and relative orientation of radical-pair partners. Orientation-resolving high-field PELDOR or DEER experiments have been performed in recent years independently by three research groups: the Prisner group in Frankfurt,[235–237] the Jeschke group in Konstanz[238] and the Möbius group in Berlin.[52,232] It is anticipated that more groups, striving for full structure determination of (bio)chemical complexes, will use orientation-resolving high-field electron–electron dipolar spectroscopy in the near future.

2.2.2.1 PELDOR

Figure 2.29 shows the principal pulse schemes typically employed for dual-frequency PELDOR (or DEER) experiments. The 3-pulse version was first introduced by Milov et al.,[214] see Figure 2.29(a). It is analogous to the SEDOR (spin-echo double resonance) sequence used in solid-state NMR to detect the coupling between two nuclear spins.[240] In this sequence, a 2-pulse Hahn echo sequence at v_a with a fixed pulse separation time τ_1 is involved to selectively detect the echo intensity of the radical A of the A–B radical pair. An additional microwave pulse at time τ_2 after the first pulse and at the microwave frequency v_b flips the B spin by 180°. If the spins A and B are coupled by dipolar interaction, the detected echo intensity becomes modulated when varying τ_2. The interaction frequency v_{AB} is obtained by Fourier transformation of the echo time trace.

To demonstrate why it so happens, we consider the somewhat simplified pulse scheme depicted in Figure 2.29(b). In this scheme the inversion π-pulse is

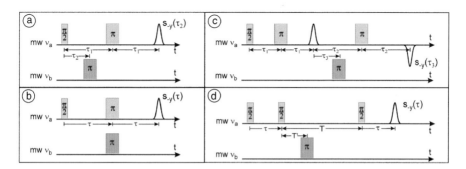

Figure 2.29 Principal dual-frequency pulse schemes of pulse electron–electron double resonance. (a) 3-pulse constant-time PELDOR;[216] (b) 3-pulse variable-time PELDOR; (c) 4-pulse constant-time PELDOR (DEER);[220] (d) 4-pulse variable-time PELDOR based on a stimulated-echo sequence.[232,239]

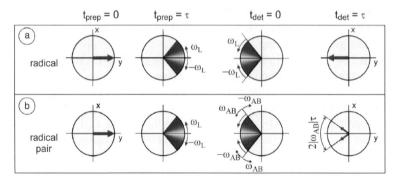

Figure 2.30 Transversal magnetization development in the rotating frame with microwave frequency of the observer spins A during the preparation (t_{prep}) and detection time (t_{det}): (a) For observer spins without dipolar coupling to the partner spins (monoradical) and (b) spins coupled via dipolar interaction in the radical pair with $\omega_{AB} = 2\pi \cdot \nu_{AB}$. During evolution time the subensemble magnetization vectors rotate with slightly different angular velocities, ω_L, due to different local fields they experience.

applied simultaneously with detection π-pulse, and the interpulse time τ is varied. The dipolar spin-echo modulations appear when the local dipolar field B_d induced at "observer" spins A by their "partner" spins B in the radical pair, gets inverted by spin flips of B that occurred between the preparation (time between the first and second pulse) and detection periods (between the second pulse and echo maximum) of the pulse sequence for spin-echo measurements at the resonance frequency of spins A. Figures 2.29(c) and 2.29(d) will be explained further below. An illustration of the echo events resulting in a dipolar modulation of the echo decay is given in Figure 2.30. It shows the evolution of transversal spin-magnetization vectors in the course of echo formation (rotating frame representation) for the 3-pulse variable-time PELDOR sequence depicted in Figure 2.29(a). When observer spins A do not experience any change of the dipolar field, their transversal magnetizations, having diverged from the $+y$-axis during the preparation time $t_{prep} = \tau$ because of different Larmor frequencies, converge during the detection period with the same frequencies. The sum magnetization becomes restored at the $-y$-axis to form an echo at the detection time $t_{det} = \tau$ (see Figure 2.30(a)). An inversion of dipolar fields, if experienced by a fraction of the observer spins that is given by the factor q_{inv}, shifts their precession frequencies during the detection period from those during the preparation period by $+\nu_{AB}$ or $-\nu_{AB}$. Therefore, their transversal magnetizations converge at $t_{det} = \tau$ to the two directions deviating from the $-y$-axis by angles $\pm 2\pi \cdot \nu_{AB} \cdot \tau$ (Figure 2.30(b)). The sum echo signal recorded as a function of τ with the microwave detection phase referenced to the $-y$-axis of the rotating frame (the "in-phase" detection), is partially modulated:

$$s_{-y}(\tau) = A_1(\tau) \cdot [1 - q_{inv} + q_{inv} \cdot \cos(2\pi \cdot \nu_{AB} \cdot \tau)] \quad (2.59)$$

where the relaxation factor $A_1(\tau)$ describes the echo decay because of transversal relaxation of the observer spins. The maximum modulation depth, $q_{inv} = 1$, corresponds to the ideal case where the spins A and B are spectrally well separated and both are fully excited by the detection and inversion pulses, respectively. This ideal case is, however, never met because the spectra of the A and B spins partially or even fully overlap. If orientational selectivity is required, the excitation bandwidth has to be inevitably reduced which leads to a shallow modulation depth, often with $q_{inv} \approx 0.01$.

The 3-pulse PELDOR sequence suffers from the spectrometer deadtime, which can reach several pulse lengths, especially at low microwave frequencies, e.g. at X-band. The effect of spectrometer deadtime can be avoided by using the 4-pulse constant-time DEER sequence introduced by G. Jeschke and coworkers[4,220] and now widely used at X-band, see Figure 2.29(c). In this sequence, an additional π-pulse at v_a is used to produce a refocused Hahn-echo, and the B-spins are manipulated between the first and the refocusing π-pulses. By proper setting of the interpulse separations the effect of the deadtime is eliminated. This method suffers, however, from relatively low echo intensity, because all A-spin manipulations take place in the xy plane of the rotating coordinate system and, thus, the initial echo intensity gets reduced by $\exp[-(2\tau_1 + 2\tau_2)/T_2]$, see Figure 2.29(c). The situation can be improved by using either 4-pulse variable-time PELDOR sequence[209] or the 4-pulse variable-time PELDOR based on the stimulated-echo sequence.[232,239]

Our strategy for high-field PELDOR spectroscopy is as follows: We use the 3-pulse stimulated-echo (SSE) sequence, in which the mw pulses are frequency adjusted to the observer spins A, and apply within the mixing period between the SSE pulses an additional mw π-pulse in resonance with the partner spins B to flip them, see Figure 2.29(d). The PELDOR-time trace is obtained measuring the SSE echo intensity at different τ values and using the reference-signal deconvolution technique. The reference-time trace is obtained by simply skipping the π-pulse at v_b. Division of the trace with the additional π-pulse by the reference trace yields the pure dipolar evolution function. When performed at stepped resonance field positions of both excitation sites within the spectra of the coupled radicals A and B, this strategy allows to find the particular field positions in the EPR spectrum that are conjugated by dipolar interaction between the A and B spins of the radical pair. The condition for achieving such a double magnetoselection gives additional correlations between the angular parameters of the dipolar frequency that define the relative orientation of the radicals A and B, thereby completing the description of the three-dimensional pair structure.

Orientational information on radicals in disordered solids is, of course, only available at a sufficient degree of orientation selectivity in the EPR spectrum. Therefore, PELDOR experiments in high-field EPR spectroscopy with adequate Zeeman magnetoselectivity appears to be a promising approach for structure determination.[232] Moreover, interferences from nuclear and hyperfine echo-modulations are normally eliminated at high fields. Nonetheless, the separation of the electron–electron dipolar couplings from other modulation or line-broadening effects becomes more demanding under high-field/high-frequency conditions. An obvious problem is the excitation bandwidth that must exceed the dipolar coupling, since both EPR lines of the dipolar doublet of a given A or B spin have to be excited. In PELDOR, the pump pulse at the second frequency v_b must excite a significant fraction of B spins that are dipolarly coupled to the observer spins A, whose resonance frequency is v_a. In high-field EPR the spectral width of the two radicals in the weakly coupled pair increases in proportion to their difference in g-values and applied Zeeman field. Fortunately, the bandwidth of the EPR resonator also increases with high mw frequency: Even for a single-mode cavity with high Q value a bandwidth of 100 MHz was achieved at 95 GHz (W-band), see Section 3.4.1 for instrumental details. This is sufficient for accommodating both v_a and v_b for donor–acceptor radical pairs in photosynthesis, but not for nitroxide spin-labelled radical pairs. W-band EPR spectra of nitroxide radicals are typically spread over a range of 400 MHz as compared to 200 MHz at X-band. Hence, dual-frequency (DF)-PELDOR at W-band is applicable for donor–acceptor ion radical pairs in photosynthesis, and detailed information of interspin distance and orientation of the charge-separated cofactors in the photocycle could be obtained,[232] see Section 5.2.1.7. For nitroxide biradicals, on the other hand, W-band PELDOR at a single mw frequency, but with additional fast field jumps (FJ-PELDOR) turned out to be an adequate solution of the bandwidth problem.[239] The pros and cons of two-frequency *versus* field-step pulsed ELDOR techniques,[239,241] as well as advantages and disadvantages of different PELDOR sequences,[4,209,241] have been discussed in the literature and, therefore, will not be repeated here. At FU Berlin, elaborate devices have been developed both for fast double-frequency PELDOR[232] and field-jump PELDOR[239] at W-band; details of the instrumentation are given in Sections 3.4.1.5 and 3.4.1.6.

2.2.2.2 RIDME

In the 3-pulse stimulated echo-sequence in resonance with the observer spins A, the partner spins B can also flip spontaneously by their longitudinal relaxation that occurs during the mixing period (interval T between the 2nd and 3rd microwave pulse). This relaxation mechanism is exploited in RIDME spectroscopy that, thus, can be considered as a single-frequency variant of dipolar spectroscopy in which T_1 relaxation substitutes the coherent mw irradiation in PELDOR to flip the B spins by 180° (π). Similar to PELDOR, the RIDME experiment yields an echo-decay time trace in which the dipolar coupling between the A and B spins becomes apparent as a modulation with the period of the inverse dipolar frequency.

Certain conditions have to be met for the RIDME experiment to operate as a function of the preparation time τ: The fixed mixing time T should be long enough to allow the longitudinal spin relaxation to flip the partner spins in the pair, but short enough to avoid a considerable reduction of the echo signal caused by the longitudinal spin relaxation of the observer spins. The fraction of spins B flipping an odd number of times is given by the inversion factor

$$q_{inv} = \frac{1}{2} \cdot [1 - \exp(T/T_1)] \qquad (2.60)$$

where T_1 is the spin-lattice relaxation time. Under conditions $T_1 \gg \tau$ and $(T_1 \cdot \nu_{AB})^2 \gg 1$, the stimulated-echo amplitude as a function of interpulse time τ is given by[233]

$$s_{-y}(\tau) = A_1(\tau) \cdot A_2(T) \cdot [1 - q_{inv} + q_{inv} \cdot \cos(2\pi \cdot \nu_{AB} \cdot \tau)] \qquad (2.61)$$

Here, the relaxation factors $A_1(\tau)$ and $A_2(T)$ describe the echo decay because of transversal and longitudinal relaxation of the observer spins. Thus, a measurement of the echo amplitude will exhibit a periodic oscillation with ν_{AB}. Unfortunately, hyperfine modulations obscure the RIDME effect. Consequently, RIDME experiments are best performed at high magnetic fields and microwave frequencies, where hyperfine modulations are strongly reduced.[232,234]

When performed consecutively at stepped field positions within the partially Zeeman-resolved high-field EPR spectrum of the two radicals of a coupled pair, RIDME is potentially capable of revealing the specific field positions at which the dipolar modulation occurs at the principal values, $\nu_{\parallel} = \nu(\theta = 0)$ and $\nu_{\perp} = \nu(\theta = \pi/2)$, of the dipolar frequency. Magnetoselection at these field positions establishes correlations between the polar angles ($\eta_{A(B)}$, $\phi_{A(B)}$) that define possible orientations of the dipolar axis in the molecular frame of the radicals in the pair.

The main advantage of RIDME experiment is its simplicity as compared to PELDOR because it is based just on the stimulated-echo sequence. It can be performed on any pulsed EPR spectrometer without needing dual-frequency or field-jump extensions. Moreover, the large modulation depths, even at high-field EPR, can be reached because the inversion factor q_{inv} depends only on the time T in the stimulated-echo sequence. The main disadvantage of the RIDME method is the partial loss of orientational selectivity. It can be a difficult task to assign the observed modulations to a particular radical in the pair, especially if the radicals are identical or their EPR spectra strongly overlap, as will be demonstrated below. Additionally, the success of a RIDME experiment is strongly dependent on the proper choice of experimental conditions, in particular of the sample temperature. The dipolar modulations are enhanced in the case of a pure T_1 relaxation process, see eqn (2.60), but the decay of stimulated-echo intensity can be governed by additional mechanisms, for example spectral diffusion.[242] Thus, one can easily slide into the awkward situation when the echo signal vanishes before the dipolar modulation has developed.

For details concerning the 95-GHz high-field PELDOR and RIDME experiments on transient radical pairs in bacterial photosynthetic reaction centers, see Section 5.2.1.7 and ref. [232].

2.2.2.3 High-Field RIDME and PELDOR on Nitroxide Radical Pairs

To demonstrate the power of the high-field RIDME and PELDOR methods also for nitroxide radical pairs, as well as to show how and what kind of information can be obtained, we present our preliminary W-band results from a model nitroxide biradical. The structure of the shape-persistent biradical B1 is depicted in Figure 2.31. This model biradical was the object of numerous previous investigations including S- and X-band PELDOR[220,243–246] and X-ray crystallography.[220] Therefore, it was chosen as a test system for our W-band RIDME and PELDOR investigations. To clarify the characteristics of the dipolar spectra, all the experiments performed on the biradical were repeated with the single-nitroxide radical R1-^{14}N, see Figure 2.22, which we had used in our nuclear ESEEM experiments.[180] *Ortho*-terphenyl was used again as a host matrix, and the experiments presented here were performed at 180 K.

Figure 2.32 shows the microwave pulse setting of the RIDME experiment. The stimulated-echo pulse sequence is applied at the field position corresponding to the g_{zz} value ($m_I = 0$) of the B1 nitroxide biradical EPR spectrum. The separation time between the second and the third $\pi/2$-pulse, T, was set to the 1/e decay time of the stimulated echo, *i.e.* $T = 33\,\mu s$. The interpulse time between the first and the second pulse, τ, was swept from 30 ns to 2000 ns in 10 ns steps. The detected echo decay $s_{-y}(\tau)$ is shown in Figure 2.32 (bottom left). The pronounced modulation appears at a frequency of 2.1 MHz, as is read out from the normalized-amplitude Fourier-transform spectrum, see Figure 2.32 (bottom right). At the chosen field position no modulations due to ^{14}N nuclei are expected, see Section 2.2.1.4. Thus, the observed modulation is ascribed to the RIDME effect.

In fact, the field–frequency RIDME spectrum of the R1-^{14}N radical reveals only several peaks in the g_{xx}, g_{yy} spectral region, Figure 2.33(a). They are all assigned to ^{14}N nuclear modulation. In addition to the modulation pattern observed in stimulated ESEEM experiment, see Figure 2.26, two peaks positioned at about 10 MHz in the g_{yy} region are detected. They result from the ^{14}N

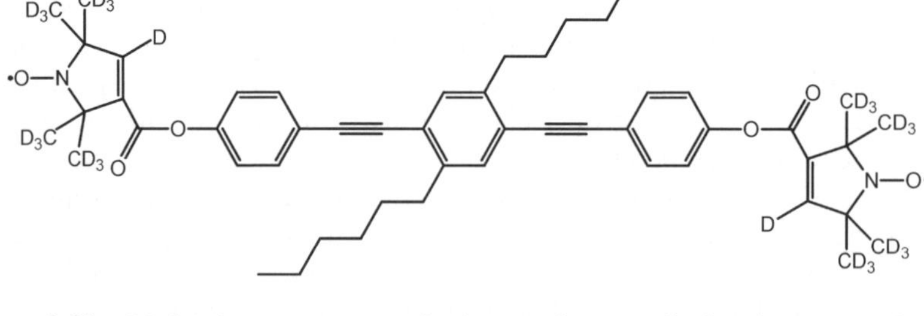

Figure 2.31 Molecular structure of the perdeuterated (headgroups) nitroxide biradical B1. The biradical was synthesized by H. Zimmermann (Max-Planck-Institute for Medical Research, Heidelberg), as has been described previously.[247]

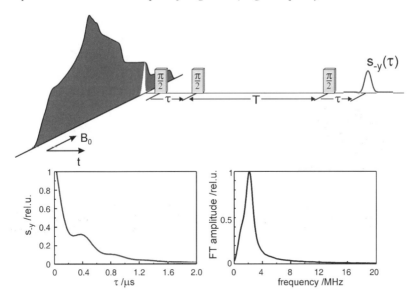

Figure 2.32 Top: W-band microwave pulse sequence for the RIDME experiment on the nitroxide biradical B1 in *ortho*-terphenyl solution at 180 K. Top: The echo-detected EPR spectrum of B1 as well as the microwave excitation bandwidth for typical microwave pulse-length settings are shown. Typical settings of the pulse-sequence parameters are $T = 33\,\mu s$, $\pi/2$-pulse length $t_p = 30$ ns. The time τ is stepped from $\tau_0 = 30$ ns to 2000 ns in 10 ns steps. Bottom left: Representative example of a dipolar modulation echo-decay trace at the indicated magnetic-field position. Bottom right: Fourier-transformed spectrum of the RIDME decay, the cutoff frequency of the high-pass filter was set to 1 MHz, see text.

nuclear combination frequencies and the powder averaging.[4] The RIDME pulse sequence produces nuclear modulations similar to the 2-pulse ESEEM experiment because the time τ is stepped. The smaller amplitudes of the modulations, compared to Figure 2.26, are explained by the loss of echo intensity due to T_1 relaxation and by faster echo decays compared to the stimulated-ESEEM experiment.

The ^{14}N nuclear modulation frequencies are also observed in the RIDME spectrum of the B1 biradical, see Figure 2.33(b). These are, however, about an order of magnitude weaker compared to the additional modulations which appear at 4.2 MHz and 2.1 MHz in the g_{xx},g_{yy} and g_{yy},g_{zz} spectral regions, respectively. These modulations are unambiguously assigned to dipolar couplings of the electron spins in the biradical. The frequency of 4.2 MHz corresponds to the parallel dipolar frequency, $v_\| = |v_{AB}(\theta=0)| = 2 \cdot v_d$, and the frequency of 2.1 MHz is assigned to the perpendicular dipolar frequency, $v_\perp = |v_{AB}(\theta=\pi/2)| = v_d$, see eqn (2.57). These values agree with $v_\| = 4.3$ MHz and $v_\perp = 2.1$ MHz observed by S-band PELDOR.[245] The insert in Figure 2.33(b) shows the T-time development of the normalized modulation amplitude S_1/S_2

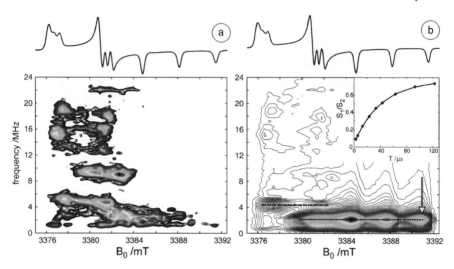

Figure 2.33 Frequency–field dependence of the modulations observed by the RIDME pulse sequence applied to 1 mM R1-^{14}N radicals (a) and B1 nitroxide biradicals (b) in *ortho*-terphenyl glass at 180 K. The structure of R1-^{14}N is depicted in Figure 2.22. The contour lines are shown as isohypses from 0.05 to 1 of the maximum FT RIDME intensity. The dotted and dashed lines in (b) indicate the spectral regions of the canonical dipolar modulation frequencies. On top, the corresponding experimental cw EPR spectra are shown. Note that cw EPR spectra are practically identical, except for a slightly broadened g_{xx} component in the spectrum of B1. The insert in (b) shows the development of the normalized modulation amplitude S_1/S_2 with time T. For details, see text.

at 2.1 MHz in the g_{zz} ($m_I = -1$) spectral position indicated by the arrow in the contour plot. The modulation amplitude was obtained by fitting the experimental RIDME decays to the equation

$$s_{-y}(\tau) = S_1 \cdot \exp(-\tau/D_1) \cdot \cos(2\pi \cdot \nu_\perp \cdot \tau) + S_2 \cdot \exp(-\tau/D_2) \quad (2.62)$$

In the ideal case, the decay time constants, D_1 and D_2, of the modulated and nonmodulated parts should be equal, see eqns (2.60)–(2.61). Moreover, the normalized modulation amplitude S_1/S_2 should approach unity for $T \gg T_1$ because it is given by

$$\frac{S_1}{S_2} = \frac{q_{inv}}{1 - q_{inv}} = \frac{1 - \exp(-T/T_1)}{1 + \exp(-T/T_1)} \quad (2.63)$$

The analysis of the experimental data, see insert in Figure 2.33(b), in terms of eqn (2.63) yields an S_1/S_2 ratio of 0.7 for $T \gg T_1$. Additionally, the normalized modulation amplitude does not tend to zero at $T = 0$. This shows that the dipolar modulation should also be observed in a 2-pulse experiment,[233] which was indeed approved experimentally. The deviation of the maximum S_1/S_2 value from unity can be explained by the different physical mechanisms influencing the amplitudes of modulated and nonmodulated decays to a different extent, for example intramolecular spin–spin dipolar interactions, fast librational motion of the nitroxide headgroups or slow rotation of the methyl groups. The different combination of these mechanisms also explains the different decay time constants D_1 and D_2, as well as their different T–time dependence observed in the experiment. However, here we want to concentrate mainly on the intermolecular dipolar interaction and, therefore, refer to refs. [242,248] for further details.

In order to obtain structural information about the biradical, the RIDME spectrum in Figure 2.33(b) has to be considered in greater detail. The important parameters are the dipolar frequencies and their positioning within the nitroxide spectrum. The amplitudes of the modulation are less important, because the mechanisms influencing the dipolar modulation depth depend on the chosen field positions of the EPR spectrum. The most important parameter is the positioning of the parallel modulation frequency, $v_{\parallel}$, within the RIMDE spectrum. Figure 2.34(a) shows the headgroup of the nitroxide overlaid with the unit sphere of the g-tensor axes system, and Figure 2.34(b) shows the structure of an artificial nitroxide radical pair, R_A–R_B. Let's consider only the nitroxide R_A as the observer radical. If the dipolar vector r_{AB} coincides, for example, with the x-axis ($\eta_A = \pi/2$, $\phi_A = 0$) the parallel dipolar modulation will show up only at the g_{xx} position in the RIDME spectrum, because only in this spectral position r_{AB} coincides with external magnetic field B_0, i.e. $\theta = 0$, see eqn (2.57). In the case of arbitrary η_A and ϕ_A angles, the parallel modulation will appear in the field position that corresponds to $g_A(\eta_A,\phi_A)$ selected by B_0, i.e. $\eta_A = \theta_A$ and $\phi_A = \varphi_A$. The consideration of the R_B radical as observer is analogous, only the angles η, ϕ have to be redefined. In RIDME, the radicals R_B

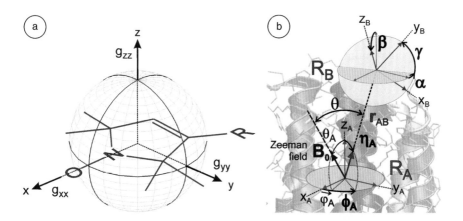

Figure 2.34 (a) Structure of the nitroxide fragment of biradical B1, definition of the principal axes system and unit sphere. (b) Definition of the radical-pair geometry parameters.

act independent of their orientations on radicals R_A, and *vice versa*, because the dipolar modulation is driven by T_1 relaxation that, to a first approximation, can be assumed to be isotropic. Thus, the resulting RIDME spectrum represents just the sum of the RIDME spectra of the individual radicals. This is a main disadvantage of the RIDME method. The RIDME spectrum of the biradical B1, Figure 2.33(b), shows that the parallel modulation $v_\parallel = 4.2$ MHz appears both in the g_{xx} and g_{yy} spectral regions. This might be interpreted by the nonequivalency of the radicals in the pair, for example ($\eta_A = \pi/2$, $\phi_A = 0$) for radical R_A and ($\eta_B = \pi/2$, $\phi_B = \pi/2$) for radical R_B. From symmetry consideration of the biradical, however, this cannot be true, *i.e.* $\eta_A \approx \eta_B$ and $\phi_A \approx \phi_B$. An additional structural information can be obtained from the positioning of the perpendicular modulation $v_\perp = 2.1$ MHz. It also appears in the g_{yy} and g_{zz} spectral regions, but from the same considerations as for the parallel modulation we cannot ascribe the g_{yy} region modulation either to the R_A or to the R_B radical. Thus, the RIDME spectrum only reveals that the z-axes of both radicals are approximately perpendicular to the dipolar vector r_{AB}.

The RIDME results for the model biradical disclose the advantages and disadvantages of the RIDME method at high magnetic fields. RIDME is very powerful in detecting the canonical dipolar modulation frequencies, $v_\parallel$ and $v_\perp$, of the nitroxide radical pair. Nitrogen nuclear modulation does not disturb the RIDME recordings due to its small amplitude compared to dipolar modulation. However, the orientational information can be extracted only partially from the RIDME spectra, even in the case of a symmetrical biradical, due to the partial loss of orientational selectivity.

To obtain a more precise information on the structure, the higher orientational selectivity of the PELDOR method has to be exploited. The strategy is not to rely on the complete 4D PELDOR experiment but rather to refine stepwise the information obtained from the RIDME spectrum. Figure 2.35 shows the PELDOR pulse sequence with the detection-pulse frequency adjusted to v_a, which corresponds to the spectral position B' optimized for maximum orientational selectivity and maximum modulation amplitude at $v_\parallel$ as observed in RIDME. The pumping π-pulse is set in the spectral position B'' where additional $v_\parallel$ modulations are detected in the RIDME experiment, in this case in the g_{xx},g_{yy} spectral region. The B'' position is stepped either by changing the frequency of the pumping microwave v_b according to $B'' = B' + (v_a - v_b) \cdot h/(g_e \cdot \mu_B)$ in the dual-frequency experiment, or by rapidly changing the external magnetic field by $\Delta B = B'' - B'$ and applying the pumping pulse at microwave frequency v_a of the detection sequence in the field-jump PELDOR experiment.

An example of the PELDOR decay trace for a B'' value anywhere between the g_{xx} and g_{yy} regions of the EPR spectrum is shown in Figure 2.35 (bottom left). A distinct modulation is observed at the dipolar frequency 4.24 MHz, as can be read out from the Fourier-transformed PELDOR spectrum, see Figure 2.35 (bottom right). The same experiment performed on the radical R1 does not reveal any modulations. Figure 2.36(a) shows the PELDOR spectrum in contour-plot representation for the different B'' settings and B' set at the g_{xx} position. A strong peak at 4.24 MHz is observed with its maximum at the field

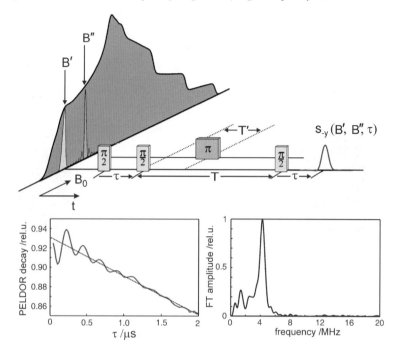

Figure 2.35 Top: Pulse sequence for the dual-frequency W-band PELDOR experiment on the nitroxide biradical B1 in *ortho*-terphenyl solution at 180 K. The echo-detected EPR spectrum of B1 as well as the microwave excitation bandwidth for typical microwave pulse-length settings are shown for the sake of clarity. Typical settings of pulse-sequence parameters are $T = 5.6\,\mu s$, $T' = 1.2\,\mu s$, $\pi/2$-pulse length $t_p = 32$ ns in the observer pulse sequence at microwave frequency ν_a. The pulse length of the additional π-pulse was adjusted to account for the mismatch of its frequency ν_b from the resonance frequency ν_a of the EPR cavity. The preparation time τ is varied from 50 ns to 2000 ns in 20 ns steps. Bottom left: Representative example of a PELDOR decay trace (the ratio of echo amplitudes recorded with additional π-pulse on and off *versus* delay time τ). Bottom right: The Fourier-transformed spectrum of the PELDOR decay trace is given, the cutoff frequency of the high-pass filter was set to 1 MHz. For details, see text.

position $B_\parallel$, which is significantly shifted towards g_{yy}. The field position $B_\parallel$ and the magnetic interaction parameters of the nitroxide, see Table 2.1, allow determination of the orientation-dependent g-values $g_{A(B)}(\theta_{A(B)}, \varphi_{A(B)})$ of those radicals that are contributing to the EPR spectrum at this field position. The selected $g_{A(B)}$ values can be plotted on the orientation-selection unit sphere of the g-axes system introduced in Figure 2.34. Figure 2.37(a) shows such a selection trace. For simplification, the ^{14}N hyperfine splitting is omitted. From symmetry considerations of the biradical structure, the same selection trace is valid for both nitroxide headgroups in B1. Thus, at this stage of our PELDOR/RIDME investigations, we can only partially reconstruct the 3D-structure of

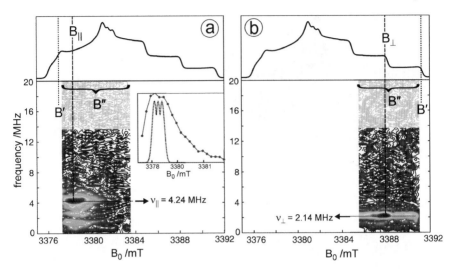

Figure 2.36 Dipolar PELDOR spectra of the B1 nitroxide biradical in *ortho*-terphenyl solution at 180 K. The observer mw frequency is fixed at a value corresponding to the resonance field B', while the pump microwave is swept through a region around the resonance frequency determined by the chosen field value B''. The contour plots show the Fourier amplitudes of PELDOR echo-decay traces. The insert in (a) shows Fourier amplitudes (dots) as the function of the magnetic field at the slice position $v_\parallel = 4.24$ MHz. Additionally, the dotted line shows the three ^{14}N hyperfine lines with widths that correspond to inhomogeneously broadened EPR linewidth of B1. For details, see text.

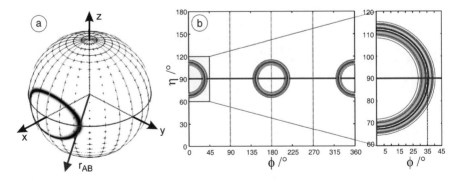

Figure 2.37 Graphical solution for the angles η and ϕ in the molecular frame of the nitroxide radical headgroups in biradical B1. (a) The corresponding selection traces are plotted on the unit sphere. There are four crossing points of the two selection traces providing four possible orientations of the dipolar axis r_{AB}, but only one direction is shown for the sake of clarity. (b) This part of the figure shows the (η,ϕ) projection of the selection traces. For details, see text.

B1: Any biradical structure with a dipolar axis defined by the angles η and ϕ, as obtained from the selection trace, will explain the experimental observations. The different combinations of η and ϕ angles can be simply read out from the contour plot of the selection–trace projection on the (η,ϕ) plane, see Figure 2.37(b).

To refine the biradical structure in a next step, additional PELDOR experiments have to be performed. As pointed out above, only the perpendicular dipolar frequencies are observed in the g_{zz} region of the nitroxide spectrum. Thus, the next experimental step is to set the observer pulse sequence at the field position B' which has a maximum z-orientation selectivity, *i.e.* at the high-field edge of the EPR spectrum, see Figure 2.36(b). The field position of the pump π-pulse is consecutively stepped inside the spectrum, and at each position the PELDOR traces are recorded. The resulting PELDOR spectrum is depicted in Figure 2.36(b). Only the peak at the frequency $v_\perp = 2.14$ MHz is observed. Moreover, the maximum PELDOR intensity is exactly at the position of the nearest ^{14}N hyperfine line. Because no frequencies higher than $v_\perp = 2.14$ MHz are observed, we can conclude that the z-axes of both nitroxide radicals of the headgroups are perpendicular to the dipolar axis. Thus, the additional selection trace, which in this case is just a ring in the xy molecular plane, can be put on the selection unit sphere, see the blue trace in Figure 2.37(a). The crossings of both selection traces give us the proper set of angles.

Figure 2.37(b) shows the multiple solutions when using the graphical method, *i.e.* by the crossings of the selection traces. Four possibilities of the dipolar-axis orientation in the nitroxide molecular system are found, namely $\eta = 90°$, $\phi = (35°; 145°; 215°; 325°)$. This leads to eight principal possibilities to build a structural model of the radical pair. These orientation ambiguities occur because the measured g-value for an orientation defined by the vector ***h*** is given by $(\boldsymbol{h} \cdot \tilde{g} \cdot \tilde{g} \cdot \boldsymbol{h})^{1/2}$, where $\tilde{g}$ is the g-tensor, and the components of the vector ***h*** are the direction cosines of the external magnetic field B_0 in the reference axes system.[3] The observed dipolar interactions follow the same rule. Thus, this ambiguity problem can principally not be solved by EPR methods alone, but needs additional information from other sources. For the nitroxide biradical B1, assuming the apparent symmetry and directing the g_{xx} principal axis along the molecular x-axis, Figure 2.34(a), we can reduce the number of possible structure variants. Actually, there are only two principally different structures, which can be described as *cis* and *trans* conformations of the biradical. These structures can by visualized by sticking together two nitroxide molecular frames, Figure 2.37(a). The dipolar (interconnecting) axis is directed from the first frame with (η, ϕ)=(90°, 35°) to the second frame, either with (η, ϕ)=(90°, 35°) or (η, ϕ)=(90°, 325°). To be able to distinguish between these two conformations, we need the help of quantum-chemical calculations of the energy-minimized structure. Principally, a set of five angles is required: the polar angles (η, ϕ) of the dipolar vector in the coordinate system of the first (observer) radical, and the three Euler angles (α,β,γ) that define the orientation of the coordinate system of the second radical with respect to the observer coordinate frame. Instead of using such an asymmetrical definition, an additional "dipolar" frame can be introduced, see

Section 5.2.1.7 and ref. [232]. In the dipolar-frame representation two sets of the polar angles (η, ϕ) for each radical are used. The fifth angle, ξ, fixes the structure with respect to rotation of both radical frames around the dipolar axis. In the case of the biradical B1, the angle ξ is not necessary to be determined: Because of the rotational freedom about the phenyleneethylene backbone, ξ is uniformly distributed between 0 and 360°.[238]

At this point, we will summarize the results of the RIDME/PELDOR experiments on biradical B1: (*i*) From $v_{\|} = 4.24 \pm 0.05$ MHz and $v_{\perp} = 2.14 \pm 0.05$ MHz the interspin distance of $r_{AB} = 2.90 \pm 0.03$ nm (29.0 ± 0.3 Å) is calculated, see eqn (2.58). The exchange interaction is negligible. (*ii*) The structure of B1 can be described as follows: The z-axes of both molecular frames are perpendicular to the dipolar axis, on average the x-axes are tilted by 35° away from the dipolar axis. (*iii*) Two conformations of the biradical, *cis* and *trans*, are principally possible and compatible with the experimental data.

These results are in fair agreement with the molecular structure used in recent high-field PELDOR (DEER) work on a similar biradical,[238] where a tilt angle of 25° was obtained. Here, however, we have to point out that in our given result a statistical distribution of the orientations was not taken into account. The insert in Figure 2.36(a) shows the PELDOR spectrum at the parallel frequency in comparison with the EPR spectrum of a single nitroxide orientation contributing at this field position. In the case of fixed relative nitroxide orientations, the PELDOR spectrum and EPR spectrum should coincide. The broadening of the PELDOR spectrum reflects the statistical deviation of the dipolar axis from the averaged position given by the (η, ϕ) angles. This broadening, however, was analyzed in greater detail. Taking, for example, a Gaussian distribution of ϕ angles, *i.e.* the rotation of nitroxide around the z-axis, a half-width of the distribution function of about 15° is determined, and the averaged tilt angle becomes about 30°.

Thus, we conclude that high-field RIDME/PELDOR experiments are able to determine the structure of the chosen nitroxide radical pair. Moreover, the orientational distribution of the nitroxide headgroups can be determined. In the example presented the RIDME experiment is used to collect preliminary information about the dipolar coupling frequency. This information is step by step refined by PELDOR experiments, until the final structure is revealed. This approach appears to be a useful strategy for high-field dipolar EPR spectroscopy on nitroxide radical pairs.

Now the question may arise: Why did high-field RIDME/PELDOR not find yet practical application to doubly spin-labelled proteins, whereas X-band PELDOR is already widely used to reveal the structures and structural changes in such systems? In fact, X-band PELDOR (DEER) became an established method for structure determination of doubly spin-labelled biological and biochemical systems.[211,222,224,226,230] The main output of such X-band experiments is the distance as well as the distance distribution of the two nitroxides in the pair. The distance constraints allow construction of structural models that yield useful insights into the structure of the protein or the superstructure of a protein complex.

Partial orientation selectivity, which sometimes appears in X-band PELDOR experiments, has to be treated carefully.[219,249,250] It does not yield orientational information but can rather lead to misinterpretations of the experimental distance results.[251,252] To further diminish the orientational artifacts, S-band PELDOR[245] or double-quantum coherence (DQC) EPR[211,221] are recommended to use. They are less sensitive to orientational selectivity. In S-band PELDOR, this is due to the smaller g-tensor spectral resolution at lower Zeeman field. In DQC, where all six pulses, frequency tuned to the observer spin A, are nonselective and intense in order to excite the whole spectral distribution of the spins, *i.e.* all the spins of the dipolarly coupled radical pair can be considered as A spins.[211] In many doubly spin-labelled systems, however, X-band PELDOR yields reliable distance results due to the flexibility of the nitroxide fragments, which at least partially destroys orientational selectivity.

For high-field PELDOR, however, the flexibility of the nitroxide labels in proteins turns out to be a main disadvantage. The example described above shows that even a small distribution of relative orientations of nitroxide fragments of the order of 10° leads to significant broadening of the PELDOR responses. In the limiting case, where the dipolar axis is uniformly distributed over the first octant of the (η, ϕ) unit sphere, see Figure 2.34, each position in the high-field PELDOR spectrum will respond with approximately a Pake pattern. The limited excitation bandwidth in combination with the increased spectral resolution at higher Zeeman fields will result in diminishing the dipolar modulation amplitude. Taking into account that even in rigid radical-pair systems the dipolar modulation seldom exceeds several per cent, these signals become undetectable by high-field PELDOR. The high-field RIDME method is still applicable in this case. However, the information that can be obtained from high-field RIDME responses is restricted to distances in this situation and, thus, is exactly the same as obtained from low-field PELDOR experiments.

The flexibility of the protein domains and their conformational changes in biological action are in themselves interesting fields of research. Unfortunately, the flexibility of the widely used nitroxide MTS spin label is often due to the flexible tether that links the nitroxide headgroup to the protein backbone. This flexibility of the spin label, however, has little, if anything, to do with the protein motion, and sophisticated temperature-dependent multifrequency EPR experiments with elaborate spectra-analysis procedures are needed to disentangle the different modes of motion in the overall dynamics of such spin-labelled systems.[211,253] Thus, further developments of the spin-labelling techniques are needed that are capable to rigidly couple the nitroxide side chain to the protein backbone. New approaches have already been presented fulfilling this requirement. They include solid-phase protein synthesis where the natural and nitroxide-containing artificial amino acids are incorporated in a step-by-step process.[254] The spin-label TOAC,[255–257] for example, which is a helicogenic nitroxyl amino acid, is by far the most popular unnatural amino acid. It was successfully incorporated in artificially made polypeptides and studied by PELDOR.[249,258] However, the six-ring structure of TOAC is sensitive to reduction/oxidation processes and acidic media and, therefore, can hardly be incorporated into functional protein systems. More promising is the use of pyrroline-type nitroxide containing amino acids[259] in combination with recombinant DNA techniques and chemical peptide synthesis.[254] Unfortunately, up to now the relatively small yield of even mono-spin labelled proteins does not yet allow for high-field dipolar spectroscopy on such objects. Hopefully, future improvements in peptide chemistry and protein manipulation techniques[260] will afford adequate potential for using such artificial amino acids in the spin triangulation of proteins by high-field RIDME/PELDOR methodology.

We summarize the perspectives of orientation-selective electron–electron dipolar high-field EPR techniques in conjunction with spin-labelling techniques:

(i) RIDME requires that at least one nitroxide radical, R_A as observer, is rigidly linked to the protein at a well-defined position. In this case one can aim at a distant nitroxide R_{B1}, preferably flexible concerning orientations and, therefore, silent in RIDME. The detected high-field RIDME spectrum would yield the interspin distance r_{AB} and the polar angles η_A, ϕ_A that describe the positioning of the dipolar vector with respect to the molecular frame of the observer nitroxide. Performing an additional RIDME measurement on the distant nitroxide R_{B2} at a different position allows a geometric constraint to be built without needing an additional measurement on a double mutant with two distant nitroxides, i.e. R_{B1}–R_{B2}. Such a measurement is, however, necessary if only the distance between the nitroxides, but no angular information is obtained, i.e. by X-band PELDOR. Moreover, the additional information about the η_A, ϕ_A angles can simplify the characterization of potential conformational changes of the protein in biological action. It will allow not only to detect changes in the distance, i.e. the radial shifts, but also changes in relative orientation, i.e. tangential shifts, which cannot be detected by low-frequency dipolar techniques.

(ii) PELDOR generally requires two rigidly fixed nitroxides. It can be applied, in conjunction with RIDME, on systems for which the full orientational information is desired. The PELDOR method can be applied to systems in which the relative angular-orientation distribution of the nitroxides in the pair does not exceed 20°. The most promising way, however, will be to use an observer nitroxide with well-resolved g-tensor and a distant partner spin of different type with small g-anisotropy.

Thus, high-field RIDME/PELDOR methods in conjunction with site-directed spin-labelling techniques have a high potential for solving geometric structure problems with less effort in molecular biology than would be required for low-frequency dipolar methods. However, the development of the new labels of nitroxide or different type, which would fulfill the requirements for high-field dipolar methods, appears to be mandatory.

References

1. *Multiple Electron Resonance Spectroscopy*; M. M. Dorio, J. H. Freed, Eds. Plenum, New York, 1979.
2. *Advanced EPR, Applications in Biology and Biochemistry*; A. J. Hoff, Ed. Elsevier, Amsterdam, 1989.
3. N. M. Atherton, *Principles of Electron Spin Resonance*, Ellis Horwood, New York, 1993.
4. A. Schweiger and G. Jeschke, *Principles of Pulse Electron Paramagnetic Resonance*, Oxford University Press, Oxford, 2001.
5. A. A. Doubinski, In: *Electron Paramagnetic Resonance*, Vol. **16**, N. M. Atherton, M. J. Davies, B. C. Gilbert, Eds., Royal Society of Chemistry, Cambridge, 1998, pp. 211.
6. K. Möbius, *Chem. Soc. Rev.*, 2000, **29**, 129.
7. J. H. Freed, *Annu. Rev. Phys. Chem.*, 2000, **51**, 655.
8. G. M. Smith and P. C. Riedi, In: *Electron Paramagnetic Resonance*, Vol. **17**, N. M. Atherton, M. J. Davies, B. C. Gilbert, Eds., Royal Society of Chemistry, Cambridge, 2000, pp. 164.
9. T. Prisner, M. Rohrer and F. MacMillan, *Annu. Rev. Phys. Chem.*, 2001, **52**, 279.
10. M. Fuhs and K. Möbius, In: *Lecture Notes in Physics*, Vol. **595**, C. Bertier, P. L. Levy, G. Martinez, Eds., Springer, Berlin, 2002, pp. 476.
11. P. C. Riedi and G. M. Smith, In: *Electron Paramagnetic Resonance*, Vol. **18**, B. C. Gilbert, M. J. Davies, D. M. Murphy, Eds., Royal Society of Chemistry, Cambridge, 2002, pp. 254.
12. M. Bennati and T. F. Prisner, *Rep. Prog. Phys.*, 2005, **68**, 411.
13. K. Möbius, A. Savitsky, A. Schnegg, M. Plato and M. Fuchs, *Phys. Chem. Chem. Phys.*, 2005, **7**, 19.
14. K. Möbius and D. Goldfarb, In: *Biophysical Techniques in Photosynthesis*, Vol. **II**, T. J. Aartsma, J. Matysik, Eds., Springer, Dordrecht, 2008, pp. 267.
15. *Appl. Magn. Reson.*; Vol. 21, No. 3-4; Möbius K., Guest Ed., 2001.
16. *Very High Frequency (VHF) ESR/EPR Biological Magnet Resonance*; Vol. 22; O. Grinberg, L. J. Berliner, Eds.; Kluwer/Plenum, New York, 2004.
17. *Magn. Reson. Chem.* Vol. 43, Special Issue; W. Lubitz, K. Möbius, K. P. Dinse, Guest Eds., 2005.
18. A. Abragam, *The Principles of Nuclear Magnetism*, Oxford University Press, Oxford, 1961.
19. A. Carrington and A. D. McLachlan, *Introduction to Magnetic Resonance*, Harper, New York, 1967.
20. J. S. van der Brink, A. P. Spoyalov, P. Gast, W. B. S. van Liemt, J. Raap, J. Lugtenburg and A. J. Hoff, *FEBS Lett.*, 1994, **353**, 273.
21. W. Lubitz and G. Feher, *Appl. Magn. Reson.*, 1999, **17**, 1.
22. M. Flores, R. Isaacson, E. Abresch, R. Calvo, W. Lubitz and G. Feher, *Biophys. J.*, 2007, **92**, 671.

23. P. J. Bratt, P. Heathcote, A. Hassan, J. van Tol, L. C. Brunel, J. Schrier and A. Angerhofer, *Chem. Phys.*, 2003, **294**, 277.
24. W. R. Hagen, In: *Advanced EPR Applications in Biology and Biochemistry*, A. J. Hoff, Ed., Elsevier, Amsterdam, 1989, pp. 785.
25. J. R. Pilbrow, *Transition Ion Electron Paramagnetic Resonance*, Clarendon, Oxford, 1990.
26. G. Feher, *Bell System Tech. J.*, 1957, **36**, 449.
27. C. P. Poole, *Electron Spin Resonance*, Wiley, New York, 1983.
28. V. A. Serezhenkov, E. V. Domracheva, G. A. Klevezal, S. M. Kulikov, S. A. Kuznetsov, P. I. Mordvintcev, L. I. Sukhovskaya, N. E. Schklovskykordi, A. F. Vanin, N. V. Voevodskaya and A. I. Vorobiev, *Radiat. Prot. Dosimet.*, 1992, **42**, 33.
29. V. E. Galtsev, O. Y. Grinberg, Y. S. Lebedev and E. V. Galtseva, *Appl. Magn. Reson.*, 1993, **4**, 331.
30. V. E. Galtsev, E. V. Galtseva and Y. S. Lebedev, *Appl. Radiat. Isot.*, 1996, **47**, 1311.
31. I. Salikhov, T. Walczak, P. Lesniewski, N. Khan, A. Iwasaki, R. Comi, J. Buckey and H. M. Swartz, *Magn. Reson. Med.*, 2005, **54**, 1317.
32. H. M. Swartz, A. Iwasaki, T. Walczak, E. Demidenko, I. Salikhov, N. Khan, P. Lesniewski, J. Thomas, A. Romanyukha, D. Schauer and P. Starewicz, *Radiat. Prot. Dosimet.*, 2006, **120**, 163.
33. H. M. Swartz, G. Burke, M. Coey, E. Demidenko, R. Dong, O. Grinberg, J. Hilton, A. Iwasaki, P. Lesniewski, M. Kmiec, K. M. Lo, R. J. Nicolalde, A. Ruuge, Y. Sakata, A. Sucheta, T. Walczak, B. B. Williams, C. A. Mitchell, A. Romanyukha and D. A. Schauer, *Radiat. Meas.*, 2007, **42**, 1075.
34. N. D. Yordanov and Y. Karakirova, *Radiat. Meas.*, 2007, **42**, 347.
35. N. D. Yordanov and K. Aleksieva, *Radiat. Phys. Chem.*, 2007, **76**, 1084.
36. H. P. Hu, G. Sosnovsky and H. M. Swartz, *Biochim. Biophys. Acta*, 1992, **1112**, 161.
37. H. M. Swartz and J. Dunn, *Adv. Exp. Med. Biol.*, 2005, **566**, 295.
38. J. Krzystek, J. Telser, B. M. Hoffman, L. C. Brunel and S. Licoccia, *J. Am. Chem. Soc.*, 2001, **123**, 7890.
39. C. Mantel, A. K. Hassan, J. Pecaut, A. Deronzier, M. N. Collomb and C. Duboc-Toia, *J. Am. Chem. Soc.*, 2003, **125**, 12337.
40. E. J. Reijerse, P. J. v. Dam, A. A. K. Klaassen, W. R. Hagen, P. J. M. v. Bentum and G. M. Smith, *Appl. Magn. Reson.*, 1998, **14**, 153.
41. O. Burghaus, M. Rohrer, T. Götzinger, M. Plato and K. Möbius, *Meas. Sci. Technol.*, 1992, **3**, 765.
42. A. W. Hornig and J. S. Hyde, *Mol. Phys.*, 1963, **6**, 33.
43. S. D. Chemerisov, O. Y. Grinberg, D. S. Tipikin, Y. S. Lebedev, H. Kurreck and K. Möbius, *Chem. Phys. Lett.*, 1994, **218**, 353.
44. D. Goldfarb and V. Krymov, In: *Very High Frequency (VHF) ESR/EPR Biological Magnetic Resonance*, Vol. **22**, O. Grinberg, L. J. Berliner, Eds., Kluwer/Plenum, New York, 2004, pp. 305.

45. A. L. Maniero, In: *Very High Frequency (VHF) ESR/EPR Biological Magnetic Resonance*, Vol. **22**, O. Grinberg, L. J. Berliner, Eds., Kluwer/ Plenum, New York, 2004, pp. 478.
46. K. Möbius, M. Plato and W. Lubitz, *Phys. Rep.*, 1982, **87**, 171.
47. K. P. Dinse, R. Biehl and K. Möbius, *J. Chem. Phys.*, 1974, **61**, 4335.
48. R. Biehl, M. Plato and K. Möbius, *J. Chem. Phys.*, 1975, **63**, 3515.
49. M. Mehring, P. Höfer and A. Grupp, *Ber. Bunsen Ges. -Phys. Chem. Chem. Phys.*, 1987, **91**, 1132.
50. B. Epel and D. Goldfarb, *J. Magn. Reson.*, 2000, **146**, 196.
51. O. Burghaus, A. Toth-Kischkat, R. Klette and K. Möbius, *J. Magn. Reson.*, 1988, **80**, 383.
52. A. Schnegg, A. A. Dubinskii, M. R. Fuchs, Y. A. Grishin, E. P. Kirilina, W. Lubitz, M. Plato, A. Savitsky and K. Möbius, *Appl. Magn. Reson.*, 2007, **31**, 59.
53. F. Gerson, *High Resolution ESR Spectroscopy*, Wiley, New York, 1970.
54. F. Gerson and W. Huber, *Electron Spin Resonance Spectroscopy of Organic Radicals*, Wiley-VCH, Weinheim, 2003.
55. H. M. McConnell, *J. Chem. Phys.*, 1956, **24**, 764.
56. J. P. Colpa and J. R. Bolton, *Mol. Phys.*, 1963, **6**, 273.
57. G. Giacometti, P. L. Nordio and M. V. Pavan, *Theor. Chim. Acta*, 1963, **1**, 404.
58. A. D. McLachlan, *Mol. Phys.*, 1960, **3**, 233.
59. M. Plato, E. Tränkle, W. Lubitz, F. Lendzian and K. Möbius, *Chem. Phys.*, 1986, **107**, 185.
60. M. Plato, W. Lubitz, F. Lendzian and K. Möbius, *Isr. J. Chem.*, 1988, **28**, 109.
61. M. Plato and K. Möbius, In: *Chlorophylls*, H. Scheer, Ed., CRC, Boca Raton, 1991, pp. 1015.
62. M. Karplus and G. K. Fraenkel, *J. Chem. Phys.*, 1961, **35**, 1312.
63. J. P. Colpa and E. de Boer, *Mol. Phys.*, 1964, **7**, 333.
64. D. Lazdins and M. Karplus, *J. Chem. Phys.*, 1966, **44**, 1600.
65. H. Kurreck, B. Kirste and W. Lubitz, *Electron Nuclear Double Resonance Spectroscopy of Radicals in Solution*, VCH Publishers, New York, 1988.
66. J. H. Freed, *J. Chem. Phys.*, 1965, **43**, 2312.
67. J. H. Freed, *J. Phys. Chem.*, 1967, **71**, 38.
68. J. H. Freed, D. S. Leniart and H. D. Connor, *J. Chem. Phys.*, 1973, **58**, 3089.
69. K. P. Dinse, R. Biehl, M. Plato and K. Möbius, *J. Magn. Reson.*, 1972, **6**, 444.
70. K. P. Dinse, K. Möbius and R. Biehl, *Z. Naturforsch. A*, 1973, **A 28**, 1069.
71. M. Plato, W. Lubitz and K. Möbius, *J. Phys. Chem.*, 1981, **85**, 1202.
72. W. Lubitz, In: *Electron Paramagnetic Resonance*, Vol. **19**, B. C. Gilbert, M. J. Davies, D. M. Murphy, Eds., Royal Society of Chemistry, Cambridge, 2004, pp. 174.
73. K. Möbius, *EPR Newsletter*, 2006, **16**, 18.

74. K. Möbius, W. Fröhling, F. Lendzian, W. Lubitz, M. Plato and C. J. Winscom, *J. Phys. Chem.*, 1982, **86**, 4491.
75. J. H. Freed, D. S. Leniart and J. S. Hyde, *J. Chem. Phys.*, 1967, **47**, 2762.
76. F. Bloch, *Phys. Rev.*, 1956, **102**, 104.
77. N. M. Atherton and B. Day, *Mol. Phys.*, 1974, **27**, 145.
78. K. Möbius, *Ber. Bunsen Ges. -Phys. Chem. Chem. Phys.*, 1974, **78**, 1116.
79. J. R. Norris, H. Scheer, M. E. Druyan and J. J. Katz, *Proc. Natl. Acad. Sci. USA*, 1974, **71**, 4897.
80. K. P. Dinse, PhD Thesis, Freie Universität Berlin, 1971.
81. J. H. Freed, In: *Mulitiple Electron Resonance Spectroscopy*, M. M. Dorio, J. H. Freed, Eds., Plenum, New York, 1979, pp. 73.
82. S. Geschwind, In: *Hyperfine Interactions*, A. J. Freeman, R. B. Frankel, Eds., Academic Press, New York, 1967.
83. K. Möbius and R. Biehl, In: *Mulitiple Electron Resonance Spectroscopy*, M. M. Dorio, J. H. Freed, Eds., Plenum, New York, 1979, pp. 475.
84. K. Möbius, W. Lubitz and M. Plato, In: *Advanced EPR, Applications in Biology and Biochemistry*, A. J. Hoff, Ed., Elsevier, Amsterdam, 1989, pp. 441.
85. K. Möbius, In: *Foundations of Modern EPR*, G. R. Eaton, S. S. Eaton, K. M. Salikhov, Eds., World Scientific, Singapore, 1998, pp. 557.
86. C. W. M. Kay, Y. A. Grishin, S. Weber and K. Möbius, *Appl. Magn. Reson.*, 2007, **31**, 599.
87. W. Lubitz, M. Plato, K. Möbius and R. Biehl, *J. Phys. Chem.*, 1979, **83**, 3402.
88. K. Möbius, *Mol. Cryst. Liq. Cryst.*, 2003, **394**, 1.
89. K. P. Dinse, M. Plato, R. Biehl, K. Möbius and H. Haustein, *Chem. Phys. Lett.*, 1972, **14**, 196.
90. R. Biehl, W. Lubitz, K. Möbius and M. Plato, *J. Chem. Phys.*, 1977, **66**, 2074.
91. J. Schlüpmann, M. Huber, M. Toporowicz, M. Plato, M. Köcher, E. Vogel, H. Levanon and K. Möbius, *J. Am. Chem. Soc.*, 1990, **112**, 6463.
92. C. W. M. Kay, F. Conti, M. Fuhs, M. Plato, S. Weber, E. Bordignon, D. Carbonera, B. C. Robinson, M. W. Renner and J. Fajer, *J. Phys. Chem. B*, 2002, **106**, 2769.
93. W. Lubitz, In: *Chlorophylls*, H. Scheer, Ed., CRC, Boca Raton, 1991, pp. 903.
94. G. Feher, *J. Chem. Soc., Perkin Trans.*, 1992, **2**, 1861.
95. H. Levanon and K. Möbius, *Annu. Rev. Biophys. Biomol. Struct.*, 1997, **26**, 495.
96. S. Weber, In: *Electron Paramagnetic Resonance*, Vol. **17**, N. M. Atherton, M. J. Davies, B. C. Gilbert, Eds., Royal Society of Chemistry, Cambridge, 2000, pp. 43.
97. F. Gerson and G. Gescheidt, In: *Biomedical EPR–Part B, Methodology, Instrumentation and Dynamics*, S. S. Eaton, G. R. Eaton, L. J. Berliner, Eds., Springer, New York, 2005, pp. 145.

98. F. Lendzian, W. Lubitz, H. Scheer, C. Bubenzer and K. Möbius, *J. Am. Chem. Soc.*, 1981, **103**, 4635.
99. F. Lendzian, W. Lubitz, R. Steiner, E. Tränkle, M. Plato, H. Scheer and K. Möbius, *Chem. Phys. Lett.*, 1986, **126**, 290.
100. F. Lendzian, W. Lubitz, H. Scheer, A. J. Hoff, M. Plato, E. Tränkle and K. Möbius, *Chem. Phys. Lett.*, 1988, **148**, 377.
101. F. Lendzian, M. Huber, R. A. Isaacson, B. Endeward, M. Plato, B. Bönigk, K. Möbius, W. Lubitz and G. Feher, *Biochim. Biophys. Acta*, 1993, **1183**, 139.
102. H. Murai, S. Tero-Kubota and S. Yamauchi, In: *Electron Paramagnetic Resonance*, Vol. **17**, B. C. Gilbert, M. J. Davies, K. A. McLauchlan, Eds., Royal Society of Chemistry, Cambridge, 2000, pp. 130.
103. A. Savitsky and K. Möbius, *Helv. Chim. Acta*, 2006, **89**, 2544.
104. S. Weber, *Biochim. Biophys. Acta*, 2005, **1707**, 1.
105. J. H. van der Waals and M. S. de Groot, In: *The Triplet State*, A. B. Zahlan, Ed., Cambridge University Press, Cambridge, 1967, pp. 101.
106. S. Yamauchi and D. W. Pratt, *Mol. Phys.*, 1979, **37**, 541.
107. S. K. Wong, D. A. Hutchinson and J. K. S. Wan, *J. Chem. Phys.*, 1973, **58**, 985.
108. F. J. Adrian, *J. Chem. Phys.*, 1974, **61**, 4875.
109. P. W. Atkins and G. T. Evans, *Chem. Phys. Lett.*, 1974, **25**, 108.
110. P. W. Atkins and G. T. Evans, *Mol. Phys.*, 1974, **27**, 1633.
111. J. B. Pedersen and J. H. Freed, *J. Chem. Phys.*, 1975, **62**, 1706.
112. P. W. Atkins and K. A. McLauchlan, In: *Chemically Induced Magnetic Polarization*, A. R. Lepley, G. L. Closs, Eds., Wiley, New York, 1973, pp. 41.
113. F. J. Adrian, *Rev. Chem. Intermed.*, 1986, **7**, 173.
114. L. Monchick and F. J. Adrian, *J. Chem. Phys.*, 1978, **68**, 4376.
115. K. M. Salikhov, Y. N. Molin, R. Z. Sagdeev and A. L. Buchachenko, *Spin Polarization and Magnetic Effects in Radical Reactions*, Elsevier, Amsterdam, 1984.
116. J. H. Freed and J. B. Pedersen, In: *Advances in Magnetic Resonance*, Vol. **8**, J. S. Waugh, Ed., Academic Press, New York, 1976, pp. 1.
117. F. J. Adrian, *Res. Chem. Intermed.*, 1991, **16**, 99.
118. A. I. Shushin, J. B. Pedersen and L. I. Lolle, *Chem. Phys.*, 1993, **177**, 119.
119. F. J. Adrian, *Chem. Phys. Lett.*, 1997, **272**, 120.
120. F. J. Adrian, *J. Chem. Phys.*, 1971, **54**, 3918.
121. J. B. Pedersen and J. H. Freed, *J. Chem. Phys.*, 1973, **58**, 2746.
122. J. B. Pedersen and J. H. Freed, *J. Chem. Phys.*, 1973, **59**, 2869.
123. A. I. Shushin, *Chem. Phys.*, 1990, **144**, 201.
124. A. I. Shushin, *Chem. Phys.*, 1990, **144**, 223.
125. F. J. Adrian, *Chem. Phys. Lett.*, 1981, **80**, 106.
126. T. J. Burkey, J. Lusztyk, K. U. Ingold, J. K. S. Wan and F. J. Adrian, *J. Phys. Chem.*, 1985, **89**, 4286.
127. A. D. Trifunac, *Chem. Phys. Lett.*, 1977, **49**, 457.
128. A. J. Hoff and J. Deisenhofer, *Phys. Rep.*, 1997, **287**, 2.

129. M. Fuhs, G. Elger, A. Osintsev, A. Popov, H. Kurreck and K. Möbius, *Mol. Phys.*, 2000, **98**, 1025.
130. P. J. Hore, In: *Advanced EPR, Applications in Biology and Biochemistry*, A. J. Hoff, Ed., Elsevier, Amsterdam, 1989, pp. 405.
131. A. J. Hoff, In: *The Photosynthetic Reaction Center*, Vol. 2, J. Deisenhofer, J. R. Norris, Eds., Academic, San Diego, 1993, pp. 331.
132. D. Stehlik, C. H. Bock and J. Petersen, *J. Phys. Chem.*, 1989, **93**, 1612.
133. R. Z. Sagdeev, W. Möhl and K. Möbius, *J. Phys. Chem.*, 1983, **87**, 3183.
134. F. Lendzian, P. Jaegermann and K. Möbius, *Chem. Phys. Lett.*, 1985, **120**, 195.
135. K. Möbius, *J. Chem. Soc., Faraday Trans. I*, 1987, **83**, 3469.
136. P. J. Hore, C. G. Joslin and K. A. McLauchlan, In: *Electron Spin Resonance*, Vol. **5**, P. B. Ayscough, Ed., The Chemical Society, London, 1979, pp. 1.
137. L. T. Muus, S. Frydkjaer and K. B. Nielsen, *Chem. Phys.*, 1978, **30**, 163.
138. H. Paul and H. Fischer, *Helv. Chim. Acta*, 1973, **56**, 1575.
139. H. Langhals and H. Fischer, *Chem. Ber. Recl.*, 1978, **111**, 543.
140. A. I. Grant and K. A. McLauchlan, *Chem. Phys. Lett.*, 1983, **101**, 120.
141. R. Baer and H. Paul, *Chem. Phys.*, 1984, **87**, 73.
142. A. N. Savitsky, M. Galander and K. Möbius, *Chem. Phys. Lett.*, 2001, **340**, 458.
143. F. Jent, H. Paul and H. Fischer, *Chem. Phys. Lett.*, 1988, **146**, 315.
144. M. D. E. Forbes, *J. Phys. Chem.*, 1992, **96**, 7836.
145. U. Kolczak, G. Rist, K. Dietliker and J. Wirz, *J. Am. Chem. Soc.*, 1996, **118**, 6477.
146. G. W. Sluggett, P. F. McGarry, I. V. Koptyug and N. J. Turro, *J. Am. Chem. Soc.*, 1996, **118**, 7367.
147. M. Kamachi, K. Kuwata, T. Sumiyoshi and W. Schnabel, *J. Chem. Soc., Perkin Trans.*, 1988, **2**, 961.
148. A. I. Shushin, *Chem. Phys. Lett.*, 1990, **170**, 78.
149. T. N. Makarov, A. N. Savitsky, K. Möbius, D. Beckert and H. Paul, *J. Phys. Chem. A*, 2005, **109**, 2254.
150. S. Jockusch and N. J. Turro, *J. Am. Chem. Soc.*, 1998, **120**, 11773.
151. D. Hristova, I. Gatlik, G. Rist, K. Dietliker, J. P. Wolf, J. L. Birbaum, A. Savitsky, K. Möbius and G. Gescheidt, *Macromol.*, 2005, **38**, 7714.
152. S. Jockusch, I. V. Koptyug, P. F. McGarry, G. W. Sluggett, N. J. Turro and D. M. Watkins, *J. Am. Chem. Soc.*, 1997, **119**, 11495.
153. A. Schweiger, *Angew. Chem. Int. Ed.*, 1991, **30**, 265.
154. H. Thomann and W. B. Mims, In: *Pulsed Magnetic Resonance: NMR, ESR, Optical*, D. M. S. Bagguley, D. M. Slingsby, Eds., Oxford University Press, London, 1992.
155. H. Thomann and M. Bernardo, In: *EMR of Paramagnetic Molecules; Biological Magnetic Resonance*, Vol. **13**, L. J. Berliner, J. Reuben, Eds., Plenum, New York, 1993, pp. 275.
156. W. B. Mims, *Proc. R. Soc. London, Ser. A*, 1965, **283**, 452.
157. E. R. Davies, *Phys. Lett. A*, 1974, **A 47**, 1.
158. P. E. Doan and B. M. Hoffman, *Chem. Phys. Lett.*, 1997, **269**, 208.

159. B. Epel, D. Arieli, D. Baute and D. Goldfarb, *J. Magn. Reson.*, 2003, **164**, 78.
160. M. Rohrer, F. MacMillan, T. F. Prisner, A. T. Gardiner, K. Möbius and W. Lubitz, *J. Phys. Chem. B*, 1998, **102**, 4648.
161. R. Klette, J. T. Törring, M. Plato, K. Möbius, B. Bönigk and W. Lubitz, *J. Phys. Chem.*, 1993, **97**, 2051.
162. J. W. A. Coremans, O. G. Poluektov, E. J. J. Groenen, G. W. Canters, H. Nar and A. Messerschmidt, *J. Am. Chem. Soc.*, 1996, **118**, 12141.
163. R. Carmieli, P. Manikandan, A. J. Kalb and D. Goldfarb, *J. Am. Chem. Soc.*, 2001, **123**, 8378.
164. B. M. Hoffman, V. J. DeRose, P. E. Doan, R. J. Gurbiel, A. L. P. Houseman and J. Telser, In: *EMR of Paramagnetic Molecules; Biological Magnetic Resonance*, Vol. **13**, L. J. Berliner, J. Reuben, Eds., Plenum, New York, 1993, pp. 151.
165. L. Mayas, M. Plato, C. J. Winscom and K. Möbius, *Mol. Phys.*, 1978, **36**, 753.
166. M. Flores, R. A. Isaacson, R. Calvo, G. Feher and W. Lubitz, *Chem. Phys.*, 2004, **294**, 401.
167. G. Soda and T. Chiba, *J. Phys. Soc. Jpn.*, 1969, **26**, 249.
168. S. Sinnecker, E. Reijerse, F. Neese and W. Lubitz, *J. Am. Chem. Soc.*, 2004, **126**, 3280.
169. D. Baute, D. Arieli, F. Neese, H. Zimmermann, B. M. Weckhuysen and D. Goldfarb, *J. Am. Chem. Soc.*, 2004, **126**, 11733.
170. S. Kababya, J. Nelson, C. Calle, F. Neese and D. Goldfarb, *J. Am. Chem. Soc.*, 2006, **128**, 2017.
171. D. Goldfarb, K. V. Narasimhulu and R. Carmieli, *Magn. Reson. Chem.*, 2005, **43**, S40.
172. R. Carmieli, P. Manikandan, B. Epel, A. J. Kalb, A. Schnegg, A. Savitsky, K. Möbius and D. Goldfarb, *Biochemistry*, 2003, **42**, 7863.
173. A. Deacon, T. Gleichmann, A. J. Kalb, H. Price, J. Raftery, G. Bradbrook, J. Yariv and J. R. Helliwell, *J. Chem. Soc., Faraday Trans.*, 1997, **93**, 4305.
174. R. J. Kalb, J. Habash, N. S. Hunter, H. J. Price, J. Raftery, J. R. Helliwell, In: *Manganese and its Role in Biological Processes; Metal Ions in Biological Systems*, A. Sigel, H. Sigel, Eds., Dekker, New York, 2000, pp. 279.
175. P. Manikandan, R. Carmieli, T. Shane, A. J. Kalb and D. Goldfarb, *J. Am. Chem. Soc.*, 2000, **122**, 3488.
176. A. Schnegg, C. W. M. Kay, E. Schleicher, K. Hitomi, T. Todo, K. Möbius and S. Weber, *Mol. Phys.*, 2006, **104**, 1627.
177. M. R. Fuchs, E. Schleicher, A. Schnegg, C. W. M. Kay, J. T. Törring, R. Bittl, A. Bacher, G. Richter, K. Möbius and S. Weber, *J. Phys. Chem. B*, 2002, **106**, 8885.
178. S. Weber and R. Bittl, *Bull. Chem. Soc. Jpn.*, 2007, **80**, 2270.
179. W. B. Mims, *Phys.Rev. B*, 1972, **5**, 2409.
180. A. Savitsky, A. A. Dubinskii, M. Plato, Y. A. Grishin, H. Zimmermann and K. Möbius, *J. Phys. Chem. B*, 2008, **112**, 9079.

181. W. L. Hubbell and C. Altenbach, *Curr. Opin. Struct. Biol.*, 1994, **4**, 566.
182. J. B. Feix and C. S. Klug, In: *Electron Paramagnetic Resonance*, Vol. **20**, B. C. Gilbert, M. J. Davies, D. M. Murphy, Eds., Royal Society of Chemistry, Cambridge, 2006, pp. 50.
183. H.-J. Steinhoff, *Front. Biosci.*, 2002, **7**, 97.
184. A. I. Smirnov, In: *Electron Paramagnetic Resonance*, Vol. **18**, B. C. Gilbert, M. J. Davies, D. M. Murphy, Eds., Royal Society of Chemistry, Cambridge, 2002, pp. 109.
185. K. Möbius, A. Savitsky, C. Wegener, M. Plato, M. Fuchs, A. Schnegg, A. A. Dubinskii, Y. A. Grishin, I. A. Grigor'ev, M. Kuhn, D. Duche, H. Zimmermann and H. J. Steinhoff, *Magn. Reson. Chem.*, 2005, **43**, S4.
186. M. Plato, H. J. Steinhoff, C. Wegener, J. T. Törring, A. Savitsky and K. Möbius, *Mol. Phys.*, 2002, **100**, 3711.
187. R. Owenius, M. Engström, M. Lindgren and M. Huber, *J. Phys. Chem. A*, 2001, **105**, 10967.
188. M. Engström, R. Owenius and O. Vahtras, *Chem. Phys. Lett.*, 2001, **338**, 407.
189. Z. Ding, A. F. Gulla and D. E. Budil, *J. Chem. Phys.*, 2001, **115**, 10685.
190. M. Pavone, A. Sillanpaa, P. Cimino, O. Crescenzi and V. Barone, *J. Phys. Chem. B*, 2006, **110**, 16189.
191. H. J. Steinhoff, A. Savitsky, C. Wegener, M. Pfeiffer, M. Plato and K. Möbius, *Biochim. Biophys. Acta*, 2000, **1457**, 253.
192. T. I. Smirnova, T. G. Chadwick, M. A. Voinov, O. Poluektov, J. van Tol, A. Ozarowski, G. Schaaf, M. M. Ryan and V. A. Bankaitis, *Biophys. J.*, 2007, **92**, 3686.
193. J. Seliger and V. Zagar, *Chem. Phys.*, 2004, **306**, 309.
194. J. N. Latosinska, M. Latosinska and J. Koput, *J. Mol. Struct.*, 2003, **648**, 9.
195. A. Lai, H. L. Flanagan and D. J. Singel, *J. Chem. Phys.*, 1988, **89**, 7161.
196. H. L. Flanagan and D. J. Singel, *J. Chem. Phys.*, 1987, **87**, 5606.
197. H. L. Flanagan, G. J. Gerfen, A. Lai and D. J. Singel, *J. Chem. Phys.*, 1988, **88**, 2162.
198. A. Bloess, K. Möbius and T. F. Prisner, *J. Magn. Reson.*, 1998, **134**, 30.
199. L. Kevan, In: *Time Domain Electron Spin Resonance*, L. Kevan, R. N. Schwartz, Eds., Wiley, New York, 1979, pp. 279.
200. J. W. A. Coremans, O. G. Poluektov, E. J. J. Groenen, G. W. Canters, H. Nar and A. Messerschmidt, *J. Am. Chem. Soc.*, 1997, **119**, 4726.
201. W. B. Mims, J. Peisach and J. L. Davis, *J. Chem. Phys.*, 1977, **66**, 5536.
202. S. P. Van, G. B. Birrell and O. H. Griffith, *J. Magn. Reson.*, 1974, **15**, 444.
203. S. V. Paschenko, Y. V. Toropov, S. A. Dzuba, Y. D. Tsvetkov and A. K. Vorobiev, *J. Chem. Phys.*, 1999, **110**, 8150.
204. A. T. Bullock and C. B. Howard, *J. Chem. Soc., Faraday Trans. I*, 1980, **76**, 1296.
205. M. F. Ottaviani, G. Martini and L. Nuti, *Magn. Reson. Chem.*, 1987, **25**, 897.

206. I. Komaromi and J. M. J. Tronchet, *J. Phys. Chem.*, 1995, **99**, 10213.
207. M. Pavone, P. Cimino, O. Crescenzi, A. Sillanpaa and V. Barone, *J. Phys. Chem. B*, 2007, **111**, 8928.
208. T. I. Smirnova, A. I. Smirnov, S. V. Paschenko and O. G. Poluektov, *J. Am. Chem. Soc.*, 2007, **129**, 3476.
209. G. Jeschke, A. Bender, H. Paulsen, H. Zimmermann and A. Godt, *J. Magn. Reson.*, 2004, **169**, 1.
210. C. P. Slichter, *Principles of Magnetic Resonance*, Harper & Row, New York, 1963.
211. P. P. Borbat and J. H. Freed, *EPR Newsletter*, 2007, **17**, 21.
212. R. N. Schwartz, L. L. Jones and M. K. Bowman, *J. Phys. Chem.*, 1979, **83**, 3429.
213. K. M. Salikhov, S. A. Dzuba and A. M. Raitsimring, *J. Magn. Reson.*, 1981, **42**, 255.
214. A. D. Milov, K. M. Salikohov and M. D. Shirov, *Fiz. Tverd. Tela*, 1981, **23**, 975.
215. A. D. Milov, A. B. Ponomarev and Y. D. Tsvetkov, *J. Struct. Chem.*, 1984, **25**, 710.
216. A. D. Milov, A. B. Ponomarev and Y. D. Tsvetkov, *Chem. Phys. Lett.*, 1984, **110**, 67.
217. A. M. Raitsimring and K. M. Salikhov, *Bull. Magn. Reson.*, 1985, **7**, 184.
218. A. D. Milov, A. G. Maryasov and Y. D. Tsvetkov, *Appl. Magn. Reson.*, 1998, **15**, 107.
219. R. G. Larsen and D. J. Singel, *J. Chem. Phys.*, 1993, **98**, 5134.
220. M. Pannier, S. Veit, A. Godt, G. Jeschke and H. W. Spiess, *J. Magn. Reson.*, 2000, **142**, 331.
221. P. P. Borbat and J. H. Freed, *Chem. Phys. Lett.*, 1999, **313**, 145.
222. P. P. Borbat and J. H. Freed, In: *Distance Measurements in Biological Systems by EPR; Biological Magnetic Resonance*, Vol. **19**, L. J. Berliner, S. S. Eaton, G. R. Eaton, Eds., Kluwer, New York, 2000, pp. 383.
223. *Distance Measurements in Biological Systems by EPR; Biological Magnetic Resonance*, Vol. **19**, L. J. Berliner, S. S. Eaton, G. R. Eaton, Eds. Kluwer, New York, 2000.
224. K. V. Lakshmi and G. W. Brudvig, *Curr. Opin. Struct. Biol.*, 2001, **11**, 523.
225. G. Jeschke, *Chem. Phys. Chem.*, 2002, **3**, 927.
226. H. J. Steinhoff, *Biol. Chem.*, 2004, **385**, 913.
227. S. S. Eaton and G. R. Eaton, In: *Electron Paramagnetic Resonance*, Vol. **19**, B. C. Gilbert, M. J. Davies, D. M. Murphy, Eds., Royal Society of Chemistry, Cambridge, 2004, pp. 318.
228. G. Jeschke, *EPR Newsletter*, 2005, **14**, 14.
229. G. Jeschke and H. W. Spiess, In: *Novel NMR and EPR Techniques; Lecture Notes in Physics*, Vol. **684**, J. Dolinsek, M. Vilfan, S. Zumer, Eds., Springer, Berlin, 2006, pp. 21.
230. O. Schiemann and T. F. Prisner, *Q. Rev. Biophys.*, 2007, **40**, 1.

231. A. V. Astashkin and A. Kawamori, In: *Biophysical Techniques in Photosynthesis*, Vol. II, T. J. Aartsma, J. Matysik, Eds., Springer, Dordrecht, 2008, pp. 325.
232. A. Savitsky, A. A. Dubinskii, M. Flores, W. Lubitz and K. Möbius, *J. Phys. Chem. B*, 2007, **111**, 6245.
233. L. V. Kulik, S. A. Dzuba, I. A. Grigoryev and Y. D. Tsvetkov, *Chem. Phys. Lett.*, 2001, **343**, 315.
234. L. V. Kulik, S. V. Paschenko and S. A. Dzuba, *J. Magn. Reson.*, 2002, **159**, 237.
235. V. P. Denysenkov, D. Biglino, W. Lubitz, T. F. Prisner and M. Bennati, *Angew. Chem. Int. Ed.*, 2008, **47**, 1224.
236. V. P. Denysenkov, T. F. Prisner, J. Stubbe and M. Bennati, *Appl. Magn. Reson.*, 2005, **29**, 375.
237. V. P. Denysenkov, T. F. Prisner, J. Stubbe and M. Bennati, *Proc. Natl. Acad. Sci. USA*, 2006, **103**, 13386.
238. Y. Polyhach, A. Godt, C. Bauer and G. Jeschke, *J. Magn. Reson.*, 2007, **185**, 118.
239. A. A. Dubinskii, Y. A. Grishin, A. N. Savitsky and K. Möbius, *Appl. Magn. Reson.*, 2002, **22**, 369.
240. D. E. Kaplan and E. L. Hahn, *J. Phys. Rad.*, 1958, **19**, 821.
241. L. V. Kulik, Y. A. Grishin, S. A. Dzuba, I. A. Grigoryev, S. V. Klyatskaya, S. F. Vasilevsky and Y. D. Tsvetkov, *J. Magn. Reson.*, 2002, **157**, 61.
242. W. B. Mims, In: *Electron Paramagnetic Resonance*, S. Geschwind, Ed., Plenum, New York, 1972, pp. 263.
243. G. Jeschke, M. Pannier, A. Godt and H. W. Spiess, *Chem. Phys. Lett.*, 2000, **331**, 243.
244. G. Jeschke, A. Koch, U. Jonas and A. Godt, *J. Magn. Reson.*, 2002, **155**, 72.
245. A. Weber, O. Schiemann, B. Bode and T. F. Prisner, *J. Magn. Reson.*, 2002, **157**, 277.
246. G. Jeschke, H. Zimmermann and A. Godt, *J. Magn. Reson.*, 2006, **180**, 137.
247. A. Godt, C. Franzen, S. Veit, V. Enkelmann, M. Pannier and G. Jeschke, *J. Org. Chem.*, 2000, **65**, 7575.
248. L. V. Kulik and S. A. Dzuba, *J. Struct. Chem.*, 2004, **45**, 298.
249. A. D. Milov, Y. D. Tsvetkov, F. Formaggio, S. Oancea, C. Toniolo and J. Raap, *J. Phys. Chem. B*, 2003, **107**, 13719.
250. A. G. Maryasov, Y. D. Tsvetkov and J. Raap, *Appl. Magn. Reson.*, 1998, **14**, 101.
251. I. V. Borovykh, S. Ceola, P. Gajula, P. Gast, H. J. Steinhoff and M. Huber, *J. Magn. Reson.*, 2006, **180**, 178.
252. P. Gajula, S. Milikisyants, H. J. Steinhoff and M. Huber, *Appl. Magn. Reson.*, 2007, **31**, 99.
253. P. P. Borbat, A. J. Costa-Filho, K. A. Earle, J. K. Moscicki and J. H. Freed, *Science*, 2001, **291**, 266.

254. C. F. W. Becker, K. Lausecker, M. Balog, T. Kalai, K. Hideg, H. J. Steinhoff and M. Engelhard, *Magn. Reson. Chem.*, 2005, **43**, S34.
255. C. Toniolo, M. Crisma and F. Formaggio, *Biopolymers*, 1998, **47**, 153.
256. V. Monaco, F. Formaggio, M. Crisma, C. Toniolo, P. Hanson and G. L. Millhauser, *Biopolymers*, 1999, **50**, 239.
257. V. Monaco, F. Formaggio, M. Crisma, C. Toniolo, P. Hanson, G. Millhauser, C. George, J. R. Deschamps and J. L. Flippen-Anderson, *Bioorg. Med. Chem.*, 1999, **7**, 119.
258. A. D. Milov, R. I. Samoilova, Y. D. Tsvetkov, F. Formaggio, C. Toniolo and J. Raap, *J. Am. Chem. Soc.*, 2007, **129**, 9260.
259. M. Balog, T. Kalai, J. Jekö, Z. Berente, H. J. Steinhoff, M. Engelhard and K. Hideg, *Tetrahedron Lett.*, 2003, **44**, 9213.
260. M. E. Hahn and T. W. Muir, *Trends Biochem. Sci.*, 2005, **30**, 26.

CHAPTER 3
Instrumentation

EPR spectroscopy, as any other spectroscopic method, has to meet three main requirements *(i)* detection; *(ii)* identification and *(iii)* characterization of the sample. That means, first of all, the method should be sensitive enough to detect the signals of interest from the system under investigation. In the case of EPR it is the paramagnetic absorption due to unpaired electron spins in radicals, triplet states and transition-metal ions. The second step is the assignment of the observed signals to one or more species in the sample or, *vice versa*, identification of the species from the observed signals. And finally, the characterization of the species and/or their environment using the specific spin-interaction parameters extracted from the observed EPR signals. Historically, EPR spectroscopy had to go a long way to fulfil the above requirements. Sensitive detection methods were developed but needed decades to reach the present standard. The systematic investigations of a large number of the paramagnetic species provided the knowledge about their spectroscopic properties and allowed to understand the molecular interactions suitable for characterizing the spin system. Finally, sophisticated data-analysis techniques were developed and approved that allow EPR spectroscopists to extract the interaction of interest from the complex spectra. In the meantime, EPR spectroscopy is approaching the level of technical development necessary to become a powerful analytical method, ready to be applied to characterize the investigated systems to hitherto unprecedented details. In the following, we briefly describe the experimental techniques used in EPR of the order of their historical development, focusing on their applications and technical aspects at high magnetic fields and microwave frequencies.

High-Field EPR Spectroscopy on Proteins and their Model Systems:
Characterization of Transient Paramagnetic States
By Klaus Möbius and Anton Savitsky
© 2009 Klaus Möbius and Anton Savitsky
Published by the Royal Society of Chemistry, www.rsc.org

3.1 Experimental Techniques

3.1.1 Continuous-Wave EPR (cw EPR)

The first EPR and NMR experiments were performed using the continuous-wave (cw) technique. Nowadays, in NMR the cw technique is completely replaced by pulsed methods. In contrast, cw EPR continues to be a powerful method applicable to a broad range of systems. Conventional cw EPR employs continuous microwave (mw) sources to irradiate the sample in the mw resonator, which is placed in an external magnetic field B_0. The external magnetic field is sine-form modulated with an amplitude below the width of the EPR line. The spectrum is recorded while sweeping the B_0 value through the resonance region with corresponding mw absorption. The change of the mw power is lock-in detected at the frequency of the field modulation. Cw EPR detects the signals from the paramagnetic species in steady-state concentrations.

The overall sensitivity of a cw EPR spectrometer is determined by the minimum detectable number of spins. For two-level systems obeying the Curie law it is given by[1,2]

$$N_{min} = \text{Const} \cdot \frac{V_S \cdot T_S}{Q_U \cdot \eta} \cdot \frac{\Delta\omega_{pp}}{\omega_0} \cdot \frac{\Delta B_{pp}}{B_{mod}} \cdot \left(\frac{\Delta f \cdot kT_{system}}{P_w}\right)^{1/2} \quad (3.1)$$

where V_S and T_S are the volume and temperature of the sample; Q_U is the unloaded quality factor of the cavity; η is the filling factor, ω_0 is the spectrometer frequency; $\Delta\omega_{pp}$ is the peak-to-peak homogeneous linewidth of EPR line in angular frequency units; ΔB_{pp} is the overall (inhomogeneous and homogeneous) EPR linewidth and $B_{mod} < \Delta B_{pp}$ is the modulation amplitude; Δf is the effective detection bandwidth, P_w the mw power incident on the cavity, and T_{system} is the effective noise temperature of the spectrometer system.

The factor $P_{system} = \Delta f \cdot kT_{system}$ gives the full effective system noise power. It consists of two contributions: $P_{system} = P_{detector} + P_{transmitter}$. The first contribution, $P_{detector}$, determines the "noise floor" of the EPR spectrometer. The noise floor is independent of the incident mw irradiation. It is generally measured by the noise figure and the amplification gain of the detection network. The noise figure describes the loss of the useful signal intensity on the way from the EPR cavity to the end detector, *e.g.*, due to losses in the waveguides and mixer elements, accompanied by the gain of noise intensity in the amplifier elements. The typical noise figure of modern EPR spectrometers is about 5–10 dB. It can hardly be improved without resorting to mw elements operating at cryogenic temperatures. The second contribution, $P_{transmitter}$, is crucial for the success of the EPR experiment, especially at high mw frequencies. The quality of the mw source is characterized in terms of amplitude noise and phase noise. The phase noise describes the width of the mw radiation in the frequency domain, *i.e.* the mw power at frequency offsets ($\omega - \omega_0$) from the carrier frequency (ω_0). The amplitude noise is the fluctuation of this mw power. The noise power, if reflected directly from the EPR cavity or phase-to-amplitude converted at the cavity resonance curve, reaches the detector and increases the noise temperature. In contrast to the noise floor, the noise contribution due to the mw source noise depends on the incident microwave power. Under certain conditions, the source noise can overcome the detector noise and limits the sensitivity of the cw EPR detection. A detailed analysis of the source noise contributions can be found in refs. [3,4].

3.1.2 Time-Resolved EPR (TREPR)

The observation of the EPR spectra of transient paramagnetic species, radicals or triplets generated by laser flashes, needed the development of EPR spectrometers with fast detection capability. Historically, two techniques were established: the first uses cw microwave irradiation and different sampling methods; the second uses pulsed microwave irradiation. Both methods have advantages and disadvantages, which may also depend on the specific ratio mw frequency/magnetic field of the spectrometer. Thus, for any investigated system one has to choose the optimal method, depending on the time resolution and sensitivity required.

Conventional EPR spectroscopy is using cw microwaves and field modulation up to 100 kHz with phase-sensitive detection. The response time of the system is typically tens of microseconds. For certain chemical and biological systems, this time resolution already allows study of radical reactivity. However, it is difficult to detect and follow spin-polarized signals under field-modulation conditions. This is mostly because of short electron-spin relaxation times that lead to Boltzmann populations of the spin states or, at least, to their partial equilibration, before the radicals can be detected. In the mid-1970s, several attempts were made to improve the time resolution of cw EPR to about 1 μs by using field modulation frequencies up to 2 MHz.[5,6] Nevertheless, because of still insufficient time resolution and experimental difficulties to cope with the small penetration depth of the modulation field and the development of low-noise mw amplifiers, the cw EPR method was superseded by the so-called "direct-detection" time-resolved EPR method, known as TREPR. In TREPR, the resonance responses of the transient radicals are detected during continuous mw excitation, but without magnetic field modulation. The transient EPR signal following the sudden radical generation, for example by a short laser pulse, is taken at a fixed external magnetic-field value directly from the mw bridge mixer or diode detector and collected using either a transient digitizer or a boxcar signal averager. The complete EPR spectrum is obtained by stepping the external magnetic field through the resonance region.

The time resolution of a TREPR spectrometer is limited by the system response time. It determines how fast the spectrometer can follow changes of the EPR absorption of the sample in the mw cavity. Any change of the perpendicular magnetization $v(t)$ will result in a change of the spectrometer output signal $S(t)$. This is given by the convolution integral of $v(t)$ with the response function $f(t)$ of the instrument: $S(t) = f(t) \otimes v(t)$. The response function of the detection

Instrumentation

instrument is given by the response characteristics of the microwave and amplification circuitry. In the absence of any additional restriction of this circuitry, the response time of the EPR instrument is determined by the bandwidth of the EPR cavity. The reflection coefficient of a critically coupled cavity (here we discuss only the common reflection-mode spectrometer) is given by

$$|\Gamma|^2 = \frac{\varepsilon^2}{4 + \varepsilon^2}, \quad \text{with} \quad \varepsilon = 4 \cdot Q_L \cdot \frac{\omega - \omega_0}{\omega_0} \quad (3.2)$$

where Q_L is the loaded quality factor of the cavity and ω_0 is the spectrometer frequency.[7] Thus, the EPR absorption signal, appearing at frequency $(\omega - \omega_0)$, will be filtered by the cavity, as given by $\sqrt{1 - |\Gamma|^2}$ in the frequency domain. After Fourier transformation of the filtering function, one obtains the response function of the cavity in the time domain

$$f = \exp[-t/T_R] \quad \text{with} \quad T_R = 2 \cdot Q_L \cdot \omega_0^{-1} \quad (3.3)$$

where T_R is the ringing time of the EPR cavity. Actually, this ringing time gives the physical limit of the EPR instrument response time. It should be mentioned that the rise time of EPR signals is not given only by the response time defined above, but also by a physical limitation for cw TREPR detection: The concentration of photogenerated transient radicals typically reaches its maximum at the end of the laser flash, when the ensemble magnetization is not yet tipped away from the z-axis by the cw microwave field.

As follows from eqn (3.3), the ringing time T_R of the EPR cavity is inversely proportional to the mw frequency of the EPR spectrometer. Fixing a typical loaded quality value $Q_L = 2000$ of a single-mode mw cavity, $T_R = 67$ ns, 6.7 ns, 2.3 ns is calculated for X-band (9.5 GHz), W-band (95 GHz) and 275 GHz EPR, respectively. Thus, the typical time resolution of X-band TREPR spectrometers is about 50–100 ns, which is much longer than the laser pulse duration of typically 5 ns. At W-band, however, the response time is below 10 ns and becomes adequate for the pulsed laser experiments. The time resolution of X-band TREPR can be improved by using cavities with low quality factor. However, normally this is not the method of choice. Decreasing the cavity quality factor leads directly to a decrease of the EPR signal[2] and, thus, to lower sensitivity of the EPR experiment.

The sensitivity of a TREPR experiment in the direct-detection mode is proportional to the ratio of the signal amplitude and the noise amplitude. The signal amplitude depends on the radical system under study, *i.e.* on the amount of mw absorption due to the magnetization of transient paramagnetic species created during or after their generation. The magnetization can be increased by optimization of the experimental conditions, for example, laser light intensity and sample size. The noise amplitude of TREPR, similar to cw experiments, is determined by three contributions: *(i)* the noise of the mw detection and amplification circuitry; *(ii)* the noise of the mw source; *(iii)* reproducibility of the experimental conditions for each laser shot, *i.e.* laser light intensity, temperature, sample bleaching, *etc.* The second contribution is more crucial for the success of the TREPR experiment. For cw EPR, this noise is less important because the cavity serves as a filter at typical frequency offsets up to 100 kHz, eqn (3.2), corresponding to the frequency of the field modulation. TREPR, however, uses broadband detection to be able to follow fast signal changes. The cavity reflects 50 percent of the incident microwave power at the frequency offset corresponding to v_0/Q_L, eqn (3.2). If the mw noise is high enough, starting from a certain excitation power, its collective contribution to the spectrometer noise amplitude will override the noise floor, then linearly increases with increasing mw amplitude. Thus, in this situation no signal-to-noise ratio improvement can be obtained by increasing the mw power, because the EPR signal depends on the amplitude of the excitation microwave as well. Improved microwave technology in recent years resulted in commercial low-noise microwave sources and power amplifiers. Nevertheless, care should be taken to choose the proper bandwidth of the detection circuitry in order to minimize the contribution of mw source noise to the overall noise amplitude.

The study of formation and decay kinetics of transient radical species requires not only the best possible time resolution but also the ability to adequately analyze the EPR signals. While TREPR provides an excellent time resolution, especially when going to high mw frequencies, it has a number of negative characteristics concerning data analysis. First, the time development of the EPR spectra is governed by the continuous interaction between the spin system and the cw microwave field. The perturbation shows up by the strong time dependence of the EPR linewidth in the time domain when the inverse of the time delay between radical formation and signal detection becomes comparable to the intrinsic linewidth of the EPR signal. This can prevent spectra detection or radical identification at early times after radical formation. Secondly, the mw radiation field also influences the time evolution of the EPR signals. In addition to the usual kinetic parameters of the chemical reaction and spin relaxation as well as of radical dynamics and CIDEP (chemically induced dynamic electron polarization) production, the time development of the EPR responses is strongly influenced by the mw field strength in the cavity. It not only dominates the signal rise, but can also produce an oscillatory behavior of the signal. As the result, the determination of rate parameters from TREPR spectra is commonly based on modified Bloch equations that, in addition to spin and chemical dynamics, also account for the perturbation of the system by the mw field. Generally, the analysis involves a nonlinear, multiparameter least-squares fit of the experimental data to the numerical solution of a set of differential equations.[8,9] The equations are based on an *a priory* model accounting for all the processes that play a role in the spin evolution. Only in rare cases is it possible to simplify the data analysis, see below. Additionally, the influence of the mw field strength often has the consequence that TREPR measurements can cover only the time period during which the spin system is far from thermal equilibrium. However, the last problem can be overruled by going to high mw frequencies, see below.

The problem of continuous perturbation of the spin system by the cw microwave irradiation that complicates the analysis of TREPR data can be avoided by applying pulsed EPR techniques. We note, however, that continuous mw irradiation of the sample also has a positive aspect: This allows TREPR and cw EPR experiments to be performed simultaneously. A small (compared with the EPR linewidth) magnetic field modulation does not perturb fast TREPR responses, but can be used for parallel phase-sensitive signal detection to observe long-lived radical reaction products that are not detectable by TREPR.

3.1.3 Pulse EPR

Pulsed EPR methods such as Fourier transform (FT) EPR and electron spin-echo (ESE) spectroscopy have become well established.[10–12] In these techniques the magnetization along the external field direction (z magnetization), which is associated with the formation and decay of transient radicals, is monitored by turning the magnetization vector in the transverse xy plane with one or more short mw pulses, followed by the detection of EPR responses in the absence of the excitation mw field.

There are two advantages of pulsed methods compared to TREPR. First, the detection occurs without incident mw power. Thus, the noise signal in pulsed experiments is determined by the noise floor of the spectrometer and by the stability of experimental conditions, i.e. it becomes independent of mw source noise. Second, the measured EPR response is directly proportional to the z magnetization existing in the moment of mw excitation. This simplifies data analysis because the excitation mw field has not to be included in the analysis model.

Besides these advantages, there are two serious limitations that are connected to each other, time resolution and sensitivity. The time resolution of pulsed techniques is determined by the time from the start of the detection mw pulse sequence to the moment at which the EPR response can be recorded. The time resolution is limited by two contributions: *(i)* the length of the pulse sequence and *(ii)* the deadtime of the EPR spectrometer.

In FT EPR, the pulse sequence consists only of one $\pi/2$ mw pulse followed by measuring the free induction decay (FID). The pulse length required for a $\pi/2$ pulse, assuming the Larmor frequency of the spin to be equal to the chosen mw frequency, is given by

$$\tau_{\pi/2} = \frac{\pi}{2} \cdot \frac{1}{\gamma \cdot B_1} \tag{3.4}$$

where γ denotes the gyromagnetic ratio of the electron spin. The B_1 field amplitude is a function of the available mw excitation power, P_w, and the power-to-field conversion efficiency, K, of the EPR resonator:

$$B_1 = K \cdot \sqrt{P_w} = C \cdot \sqrt{\frac{\mu_0 \cdot Q_L}{\omega_0 \cdot V_C}} \cdot \sqrt{P_w} \tag{3.5}$$

where C is a constant determined by the cavity; Q_L and V_C are the loaded quality factor and effective volume of the cavity. In a real experiment, however, the pulse length is also strongly influenced by the bandwidth of the EPR cavity. The B_1 field is not created immediately after switching on the mw power, but grows in with the ringing time constant T_R of the cavity, see eqn (3.3). Thus, for an X-band spectrometer equipped with a cavity of loaded quality factor $Q_L = 2000$, it is impossible to create pulses shorter than about 60 ns, even when enough mw power is available. At low mw frequencies, this problem can be overcome by lowering the quality factor of the cavity by overcoupling.[13] This, unfortunately, results in fairly low power-to-field conversion efficiency as well as in reduced sensitivity. At higher mw frequencies this limitation becomes less critical. Thus, at W-band and $Q_L = 2000$, pulses as short as 6 ns can be generated. Moreover, the conversion efficiency is much higher at high mw frequencies, because the B_1 field is concentrated in a smaller volume of the cavity. To create a $\pi/2$ pulse of 10 ns length at X-band, about 1 kW mw power incident on a single-mode cavity is required ($Q_L \approx 100$), whereas at W-band about 500 mW is already sufficient ($Q_L \approx 2000$). At 360 GHz this pulse length can be achieved with only 8 mW incident on the single-mode cavity.

The EPR response cannot be measured immediately after switching off the mw power because of the cavity ringing. Thus, one has to wait for a certain time until the ringing mw power has decayed approximately to the level of the expected EPR response. This time is called the deadtime of the spectrometer. For X-band, it typically amounts to about 100 ns. At higher frequencies, for instance W-band, the deadtime is typically one order of magnitude shorter, owing to the lower mw excitation power required and the faster cavity ringing.

The deadtime of an EPR spectrometer not only limits the time resolution of pulsed experiments, but also limits its applicability to certain radical systems. The FID is governed by the linewidth of the EPR signal. The rate of exponential damping of the FID signal $(1/T)$ is related to the full width at half-height of an EPR line, $\Delta B_{1/2}$; for instance for a Lorentzian line $\Delta B_{1/2} = 2/\gamma \cdot T$. Thus, for the FID to decay slower than the spectrometer deadtime, the X-band FID technique cannot be used for paramagnetic systems with linewidths larger than 0.1 mT. In contrast to FID, the decay of an ESE response is only determined by the homogeneous linewidth of an EPR transition, *i.e.* by the T_2 relaxation time. On the other hand, an inhomogeneous broadening of the EPR transition is required for the echo formation.

Pulsed EPR probes the full magnetization of the sample, and not only transient contributions as TREPR does. If some persistent or long-lived, on the time scale of the experiment, paramagnetic species are generated or initially present in the sample, their EPR responses are also detected by pulsed EPR. In this case, to obtain solely the EPR spectra of the transient radicals, one has to perform additional measurements to get rid of background signals.

3.2 Historical Overview of High-Field/High-Frequency EPR Spectrometers

The great leap forward in microwave and electronic technology, triggered by the development of radar techniques during World War II, allowed construction of a new generation of spectrometers with the sensitivity and resolution required to detect magnetic resonance effects. In 1944 towards the end of the war, the first experiments of E.K. Zavoisky[14–17] at Kazan University revealed electron spin resonance effects in solids. At first he used frequencies of about 30 MHz and fields of about 1 mT. He observed a magnetic-field-dependent paramagnetic absorption in pure $CrCl_3$ and $MnSO_4$ samples, but also in H_2O-diluted samples. Then, using higher frequencies, 140 MHz and 3 GHz, he demonstrated that the absorption maximum shifts to higher magnetic fields, 5 mT and 108 mT, respectively. The frequency/field ratio of the absorption maximum, about 28 MHz/mT, corresponds to the electron Larmor frequency in the applied external magnetic field, since the investigated ions are characterized by a Landé factor of about $g = 2$. Practically at the same time, in 1945, F. Bloch, W. Hansen and M.E. Packard[18,19] at Stanford University as well as E.M. Purcell, H.G. Torrey and R.V. Pound[20,21] at Harvard University succeeded to detect the proton magnetic-resonance absorption in paraffin at 7.8 MHz in the field of 185 mT, and in paraffin and water at 30 MHz in the field of 710 mT. The frequency/field ratio corresponds to the proton Larmor frequency in the applied magnetic field, determined by the gyromagnetic ratio of the proton (42 kHz/mT). These 1944/1945 experiments opened the field of electron and nuclear magnetic resonance.

Since their discovery, both EPR and NMR have undergone continuous improvement due to new developments of spectroscopic instrumentation and methodology. In the beginning, the development of EPR and NMR went side-by-side. During the first decade after discovery, a large number of spectrometers were constructed at various universities and research laboratories. Parallel to the laboratory developments, commercialization of the magnetic resonance spectrometers had started. R.H. Varian (founder of Varian Associates, Palo Alto) had tracked the development of NMR at Stanford and Harvard during World War II, and he hired M.E. Packard, a key NMR researcher, to become head of Varian's NMR project. The first three 30-MHz NMR spectrometers were delivered in 1952.[22] In 1954, Varian Associates started the development of EPR spectrometers. The first six X-band EPR instruments were shipped in 1956.[23] In the years of foundation, the same technical problems had to be solved in NMR and EPR. However, very soon the main difference concerning the spectrometer performance in terms of sensitivity and resolution was recognized. It turned out that for NMR spectral resolution is the key criterion of performance. NMR lines of samples in solution are usually extremely sharp, the linewidth often being less than 0.1 Hz. In order to distinguish closely spaced peaks, the magnetic field inhomogeneity and the instability of the radio-frequency source have to fulfil the stringent requirement to be of the order of 10^{-8}. The narrowest EPR line, on the other hand, is about

30 kHz (10 μT) and, hence, the requirements for field inhomogeneity and frequency instability are not so strict, *i.e.* 10^{-5} to 10^{-6}. Thus, detection sensitivity, prior to spectral resolution, becomes the key criterion for the performance of an EPR spectrometer.

The general formula to solve sensitivity problems of both EPR and NMR is to go to higher resonance frequencies and correspondingly higher magnetic fields. This turned out to be fully true for NMR but, in practice, only partially true for EPR. The development of high-frequency NMR becomes limited by the quality and availability of superconducting magnets. In EPR, it is the high resonance frequency that provides the major technological challenges. This becomes obvious if we plot the magnetic field strength *vs.* the time at which the corresponding NMR or EPR spectrometer was realized, see Figure 3.1. The field and frequency of the commercial NMR instruments increase monotonically with the years. In the meantime, 950-MHz high-resolution NMR spectrometers operating at 22.3 T are commercially available. A commercial high-field EPR spectrometer operating at 94 GHz (W-band) was introduced by Bruker Biospin in 1996, more than 30 years after Varian Associates had presented a 35-GHz (Q-band) EPR spectrometer in 1962. Thus, the larger part of EPR applications at high-frequency/high-field conditions was done using laboratory-built equipment. In the following we will consider the history and techniques of high-field EPR in more detail.

One more difference between EPR and NMR developments has to be pointed out: the transition from cw to pulsed instrumentation. The era of time-domain NMR began already in 1966 when R.R. Ernst and W.A. Anderson at Varian Associates, Palo Alto, introduced Fourier transform NMR.[24] This initiated the rapid development of pulsed NMR techniques that allow a large number of sophisticated experiments in liquids and solids to be done that cannot be performed by cw techniques. Interestingly, the first commercial FT

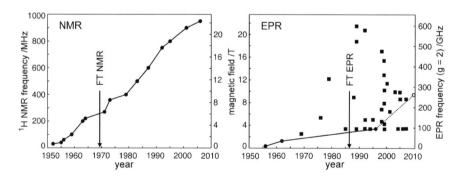

Figure 3.1 Instrumental development of NMR and EPR spectrometers towards high magnetic fields. The interconnected filled cycles show the year of the corresponding spectrometer becoming commercially available. The years of introduction of pulse Fourier-transform (FT) instruments are marked by arrows. The filled squares indicate the laboratory-built high-field EPR instruments, as described in the literature for the first time, see Table 3.1.

NMR spectrometer was not introduced by Varian Associates, but by the competitive company Bruker Physik, Karlsruhe, in 1969. In contrast, the utilization of the time domain in pulse EPR took many more years. The first commercial pulsed EPR spectrometer became available from Bruker Biospin in 1987.[25] The reason for this delay is the retarded development of fast control and data acquisition systems as well as of high-power microwave sources required for pulse EPR. The electron-spin relaxation times are orders of magnitude shorter than those of nuclear spins. Moreover, the EPR spectrum is spread over a much broader frequency range compared to an NMR spectrum. Thus, first to excite and afterwards to detect EPR transitions a nanosecond time resolution of control and acquisition systems is required. The first FT NMR experiments were done with only 2 ms time resolution,[24] which was technically possible since the 1960s. The systems fulfilling the FT EPR requirements appeared only in the beginning of the 1980s.

3.2.1 First Generation

In 1960, J.B. Mock from Bell Telephone Laboratories[26] described a transmission-type EPR spectrometer operating in the frequency range of 50 to 150 GHz using tuneable backward-wave oscillators as the microwave sources. The magnetic field up to 2.4 T was supplied by a modified electromagnet. From the instrumental point of view, this spectrometer is more related to the microwave absorption spectroscopy. The mw power was transmitted through the sample in a rectangular waveguide and detected by a crystal detector. A sensitivity of 2×10^{15} spins/mT in a 1-Hz bandwidth was achieved.[27] The spectrometer was used for direct measurements of linewidth and resonance frequency of zero-field lines of Mg^{2+} in $MgWO_4$ single crystals.

The first "real" EPR measurements at mm-wavelength were described by P.R. Elliston and coworkers[28] in 1963. They used a simple reflection-type system operating at 64.2 GHz with no cavity incorporated. The heart of the spectrometer system was a balanced harmonic generator, employing a "magic tee". The same device was used as the detector of the harmonic signal from the sample placed in the short-cut waveguide. A spectrometer sensitivity of 9×10^{16} spins/mT at 1 Hz bandwidth was determined using $CuSO_4 \cdot 5H_2O$ as a sample at room temperature. The spectrometer was primarily intended for antiferromagnetic resonance studies.

In 1969, simultaneously two high-field EPR spectrometers were introduced, both working at about 70 GHz. E.F. Slade and D.J.E. Ingram[29] described the spectrometer setup that utilized, for the first time, a superconducting magnet. The reflection-type spectrometer was equipped with a klystron, a cylindrical TE_{011} cavity and an indium-antimonide detector. A sensitivity of 5×10^{10} spins/mT was measured using a phosphorus-doped silicon crystal. Ingram also proposed a modified 140 GHz spectrometer version, using a harmonic generator to produce the mw power.[30] The abandonment of the cavity and the low mw power available from the harmonic generation reduced the sensitivity of the spectrometer considerably below that of the 70-GHz version and, hence, this spectrometer version did not find practical application. The 70-GHz spectrometer of H. van den Boom,[31] similar to that of Slade and Ingram, used a klystron as a mw source, an oversized rectangular TE_{105} resonator and a reflection detection scheme with reference arm. The sensitivity of 5×10^{12} spins/mT was rather low, due to the oversized cavity and standard crystal detector.

A Fabry–Perot resonator for millimeter and submillimeter EPR was introduced by I. Amity[32] in 1970. In the same year, H.C. Box and coworkers[33] presented the results of 70-GHz EPR investigations of paramagnetic radical cations produced by ionizing radiation in single crystals of organic substances. Their spectrometer utilized heterodyne detection and a cylindrical TE_{011} cavity in transmission mode. Soon after, in 1973, Y. Alpert and coworkers[34] measured zero-field splittings of the high-spin ferric ion in a frozen solution of acid methemoglobin from the frequency dependence of the g-value in the frequency range of 70 to 400 GHz. The sample was placed in the oversized waveguide of the transmission type spectrometer equipped with a klystron and BWO (backward wave oscillator) as the mw sources and a carbon bolometer as the detector. The superconducting coil produced a magnetic field of up to 5.7 T.

All spectrometers described above can be considered, at least to some extent, as high-frequency/high-field EPR instruments. They were, however, designed for measurements of zero-field splittings of transition-metal ions or for samples containing high concentrations of the paramagnetic species at low temperatures. The absolute sensitivity and the concentration sensitivity were too low to allow for applications to (bio)chemical samples, especially at elevated temperatures. Thus, one can consider these spectrometers as high-field EPR spectrometers of the first generation.

3.2.2 Second Generation

In 1971, Ya.S. Lebedev from the Institute of Chemical Physics in Moscow started a long-term research and development project with the aim to realize a high-frequency/high-field EPR spectrometer with high sensitivity, resolution, stability and flexibility for broad applications in physics, chemistry and biology.[35–37] The concept of a 140-GHz EPR instrument operating at 5 T was developed. The construction started in 1973 in close cooperation with the Physical-Technical Institute in Donetsk (Ukraine). In the following years, several prototypes of the spectrometer were constructed. Different microwave sources including klystrons, BWOs, diffraction generators and solid-state oscillators were tested. Single-mode resonators, oversized cavities and Fabry–Perot resonators, as well as different detectors, were compared in their performance. The first working version of the 140-GHz cw EPR spectrometer, finished in 1979, reached a sensitivity of 4×10^8 spins/mT at 1 Hz bandwidth, which was about three orders of magnitude higher than for X-band EPR at that time. In parallel with the continuous development of EPR instrumentation, the Lebedev group performed a series of EPR experiments approving the advantages of high-field/high-frequency EPR for application to physical, chemical and biological systems. Thus, it is legitimate to consider the pioneering work of Ya.S. Lebedev and his coworkers as the starting of the second generation in the history of high-frequency/high-field EPR development.

Inspired by the opportunities high-field EPR offers, several research groups across the world have started with the development of their own EPR spectrometers dedicated to their specific research objects. The construction of spectrometers working at high frequencies went in parallel with the implementation of sophisticated electronic instrumentation well suited for EPR techniques. In Europe, the Berlin[38,39] and Leiden[40,41] EPR groups first focused on the construction of sensitive W-band (95 GHz) EPR and on the extension of W-band EPR spectrometers to ENDOR capabilities. Cw ENDOR was realized in Berlin in 1988.[42] The first pulsed EPR experiments at W-band were performed in Leiden in 1989.[40] Pulsed ENDOR was simultaneously developed around 1994 in both laboratories.[43,44] The research group of J.H. Freed at Cornell University was the first to push EPR instrumentation to far-infrared technology, and pioneered the use of quasioptical mw techniques in EPR. Details of their 250 GHz EPR spectrometer were published in 1988.[45] One year later, the group of L.-C. Brunel at the Grenoble High-Magnetic-Field Laboratory realized an EPR system operating up to 525 GHz and magnetic fields up to 20 T.[46] By now, the complete list of laboratory-built high-field/high-frequency EPR spectrometers has become rather long. Therefore, we have tried to summarize the key aspects of instrumental development in the form of Table 3.1. The main technical concepts of high-frequency EPR will be discussed in the following sections. Before doing so, we make several comments on commercialization of high-frequency EPR instruments.

The commercialization of scientific instrumentation is a necessary step to make the technique available to a wide scientific community. This general statement also applies to high-field EPR. The EPR community is not limited to a relatively small number of physicists and physical chemists able to construct, maintain and operate such complex spectrometers. At the beginning of the 1990s, a commercial high-field EPR system was produced in Donetsk using the design of the instrument developed in Lebedev's group in Moscow. It included a complete spectrometer or parts of it, for example the mw bridge, designed for cw or pulsed EPR at 95 GHz and 140 GHz. Several such mw systems were installed, for example in the Weizmann Institute, Rehovot,[53] University of Leiden,[85] Argonne Natl. Laboratory,[86] Albert Einstein College, New York,[58] and UC Davis.[59,60] The widely used W-band EPR spectrometer, operating in cw (Elexsys E600) and pulsed (E680) versions, have been supplied by Bruker Biospin, Rheinstetten since 1996.[50,87] The initial problem of low mw power was solved by Bruker in 2004 by introducing an additional power module.[88] This allows pulsed EPR, ENDOR and ELDOR measurements with $\pi/2$ pulses of about 30 ns to be performed. The new project to develop 263 GHz/9.5 T EPR was announced by Bruker Biospin in 2007. This spectrometer utilizes a quasioptical design for the mw bridge and will be capable of cw and pulse modes of operation.

Table 3.1 Summary of laboratory-built high-field/high-frequency EPR spectrometers.

Frequency/GHz	EPR mode	Detection scheme	MW bridge scheme	Transmission line	Resonator	MW source	Year	ref.
150	cw, pulsed	HO	WG	OWR	TE011, FP	K, SS, BWO, IMPATT	1977	35–37,47,48
95	cw, pulsed	HE	WG	OWR	TE011, FP	K, SS	1985	38,39,42,43
95	cw, pulsed	HE	WG	OWR, OWC	FP	IMPATT	1989	40,44,49
95	cw, pulsed	HE	WG	OWR	TE011	G	1992	50
95	cw	HO	WG	OWR, DWG	TE013	G	1994	51,52
95, 140	cw, pulsed	HE	WG	OWR	TE011	SS	1999	53
95	cw, pulsed	HE	OQ	CWG	FP	SS, EIKA	2004	54
95	cw, pulsed	HE	WG	OWR	TE011	SS	2006	55
95	cw	HE	WG	OWR	TE011, LGR	G	2007	56
30–120	cw	HO	WG	WR	TE011	MVNA	1999	57
130	cw	HO	WG	OWR	TE011	G	1998	3
130	cw	HE	WG	OWR	TE011	G	2002	58
130	cw	HE	WG	OWR	TE011	G	2005	59,60
95–475	cw	HE	QO	CWG	FP	G	1998	3
90, 180	cw	HE	OQ	CWG	FP	G	1998	4
140	cw, pulsed	HE	WG	OWR	FP	G, EIK	1992	61,62
140	cw, pulsed	HE	WG	OWR	TE011, FP	G, GTR	1995	63
180	cw, pulsed	HE	QO	CWG	TE011	G	2001	64
220	cw	HO	QO	LS	FP	G	1999	65
120, 240	cw, pulsed	HE	QO	CWG	TE011	G	2007	66
120, 240	cw, pulsed	HE	QO	CWG	FP, NRS	G	2005	67
275	cw, pulsed	HE	QO	CWG	TE011	G	2003	68,69
285	cw	HO	WG	OWC	NRS	G, SS	1999	70,71
360	cw, pulsed	HE	QO	CWG	FP	G, O	1999	72,73
160–525	cw	HO	QO	CWG	NRS	FIR	1989	46
240, 316	cw	HO	QO	OWC	WGD	FIR	2000	74
95–3000	cw	HO	QO	OWC,	NRS, FP	G, FIR	1999	75–77
604	pulsed	HE	QO	OWC, DWG	FP	FIR	1999	78
170, 250	cw	HO	QO	LS, CWG	FP	G	1988	45,79,80
250–7000	cw	HO	QO	OWC	NRS	G, BWO, FIR	2003	81
38–889	cw	HO	QO	OWC	NRS	G, GTR	1998	82–84

Abbreviations used: Abs. – absorption mode, HO – homodyne; HE – heterodyne; WG – waveguide; QO – quasioptical; OWR – oversized rectangular waveguide; OWC – oversized cylindrical waveguide; DWG – dielectric waveguide; CWG – corrugated waveguide; FP – Fabry-Perot; LGR – loop-gap resonator; WGD – whispering gallery dielectric resonator; NRS – non resonant structure; K – klystron; G – Gunn diode (phase-locked or varactor-tuned); GTR – gyrotron; EIK – extended interaction klystron; EIKA – extended interaction klystron amplifier; FIR – far-infrared laser; SS – solid-state oscillator (based on low-frequency transistor oscillators with active multiplication, typically by IMPATT component); O – orotron; IMPATT – impatt oscillator; MVNA – vector network analyzer.

3.3 Technical Aspects of High-Field/High-Frequency EPR

3.3.1 Sensitivity Considerations

The full exploration of the resolution capability in high-field EPR requires adequate detection sensitivity of the spectrometer. In principle, the absolute sensitivity should significantly increase with increasing Zeeman field and corresponding mw frequency ω_0. The rate of energy absorption near EPR resonance, $R(\omega_0)$, is proportional to the population difference between α ($m_s = -1/2$) and β ($m_s = 1/2$) electron-spin states, the probability of inducing a transition between these states and the energy absorbed or emitted in a transition. These factors are proportional to $\frac{\hbar \cdot \omega_0}{2 \cdot kT}$ (for $\hbar \cdot \omega_0 < kT$, where k is the Boltzmann constant, T the temperature), ω_1^2 and $\hbar \cdot \omega_0$ resulting in $R \propto \omega_0^2$. In reality, however, technical factors also influence the sensitivity, *i.e.* the quality factor Q of the resonator and the filling factor η, see eqn (3.1). If one assumes that the mw source, components, detection and amplification stages are frequency independent in their noise and performance (which, in practice, is unrealistic), simple expressions for the frequency dependence of the cw EPR detection sensitivity can be deduced in terms of the minimum number of detectable electron spins, N_{min}. For constant incident mw power and unsaturated EPR lines, one obtains:[1,2]

(i) $N_{min}/V_S \propto \omega_0^{-3/2}$, for the minimum detectable spin concentration, when the sample size is scaled by the same factor as the cavity dimensions,
(ii) $N_{min} \propto \omega_0^{-9/2}$, for the minimum number of detectable spins when the sample size is not varied, *i.e.* when the sample volume V_S is constant and the filling factor is proportional to V_S/V_C with the cavity volume being proportional to ω_0^{-3}.

Case (i) holds for liquid or powder samples, case (ii) applies to small crystals. For a constant magnetic field energy density, *i.e.* for the same ω_1 at the sample, the proportionality reduces to $\omega_0^{-3/4}$ and $\omega_0^{-15/4}$ for the case of a constant ratio V_S/V_C and a constant V_C, respectively.[1,2] (The exact frequency factors have been reconsidered in ref. [180].) In theory, therefore, from the view point of sensitivity it should be advantageous to perform EPR at frequencies as high as possible. In practice, however, existing high-field EPR spectrometers often fall short of the predicted sensitivity.[37] This is because the spectral purity and stability of the mw source and its power output, the noise figures and insertion losses of components and detector, as well as the detection bandwidth have to be included in the consideration.

The choice of an optimum mw frequency and Zeeman field is, of course, dependent on the detectable spin concentration for a given sample. In reality, however, one always needs to compromise between optimum sensitivity and desired information to be obtained from the EPR spectrum. For example, for rigid-limit spectra one must consider the origins of the overall EPR linewidth. In particular, it is important to what extent g-strain (*i.e.* the distribution of g-values arising from an inhomogeneity of the local environment of individual spins) determines the inhomogeneous EPR linewidth in relation to other sources of line broadening, for instance unresolved hyperfine couplings. As the Zeeman field increases, the g-strain contribution can become dominant, and a further increase of the field will neither enhance the EPR resolution nor the sensitivity of cw EPR detection.

In pulse EPR, one has to consider the spectral features in connection with the mw excitation bandwidth. Increased g-tensor resolution at high fields spreads the EPR spectrum over a wider interval of the magnetic resonance field and, consequently, the spectral density in the spectrum drops. Thus, taking into account the power-limited excitation bandwidth of mw pulses in high-field EPR, one will observe an echo or FID response proportional to the reduced spectral density at a fixed magnetic-field value within the spectrum. This may result in a drastic loss of detection sensitivity. Moreover, the prolongation of T_1 of the spin system with increasing magnetic field may enforce a reduction of the repetition rate of pulsed detection, thereby lowering the signal-to-noise ratio.

For the detection of the transient species the situation is even more complicated, especially if spin-polarized states are involved. The magnetic-field-dependent polarization mechanisms can lead either to an increased enhancement factor or to its decrease with rising ω_0. On the one hand, this may help to suppress the polarization effects in the EPR spectrum, for example, from the triplet mechanism that is dying out at high Zeeman fields. On the other hand, this makes general sensitivity considerations impossible without quantifying the extent of spin polarization of the system under investigation.

3.3.2 Detection Schemes

There are five basic requirements for an EPR spectrometer: *(i)* A strong homogeneous and stable magnetic field is needed. *(ii)* Suitable low-noise microwave sources must be available. *(iii)* The microwave power must be transferred with minimal losses to a resonator with the sample and from the resonator to the detector by means of a suitable mw transmission line. *(iv)* The resonator must concentrate the incident microwave radiation onto the sample and allow detection of the small amount of energy absorbed when the EPR resonance condition is met. *(v)* The detector must be able to measure, with high signal-to-noise ratio, the variation of the mw power level in the case of EPR absorption. In Figure 3.2(a) the simplest way, in which the necessary components can be assembled to form an EPR spectrometer, is shown. In this transmission-type homodyne system, a certain amount of mw power in the resonator is coupled out and fed directly to the detector. The sensitivity of such a system is optimized when one half of the incident mw power is absorbed in the resonator, one quarter is reflected and one quarter is transmitted. Figure 3.2(b) shows the operation principle of a reflection-resonator homodyne EPR spectrometer. For optimum sensitivity, the resonator is exactly matched to the transmission-line waveguide and absorbs all incident mw power. In the case of

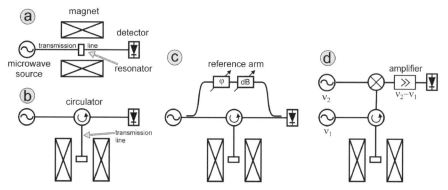

Figure 3.2 Schematic diagrams of EPR spectrometers with (a) homodyne detection and transmission-mode resonator (b) homodyne detection and reflection-mode resonator; (c) homodyne detection with reference arm and reflection-mode resonator; (d) heterodyne detection with reflection-mode resonator.

EPR absorption, the matching between resonator and waveguide changes, and the same amount of microwave power as absorbed by the sample reaches the detector. The excitation and absorption microwaves are decoupled by the circulator, which ensures that a signal fed into one of its ports leaves the adjacent port only. The optimum sensitivity for this reflection-resonator scheme is a factor of four higher than for the similar cavity operating in transmission mode. If it is required that a reflection-resonator spectrometer provides information about EPR absorption and dispersion separately, a reference (or "bucking") arm should be included in the system, see Figure 3.2(c). By using directional couplers, an additional mw signal can be split off from the microwave source and fed directly to the detector for interferring with the mw signal reflected from the resonator and carrying the EPR information. The phase and magnitude of this bucking signal can be adjusted by a phase shifter and attenuator to obtain the required degree of suppression of either the absorption or dispersion signal. The principles, limitations and practical realizations of the different homodyne EPR detection schemes in Figures 3.2(a)–(c) are described in the monographs of T.H. Wilmshurst,[7] C.P. Poole[2] and D.J.E. Ingram,[30] to which we refer for further details. The heterodyne detection with reflection-mode resonator, see Figure 3.2(d), does not play an important role in X-or Q-band EPR because of the available low-noise mw signal amplifiers, but becomes very important in high-frequency EPR where such low-noise signal amplifiers (above 100 GHz) are not yet available. The heterodyne detection will be discussed further below.

All the considerations concerning the basic homodyne detection schemes at X- or Q-band generally hold also for high-frequency EPR. However, the degradations in the performance of specific spectrometer parts at high mw frequencies put limitations on the spectrometer design and require a careful choice of the components. In the following sections we will describe the microwave sources, resonators, transmission lines and magnet systems that have been successfully tested and employed in high-field spectrometer designs of various laboratories. Here, we review different detection techniques at high mw frequencies that are directly connected with the performance of detector. For additional information we refer to review articles dealing in detail also with aspects of instrumentation development of high-frequency EPR.[37,89–96]

The microwave detector is employed in an EPR spectrometer to convert the microwave EPR signal to lower frequency (rf or DC) that can be handled by standard recording devices. The detector for cw EPR has to fulfil the following requirements: *(i)* It should have a high sensitivity combined with small intrinsic noise, *i.e.* the signal-to-noise ratio determines the smallest detectable signal. *(ii)* Its dynamic range should be large enough to allow for linear registration of small as well as large signals. For TREPR spectroscopy, additionally the response time of the detector is of importance. In pulsed EPR spectroscopy, only a short response time and high dynamic range of the detector would enable fast detection and short deadtime of the spectrometer. The conventional homodyne spectrometers employ either diode detectors or bolometers for retrieving the EPR signal from the background noise.

Bolometers are thermal radiation detectors. These incoherent detectors are suitable to detect mw power in the very broad frequency range from GHz to THz. The radiation heats an absorbing element, and a resistance thermometer is used in a compensation circuit to sense the resulting temperature change. The ultimate sensitivity of a bolometer is determined by the thermal fluctuations of the absorbing element which are described by $NEP^2 = 4 \cdot kT^2 \cdot G$. Here, NEP is the noise-equivalent power (measured in $W/\sqrt{Hz}$), T is the operating temperature (measured in K), and G is the thermal conductance (measured in W/K) between the absorbing element and the temperature bath. The sensitivity of the bolometer can be increased by reducing either the temperature T or the thermal conductivity G. However, since the absorber has a heat capacity C (measured in J/K), it will have a thermal time constant of $\tau = C/G$. If G is reduced to improve the sensitivity, C must also be reduced to ensure that τ is kept much shorter than the characteristic modulation time of the detected signal. The best bolometers applied nowadays in astrophysics approach the NEP level of $10^{-18} W/\sqrt{Hz}$. These bolometers, however, have to be cooled to mK temperatures and have a response time >1 ms.[97] The NEP value should be compared with the thermal noise power at the operating temperature that determines the smallest possible EPR signal power that can be detected. The thermal noise power of an ohmic resistance is given by $\Delta f \cdot k \cdot T = 4 \times 10^{-21}$ W ($= 4 \times 10^{-18}$ mW $= -174$ dBm) at 293 K and 1 Hz bandwidth. Thus, the noise of the best available bolometers still exceeds the kT noise by as much as a factor of 250. The standard commercial bolometer systems, for example InSb hot-electron bolometers operating at 4.2 K, have a NEP of about $10^{-13} W/\sqrt{Hz}$ with a time constant of about 1 µs. This response time is short enough to allow for sensitive cw EPR detection with high-frequency (≈ 100 kHz) field modulation or TREPR with direct-detection of transient radicals with lifetimes

> 1 μs, but is too slow for high-performance pulsed EPR spectrometers with sub-μs pulse trains.

The diode detectors, typically of Schottky-barrier type, are characterized by a fast response time of about 1 ns. They do not require cryogenic cooling and can be used to detect the mw power up to the THz region. Under low-power operation the diode detectors have a NEP of about 1.5×10^{-12} W/$\sqrt{\text{Hz}}$ at 150 GHz, rising to 2×10^{-11} W/$\sqrt{\text{Hz}}$ at 800 GHz.[98] Thus, simple homodyne systems, Figures 3.2(a) and (b), with a detection based on Schottky diodes are not optimal in terms of detection sensitivity, and bolometer detection is to be preferred. Moreover, as the input power increases, for example by the bucking signal of the reference arm, excess noise is generated in the diode. This noise typically has a $1/f$ power spectrum, where f is the output frequency, and is known as flicker noise. Flicker noise becomes increasingly important in cw EPR with field modulation in the kHz region.

The way to overcome the noise of the diode detector at mw frequencies up to Q-band is to amplify the EPR signal by a low-noise mw preamplifier prior the detection. The development of mw low-noise amplifiers (LNA) based on GaAs-FET technology (gallium arsenide field effect transistors) in the 1980s lead to a significant improvement of spectrometer sensitivity. Nowadays, these low-noise amplifiers offer noise figures below 1 dB and small-signal gains up to 40 dB at moderate saturation characteristics (typically 10 dBm at 1 dB compression point). The performance, however, degrades when going to high frequencies above 40 GHz. Beyond this frequency, LNAs based on heterostructure FET technology can be used. Heterostructure transistors are also called "hot-electron mobility transistors" (HEMT). Until the mid-1990s, the technical frequency limit of these HEMT amplifiers was about 60 GHz. The rapid development in manufacturing techniques, for instance the development of monolithic microwave integration circuits (MMIC), as well as the introduction of new semiconductor materials widened the frequency range of the HEMT amplifiers. In the meantime, the HEMT LNAs are commercially available up to 100 GHz. The first application of the HEMT LNA in a high-field EPR spectrometer was reported by the Illinois group.[52] They incorporated a 95-GHz LNA with 22 dB gain and 5 dB noise figure (NF) in their cw W-band homodyne mw bridge. An improvement of the EPR signal/noise ratio by a factor of 3 (10 dB reduction of the detector NF) was observed when employing a diode detector with LNA compared to the diode detector without LNA.

In our laboratory, we also tried to integrate a 95-GHz LNA (18 dB gain, 4.6 dB NF) in our heterodyne W-band bridge, see Section 3.4.1.1. The sensitivity of cw and TREPR detection was found to be improved by a factor of 2 (6 dB reduction of NF). Unfortunately, the amplifier leads to a degradation of the pulsed EPR performance. The quite low output saturation power, typically 0 dBm around 95 GHz, results in a longer deadtime of the spectrometer. The 95-GHz LNA is used in the commercial W-band EPR spectrometer of Bruker Biospin.[88] Recent developments in HEMT technology show the possibility to build low-noise amplifiers for mw frequencies up to 600 GHz.[99,100] The version of an amplifier for 270 GHz described recently[101] showed a noise figure of 7.5 dB with a gain of 11.4 dB.

A more complex solution of the noise problem, that has been used in high-frequency/high-field EPR, is the application of the heterodyne-detection scheme with an intermediate frequency (IF) sufficiently large to eliminate $1/f$ noise from the Schottky diode. In this scheme, see Figure 3.2(d), the Schottky diode is used as a mixer multiplying the EPR mw signal with the signal of the local oscillator (LO), operated at a somewhat different frequency, to produce the IF signal power. The IF signal can be further amplified by a standard low-noise amplifier to overcome the noise of the final detector, typically also a diode or a mixer-type detector. The single-side-band noise figure of the heterodyne system depicted in Figure 3.2(d) can be as small as 10 dB.

The disadvantage of a heterodyne system with a main mw source and a local oscillator operating independent from each other is its sensitivity to the uncorrelated phase noise of both sources. This problem can be solved either by phase locking both sources to a master oscillator (as is realized in our 360-GHz EPR spectrometer, see Section 3.4.2.2) or by using, as the LO, the up-converted signal from the main source and mixing it with the signal of the low-frequency oscillator (as is realized in our 95-GHz spectrometer, see Section 3.4.1.1). An additional advantage of the heterodyne system is that amplification and phase manipulation of the EPR signal can be performed at low IF frequency using high-performance low-cost components. The detection bandwidth of the heterodyne systems depends on the choice of the IF frequency, and values of ±500 MHz required for fast pulsed EPR can be easily achieved. Moreover, the high dynamic range of the diode-mixer based systems, which is determined by a rather high LO power, makes them well suited for high-frequency pulsed EPR spectrometers. Because of this, superconducting-insulating-superconducting (SIS) mixers are not applied due to their saturation at low power levels.[102]

Around a chosen mw frequency, the operation bandwidth of the heterodyne systems based on diode mixers is rather limited. This can be improved by using cryogenically cooled InSb mixers in a combination with a homodyne detection scheme.[4] They offer good noise performance over a wide frequency band and have much lower $1/f$ noise compared to InSb bolometers. The major disadvantage is that the InSb mixers have a detection bandwidth of only 1 MHz, the same as the InSb bolometers have. Therefore, they are unsuitable for pulsed EPR applications.

To sum up, the heterodyne detection scheme based on diode mixers offers the best cw, time-resolved and pulsed EPR performance, but considerably complicates the spectrometer design. It requires additional microwave components either of waveguide of quasioptical type. For very high frequencies, for example 360 GHz, the use of two expensive high-frequency microwave sources becomes necessary, because of the small source output power. The homodyne detection scheme based on InSb mixers/bolometers has become an excellent choice for broadband high-frequency cw EPR spectrometers. The rapid development of low-noise high-frequency amplifiers will significantly improve the homodyne detection sensitivity in the future.

3.3.3 Microwave Sources

Most millimeter and submillimeter mw sources for EPR spectroscopy can be divided into two groups: vacuum-tube oscillators and solid-state oscillators. Vacuum-tube oscillators such as klystrons, BWOs, gyrotrons, extended interaction oscillators (EIOs) and magnetrons have been, and still are used as coherent mm-wave sources. They are, however, increasingly replaced by solid-state oscillators, which are smaller, easier to operate and reach excellent performance with increased reliability.

Criteria for selecting a mw source for EPR purposes include noise performance, output power, tuning bandwidth, durability, convenience and costs. The solid-state sources are often preferred because of relatively low cost and compact encasement. This simplifies integration into the EPR microwave bridge with its compact array of components. Moreover, solid-state sources do not require high control voltages, additional cooling and are generally less sensitive to stray fields from the cryomagnet. They deliver, however, only low output power compared to vacuum-tube devices. EPR solid-state sources generally employ negative-resistance devices based on IMPATT (GaAs, Si) or Gunn diodes (GaAs, InP). Gunn-diode sources are used in the frequency range up to 150 GHz providing 100 mW (20 dBm) of cw radiation at 95 GHz and 20 mW (13 dBm) at 140 GHz. The InP Gunn diodes are more efficient than GaAs diodes. Above 100 GHz the GaAs IMPATT device has approximately the same efficiency as a Gunn diode, and operates up to 150 GHz. Silicon IMPATT devices provide the highest cw power levels at frequencies up to 220 GHz where about 50 mW (17 dBm) can be obtained.

The disadvantage of free-running solid-state sources is rather poor noise and frequency jitter characteristics, which require phase-lock loops to a reference oscillator or external cavity stabilization. For example, the free-running GaAs Gunn diode at 95 GHz has a phase noise of about -40 dBc/Hz at 10 kHz frequency offset (-70 dBc/Hz at 100 kHz) which is the typical range of cw EPR with field modulation. Phase locking to a 100-MHz low-noise quartz oscillator allows improvement in the noise performance. In practice, the noise performance of a 100-MHz oscillator up-scaled to 95 GHz can be reached, *i.e.* -100 dBc/Hz at 10 kHz and -130 dBc/Hz at 100 kHz. This phase-noise level fulfils all requirements for modern cw EPR.[4] Phase locking or cavity stabilization, however, limits the sweepability of the mw sources that is necessary, for example, to control the matching of the EPR resonator. The diode devices can be used as mw sources, amplifiers and active multipliers. This property allows realization of different designs to obtain the desired EPR operation frequency, either by multiplying the low-frequency microwaves or by combining two mw fields by a balanced mixer with further mw amplification. The diode devices can be expected to operate continuously for over 100 000 h, thus offering a longer lifetime than vacuum-tube oscillators.

For frequencies above 140 GHz it is necessary to employ one or more stages of frequency multiplication using varactor-diode harmonic multipliers. While harmonic multiplication can be very efficient in frequency doubling or even tripling, the efficiency drops rapidly at higher harmonic numbers because of the difficulty in properly terminating the harmonics below the desired one. Typically, power levels of 200 mW at about 90 GHz are available prior to multiplication. Output powers of 20 mW and 5 mW can be obtained after doubling to 180 GHz[64] and tripling to 280 GHz,[68] respectively. At 360 GHz, a mw power of 1 mW is obtained after tripling the 120 GHz microwave.[72] Such low power levels turned out to be already satisfactory to perform pulsed EPR experiments. Using single-mode TE_{011} cavities, $\pi/2$ pulse lengths of 60 ns (180 GHz) and 100 ns (275 GHz) have been reached.[64,68] The rapid development in the diode oscillator, amplifier and multiplier techniques[99,103–105] raises hope that in the near future the output power of solid-state devices can be increased to a level of pulsed high-frequency EPR performance that is comparable with that of X- or Q-band pulsed EPR spectrometers.

Vacuum-tube oscillators inherently produce higher mw powers than solid-state oscillators. For example, reflex klystrons are capable to produce up to 1 W cw radiation at frequencies up to 220 GHz. Due to their excellent phase-noise characteristics, klystrons found wide application in the first versions of high-field EPR spectrometers.[39] Unfortunately, since the beginning of the 1990s the production of reflex klystrons above 80 GHz was continuously curtailed and finally stopped. The resulting high price together with the limited lifetime of klystrons (5000 h typical) had the consequence that reflex klystrons are no longer considered as mw sources for high-field EPR.

BWOs are suitable sources for mw radiation up to 2 THz.[106] At 100 GHz they can generate more than 10 W in cw, and more than 10 kW in pulsed mode.[107] At higher mw frequencies the output power of BWOs declines drastically to about 1 mW at 1000 GHz. They are commercially available for several frequency bands and tuneable over the whole range of a band. Only a few applications of BWOs in high-field EPR were described in the literature.[81,108] Probably, this is because the output characteristics in terms of phase and frequency jitter are determined by the stability of the high-voltage BWO power supply. Thus, either very stable power supplies in the kV region (stabilized to 10^8) or phase-locking techniques are needed to comply with the frequency-stability and phase-noise requirements for cw EPR.

The medium-power gyrotrons (cyclotron-resonance masers) developed at Fukui University in Japan have been used for high-field EPR.[83,84] They supply more than 100 W cw power at frequencies below 440 GHz at fundamental operation of the electron cyclotron resonance, and about 10 W below 880 GHz in harmonic operation. However, gyrotrons require the same external magnetic fields as EPR does for $g=2$ systems at this resonance frequency, *i.e.* 16 T for 440 GHz. This narrows the general applicability of these gyrotron devices in high-field EPR. At MIT in Cambridge the research group of R.G. Griffin had developed a 140-GHz microwave bridge for high-field EPR employing a gyrotron with 1–10 W output power.[63] The gyrotron was also used in ELDOR experiments in which the 10 W output power was gated with a diode array switch.[109] The Griffin group improved the gyrotron design related to stability and maintenance and operates them routinely in cw mode at 140 and

250 GHz.[110] To use them in pulse mode for EPR and ENDOR faces problems with the present diode array switch. It requires development of novel switching devices that can handle the available cw output power levels of the gyrotrons and allows for switching on and off with sufficient speed. Besides in high-field EPR and ENDOR, the Griffin group applies cw gyrotron oscillators at power levels of 25–100 W in DNP-enhanced NMR studies at frequencies up to 460 GHz operating at the second harmonic of the cyclotron frequency, $\omega - 2\omega_c$,[111–113] (DNP: dynamic nuclear polarization).

A conceptually different type of vacuum-tube high-frequency microwave source, the Orotron, was introduced to high-field EPR spectroscopy by the Möbius group at FU Berlin.[73] It was designed as the pulsed 360-GHz source in the quasioptical mw bridge feeding a Fabry–Perot resonator in the 14-T cryomagnet. Pulsed Orotron operation is achieved by inserting an additional gate electrode into the vacuum-tube device that is connected to a high-voltage pulsing unit to control the electron-beam current. The generated pulses at 360 GHz have pulse lengths from 100 ns – 10 μs and a pulse power of about 30 mW. We discuss the pulsed Orotron and first 360 GHz FID results from a test sample further below in Section 3.4.2.9.

Recently, the extended interaction klystron amplifier (EIKA) was successfully introduced as mw source in pulsed W-band EPR[54] by the Freed group at Cornell. Their spectrometer achieves 3–5 ns π/2 pulses using 1 kW output power of the EIKA quasioptically transmitted to a Fabry–Perot resonator.

Beyond 400 GHz far-infrared (FIR) lasers offer high sub-mm microwave power at spot frequencies and, indeed, have been successfully used for high-field EPR.[46,75,78,114] FIR laser sources suffer, however, from high amplitude- and phase-noise levels, which limit their applicability in cw EPR, and from the lack of tunability. In the pulsed mode, the pulse-sequence formation and variation of interpulse distances require special efforts[78,115] that restrict the application of FIR lasers as sub-mm mw source to special cases.

3.3.4 Resonators

The resonator is a crucial part of an EPR spectrometer having a large impact on sensitivity and bandwidth. The resonator amplifies the microwave fields, localizes the field maximum at the sample and picks up the EPR response and couples it to the receiver. The main characteristic feature of a resonator is its loaded quality factor Q_L. It consists of two parts: the unloaded quality factor Q_U, which is determined by the losses in the resonator and sample, and the radiation (coupling) quality factor Q_R with $1/Q_L = 1/Q_U + 1/Q_R$. Beside the Q_L-value, the filling factor η, which quantifies the ratio of the energy of the mw magnetic field at the sample to that integrated over the resonator volume, is of particular importance for EPR sensitivity. In the case of a critically coupled resonator ($Q_U = Q_R$, $Q_L = Q_U/2$), the concentration sensitivity is proportional to ($Q_U \cdot \eta \cdot \omega_0$), which has to be maximized for cw EPR experiments. The conversion factor, K, which relates the B_1 field to the square root of the incident power, $B_1 = K \cdot \sqrt{P}$, is proportional to $(Q_L \cdot \omega_0^{-1} \cdot V_C^{-1})$, where V_C is the effective volume of the resonator. This factor has to be optimized for pulsed EPR, since a high B_1 field is required to obtain short mw pulses. The response time, $T_R = 2Q_L \cdot \omega_0^{-1}$, determines how fast the resonator reacts on changes of mw power either due to EPR absorption or incident radiation. This response time has to be considered in both pulsed and time-resolved EPR. Beyond the above quantities, an additional feature gains importance when going to high mw frequencies: The resonator for high-field EPR has to be frequency tunable because of large frequency shifts due to the sample. High-performance mw sources at high frequencies operate only in a narrow frequency region or even at a locked frequency. Therefore, it is mandatory to match the resonance frequency of the resonator to the frequency of the mw source, and not the other way round as is typically done in X-band EPR.

3.3.4.1 Single-Mode Cavities

Single-mode resonators, commonly termed "cavities", clearly provide the highest absolute sensitivity and a maximum B_1 field at the sample. Therefore, particular efforts have been devoted to the development of efficient millimeter-wave single-mode cavities. Cylindrical cavities operating in TE_{011} mode were found to be excellent resonance structures for high-field EPR. Such cavities are successfully used in high-sensitivity EPR spectrometers operating up to 275 GHz, see Table 3.1. The simple geometry of a cylindrical cavity in conjunction with its suitable dimensions up to mm wavelengths allows different probehead configurations for a large variety of EPR experiments including ENDOR, light-excitation or continuous flow-system[116] schemes to be realized.

The typical TE_{011} cavity designed for cw EPR consists of a thin-walled metal cylinder of diameter approximately equal to the operation wavelength, one or two metal plungers movable along the cylinder axis to allow for frequency tuning of the cavity by changing the cavity length, see Figure 3.3(a). The cavity is coupled to the rectangular waveguide of the mw transmission line by a circular coupling hole (iris) placed laterally or at an end plate. There are several

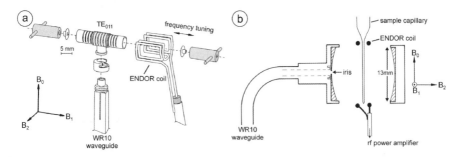

Figure 3.3 (a) Exploded view of the ENDOR version of the W-band cylindrical TE_{011} cavity. (b) ENDOR version of the W-band Fabry–Perot resonator. For details, see ref. [39].

possibilities for realizing variable coupling, which is necessary to achieve optimal coupling of the cavity to the waveguide. The simplest one uses an iris diameter corresponding to the critically coupled cavity–waveguide configuration of the empty cavity. If insertion of the sample reduces Q_U, the coupling quality factor Q_R can also be reduced to match Q_U either by moving a tiny piece of metal, for instance in ball form, or a sheet of dielectric material closer to the iris[39] or by changing the dimensions of a waveguide transformer. An alternative coupling mechanism requires that the diameter of the coupling hole is large enough to provide $Q_R > Q_U$ for the empty cavity (overcoupling). The Q_R-value can be continuously reduced by rotating the cavity around a vertical axis that goes through the waveguide and the coupling hole.[53] This coupling mechanism is particularly interesting for cavities operating above 200 GHz, because it allows an iris with relatively large diameter to be used, for instance 0.34 mm at 275 GHz.[68]

Table 3.2 summarizes the theoretical performance data of empty TE_{011} cavities in the frequency range from 95 to 360 GHz. The maximum Q-value is obtained if the cavity length, H_C, and the cavity diameter, D_C, are equal.[2] For comparison, 9.5 GHz (X-band) data are also given. The advantages of single-mode TE_{011} cavities are clearly seen. The cavities afford a value of $Q_U \cdot \eta \approx 1000$ up to 360 GHz. In contrast to X-band, the short voltage response time of $T_R = 14$ ns at W-band with the highest possible Q_U-value allows the use of the TE_{011} cavity in pulsed mode without overcoupling, i.e. without lowering the loaded quality factor Q_L by increasing Q_R. Short mw pulses can be produced with relatively low incident mw power due to the effective conversion of mw power to B_1 field strength. Theoretically, it should be feasible to produce a 360-GHz $\pi/2$ pulse length as short as 20 ns with only 2 mW incident mw power on a TE_{011} cavity. The values actually reported for different TE_{011} variants are very close to the calculated ones for the various mw frequencies. At W-band, quality factors Q_U of about 6000 have been achieved.[39,44,50,117] Q_U values of 4000 and

Table 3.2 Comparison of theoretical performance data of a cylindrical TE_{011} cavity for different mw frequencies.

Frequency GHz	$D_C = H_C$ mm	$Q_U{}^a$	T_R ns	$D_S{}^b$ mm	$V_S{}^c$ µl	$Q_U \cdot \eta$	K $mT/\sqrt{mW}$	$P(t_{\pi/2})^d$ mW
9.5	41.6	27 000	450	8.80	2500	5400	0.11	18 000
95	4.16	8600	14	0.88	2.5	1720	1.86^e	57
180	2.20	6200	5.5	0.47	0.38	1240	4.13	12
275	1.44	5000	2.9	0.30	0.10	1000	7.02	4
360	1.10	4400	1.9	0.23	0.05	880	9.83	2

^{a}calculated taking into account the resistive losses in the cavity walls with the resistivity of gold at room temperature $\rho = 2.2 \times 10^{-8}\,\Omega\,m^{-1}$.
^{b}for sample size corresponding to $\eta = 0.2$ and the magnetic-to-electric filling factor ratio of $\eta/\eta_E = 10$.
^{c}for sample diameter D_S.
^{d}microwave power to produce the maximum B_1 field required for 20 ns $\pi/2$ microwave pulses.
^{e}a value of 1.68 $mT/\sqrt{W}$ for $Q_U = 7800$ is reported.[118]

3000 have been reported for 180 GHz[64] and 275 GHz[68] cavities, respectively. The experimental conversion factors are also approaching the calculated ones. For example, at W-band a $\pi/2$ pulse length of 18 ns was obtained with 65 mW incident mw power on the cavity,[117] see Section 3.4.1. And 100 ns pulses could be achieved at 275 GHz with only 1 mW incident power.[68]

Single-mode cavities provide excellent performance data especially for pulsed experiments. When increasing the frequency to sub-mm wavelengths, however, they are becoming too small for sample handling. This restricts the admissible volume of the sample. For highly concentrated samples, this may even be advantageous. For biological systems close to *in-vivo* conditions, however, the spin concentrations are rather limited. Moreover, biological samples often contain water as solvent. Water exhibits a high dielectric permittivity (ε', ε'' defining the loss tangent: $\tan\delta = \varepsilon''/\varepsilon'$) which is strongly dependent on the frequency of the electric field experienced by the sample in the mw cavity. This leads to dielectric shifts of the resonance frequency (ε' effects) and dielectric absorption losses (ε'' effects). At 100 GHz, for water at room temperature the permittivity values are $\varepsilon' \approx 8$, $\varepsilon'' \approx 18$, falling off to higher frequencies.[119] The damping loss lowers the Q_U of the cavity and reduces the EPR sensitivity. The way to overcome this problem in a cylindrical cavity is to reduce the diameter of the sample capillary, but this is paid for by a reduced filling factor. Optimization of water sample size at 298 K in terms of $Q_U \cdot \eta$ results in capillary diameters of 0.15 mm at W-band and 0.09 mm at 275 GHz,[120] corresponding to sample volumes of 70 nl and 10 nl, respectively. This puts the concentration requirements to the mM region, which is often difficult to reach in biological samples. The solution to this problem is to use mw resonators of different type, and Fabry–Perot resonators are often a good option.

3.3.4.2 Fabry–Perot Resonators

Figure 3.4 shows the calculated sensitivity of a 140-GHz EPR spectrometer with different mw resonators for different dielectric losses of the sample.[37] In the region of $\tan\delta > 1$, Fabry–Perot resonators are expected to have a better concentration sensitivity, while for $\tan\delta < 1$ TE$_{011}$ cavities are the better choice. Amity[32] was the first in the high-field EPR community who appreciated the additional advantages to be gained by using open resonator structures. Advantages are, for example, high Q-factor, convenient resonator size especially in the sub-mm region, easy implementation of field-modulation coils. This also holds for the implementation of ENDOR coils as well as devices for light excitation of the sample.[39]

In principle, microwave high-Q Fabry–Perot resonators can be constructed from any two flat conducting reflectors with accurately aligned parallel faces. This simplest configuration suffers, however, from two main limitations: the diffraction losses on the reflector edges and the requirement of an extremely precise alignment of the mirrors and incoming mw beam. Replacement of one or both reflectors by a concave mirror (usually spherical) have the advantage of

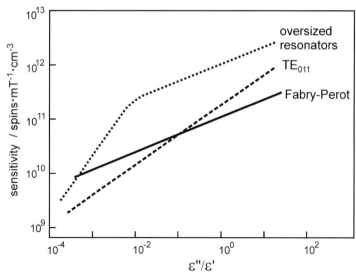

Figure 3.4 Computed cw EPR sensitivity of a 140-GHz spectrometer with the different resonators *vs.* sample dielectric losses[37] quantified by $\tan\delta = \varepsilon''/\varepsilon'$.

focusing the mw field into a smaller volume, thereby reducing diffraction and alignment problems, provided that the reflectors are of sufficiently large aperture. The stability criteria for biconcave FP resonators and their field distribution and performance characteristics have been analyzed in depth,[121,122] and experimental measures to avoid mode degeneracy occurring in exact confocal mirror arrangements are described in the literature.[123] FP resonators have been constructed and tested in almost all laboratories developing mm high-field EPR spectrometers, see Table 3.1. The resonance frequency of the FP resonator can be tuned by changing the distance between the mirrors. The mw coupling can be achieved either quasioptically or by waveguide methods. The mw waveguide coupler comprises a fundamental-mode rectangular waveguide fixed to the back of the reflector that is furnished with a coupling hole. The thickness and diameter of the iris hole have to be chosen for critical coupling of the TE_{01} waveguide mode to the TEM_{00q} mode of the FP resonator. Critical coupling over a wide range of resonator Q_U values is achieved by a variable coupling mechanism.[39] Because, the fundamental TEM_{00q} modes have Gaussian profile, they can also be coupled to the output of a corrugated waveguide through the partially reflective planar mirror of a plane–concave FP configuration.[4,54,72,80] This mirror can either be a metallic wire mesh or a dielectric window. The compatibility of FP resonators with quasioptics allows a number of polarization encoding and detection schemes to be designed. The most important reflection-induction scheme will be described further below.

For EPR, the typical FP resonator operates in the TEM_{00q} mode with the number of half-wavelengths in the resonator, q, around 6. This mode is axially symmetric, and the field variations parallel to the cylindrical axis correspond to

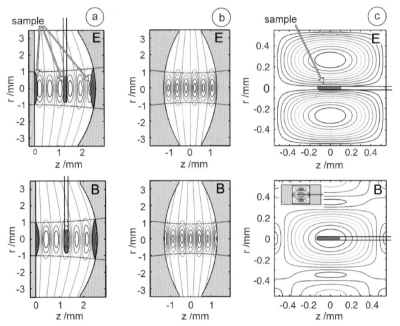

Figure 3.5 Contour plots of the electric (E) and magnetic (B) mw field amplitudes of the TEM$_{006}$ mode in a half-symmetric plane–concave (a) and symmetric concave–concave (b) Fabry–Perot resonator. The dimensions are identical with the actual FP resonator in the 360-GHz EPR spectrometer: $R = 8.06$ mm, $d = 2.6$ mm.[72,124] The electric and magnetic field amplitude distributions in a 360-GHz TE$_{011}$ single-mode cavity are shown in (c). The insert in the bottom-right picture shows the cavity in scale of the FPs (a) and (b).

the number q. The B_1 field profile in a plane perpendicular to the cylinder axis has a Gaussian shape. Typically, near-confocal arrangements of the reflectors are used, *i.e.* the radius, R, of reflector curvature is nearly equal to the distance, d, between the mirrors. The exact confocal arrangement of the mirrors ($R = d$) would lead to degeneracy of the higher modes in the resonator. Thus, the B_1 profile width, called "beam waist", is slightly depending on the position between the two mirrors. The highest B_1 field appears in the center of the resonator in a concave–concave reflector configuration, and close to the planar reflector in the plane–concave configuration. Figure 3.5 shows the contour plots of the magnetic and electric fields for the TEM$_{006}$ FP resonator in the plane–concave and concave–concave configurations optimized for operation at 360 GHz.[124] The possibilities to place the sample in the region of maximum B_1 field but minimum electric field are indicated. The advantages of FP resonators compared to single-mode cavities are obvious: It is possible to place the sample either in the middle of the FP resonator or on one of the mirrors. Moreover, the radius of the cross section with high B_1 field is approximately equal to the wavelength. This increases the useful sample volume without going too much into the region of significant electric field amplitude.

Ideally, the quality factor of a FP resonator is determined solely by resistive losses in the mirrors at a particular frequency, ω_0. The unloaded quality factor is given by

$$Q_U = \frac{d}{2} \cdot \left(\frac{\mu_0 \cdot \omega_0}{2 \cdot \rho}\right)^{1/2} \tag{3.6}$$

where ρ is the resistivity of the mirror material and d the mirror separation.[125] The resonance condition for the TEM_{00q} mode is

$$\frac{2 \cdot d}{\lambda_{00q}} = q + \frac{1}{\pi} \cdot \cos^{-1}\left(1 - \frac{d}{R}\right) \tag{3.7}$$

where $\lambda = 2 \cdot \pi \cdot c/\omega_0$.[126] A value $Q_U = 22\,000$ can be reached for a W-band TEM_{006} FP resonator with gold mirrors ($R = 20$ mm). This is more than a factor of 2 higher than for a cylindrical TE_{011} cavity. To compare the theoretical concentration sensitivities for the FP resonator and the cylindrical cavity, the product $Q_U \cdot \eta$ has to be compared. Taking a nonlossy sample in a 0.8-mm capillary, filling factors of 0.003 and 0.2 are calculated for the FP resonator with TEM_{006} mode and a TE_{011} cavity, respectively. This results in $Q_U \cdot \eta = 66$ and 1720. Thus, for nonlossy samples the cylindrical cavity provides 25 times higher sensitivity than the FP resonator. This unfavorable factor can be drastically decreased if a flat cell in combination with the FP is utilized.[127] The filling factor of a 0.9-mm thick flat cell in a W-band FP resonator becomes 0.016 and the corresponding sensitivity factor $Q_U \cdot \eta = 350$. For lossy samples, however, FP resonators allow the use of much larger samples than cylindrical cavities would allow, thereby offering high concentration sensitivity, see Figure 3.4.

In practice, TE_{011} cavities can approach the theoretical limits of the quality factors. The experimental quality factors of Fabry–Perot resonator, however, are reduced due to diffraction losses and losses from the coupling iris or additional resistive losses in the coupling mesh. The different systems can be compared using the "finesse" of the resonance structure. This parameter is defined as Q_L/q and, thus, is independent of the mode number q. From $Q_U = 22\,000$ for $q = 6$ a theoretical limiting finesse of about 1800 is deduced. At W-band, for a FP resonator with iris coupling it is possible to realize a finesse of about 1200 by using large-diameter mirrors of 19 mm.[39] For the mesh coupling, the experimentally observed finesse decreases to about 500,[4] due to extra losses in the mesh. Going to higher frequencies continuously reduces the experimental finesse. Values of 260 at 180 GHz,[4] 130 at 360 GHz[72] and 40 at 600 GHz[78] were reported.

The Fabry–Perot resonators exhibit smaller conversion factors compared to single-mode cavities because of their larger volume. The total magnetic field energy is distributed between several nodal zones compared to only one in a single-mode cavity. To produce the same B_1 field at the sample position, the FP resonator in TEM_{006} mode requires at least a factor of 6 (in practice above 10) more incident mw power compared to the TE_{011} cavity. Therefore, FP resonators are less suitable for pulsed EPR experiments, excluding the rare cases when mw sources with sufficiently high power are available.[54,62,73] On the other hand, the larger volume of the FP in combination with its open structure makes the system more flexible compared to single-mode cavities. For example, the ENDOR coils can be easily introduced, see Figure 3.3(b). Different light-excitation schemes including optically detected EPR[128] are easy to realize. The full 3-axes rotation of single-crystal samples becomes possible by introducing a goniometer into the W-band FP resonator,[129] see further below. Furthermore, FP resonators can also be used in transmission mode. A transmission-mode FP resonator, coupled by two conical feedhorns to the quasioptical mw beam, was employed in the first version of a 250 GHz EPR spectrometer at Cornell.[45] Crossed-waveguide FP resonators with high mutual isolation were described for 95 GHz[130] and 140 GHz.[62] The orthogonality of the detection and excitation waveguide branches allowed an effective isolation of about 40 dB between receiver and transmitter to be achieved without using circulators or their quasioptical analogues. Thus, principally FP resonators offer more opportunities for realizing different types of EPR experiments, however, at the price of lower sensitivity and smaller B_1 fields for a given mw source.

3.3.4.3 Loop-Gap Resonators

Loop-gap resonators (LGR) in the frequency range of 2 to 35 GHz were introduced as a powerful alternative to single-mode resonators.[131] The LGR structure is determined by the geometry of the conducting loop and the isolating gaps, at which the ends of the loop approach each other. The loop is an inductive element surrounding the sample, while the gaps are capacitive elements. The dimensions of LGRs are typically 1/10 to 1/3 of the resonance wavelength. Compared to single-mode cavities, LGRs offer several attractive features such as a large mw power-to-field conversion factor, high filling factor and excellent mw field homogeneity over the sample. The quality factor of a LGR is typically one order of magnitude smaller than that of single-mode cavities. This is advantageous for pulsed and time-resolved EPR applications, especially at low mw frequencies, due to an increased resonator bandwidth, thereby shortening the response and deadtime. In principle, LGRs can be designed for any mw frequency. However, at high frequencies the dimensions of the LGR structures become so small that the machining precision becomes the limiting factor for the practical realization of LGRs. Recently, the fabrication of a LGR resonator for W-band EPR was reported.[132] It is designed for and realized by electric-discharge machining in the form of a 5-loop-4-gap structure. The resonator height is 2.5 mm and includes five equal loops with a diameter of 0.65 mm, connected by gaps of only 0.083 mm thickness. The comparison with a TE_{011} cavity showed that the signal intensities are about the same when the B_1 values at the sample are the same. The conversion factors of both resonators were also found to be similar. Thus, the only benefit of the

LGR structure at high frequencies is the greatly increased bandwidth ($Q_U = 1500$ at 95 GHz) relative to the TE$_{011}$ cavity ($Q_U = 8600$ at 95 GHz). This would be useful to reduce the impact of mw source phase noise in cw EPR, and to reduce the deadtime in pulsed EPR.

The loop-gap structures suffer from specific disadvantages at high frequencies: The resonance frequency of a LGR strongly depends on the manufacturing quality, and the insertion of typical sample capillaries shifts the resonance frequency of a LGR by up to 1%. Hence, a wide-range frequency-tuning mechanism would be desirable that, however, is difficult to realize taking into account the tiny structural dimensions at W-band or even higher frequencies.

3.3.4.4 Dielectric Resonators

Besides the established benefits of loop-gap resonators in X-band EPR, single-mode dielectric ring resonators present remarkably high quality factors. The unloaded quality factor Q_U is mainly determined by the dielectric loss in the resonator walls and, thus, is determined by $\tan\delta = \varepsilon''/\varepsilon'$ of the dielectric material. Properly chosen dielectric materials may increase the Q_U value to above 10 000 at X-band and room temperature. The resistive losses in the metallic radiation shield around the dielectric resonator are typically negligible at X-band or lower frequencies. Therefore, dielectric ring resonators are widely used in L- to X-band cw and pulsed EPR, for example the commercially available Bruker probehead (Bruker ER 4118x-x FlexLine Resonators) incorporates a dielectric ring resonator. The unique features of dielectric resonators stimulated the development of their high-frequency versions.[133] Two variants were suggested and developed by the Martinelli group in Pisa:[74,118,133–135] the dielectric ring resonators working in travelling-wave resonant modes (the whispering-gallery mode, WGM resonators) and the nonradiative dielectric tube (DT) resonators operating in the TE$_{011}$ mode.

The WGM resonators were successfully tested between 190 and 316 GHz.[74,134] The resonance structure consists of two stacked polyethylene discs of 15 mm diameter, properly coupled to the dielectric waveguide. The coupling can be varied by changing the relative position between waveguide and resonator. The resonance frequency of the WGM resonator can be tuned by a conducting disk placed parallel to the resonator and moved axially. The sample can be loaded between the two polyethylene discs. The active part of the resonator, where the interaction of the mw radiation with the sample occurs, is the volume between the two discs, bounded by the external rim. An unloaded quality factor of about 10 000 was measured. For an active sample volume of 4 μl at 285 GHz, a filling factor of 0.06 was estimated. This results in a sensitivity factor $Q_U \cdot \eta \approx 600$, coming close to the value of a single-mode cavity. The mw power-to-field conversion factor of $0.4\,\text{mT}/\sqrt{\text{W}}$ is comparable to that of a FP resonator because of similar resonator volumes. Thus, in principle the WGM resonator fulfils the requirements for high-field EPR. However, it has a significant disadvantage: Because of the travelling-wave nature, the WGM resonators do not have any region where the electric field is stationary vanishing. Therefore, the WGM resonator is very sensitive to dielectric loss in the sample, which in any position, is exposed to both electric and magnetic mw fields. In practice, this makes WGM resonators applicable only to samples without significant dielectric losses and, thus, excludes most biological systems, which notoriously contain water.

The dielectric tube or ring resonators are routinely used in X-band EPR. The characteristic dimensions of the tube are determined by the dielectric constant, ε', of the material. At X-band, a sapphire ($\varepsilon' = 9.4$) tube with outer diameter ≈ 10 mm, inner diameter ≈ 5 mm and length ≈ 20 mm provides high conversion, quality and filling factors for EPR samples of suitable dimensions. At high mw frequencies it becomes necessary to decrease the tube dimensions. The outer diameter of the tube, assumed to be resonant in the TE_{011} mode, is given by $D_{out} = 7.66 \cdot c/\omega_0 \cdot \sqrt{\varepsilon'}$,[118] which yields 1.2 mm for 95 GHz and $\varepsilon' = 9.4$. Thus, the inner diameter has to be about 0.6 mm and is comparable with the loop diameter of a LGR.[132] In practice, this limits the applicability of DT resonators to mw frequencies not exceeding 150 GHz. A prototype DT resonator for W-band was recently presented.[118,135] It consists of an arbitrarily long quartz ($\varepsilon' = 3.8$) tube (1.6 mm outer and 0.8 mm inner diameter) placed between two plane aluminum mirrors ≈ 1.5 mm apart. The resonance frequency can be varied by changing the distance between the metal mirrors. An unloaded quality factor of 5500 was calculated taking into account the aluminum conductivity only. A lower value of 1500 was measured due to finite dielectric losses in the Suprasil quartz used as dielectric material. A conversion factor of $1.8 \, mT/\sqrt{W}$ was estimated for the configuration with the experimentally determined Q_U. The high B_1 field achieved in combination with the low quality factor, as well as the simple and cheap realization when compared to loop-gap resonators makes the DT resonators a promising alternative to single-mode cavities and LGRs for pulsed EPR applications up to 150 GHz.

3.3.4.5 Nonresonant Systems

In cw EPR at frequencies above 100 GHz there is no great advantage in using a resonator for nonlossy samples provided that a sufficient quantity of the sample is available. The concomitant loss in concentration sensitivity can be partially compensated by using much larger sample volumes, up to 1 ml, so that the absolute sensitivity remains comparable to that of cavity-based mw systems, see Figure 3.4. An obvious benefit of large sample volumes is the ability to carry out complex biochemical manipulations of the sample. Large protein volumes can be characterized, prior to the high-field EPR study, by optical or X-band EPR spectroscopy without changing the sample holder.[71]

The nonresonant sample holders can be integrated in both waveguide and quasioptical designs. They offer simple sample handling and are suitable for different types of cw and time-resolved EPR experiments, for example those that require light excitation, sample rotation, thin-film positioning parallel or perpendicular to the external magnetic field. The nonresonant systems provide relatively small B_1 fields at the sample and, therefore, are not acceptable for pulsed EPR. The small B_1 for a given mw power, however, turns out to be advantageous for cw EPR experiments at low temperatures, making the spin system less susceptible to saturation effects. Figure 3.6 gives an overview of different nonresonant probehead configurations. They were constructed by the Tallahassee EPR group for a spectrometer operating at 120 and 240 GHz, utilizing wideband operation of nonresonant probeheads.[67] A concentration sensitivity of 10^{13} spins/(mT cm^3) at 240 GHz was measured using a polymerized nitroxide radical solution in polystyrene. At 90 GHz, a value of 10^{12} spins/(mT cm^3) was reported for a nonresonant FP probehead with removed coupling mesh.[4] However, the absolute sensitivity of 7×10^{11} spins/mT at 240 GHz and 10^{10} spins/mT (1 Hz bandwidth) at 90 GHz is about two orders of magnitude lower than the sensitivity with resonant probeheads.

It is pointed out that also in recent 360-GHz cw EPR experiments on frozen-solution flavin radicals a nonresonant probehead consisting of a FP resonator with removed plane coupling mesh was used.[136] And even the first 360-GHz cw ENDOR results were obtained by using this nonresonant probehead with an additional NMR rf coil around the sample (polycrystalline nitroxide radicals) placed on the spherical FP mirror.[137]

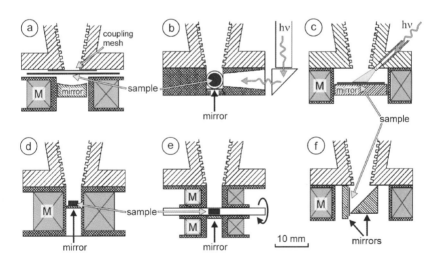

Figure 3.6 Examples of nonresonant sample-holder designs for high-field EPR: Fabry–Perot resonator, nonresonant with removed mesh (a); light-accessible probeheads (b,c); cw EPR probehead for solid or powdered samples (d); probehead for single-crystal experiments with one-axis rotation (e); thin-film probeheads for film mounted ⊥ (c) and ∥ (f) to the external magnetic field. The modulation coils are denoted by M. Figure adapted from ref. [67].

3.3.5 Microwave Transmission Lines

In a conventional EPR spectrometer the mw power from the mw bridge is transferred to the sample-containing resonator by means of a mw transfer line. In reflection mode, the same line is used to transfer the mw response from the resonator to the detector. Therefore, care has to be taken to keep the insertion loss of the transfer line as small as possible. High losses of the transfer line lead to a reduction of the mw excitation power available at the resonator. They also diminish the EPR signal power in relation to the overall noise power of the spectrometer, *i.e.* they decrease the spectrometer sensitivity.

In an X-band EPR spectrometer the typical length of the transfer line is about 1 m. This line is built in the form of a fundamental-mode rectangular waveguide or coaxial cable. At 9.5 GHz, the WR90 rectangular TE_{10} mode waveguide attenuates the mw power by about 0.1 dB/m. Typical losses of modern semirigid or flexible low-loss coaxial cables are close to 1dB/m. Both transfer-line realizations do not contribute much to a degradation of the spectrometer performance away from optimum conditions.

Most high-field/high-frequency EPR spectrometers require approximately the same transfer-line length as at X-band. A typical high-frequency spectrometer, equipped with a superconducting magnet, needs about 1 m line length in the case of a top-mounted probehead. This length can be reduced to about 0.5 m if a bottom-mounted probehead is used. Coaxial cables as transfer line become inapplicable at high mw frequencies. Above 26 GHz the insertion losses exceed 2 dB/m and reach 6 dB/m at about 70 GHz. The fundamental-mode rectangular waveguides also lose attraction when going to higher mw frequencies. For instance, at 140 GHz the WR7 waveguide (copper, room temperature) attenuates by 5.6 dB/m. Thus, other solutions have to be found to build a low-loss transfer line for frequencies higher than 90 GHz.

The necessity to improve transmission-line technologies in millimeter and submillimeter wavelength regions appeared in the 1950s in connection with the rapid development of radar techniques and radio astronomy. Additional incentives came from mw spectroscopy, hot-plasma diagnostics as well as from the idea of long-distance energy transfer by high-frequency microwaves. All these techniques require exceedingly small transmission losses of microwave power over meter distances. In the following years, the efforts of scientists and engineers resulted in numerous solutions of the transmission-line problems. Below, we consider different types of transfer lines and their application in high-field EPR spectrometers.

In metallic waveguides the electric and magnetic fields are confined to the space within the waveguides, and no mw power is lost by radiation. Dielectric losses due to the air inside the waveguide are negligible. However, some mw power is lost due to electric mw currents in the conducting walls of the waveguide, producing ohmic losses. At low mw frequencies the ohmic loss is usually very small. With increasing frequency these losses increase, reaching unacceptable levels ($>$3dB/m) above 100 GHz in single-mode rectangular waveguides machined from copper. An obvious solution is to use oversized waveguides with cross sections considerably larger than normally used for the specific frequency band. Increasing the waveguide cross section leads to decreasing wall currents. For example, at 95 GHz the microwave is attenuated in the single-mode waveguide WR10 by 2.5 dB/m. In the WR28 waveguide (2.8 times larger linear dimensions), the fundamental mode TE_{01} has a loss of 0.55 dB/m, in WR90 waveguide (9 times larger linear dimensions) the loss further decreases to a theoretical value of 0.18 dB/m. Thus, in theory, oversized waveguides have drastically lower attenuation than standard ones, in practice, however, parasitic excitation and propagation of high-order mw modes lead to additional losses. Moreover, surface-resistance values in the mm and sub-mm region are several times higher than expected from tabulated DC values of the resistance of the wall material, resulting in waveguide losses in excess of what is theoretically expected. Additional losses arise from the two tapered transitions required to transfer the microwave from single-mode to oversized waveguides and back. All in all, oversized rectangular transmission lines find practical application up to 140 GHz. Various laboratories opted for oversized rectangular waveguides for their high-frequency EPR spectrometers, for example at W-band,[39,49,53] achieving for their transmission lines losses between 1.5 dB[39] and 3 dB.[49] For a 140-GHz EPR/ENDOR spectrometer, insertion losses of 8–9 dB for a transmission line consisting of a combination of oversized and single-mode waveguides, tapers and bends were reported.[61]

Also, oversized circular waveguides were considered as an alternative to oversized rectangular waveguides. For small oversizing, however, the transfer losses of the fundamental circular TE_{11} mode are comparable to those of the rectangular TE_{10} mode. Total losses of 3 dB at 95 GHz were reported for a 1.5-m transfer line that includes an oversized TE_{11} circular waveguide of 4 mm diameter and two rectangular-to-circular tapered mode transitions.[40] For large oversizing, the TE_{01} mode becomes the mode of lowest loss. But imperfections of the cylindrical geometry can easily cause mode conversion that increases the transfer losses.

An alternative approach was made by incorporating a dielectric waveguide in a W-band EPR spectrometer.[52] The flexible waveguide is made from PTFE, and insertion losses of <1 dB/m were observed.

For "ultra" high-frequency EPR spectrometers employing quasioptical elements in their mw bridge, several methods have been tried for low-loss transmission of the microwaves between the quasioptical bridge and the probehead. These methods include large oversized cylindrical waveguides,[75,78] lens trains[45] and dielectric tubes.[78] Such methods did not find wide applicability in sub-mm EPR spectrometers because of the difficulty to couple the Gaussian mw beam to the transfer line, distortions of the beam properties due to the mode conversions in cylindrical pipes, high insertion losses of dielectic waveguides, and high insertion losses of lens trains. The method that has become most commonly used is the corrugated cylindrical waveguide, see Table 3.1.

The advantages of such corrugated waveguide transmission lines are: high coupling efficiency between the Gaussian beam and the HE_{11} mode in the corrugated waveguide, very low propagation losses of the HE_{11} mode. Typically, the corrugations in such a waveguide are rectangular with $\lambda/4$ depth, spaced at a distance of about $\lambda/2$, where λ is the central wavelength of the mw beam. The maximum low-loss frequency is related to the corrugation width and period, the minimum low-loss frequency is related to the corrugation depth and pipe diameter. In the corrugated waveguide, the HE_{11} mode can propagate with any polarization, from linear to circular, with negligible crosspolarization. More details including operation principles can be found, for example, in ref. [138].

The corrugated cylindrical waveguide transmission lines are, in principle, applicable for any high-frequency spectrometer operating at $\geq 90\,GHz$.[4] For mw bridges in rectangular waveguide designs, however, they are less attractive because expensive transitions are required to taper the fundamental rectangular propagation mode TE_{10} to the cylindrical HE_{11} mode of the corrugated waveguide. At frequencies above 140 GHz, as well as at lower frequencies for spectrometers based on quasioptical elements, the cylindrical corrugated waveguides are now commonly used. Their typical insertion losses in high-field EPR spectrometers are about 1 dB.

3.3.6 Magnet Systems

The magnet system for an EPR spectrometer has to fulfil at least three conditions: *(i)* the inhomogeneity of the static magnetic field over the EPR sample should be smaller than the EPR linewidth of the system under study, *(ii)* the field instability has to be much smaller than the EPR linewidth, otherwise additional noise will be generated, *(iii)* the magnetic field has to be sweepable over the full range of the EPR spectrum. At lower fields, iron-core electromagnets perfectly fulfil these requirements. Unfortunately, the iron saturation magnetization of about 2.2 T (corresponding to 60 GHz EPR frequency for $g=2$ samples) does not allow construction of iron-core magnets suitable to produce static fields much over 2 T without drastic increase of the magnet size and electric power consumption.[139]

For mm and sub-mm high-field EPR applications, sweepable superconducting magnets normally provide the required Zeeman fields, B_0. Present-day Nb-based superconducting magnet technology approaches its limit of B_0 fields at about 22 T. The performance of high-field cryomagnets critically depends on the specifications of the superconducting wires. NbTi multifilament wires are typically used for solenoids for field strengths of up to about 9.5 T.[140] For the 360-GHz EPR spectrometer at FU Berlin, a cryomagnet with maximum field strength of 14 T is used.[72] For such high fields multifilament superconduction wires based on Nb_3Sn are used for the inner sections of the cryomagnet.[140] Up to 14 T, the superconducting magnet systems guarantee a field inhomogeneity of <5 ppm within a 1-cm diameter sphere at warm-bore diameters of >80 mm. The field homogeneity is sufficient for EPR applications due to the relatively large intrinsic EPR linewidths (>10 µT) and small sample sizes (<5 mm largest dimension of the active sample volume). The wide bore of the magnet allows insertion of EPR probeheads and cryostats with easy-to-handle dimensions. Fine tuning of the magnetic field around the preset B_0 value for a given g-factor of the sample molecule is obtained by the integrated sweep-coil system, either superconducting or operating at room temperature. The additional superconducting coil integrated in the magnet system allows for a sweep range of ±0.1 T without substantial increase of liquid-helium consumption. Additional water-cooled room-temperature sweep coils are either part of the magnet construction[87] or realized in the form of a solenoid insert into the warm bore of the cryomagnet.[52,55,60] This variant decreases, of course, the dimensions of the cryostat and EPR probehead. The reported sweep range of room-temperature inserts is limited to about ±0.04 T.[55,60,87]

Current superconducting magnet systems have the disadvantage of a rather high consumption of liquid helium. Another disadvantage is the difficulty to monitor the exact magnetic-field value during a running EPR experiment. The first problem can be solved by the recently introduced cryogen-free superconducting magnet systems. Such systems are commercially available nowadays with suitable room-temperature bores and magnetic fields up to 16 T. They are based on mechanically cooling technology, i.e. they do not require any liquid coolant for operation. Very recently, cryogen-free magnets were employed in NMR at 400 MHz, 9.4 T[141] and high-field EPR up to 12 T.[66,142] The magnet system for EPR could be stabilized to better than 0.3 ppm/h, which is sufficient for many EPR experiments. A difficult problem exists so far with active cryocooling systems: They are still generating unwanted mechanical vibrations, which require strict mechanical isolation of the EPR mw components and detection circuitry from the magnet. It is anticipated that future improvements of the mechanical cooling systems will solve this problem.

The problem of exact magnetic-field monitoring in superconducting magnet systems is still unsolved. High-resolution EPR experiments require reproducible and stable magnetic-field values that, however, are difficult to maintain at high fields without feedback loops. At lower fields, e.g., in X-band or Q-band EPR, NMR gaussmeters (magnetometers) are used for measuring absolute magnetic-field values. The high homogeneity of the magnetic field inside the pole gap of an iron-core magnet allows the gaussmeter probehead, which contains the proton-spin system (typically H_2O, or a ^1H-rich solid) to be placed outside the EPR cryostat, and reliable information is obtained about the relative field values when stepping the magnetic field. The absolute magnetic-field value at the sample is obtained by taking into account the systematic field shift between the NMR probehead and EPR sample positions. This method is difficult to realize with superconducting magnets due to the large field gradients outside the homogeneous field region occupied by the EPR probehead. Commercially available NMR gaussmeters for high-field applications use the cw NMR signal of H_2O or D_2O. Characteristic sample-volume dimensions of 4 mm diameter and length require a field homogeneity better than 100 ppm/mm. Moreover, the NMR probehead is operating only at ambient temperatures, which prohibits its positioning inside the EPR cryostat. Recently, a gaussmeter based on the FT NMR technique was successfully applied to monitor the magnetic field during an EPR experiment at 5 T.[143] The instrument allows measurement of field sweeps of ±0.4 T with a resolution of 10^{-5} T (2 ppm). The NMR probe is placed outside the EPR cryostat, but has to stay in the homogeneous field region. This limits the applicability of the gaussmeter to special cryostat and magnet designs. The development of a solid-state NMR gaussmeter using aluminum NMR of small Al particles of about 5 μm diameter looks rather promising.[144] This gaussmeter was used for monitoring the magnetic field in wiggler magnets at 4.2 K[144] in the field region from 2 T to 7 T. The small sample volume of 1 mm diameter and 3 mm length allows for field measurements with an accuracy of ±20 ppm inside field gradients of up to 5 mT/mm. These specifications are favorable in view of potential applications to EPR superconducting magnet systems. For coarse (1000 ppm) magnetic-field monitoring, Hall-effect sensors can be used.[66]

In the field region around $g = 2$, EPR standard samples are generally applied as a reference for measurements of magnetic-field values. We are using the six-line EPR signal of Mn^{2+} in MgO recorded simultaneously with the EPR spectrum of interest to calibrate the magnetic-field axis for g-value measurements. The narrow EPR lines of the six hyperfine components of Mn^{2+} ($I = 5/2$) are positioned around $g(Mn^{2+}) = 2.00101 \pm 0.00005$ with line separations according to the hyperfine splitting of 8.710 ± 0.003 mT, covering a field range of about 50 mT.[39] The magnetic-field axis is calibrated with an accuracy of ±0.02 mT, which corresponds to ±5 ppm at 3.4 T (W-band EPR). A similar procedure is utilized in the calibration protocol for the magnetic field sweep of the commercial W-band EPR spectrometer from Bruker Biospin.[87]

In the near future, high-field/high-frequency EPR spectrometers will probably be extended to above 20 T/600 GHz taking advantage of the rapid development of superconducting magnet technology for high-field NMR spectroscopy. At present, 14 T cryomagnets are about the limit for university-based EPR spectrometers. Beyond 14 T, only specialized national or international facilities offer the dedicated magnet systems suitable for EPR spectroscopy above 400 GHz. Bitter-type electromagnets can provide static magnetic fields up to 35 T (1 THz EPR for $g = 2$). They require 20 MW electric power and massive water cooling. Beyond 35 T, the hybrid magnets, *i.e.* a combination of Bitter and superconducting magnets, can push the static field up to 45 T (corresponding to 1.25 THz EPR). A very promising new research area is opened by the repetitive-pulse magnets that can provide fields up to 80 T, but only over periods of several tens of milliseconds. In a pulsed coil magnet delivering fields up to 50 T for 15 ms, 2-GHz proton NMR was performed in 2005.[145,146] This is certainly exciting news for the magnetic-resonance community. A detailed description of very-high-field magnet systems is given in a recent review article.[147]

3.4 High-Field Multipurpose Spectrometers Built at FU Berlin

3.4.1 The 95-GHz Spectrometer

The W-band high-field EPR spectrometer at FU Berlin was designed as a multipurpose instrument for cw, pulsed and time-resolved EPR, as well as for

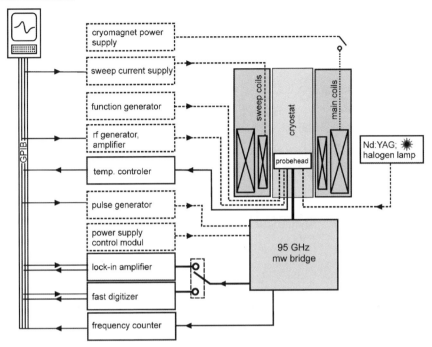

Figure 3.7 Block diagram of the W-band EPR spectrometer at FU Berlin with main components as well as control electronics for standard EPR experiments.

double-resonance experiments (ENDOR, ELDOR). The instrumental development started around 1983. The significant steps forward in the performance and capabilities of the spectrometer up to its present version have been taken predominantly by diploma and PhD students as well as postdoctoral coworkers of the Möbius group. In particular we mention (in chronological order) E. Haindl,[38] O. Burghaus[39,42,127,148,149] T. Götzinger,[150] M. Rohrer,[39,151–153] R. Klette,[129,154] T.F. Prisner,[43,96,155] A. Schnegg,[156] M. Fuhs[130,157,158] A. Savitsky[116,159,160], and Yu. Grishin.[161]

A block diagram of the present version of the EPR spectrometer is shown in Figure 3.7. The EPR spectrometer consists of the following main parts: the microwave bridge, the magnet system, the probehead, the cryostat, the control electronics. The mw bridge combines the transmitter, which produces the mw power for continuous or pulsed excitation of the sample, with the receiver for detecting the EPR signal. The external magnetic field is provided by a superconducting magnet that is sweepable to match the EPR resonance condition. The EPR probehead includes the mw cavity, sample holder, mw tuning mechanics, field-modulation coils for cw EPR, rf coils for ENDOR, *etc.* The cryostat is used to set the temperature of the sample. The control electronics govern the specific schemes needed for the various types of the experiments.

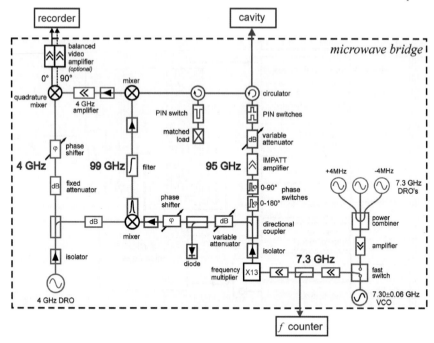

Figure 3.8 Block diagram of the W-band EPR microwave bridge.

3.4.1.1 Microwave Bridge Design

Figure 3.8 shows the detailed block diagram of the current version of the EPR mw bridge developed at FU Berlin. The heterodyne detection bridge works at an intermediate frequency (IF) of 4 GHz, which is provided from the dielectric-resonator oscillator (DRO) (*Miteq, E-HT series*) and that allows convenient manipulation of phase and amplification. The signal-oscillator output power at 95 GHz is used to irradiate the sample. When the EPR resonance condition is met, the reflected power from the cavity is mixed with the 99-GHz local-oscillator signal by a biased balanced mixer (*Elva-1, BM-10/95*). This 99-GHz signal is obtained by up-conversion, using a balanced mixer (*Farran Technology, BMF-10S*), of the 95-GHz signal-oscillator power with the 4-GHz DRO, to provide 4-GHz IF. This 4-GHz power is amplified with a low-noise preamplifier (*Miteq, AFD3 series*, +35dB) and then down-converted with the same 4 GHz DRO, as local oscillator, using either a quadrature IF mixer (*Miteq, IQZ0046*) or a double-balanced mixer (*Miteq, M0204*). The resulting quadrature- or single-phase signal is guided either to a lock-in amplifier or, after optional further amplification by a home-built balanced video amplifier (+18 dB gain), to a fast digitizing scope (*Agilent, 54830B*).

A specially designed 4-channel solid-state oscillator source (*Elva-1*) is used to obtain the 95-GHz mw power for the sample excitation. It utilizes four

Instrumentation 163

Table 3.3 Specifications of the W-band EPR microwave bridge at FU Berlin.

Parameter	Value	Units	Conditions
Overall receiver noise figure[a]	10 ± 1	dB	without protection
	11 ± 1		with protection
Overall receiver gain	$+20$	dB	double balanced mixer
	$+16$		quadrature IF mixer
	$(+18)$		additional video amplifier
Receiver bandwidth	>500	MHz	
	250		with video amplifier
Operation frequency	95 ± 0.3	GHz	
Frequency stability	<20	kHz/h	DRO
	<1	MHz/h	VCO
FM noise at 10 kHz	-100	dBc/Hz	DRO $\pm$ IMPATT[b]
	-70		VCO $\pm$ MPATT[b]
AM noise at 10 kHz	-135	dBc/Hz	(DRO/VCO) $-$IMPATT[b]
	-140		(DRO/VCO) $+$ IMPATT[b]
Output power available[a]	20	dBm	$+$IMPATT[b]
(max. on circulator)	0		$-$IMPATT[b]
Power attenuation range	0 to -130	dB	
PIN switching time	<2	ns	
Pulse-forming PIN attenuation[a]	-60	dB	
Detector protection attenuation[a]	-32	dB	
Phase-switching error	<0.2	degrees	within 95 ± 0.1 GHz
Switching time DRO/VCO	50	ns	

[a] The characteristics are measured within the operation frequency of 95 ± 0.3 GHz.
[b] $+$IMPATT: amplifier is on, $-$IMPATT: amplifier is off.

low-frequency oscillators, three of them are fixed-frequency, free-running, temperature-stabilized DROs, and one is a voltage-controlled oscillator (VCO) operating at 7.3 GHz that can be swept by ± 60 kHz. The frequency of this mw field is further up-converted by a factor of 13 using an IMPATT-active frequency multiplier (*Elva-1, IAFM-10*). An additional gain of the 95 GHz power could be obtained by a two-stage cw IMPATT amplifier (*Elva-1, IILA-10/18/95*) in the transmitter channel. The two-stage reflective pin switches (*Elva-1, FPS-SPST-10/95/2*) and two pulsed phase shifters, 0–90° and 0–180°, (*Elva-1, FPM-10/95/XX*) in the transmitter channel allow for the formation of pulses and full phase-cycle manipulation of the mw power for pulsed EPR measurements. To protect the detection part in the pulsed mode, an absorptive defence device (*Elva-1, SPST-10/95/25*) is installed in the detection channel. The key characteristics of our mw bridge are summarized in Table 3.3.

3.4.1.2 Magnet and Cryostat

In our 95-GHz spectrometer the static magnetic field is supplied by a superconducting magnet (*Cryomagnetics*) with a B_0-field strength of up to 6 T and a warm-bore diameter of 114 mm. The field homogeneity of 1 ppm (about 3 μT at

3.4 T) over a spherical volume of 10 mm diameter is sufficient for EPR applications. The magnetic field can be swept up to ±0.1 T (in 1 µT steps) by controlling the current in additional superconducting sweep coils inside the magnet dewar. Because of a slight nonlinearity of the field sweep (quadratic law, the deviation from linearity is about 0.1 mT at the 30 mT sweep range), the EPR signal of a standard Mn^{2+} powder sample (0.02% in MgO) placed inside the EPR cavity (in the TE_{011} cavity one of the tuning pistons is coated with a thin layer of the Mn^{2+} powder) is simultaneously recorded for exact field calibration.

To keep the microwave transfer losses from the mw bridge to the cavity and back as small as possible, the EPR probehead is mounted on the bottom flange of the cryomagnet. This construction excludes the application of commercial bath or gas-flow cryostats of dewar type. Therefore, we constructed a special cryostat with both sides open. It consists of a vacuum-isolated stainless-steel tube inserted into the magnet warm bore, the top flange allows for gas transfer or cryogenic liquid injection.

The nitrogen gas-flow mode is used to control the sample temperature in the range of 90 to 360 K. The temperature of the probehead is stabilized to ±0.1 K of the set temperature by a feedback loop controlling the gas flow rate and the gas temperature. The exact sample temperature is measured electrically by using a temperature-calibrated GaAlAs diode mounted close to the cavity in the probehead.

In the injection mode, either liquid nitrogen or helium is continuously injected into the probehead volume by means of the flexible transfer line, while the vapor is pumped off. The temperature is controlled by setting the injection rate and the underpressure in the probehead volume. The temperature of the sample can be stabilized to ±1 K around the set temperature in the range between 70 to 90 K when using liquid nitrogen, and between 30 to 70 K when using liquid helium. The long-term temperature stability is better than 1 K/h.

3.4.1.3 Probeheads

When striving for high spectrometer sensitivity at mm wavelengths, incorporation of a microwave resonance structure in the probehead is indispensable. There is no *a priori* preference for a single-mode cavity, which is commonly used in X- and Q-bands, or a multimode Fabry–Perot (FP) resonator, which is commonly used in the submillimeter and optical regions. The choice between these two resonance structures depends strongly on physical and technical considerations related to the type of EPR experiment to be performed. Consequently, both FP resonators and cylindrical single-mode cavities have been designed for our 95-GHz spectrometer.

Fabry–Perot Probehead

Fabry–Perot resonators have been constructed and tested in several high-field EPR laboratories (see Section 3.3.4.2). Most experimental setups use cryomagnets and FPs with vertical B_0 and resonator axes. This configuration, however, is inconvenient for sample access, light irradiation and extension to ENDOR. Hence, we have placed the axis of the FP resonator perpendicular to the B_0 axis.[39] Such a configuration allows vertically or horizontally mounted sample capillaries to be used. The simultaneously used Mn^{2+} standard sample can be adjusted in its position within the FP. It consists of two concave spherical mirrors (diameter 13 mm and 19 mm with corresponding curvature radii of 15 mm and 20 mm) in an approximately confocal arrangement (radius of curvature equal to mirror distance). The mirror distance can be tuned by a fine thread to adjust the resonator frequency to the mw source frequency (10 MHz frequency variation for 1 μm distance variation). In most EPR and ENDOR experiments, the fundamental TEM_{00q} with q set from 6 to 9 is preferred. For optimizing the detection sensitivity, the microwave coupling between resonator and waveguide is important. Sufficient coupling dynamics of the FP is achieved by moving a thin dielectric plate (MACOR) partially over the iris in one of the mirrors (for details of construction, see ref. [39]).

The basic realization of the FP probehead allows only rotation of the sample around the sample axis with an accuracy of ±1°. In a next step, a FP probehead for single-crystal investigations with rotation around three axes by means of a goniometer was developed, Figure 3.9. It allows rotation of the sample by 240° around the resonator axis and by 360° around the other two axes with an angular accuracy of better than 0.5°.[154] The FP probehead is equipped with two orthogonal quartz fiber bundles for uniform light excitation while rotating the single-crystal sample.

Partial one-dimensional ordering of biomolecular complexes can be achieved by incorporating them in specially prepared thin films.[162] In order to obtain good orientation selection with multilayers on plane surfaces, we have developed two novel plane–concave Fabry–Perot resonators with the external magnetic field either parallel or perpendicular to the plane mirror,[158] see Figure 3.10. The plane mirror can be shifted to adjust the resonance frequency. The mw beam-waist diameter on the plane mirror, on which the multilayer sample is attached, is about 4 mm and, correspondingly, the size of the multilayers for optimal sensitivity should be about 12.5 mm^2.

Fourier-transform (FT) spectroscopy with stochastic excitation of the spin system offers several advantages compared to the usual magnetic resonance methods with coherent excitation, as is demonstrated by its success in NMR[163] and X-band EPR.[164] To realize the demanding experiment of stochastic high-field/high-frequency FT EPR at 3.4 T/95 GHz, using broadband stochastic microwave excitation with a bandwidth of as large as 250 MHz, a dedicated resonator probehead was developed in our laboratory.[130,157] This probehead was realized as a bimodal Fabry–Perot resonator with two spherical gold-plated brass mirrors (diameter 19 mm, curvature $R = 15$ mm). The induction-type FP resonator operates in the TEM_{008} mode with the detection arm rotated by 90° with respect to the excitation arm. The FP resonator is overcoupled in order to decrease the quality factor to 200 and, thereby, to obtain the required large bandwidth. The mw decoupling of excitation and detection of such a

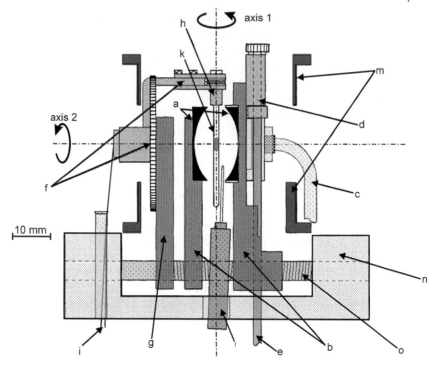

Figure 3.9 W-band Fabry–Perot resonator with goniometer with mounted single crystal (shown orientation $B_0 \parallel$ crystal c-axis. (a) mirrors (diameter 19 mm, distance *ca.* 10 mm); (b) mirror supports (synchronously movable for frequency tuning); (c) WR10 waveguide; (d) microwave coupling unit with (e) drive shaft; (f) goniometer arm with gear for rotation of the arm about the resonator axis; (g) arm support (movable for adjusting the sample in the center of the resonator); (h) rotatable sample holder; (i) thread passing through arm (f) to sample holder (h) for sample-axis rotation; (k) quartz capillary with single crystal; (l) independently adjustable quartz capillary with Mn^{2+} standard sample; (m) modulation coils; (n) holder; (o) guide rod with right/left thread for symmetrical mirror adjustment (frequency tuning). Rotation axes: axis 1, gear-driven rotation about the resonator axis; axis 2, thread-driven rotation about the sample capillary axis. For details, see refs. [129,154].

resonator is usually not better than 20 dB. It is possible, however, to increase the decoupling by adjusting conducting screws (brass, diameter 2 mm) in the center plane of the resonator. These screws work in analogy to the "paddles" in the bimodal cylindrical cavities used in X-band.[165,166] For a detailed description of the stochastic W-band FT EPR experiment, we refer to refs. [130,157].

TE_{011} Cavity Probehead
The disadvantages of FP resonators in terms of low filling and conversion factors become critical for pulsed high-field EPR experiments with only rather

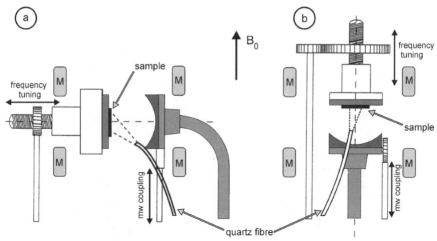

Figure 3.10 W-band plane–concave Fabry–Perot resonators with light excitation, (a) B_0 parallel and (b) perpendicular to the plane mirror. The resonators are operated in the $TEM_{005-007}$ modes. The frequency adjustment and mw coupling mechanisms are indicated. The modulation coils are denoted by M. For details, see ref. [158].

moderate mw power available at high frequencies. This was our motivation behind the construction of W-band EPR and ENDOR probeheads with cylindrical single-mode cavities. The TE_{011} mode has high filling and conversion factors, and its resonance frequency can easily be tuned by changing the cavity length. The development of cylindrical single-mode cavities for W-band started in our group in 1988.[150] The first-generation construction design proved to be very reliable in terms of frequency and coupling stability. This design concept is still applied for fabricating new generations of W-band cylindrical cavities without major modifications.

Figure 3.11 shows exploded views of the W-band EPR probehead. The main elements are the cylindrical TE_{011} cavity optimized for 95 GHz, the frequency-tuning mechanism, the mw coupling mechanism and the field-modulation coils. They are mechanically isolated from the cavity in order to avoid microphonics. At 10 kHz they supply up to 2 mT of field modulation amplitude at the sample. The cavity diameter is 4.16 mm. The cavity length can be adjusted to 4.16 ±1.5 mm (corresponding to TE_{011} mode resonance frequencies from 92 to 104 GHz for an empty cavity). This is done by changing the separation of the two sliding pistons in the cavity by means of a gear drive of high mechanical precision, Figure 3.11(b). The tuning drive consists of two cam followers 180° out of phase, thereby guaranteeing that the maximum of the mw field stays in the center of the cavity. This is an important requirement for experiments with small single-crystal samples and light excitation. For the different capillary diameters (ID 0.1 to 1.0 mm), different pairs of pistons are used. On the top of the coupling needle (DELRIN) a small silver ball of 0.7 mm diameter is

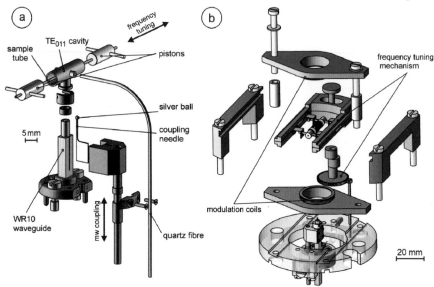

Figure 3.11 (a) Exploded view of the 95-GHz EPR TE$_{011}$ optical transmission cavity with microwave-coupling mechanism and light-access chimneys (with minimum mw radiation losses) to insert the quartz-fiber light pipe. (b) Exploded view of TE$_{011}$ (ENDOR variant) probehead with frequency-tuning mechanism and field-modulation coils. For details, see refs. [39,150].

painted, using a suspension of fine silver flakes (about 8 μm). The coupling needle can be moved in the upper part of the W-band (WR10) waveguide by means of a vertically sliding mechanism with a high-precision thread drive, Figure 3.11(a). The iris coupling hole in the middle of the cavity wall has a diameter of 0.8 mm, the wall thickness around the iris is less than 0.1 mm.

For different EPR experiments it was necessary to construct different types of cavities. Their characteristics are summarized in Table 3.4, and commented on in the following:

For pulsed W-band EPR experiments, in which a high B_1 field at the sample position is required, a special TE$_{011}$ cavity with high Q is used. The cavity body was machined from gold (99.9%). Additionally, the brass pistons for the frequency adjustment of the cavity were gold plated (100 μm) at the front side. A room-temperature unloaded $Q_U = 7400$ was measured without a sample. This is close to the calculated Q_U value of 8600 when accounting for the ohmic losses in the cavity walls only. At 90 K, $Q_U = 12000$ was determined, which is consistent with the room-temperature value taking into account the known temperature dependence of gold resistivity.

For those EPR experiments in which light irradiation of the sample is required, the standard cw cavity is replaced by an optical-transmission cavity, Figure 3.11(a). It has two symmetrically placed 0.6-mm holes in the cavity wall

Table 3.4 Summary of the performance data of TE_{011} cylindrical W-band cavities constructed at FU Berlin.

	EPR	EPR	ENDOR	FJ-PELDOR
Material	Bronze, gold coated	Gold 99.9%	Bronze, gold coated	Ti/Al/V, gold coated
Q_U (293 K)	5000	7400	4000	3000
Conversion factor^a in $mT/\sqrt{W}$	1.4	1.9	1.2	1.1
$t_p(\pi/2)^a$ /ns	24	18	28	32
Filling factor^b	0.2	0.2	0.2	0.2

^{a}measured for +18 dBm (65 mW) incident mw power on the cavity, point sample in quartz capillary, room temperature.
^{b}for nonlossy samples, taking the ratio of magnetic to electric filling factors $\eta/\eta_E = 0.1$, sample capillary I.D. 0.88 mm.

for the light access furnished with mw chimneys and holding a 0.8-mm thick quartz fiber.[152]

For ENDOR experiments, the cw variant of the probehead had to be modified.[151] The standard cw cavity is replaced by a gold-plated bronze TE_{011} cavity, the body of which is slotted with regular slots of 0.3 mm width and 0.6 mm separation to reduce rf-induced eddy currents. The slots slightly reduce Q_U to a typical value of 4000. We use two variants of the ENDOR cavity, one with and the other without light-access holes. The bronze pistons are replaced by pistons machined from MACOR ceramics with gold-plated end faces. The rf ENDOR-coil holder is mounted on the top of the upper field-modulation coil. The rf saddle coils around the cavity center produce an rf magnetic field perpendicular to both the external and mw magnetic fields.[39]

In field-jump PELDOR experiments,[161] the field-modulation coils are replaced by field-jump coils of the same geometry. The cavity with dimensions that are adapted from the ENDOR variant is machined from a titanium/aluminum/vanadium alloy and then gold plated.[161] The ENDOR pistons are used for frequency tuning. The unloaded quality factor of such a cavity configuration is about 3000.

The mw power is transferred from the bridge to the cavity and back by means of a straight oversized waveguide WR28 of 500 mm length. This allows the transfer losses to be kept below 1.5 dB at room temperature.

3.4.1.4 EPR and ENDOR Performance

For small single-crystal samples placed in the TE_{011} cavity, the cw EPR sensitivity is about 10^8 spins/mT at 1 Hz detection bandwidth, *i.e.* about two orders of magnitude higher than at standard X-band frequencies.[50] This number corresponds, for example, to a minimum detectable concentration of a nitroxide spin label in frozen solution of 0.1 μM (200 K, sample tube ID 0.6 mm, active sample volume 1 μl). In the case of Fabry–Perot resonators, a cw EPR sensitivity of about 4×10^9 spins/mT at 1 Hz detection bandwidth was experimentally determined[39,158] for small samples.

For direct-detection time-resolved EPR measurements, the time resolution of the heterodyne detection channel of about 2 ns and the short ringing time of the EPR cavity, T_R, of about 6 ns (loaded quality factor $Q_L = 2000$) provide an overall time resolution that is an order of magnitude higher than for X-band TREPR.

In contrast to pulsed X-band EPR spectrometers with their high-power TWT mw amplifiers, the mw power required for pulsed W-band experiments is drastically smaller due to the small cavity dimensions. The shortest measured $\pi/2$-pulse length of 18 ns can be achieved with a 95-GHz incident power on the gold TE_{011} cavity of only 65 mW. Another advantage of a pulsed W-band EPR spectrometer is its short deadtime. This is defined as the time interval from the moment when the mw power is switched on or off to the moment when the EPR signal can be recorded without any distortions. At W-band, the deadtime is more than an order of magnitude shorter than at X-band. This is due to the much lower mw power required for sample excitation at 95 GHz and the higher ringing frequency (shorter ringing time) of the W-band cavity. For our W-band spectrometer configuration with quadrature detector, the deadtime was measured to be $10\,\text{ns} \cdot Q_L/1000$ when using the full excitation mw power. It is determined by the power-saturation characteristics of the mw detection circuit.

For proton ENDOR, the rf circuit is tuned to and matched for 140 MHz (with 40 MHz bandwidth), corresponding to the proton Larmor frequency at 3.4 T. An rf power of 300 W in cw and 2.5 kW (*Amplifier Research, 1000LM8*) in pulse mode is available. It yields 2–6-µs long rf π-pulses in the proton ENDOR experiments. ENDOR on other nuclei can also be performed by using a short-circuit rf coil optimized for maximum current at 40 MHz central frequency (with 30 MHz bandwidth).

3.4.1.5 Field-Jump PELDOR

The pulsed magnetic field at the sample inside the TE_{011} EPR cavity is generated by a pulsed electric current through a pair of Helmholtz coils fixed outside the cavity, which replace the field-modulation coils (Figure 3.12). Each coil

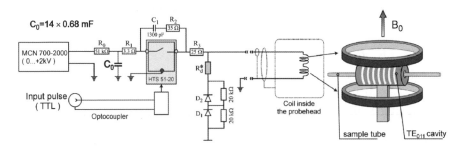

Figure 3.12 (a) Basic electric circuit of the current-stepping device and (b) probehead with the field-jump Helmholtz coils and slotted W-band cavity. For details, see ref. [161].

consists of three turns of insulated copper wire ($\varnothing$ 0.8 mm) wound around a cylindrical holder ($\varnothing$ 22 mm). The separation between the coils is 11 mm. They are electrically connected in series, and the measured inductance, $L = 0.9\,\mu\text{H}$, agrees with that calculated for this coil geometry. The calculated field/current conversion coefficient is 2.37 G/A. Thus, for the maximum field of 160 G, a pulsed current of 70 A is required.

The pulsed current generator, shown in Figure 3.12, is based on two commercial electronic devices: *(i)* The high-voltage DC source (*F. u G. Elektronik, HCE 35-2000*), supplies a stabilized voltage from 0 to 2000 V with 10^{-4} precision and a maximum output current of 15 mA. *(ii)* The fast high-voltage FET semiconductor switch (*Behlke Electronic, HTS 51-20*) with the following specified characteristics: 5000 V DC/ 200 A, turn-on rise time 35 ns, turn-off fall time 30 ns, minimum pulse duration 200 ns. The pulsed current generator in combination with stand-alone coils allows a maximum magnetic field pulse amplitude of ± 20 mT with a fall- and rise-time of about 40 ns and a repetition rate of 2 kHz when producing a 400-ns field pulse to be obtained.

To suppress eddy currents induced by the magnetic field pulses, all conductive elements in the probehead are replaced by parts manufactured from insulating materials. The slotted cylindrical TE_{011} cavity is manufactured from Ti-6Al-4V alloy. Compared to brass, this alloy has a 25 times higher ohmic resistivity and a four times smaller temperature coefficient of the resistance. For a sufficiently high unloaded Q-value of the high-resistive cavity, its inner surface was gold plated (about 4 μm). This resulted in $Q_U = 3000$ and, consequently, no significant loss of EPR sensitivity and mw B_1 field was found for this cavity as compared to the prototype brass cavity. The screening field from eddy currents in the cavity was found to decay with a time constant of 110 ± 10 ns in the Ti-6Al-4V cavity, *i.e.* an order of magnitude faster than in the brass cavity of the same geometry. The eddy-field amplitude at 500 ns after the turn-on field step is $3 \times 10^{-3} \cdot \Delta B$ that, for a 16-mT field jump, corresponds to 50 μT. Thus, to satisfy the requirements for FJ–PELDOR experiments (to achieve a stable peak-field value and to avoid residual fields after switching off) with the Ti–6Al–4V cavity, a duration of the field jump of about 700 ns and an after-pulse delay of 500 ns are estimated that determine a minimum stimulated-spin-echo (SSE) mixing time interval T of about 1 μs. This pulse pattern already allows FJ-PELDOR experiments in high magnetic fields to be made for the determination of nanometer distance and relative orientation of nitroxide spin labels in disordered samples.[161]

3.4.1.6 Dual-Frequency PELDOR

The 95 GHz microwave bridge is equipped with a fast (<50 ns) microwave switch, which allows selection between one of the three fixed-frequency DROs and the variable-frequency VCO operating around 7.3 GHz. It is placed in front of the frequency multiplier (×13) followed by an IMPATT power amplifier. The frequency operation range is determined by the bandwidth of the power amplifier, it is 600 MHz around 95.0 GHz in our configuration. The output frequencies of about 7.3 GHz are measured with the 1-kHz resolution of the frequency counter. This corresponds to a precision of 13 kHz at 95 GHz after multiplication of the low-frequency mw output by the factor 13. The dual-frequency pulsed PELDOR experiment includes a three-pulse SSE subsequence applied at the fixed frequency v_1 (DRO). The external magnetic field is set to a value B' corresponding to resonance of the "observer" radicals at this frequency. An additional microwave pulse with the flip angle π is applied to the "partner" radicals at the variable frequency v_2 (VCO). This pulse is set within the mixing time interval T of the SSE sequence. The frequency v_2 corresponds to the resonance field $B'' = B' + (v_1 - v_2) \cdot h/(g_e \cdot \mu_B)$.

A systematic control of the TE_{011} EPR cavity parameters is inevitable for the success of the dual-frequency PELDOR experiment because the amplitude of the dipolar modulation critically depends on the turning angle of the pumping mw pulse. Because the mw cavity has to be overcoupled to obtain a larger bandwidth, the critically coupled loaded quality factor, the mw coupling factor and the optimal pulse length have to be predetermined accurately. Knowing these parameters, the optimal π-pulse length can be set for mw excitation at v_2. Typical values for the critically coupled cavity are $Q_L = 2000$, $t_p(\pi/2) = 30$ ns. Overcoupling the cavity to $Q_L = 900$ reduces the $\pi/2$-pulse length to $t_p(v_1) = 36$ ns, but allows for π-pulse length of 380 ns applied 280 MHz from the cavity resonance frequency, *i.e.* $|v_1 - v_2| = 280$ MHz (corresponding to a magnetic field difference of 10 mT).[160]

3.4.2 The 360-GHz Spectrometer

The instrumental development of the 360-GHz EPR spectrometer at FU Berlin was started in 1995. Several PhD students and postdoctoral coworkers of the Möbius group were involved in this development, in particular M. Fuchs,[72,124,167] A. Schnegg,[137,168] Yu. Grishin[73,137] and T.F. Prisner.[72] During the last decade several design improvements of the spectrometer were realized including the pulse EPR and cw ENDOR extensions. The most notable difference of the 360-GHz ($\lambda \approx 0.8$ mm) spectrometer in relation to our W-band spectrometer is the use of quasioptical microwave components and a corrugated waveguide in the transmission line to the probehead (see below).

3.4.2.1 *Quasioptical Microwave Propagation*

The microwave propagation losses in normal metallic waveguides rise rapidly at high frequncies, making standard waveguides unusable above 150 GHz except for very short distances of a few cm. Dielectric fibers and similar "light-pipe" structures can work very well for frequencies in the visible and near-visible spectral region. Unfortunately, they have significant dielectric losses at millimeter waves leading to high damping of the propagated wave. Free space is a low-loss alternative. The damping of millimeter waves in air due to molecular absorption is quantified in dB/km, in contrast to dB/m or even dB/cm in classical waveguide systems. The techniques for the propagation of electromagnetic waves over free space are well developed for radiation with wavelengths of less than a micrometer. Established techniques exist for optical systems for which the characteristic dimensions are many thousand times larger than the wavelength. These systems can be designed and analyzed using the traditional methods of geometrical-optics with only rare resort to wave optics. In order to treat millimeter-wave systems in the geometrical optics way, one would have to handle optical elements at least one meter in diameter. Gaussian quasioptics offers a solution to this problem. Quasioptics can be considered as a specific branch of microwave science and engineering. This term is used to characterize methods and tools devised for handling electromagnetic waves propagating in the form of narrow directed beams, whose width w is greater than the wavelength λ, but smaller than the cross section size, D, of the limiting apertures and guiding structures: $\lambda < w < D$. Normally, $D < 100 \cdot \lambda$, but also devices as small as $D = 3 \cdot \lambda$ can be analyzed using quasioptical principles. In contrast to geometrical optics, which requires $D > 1000 \cdot \lambda$, quasioptics consider both ray-like and diffraction-causing optical phenomena when tracing the beams as they interact with the surfaces of optical elements and the matter in the pathway.

The simplest form of an electromagnetic wave propagating through free space is the Gaussian beam, which is shown in Figure 3.13. The fundamental-mode beam has a Gaussian distribution of the electric field amplitude perpendicular to the z-axis of propagation[169,170]

$$|E(z,r)| = |E(0,r)| \cdot \exp(-r^2/w^2) \qquad (3.8)$$

where r is the distance from the propagation axis and w is the beam radius. The beam radius w will have a minimum value w_0 at a specific place along the beam axis, which defines the beam waist. For a beam of wavelength λ at a distance z along the beam, as measured from the beam waist, the variation of the beam

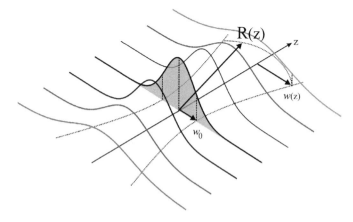

Figure 3.13 Electric-field amplitude profiles and beam parameters of a Gaussian beam propagating along the z-axis. The beam parameters are: the beam waist w_0, the beam radius $w(z)$, and the radius of curvature $R(z)$.

radius is given by

$$w(z) = w_0 \cdot \left[1 + (z/z_0)^2\right]^{1/2} \quad (3.9)$$

and the radius of curvature R of the wavefronts comprising the beam is

$$R(z) = z \cdot \left[1 + (z_0/z)^2\right] \quad (3.10)$$

where $z_0 = \pi \cdot w_0^2/\lambda$ is the confocal distance (also called the Rayleigh length). Quasioptical elements for propagation of Gaussian beams are waveguides, lenses or mirrors to refocus the Gaussian beam, and antennas or feed-horn systems to transform the waveguide modes to Gaussian modes of propagation. The actual diameters of quasioptical elements are determined by the boundary values of still tolerable diffraction and beam truncation effects. An element with a single aperture $D = 2 \cdot w$ has a coupling efficiency of 98.8%, *i.e.* a beam-power loss of -0.05 dB when transmitting a fundamental-mode beam.

The propagation of radiation in a Gaussian beam in free space is independent of its polarization. Thus, generation of beams with different polarizations and manipulation of them with polarization-dependent elements is possible. For example, wire grids with very thin conducting filaments that are separated by distances less than a wavelength, are very effective polarizers. For radiation with the electric field in the directions of the wires currents are induced, and the grid acts as a reflector. A perpendicularly oriented electrical field does not induce currents, and the radiation passes the grid without attenuation. Proper manipulation of the Gaussian beam polarization allows construction of the quasioptical analogs of corresponding waveguide-based microwave elements such as circulators, directional couplers, *etc.*, which become either extremely lossy or even impossible to realize for frequencies above 150 GHz. Detailed descriptions of such quasioptical elements can be found in the monographs by Lesurf[170] and Goldsmith.[169] The engineering and design concepts of quasioptical EPR systems are considered in detail by Gulla and Budil,[171] and a transfer matrix method for characterizing and optimizing the performance of quasioptical EPR sample resonators is described by Earle *et al.*[172]

3.4.2.2 Microwave-Bridge Design

The general design of the 360-GHz microwave bridge is depicted in Figure 3.14. There are three mw sources for sample excitation available. The pulsed Orotron source will be described below in Section 3.4.2.9. In the cw configuration,[124] the microwave at 360.03 GHz is generated either by a phase-locked tripled 120-GHz Gunn source (*Farran Technology*) or by a quadrupled 90-GHz Gunn oscillator (*Radiometer Physics*). The second source can be phase locked to a 10-MHz oscillator or swept between 358.8 GHz and 361.2 GHz, which allows

Instrumentation

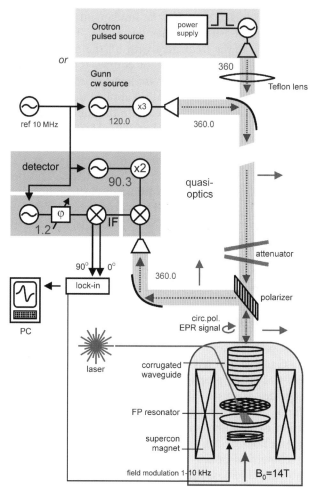

Figure 3.14 The 360-GHz EPR setup. The cw Gunn source (green) can be exchanged for the pulsed Orotron source (green) by replacing the first off-axis elliptical mirror for a Teflon lens. In the beam path of the quasioptical bridge (orange), beam propagation directions are indicated by dashed arrows (red) and polarization directions by solid arrows (green). Inside the magnet (grey), the corrugated waveguide, the modulation coil and the semiconfocal Fabry–Perot resonator with mesh coupling are indicated. The EPR-induced microwave signal reflected from the polarizer grid is focused onto the horn of the subharmonic mixer detector (blue). Light excitation is introduced into the probehead via a light guide (red).

the precise frequency tuning of the resonator containing the sample. The output power at 360 GHz is about 1 mW for both sources. For detection a heterodyne mixer scheme is employed. The central component of the receiver (*Farran Technology*) is a subharmonic mixer detector with a detection bandwidth of 100 MHz. In contrast to a fundamental mixer, in a subharmonic mixer the incoming mw signal is mixed not with the fundamental frequency of the local oscillator (LO) input but with a higher harmonic. The local oscillator can, therefore, operate at a lower frequency with all the advantages of lower-frequency components. The disadvantage, of course, is the less efficient mixing process. The LO is provided by a phase-locked, doubled 90.3-GHz Gunn oscillator. The necessary reference signal is supplied directly from the 10-MHz source of the transmitter control module. With the subharmonic mixer running at the 2nd LO harmonic, the EPR signal at 360 GHz is down-converted with a LO frequency of 361.2 GHz, so the resulting IF frequency is 1.2 GHz. This is then passed through a low-noise amplifier with 35 dB gain and a video filter. The noise figure of the complete receiver module is 15 dB. This performance is comparable to that of the most sensitive InSb hot-electron bolometer detectors nowadays available. The signal at 1.2 GHz is down-converted by a quadrature IF mixer (*Anaren Microwave*). The LO for this mixer is provided by a 1.21 GHz dielectric resonator oscillator (DRO) that again is phase locked to the 10-MHz master oscillator. Since in a quadrature mixer both in-phase and out-of-phase signal components are generated simultaneously, both the absorptive and dispersive signal components can be detected in a single experimental run.

3.4.2.3 Quasioptical Components

In our 360-GHz spectrometer[72,124] we employ a quasioptical transmission line setup in which a Gaussian beam is launched into free space via a corrugated horn antenna. Where free-space propagation is not feasible, cylindrical corrugated oversized waveguides are used. Conventional smooth-walled oversized waveguides are inferior to corrugated ones because of excitation of higher-order modes, generation of standing waves at transitions from and to fundamental-mode waveguide sections, and distortion of the propagated beam polarization.[173] In analogy to geometrical optics, to refocus a diverging beam one can employ either curved mirrors or lenses. Lenses can be the origin of standing waves in the system and cause dielectric losses. Therefore, all focusing elements in our quasioptical transmission line therefore are metallic off-axis elliptical mirrors.

To couple the microwave to the Fabry–Perot sample resonator, the output beam needs to be focused to a beam waist of 1.0 mm. This is achieved by a corrugated tapered waveguide section, made of gold-plated copper, at the end of the cylindrical waveguide. The corrugated waveguide itself consists of 100-mm long electroformed sections soldered to an overall length of 1080 mm. The diameter is 23 mm, being reduced by the taper over a length of 20 mm to 3 mm. The corrugation depth is 0.6 mm and the slot/metal pitch 0.25 mm. The material of the main waveguide section is German silver, which is preferred to copper since it has a lower coefficient for heat conduction.

3.4.2.4 Induction-Mode Operation

The linear polarization of the Gaussian beam is utilized in an induction-mode detection scheme first described by Teaney *et al.*;[165] later examples of this detection scheme at high magnetic fields were introduced for transmission mode by Prisner et al.[62] and for reflection mode by Smith et al.[4]

The Gaussian mw beam launched by the transmitter horn antenna has an initial beam waist w_0 of 1.48 mm and is linearly polarized. As is indicated in Figure 3.14, the first off-axis mirror focuses the widened beam onto the upper end of the corrugated waveguide to a beam waist of 7.24 mm. On its way the beam passes through a wire-grid polarizer that is oriented with an angle of 45° with respect to the optical axis. The free-standing wires (12 wires/mm, 25 µm diameter) will allow mw radiation that is polarized perpendicular to the direction of the wires to pass since in this direction no currents are induced, while radiation polarized along the wire direction is effectively reflected. Behind the polarizer, the residual radiation with polarization along the wires is attenuated by about 20 dB compared to the unattenuated component. By mounting a second grid on a rotation stage in front of the first, one obtains a polarizer/analyzer setup that acts as a variable attenuator with a dynamic range of 20 dB. The rotation stage is slightly tilted with respect to the optical axis to avoid the buildup of standing waves in the system.

In the cylindrically symmetric Fabry–Perot resonator the excitation mw represents a composition of two circularly polarized modes with opposite polarization. Of those two modes only one interacts with the spin system and, therefore, becomes partially attenuated upon magnetic resonance absorption. Recombined with the unattenuated component, this yields elliptically polarized mw radiation that is reflected back into the corrugated waveguide. When hitting the wire-grid polarizer, only the component orthogonal to the excitation mw is reflected onto the receiver antenna, while the excitation power itself is passed to the transmitter tripler and attenuated there.

Essential in this context is the use of the cylindrically symmetric Fabry–Perot resonator as a bimodal cavity. Since all directions of polarization in the resonator are degenerate, both the excitation microwave and the EPR induction component, polarized orthogonally to the excitation component, have to be supported by the resonator structure. This also applies to the oversized waveguide transmission line. The use of a single-mode cavity, despite many advantages, is not possible within an induction-mode scheme.

The maximum obtainable isolation of the excitation from the detection arm is ultimately limited by the crosspolarized microwave component that is induced by the off-axis mirrors. This effect is partly corrected by the polarizer grid that is passed after the first mirror. Any other misalignment in the setup will also introduce crosspolarization. Overall, this induction-mode setup provides an attenuation of the excitation power with respect to the EPR signal power of 20 to 30 dB.

It should be stressed that one of the main principles underlying the current spectrometer design was to realize an operating setup with as few quasioptical components as possible. This should ease proper alignment procedures, minimize the generation of standing waves and allow for low insertion losses. While, due to the induction mode setup the overall insertion loss from transmitter horn to receiver horn cannot be directly measured, a conservative estimate based on the known performance of the separate components leads to a 5-dB insertion loss.

3.4.2.5 Magnet and Cryostat

The superconducting magnet is a Teslatron H system (*Oxford Instruments*). It can sustain a magnetic field of up to $B_0 = 14\,\text{T}$. The experiments are typically run with a central magnetic field $B_0 = 12.846\,\text{T}$, the resonance field for the free-electron g-value and a microwave frequency of 360 GHz. The lowest accessible g-value with a 14-T central field is $g = 1.84$.

Integrated into the main coil assembly is a superconducting sweep coil with a sweep range of $\pm 100\,\text{mT}$. In this way the magnet can be swept with a sweep rate of up to 70 mT/min while the main coil remains in persistent mode. The homogeneity of the magnet is specified to 3 ppm in a 10-mm sphere. This was verified with an NMR Gaussmeter with a deuterium probe that was moved along the symmetry axis of the magnet. The linearity of the field sweep has also been tested with an NMR gaussmeter for a full sweep. The sweep is not linear and after a half-cycle, a remanent field of 4 mT can be observed. After a full cycle the field offset was 0.05 mT. This necessitates the use of a standard sample for magnetic-field calibration. We normally use Mn^{2+}/MgO dissolved in a polystyrene film as the standard sample whose magnetic interaction parameters are known with high precision.[39]

The warm bore of the cryomagnet has a diameter of 88 mm. This allows a cryostat with the rather large diameter of 62 mm to be employed to accommodate the probehead. The cryostat is a helium-cooled static-flow cryostat (*Oxford Instruments, CF1200*) that offers a temperature range of 3.8 K to 300 K. In a static-flow design, the probehead is not cooled directly by the He vapor. Instead, the probehead space is closed off and filled with a buffer gas (we use argon) that in turn is cooled by a copper heat exchanger shield that is in direct contact with the helium. This setup avoids stability problems that often arise in direct-flow cryostat designs.

3.4.2.6 Probeheads

Several probeheads have been developed for EPR and ENDOR measurements at 360 GHz.[124,137] The ENDOR version is shown in Figure 3.15. Both EPR and ENDOR probeheads are based on the plane–concave Fabry–Perot resonator operated in the TEM_{006} mode. With a resonant mirror distance $d = 2.58\,\text{mm}$ and curvature radius $R = 8.06\,\text{mm}$ one obtains a confocal distance of 3.76 mm. A typical value for the finesse of the resonator with a sample is $F = 160$, giving for the loaded quality factor $Q_L = 800$. Coupling to the tapered end of the corrugated waveguide is achieved through the flat mirror, which is a highly reflective metallic mesh (typically 30 wires/mm). The mesh consists of electroformed copper and is stretched to an exchangeable mesh holder, where it is fixed onto a circular frame. Since the mesh holders are exchangeable, the mesh

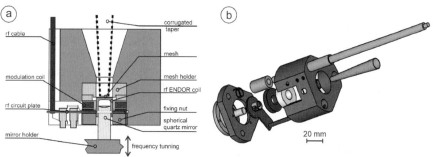

Figure 3.15 Cross-sectional (a) and exploded view (b) of the 360-GHz cw EPR/ENDOR probehead. From left to right: spherical Fabry–Perot resonator mirror (gold-plated quartz, yellow), springs (phosphor bronze, black), single-loop ENDOR coil (light blue), ENDOR rf leads and circuit (blue), tuning-rod sliding bushes (bronze, grey), modulation-coil holder (red), coupling mesh (flat Fabry–Perot mirror) holder (white), main body (TUFNOL, brown) with taper-centering screws (yellow), corrugated waveguide-taper (orange), tuning rods (copper/epoxy resin glass fabric, green). For details, see ref. [137].

can be replaced by another mesh with different reflectivity, thus allowing for variable coupling for different samples. Tuning is achieved by translating the spherical mirror along three guide rods via a micrometer screw on top of the corrugated waveguide outside the magnet. The field-modulation coils surround the spherical mirror. The modulation amplitudes can be raised up to 3 mT without generating excessive microphonics. A quartz fiber integrated into the probehead allows for photogeneration of radicals or triplet states inside the resonator by laser excitation.

The overall detection sensitivity of the spectrometer equipped with the FP resonator has been measured to be 1.5×10^{10} spins/mT at 1 Hz detection bandwidth by evaluating the signal-to-noise ratio of a known number of spins in a Mn^{2+} sample.

In an alternative probehead configuration without resonator the copper mesh is removed, which reduces the spectrometer sensitivity by a factor of 40. To compensate for the decreased sensitivity, the amount of sample has to be increased by the same factor. However, since the combination of low-noise mw components, the induction-mode detection scheme and the high Zeeman B_0 field provides an excellent overall detection sensitivity, an increase of the necessary amount of sample from 0.5 µl (typical for an organic radical in a 0.1 mM sample with resonator) to 20 µl (without resonator) is affordable in many cases.

3.4.2.7 ENDOR

Figure 3.16 shows a cross-sectional and an exploded view of the 360-GHz EPR/ENDOR probehead together with the resonant rf circuit for ENDOR experiments. The sample is placed on the gold-coated quartz mirror and irradiated by

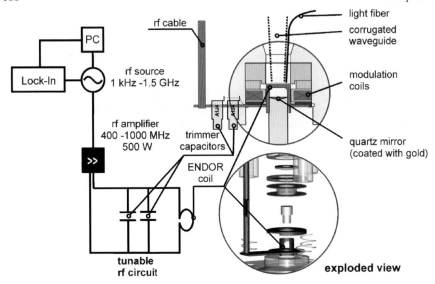

Figure 3.16 Tunable rf circuit (*left*) together with a cross-sectional and an exploded view of the 360-GHz/550-MHz EPR/ENDOR probehead (*right*). The resonant rf circuit consists of a single-loop coil and two trimmer capacitors, which provide frequency tuning and matching. The rf is generated by a sweepable generator, a power amplifier and a single-turn coil, which generates a B_2 field perpendicular to the microwave field B_1. The sample is placed on the gold-coated spherical quartz mirror. Note that for the EPR/ENDOR experiment the mw coupling mesh as the flat mirror of a Fabry–Perot resonator was removed. For details, see refs. [137,168].

a mw field B_1 and a perpendicular rf field B_2 that is produced in a single-loop ENDOR coil. The ENDOR coil is part of the resonant circuit (center frequency 547 MHz; bandwidth 60 MHz) which can be frequency tuned and impedance matched by two trimmer capacitors mounted on the rf circuit plate close to the ENDOR coil. The rf power is provided by a digital frequency generator (*Hewlett Packard, 8648B*) and amplified by a 500-W power amplifier (*Amplifier Research, 500HB*). To reduce standing waves in the rf transmission line, a high-power rf circulator is used. The maximum B_2 generated by the ENDOR coil is 0.8 mT at 20 W incident rf power, for which unwanted heating effects are avoided. ENDOR spectra are typically recorded at a fixed external field B_0 and mw field B_1, while sweeping the frequency of the rf field B_2. The B_2 field is frequency modulated, and the ENDOR signals are recorded by using a lock-in amplifier. For more details of the 360-GHz ENDOR experiments, see ref. [137].

3.4.2.8 Transient EPR Bridge with Reference Arm

If, in a standard waveguide microwave bridge with reflector-type cavity, information is required about the phase of the EPR signal, for example to separate absorption and dispersion contributions or to discriminate between emissive and absorptive lines in complex spin-polarized EPR spectra, a reference (or "bucking") arm has to be included in the bridge system with variable phase shifter and attenuator. The reference arm is fed by the main mw source via a directional coupler. Before reaching the mw detector the microwaves reflected from the cavity in the signal arm and the microwaves from the reference arm are combined again whereby the relative phase and amplitude of both coherent fields can be adjusted to give whatever degree of superposition of the two waves is required.[7]

In analogy, to obtain information on the phase of a 360-GHz EPR signal, when operating with a reflection-type resonator in a quasioptical microwave bridge, it is necessary to include a reference signal arm in the optical pathway. In our 360-GHz high-field EPR studies of photosynthetic electron-transfer systems, such phase information is needed for analyzing non-Boltzmann EPR spectra of spin-polarized transient radicals and radical pairs with lines in emission and enhanced absorption. Figure 3.17 shows the 360-GHz heterodyne microwave bridge with reference arm (*Radiometer Physics*) that was designed for time-resolved EPR (TREPR) experiments with induction-mode detection of spin-polarized transient reaction intermediates employing the Fabry–Perot resonator as the probehead.[168] The excitation mw beam (360.03 GHz) is generated by a phase-locked tripled 120.01-GHz Gunn diode source. The reference beam (361.24 GHz) is generated by a phase-locked quadrupled 90.31-GHz Gunn source. The Gaussian beams from the main mw source (360.03 GHz) and from the reference mw source (361.24 GHz) are both split into two parts by a beam splitter consisting of a metallic wire grid tilted by 45° with respect to the

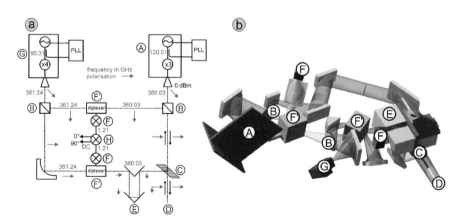

Figure 3.17 (a) Block scheme of the 360-GHz induction-mode mw bridge with reference arm. The main elements are A) 360.03-GHz and G) 361.24-GHz phase-locked mw sources, B) beam splitter, C) polarization grid, D) FP resonator, E) phase shifter, F') Fabry–Perot beam diplexer, F) and H) mixers. The frequencies and mw beam polarizations are noted. (b) Three-dimensional construction scheme of the mw bridge with reference arm. The mw Gaussian beams with corresponding diameters are shown. For details, see ref. [168].

linearly polarized beams from the horn antennas of the respective mw sources. The signal from the reference arm acts as a local oscillator for the mixer system for down-conversion of the sub-mm microwaves to an easy-to-handle L-band frequency (1.21 GHz). The power reflectivity of the beam splitters is chosen to allow for the necessary level of local oscillator power for the mixer system. Corresponding parts of the signal and reference beams are combined by a folded Fabry–Perot ring-resonator diplexer[170] consisting of two curved mirrors and two wire grids of reflectivity and transmissivity of 0.5 at 360 GHz. The distance of the grids is adjustable via a micrometer. The function of the diplexer is to couple both mw beams efficiently onto the mixer of the heterodyne receiver system. Hence, the output of the diplexer is fed into the horn antenna of a diode mixer that produces the intermediate-frequency signal of 1.21 GHz. The other part of the 360.03-GHz beam passes through the polarization grid and corrugated waveguide down to the FP resonator of the probehead in the B_0 field. At resonance, the parallel component of the mw field carrying the EPR signal is reflected at the polarization grid. The phase of the induced EPR signal beam can be adjusted by using a quasioptical variable-phase shifter consisting of two roof mirrors whose separation can be tuned via a micrometer. The properly phased beam is then combined with the other part of the 361.24-GHz beam by means of an analogous folded Fabry–Perot diplexer and transmitted onto the entrance horn of the diode mixer. The resulting 1.21-GHz microwave carrying the EPR signal is then quadrature mixed with the 1.21-GHz reference microwave down to dc voltage to obtain the EPR signal. Beyond the desired phase sensitivity of the EPR signal, the advantage of this bridge construction is that the detection becomes independent of the phase noise of the reference mw source. Our first phase-sensitive EPR experiments on a test sample have approved the soundness of the described design concept of the transient 360-GHz EPR microwave bridge with reference arm.

3.4.2.9 Pulsed Orotron Source

For pulsed EPR operation at 360 GHz the cw Gunn diode source in the spectrometer can be exchanged for a novel pulsed 360-GHz vacuum-tube source called an Orotron (*Gycom*), see Figure 3.18. The pulsed Orotron was tailor-made for our 360-GHz EPR studies in a joint German–Russian pilot project supported by the DFG.[73] The acronym Orotron was originally introduced as an abbreviation of the Russian words describing a device with an open resonator and a reflecting diffraction grating. Conceptually, the Orotron is a nonrelativistic free-electron laser using the stimulated Smith–Purcell radiation of a flat electron beam interacting with a periodic grating structure that acts as one of the mirrors in an oversized high-quality open Fabry–Perot resonator structure to achieve feedback for high output power and frequency stability. The electromagnetic field present in the open resonator of the device bunches the electron beam, and this bunching leads to coherent oscillation.[174] Such open resonator generators were independently developed by F.S. Rusin and

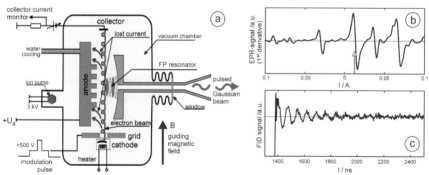

Figure 3.18 (a) Layout of the internal structure of the pulsed Orotron. The feature distinguishing it from a cw Orotron is the additional high-voltage gating grid close to the cathode. Application of high-voltage pulse trains to the grid leads to a modulation of the electron-beam current and concomitant modulation patterns of the generated microwave power. (b) cw 360-GHz EPR spectrum of a polycrystalline perylenyl ion sample. The abscissa is calibrated by the B_0 sweep-field current. (c) FID following a 1-μs mw pulse at a field position arrow-marked in (b). For details, see ref. [73].

G.D. Bogomolov in Moscow, USSR,[175] and by K. Mizuno and S. Ono in Sendai, Japan.[176] The Russian group coined the name Orotron to stress the open-resonator feature of this oscillator, whereas the Japanese group introduced the name Ledatron to stress that the existence of two different modes of electromagnetic fields, the grating surface wave and the FP resonator wave, interacting with the electron beam is essential for the device to create sufficient electron bunching for mw radiation ("Leda" refers to the mother of the twins Castor and Pollux in Greek mythology).

The present design extends the cw Orotron by a gate electrode and a high-voltage pulsing unit to control the electron-beam current. The generated pulses at 360 GHz have pulse lengths from 100 ns – 10 μs and a pulse power of up to 30 mW. The output in a broader frequency band between 320 and 380 GHz ranges from 20 up to 60 mW. Within a 10-ms time slot, incoherent pulse trains of arbitrary duration can be generated. The pulsed Orotron has been incorporated in the quasioptical microwave bridge of our heterodyne induction-mode 360-GHz EPR spectrometer. The first free-induction decay (FID) measurements at 360 GHz and a magnetic field of 12.9 T on a polycrystalline perylenyl-ion sample are considered to be very promising for future applications and extensions of Orotron-EPR spectroscopy under high-frequency/high-field conditions. In the following, some details of the novel pulsed Orotron source will be given that may be of interest to the high-frequency/high-field EPR community.

The wavelength of the diffraction radiation and its power are determined by the distance d between the rulings (spatial period length of the grating), the velocity v of the electrons in the beam, and the angle θ between the incident beam and the emitted plane wave. If the phase velocity (in the direction of the electron beam) $v_{ph} = f \cdot d$ of the diffraction radiation at frequency f is made equal to the electron velocity v, the synchronization condition

$$v = v_{\text{ph}} \tag{3.11}$$

is fulfilled, *i.e.* the time the electrons need to travel a distance d equals the time period $1/f$ of the surface wave. Then the fundamental wavelength λ of the diffraction radiation is given by[177]

$$\lambda = d \cdot (\beta^{-1} - \cos\theta) \tag{3.12}$$

where $\beta = v/c$, and c is the speed of light. Hence, for $\theta = 90°$ (radiation propagating perpendicular to the electron-beam direction), the synchronization condition leads to

$$\lambda = \beta^{-1} \cdot d \tag{3.13}$$

i.e. the wavelength of the diffraction radiation is solely determined by the grating constant d and the electron velocity v. The radiation is polarized with the electric field vector perpendicular to the grooves of the grating. The Smith–Purcell radiation is incoherent and basically very weak. In the Orotron, however, it is turned into coherent and rather powerful radiation by virtue of the Fabry–Perot resonator with high quality factor Q as a feedback device, see Fig 3.18(a).

In conventional Orotrons, a cw electron beam is generated by a cathode close to one side of the metallic grating that forms the anode to accelerate the electrons by a high voltage U_a. A strong longitudinal magnetic field B guides the electron ribbon beam close and parallel to the surface of the metallic grating, which partly covers the plane mirror of the FP resonator. B has to be high and homogeneous enough to guarantee that only a fraction of the electrons hits the water-cooled grating anode, and the other fraction reaches the collector. The concave FP mirror reflects the Smith–Purcell radiation back onto the electron beam. The FP resonator standing-wave mode interacts with the electrons via its transversal electric-field component along the direction of electron motion. If positive feedback conditions are fulfilled, as specified in eqn (3.13), the amplification gain exceeds the losses in the open FP resonator and grating structure, and the build-up of large-amplitude electric-field oscillations in the vicinity of the electron beam causes the electrons to bunch. Electron bunching is the prerequisite for amplifying the weak Smith–Purcell radiation to a strong mw output radiation polarized perpendicular to the mirror axis of the FP resonator. In this situation, the Orotron will radiate coherently at a frequency near one of the resonance frequencies of the FP-resonator modes. The radiation exits the FP resonator via a coupling slit in the concave mirror to enter the output waveguide, which is equipped with an appropriate horn antenna for emitting a Gaussian mw beam.

The theory of the Orotron[175,178,179] provides a series of requirements for Orotron operation, for instance the admissible thickness of the electron-beam

Instrumentation

ribbon. Moreover, the conditions $d \ll \lambda$ and $\lambda \ll R$ (radius of the FP mirrors) have to be met to minimize diffraction losses, and the B field has to be strong enough to confine the electron motion to essentially one dimension parallel to the grating surface and perpendicular to the rulings. Additional conditions for the resonator length (to be tuned mechanically by moving one mirror), electron velocity (to be tuned electrically by the anode voltage U_a) and minimum electron current have to be fulfilled[73] before Orotron radiation starts to be emitted at a certain frequency (or wavelength). The required anode voltage for Orotron oscillation increases quadratically with frequency and is given by (for U_a in kV):

$$U_a = \frac{m \cdot c^2}{2e} \cdot (d/\lambda_n)^2 \approx 255 \cdot (d/\lambda_n)^2 \qquad (3.14)$$

For example, for $\lambda_n = 0.8$ mm (360 GHz) and $d = 0.1$ mm, $U_a \approx 4$ kV.

The metallic grating (10×10 mm^2) is a multicomb volume grating with 50% transparency ($d = 0.1$ mm, rod height $= 0.7$ mm) embedded in the plane mirror of the FP resonator. The actual structure of the rods is that of fins with vertical slits, forming a grid themselves. Hence, the electron beam is split into many parallel beams, thereby increasing the interaction between beam and grid, *i.e.* the Orotron efficiency. The concave mirror (diameter 15 mm) is normally separated from the plane mirror by about 7 mm, but the resonator length l can be mechanically adjusted from 5 to 10 mm. For guiding the electron ribbon beam over the grating, thereby just grazing the active part of its surface without touching it, a specific arrangement of superstrong rare-earth permanent magnets was developed that provides a homogeneous field $B = 1.25$ T over the grating. The magnet configuration was designed in such a way to allow the field to fall off to almost zero at a radial distance of 30 cm. This is an essential requirement for the 14-T cryomagnet of the 360-GHz high-field EPR spectrometer, see below. The overall weight of the Orotron including the permanent magnet is about 25 kg. Its outside form is that of a cylinder with 14 cm diameter and 18 cm length.[73]

The main innovation for constructing a sub-mm Orotron with ns pulsed radiation is the insertion of an additional gating grid between the cathode and the anode. The gating grid is located as close as 0.1 mm apart from the cathode (see Figure 3.18(a)). Application of a negative grid voltage ($U_g \approx -200$ to -300 V) blocks the electron-beam current. In contrast, application of a positive voltage ($U_g \approx +100$ to $+500$ V), builds up the electron current and, when the minimum current is reached, Orotron oscillations get started. Thus, the magnitude of the electron current is controlled by the magnitude and time profile of the grid modulation voltage, *i.e.* the characteristics of the Orotron radiation now critically depend also on the modulation pattern of the gating voltage.

The microwave output of the Orotron is coupled from the top of the EPR cryomagnet into the quasioptical transmission line of the 360 GHz mw bridge described above. Instead of using the cw Gunn-diode source, the Orotron output is focused by a Teflon lens (*Radiometer Physics*) onto the corrugated waveguide to the probehead (see Figure 3.14). The lens has a special groove coating to minimize reflections and standing waves. Like in the cw system, the linear polarization of the Orotron output is used in an induction-mode detection scheme to separate the exciting microwaves from the reflected EPR-signal carrying microwaves. Despite the close proximity of the Orotron and the 14 T cryomagnet, it is mandatory that their magnetic fields do not significantly perturb each other. Therefore, the beam-guiding B field of the Orotron is effectively shielded by properly arranged segments of the superstrong rare-earth magnetic material. This shielding leads to a stray field outside the Orotron that is substantial only along the symmetry axis of the Orotron magnet, but drops off strongly in the perpendicular (radial) direction along the mw horn antenna (from $B = 1.2$ T inside the Orotron to $B < 1$ mT at 30 cm distance).

The test sample used in the first EPR measurements with the pulsed Orotron was a polycrystalline perylenyl-ion sample, (perylenyl)$_2$(PF$_6$)$_{1.1}$ × 0.8CH$_2$Cl$_2$), with a size of $0.2 \times 0.2 \times 0.2$ mm^3.[73] The sample was mounted onto the curved Fabry–Perot mirror. With the spectrometer depicted in Figure 3.14 it was possible, for the first time, to detect EPR FIDs at 360 GHz. Figure 3.18(b) shows, for comparison, the cw EPR spectrum of the perylenyl sample at room temperature, as recorded with the 360-GHz cw EPR setup.[72] The abscissa is calibrated by the current in the sweep coil for varying the B_0 magnetic field. Figure 3.18(c) shows a single-shot signal of the same sample following a 1-μs pulse from the Orotron. The signal was detected with the pulsed EPR setup (Figure 3.14) at a sweep-coil current of 0.015 A (the field-current calibration constant is (13.5 ± 0.2) mT/A). Directly after the Orotron pulse, an exponentially decaying mw signal is observed. This signal was identified by its B_0 dependence as the FID of the EPR transition at 0.015 A. The oscillations on the signal decay are due to a residual frequency mismatch between the Orotron frequency and the LO frequency of the heterodyne detection mixer. Our preliminary experiments show that the coupling of the Orotron source to the quasioptical spectrometer works and the EPR sensitivity is high enough to detect FIDs. Improved Orotron modulator characteristics for pulse trains will allow for two-pulse spin-echo experiments at 360 GHz.

To conclude this section: Many options for exploiting the strength of high-field EPR spectroscopy in terms of spectral and temporal resolution critically rely on the availability of powerful cw and pulsed mm and sub-mm microwave sources. Our present experience with a dedicated tailor-made pulsed Orotron leads us to believe that such a mw source is well suited for state-of-the-art high-field EPR and ENDOR experiments at frequencies up to 360 GHz. In comparison to other vacuum-tube based far-infrared sources, such as EIOs and gyrotrons, the price/performance ratio of an Orotron is certainly very attractive.

References

1. G. Feher, *Bell System Tech. J.*, 1957, **36**, 449.
2. C. P. Poole, *Electron Spin Resonance*, J. Wiley, New York, 1983.

3. E. J. Reijerse, P. J. van Dam, A. A. K. Klaassen, W. R. Hagen, P. J. M. van Bentum and G. M. Smith, *Appl. Magn. Reson.*, 1998, **14**, 153.
4. G. M. Smith, J. C. G. Lesurf, R. H. Mitchell and P. C. Riedi, *Rev. Sci. Instrum.*, 1998, **69**, 3924.
5. B. Smaller, J. R. Remko and E. C. Avery, *J. Chem. Phys.*, 1968, **48**, 5174.
6. P. W. Atkins, K. A. McLauchlan and A. F. Simpson, *J. Phys. E*, 1970, **3**, 547.
7. T. H. Wilmshurst, *Electron Spin Resonance Spectrometers*, Adam Hilger, London, 1967.
8. A. N. Savitsky and H. Paul, *Appl. Magn. Reson.*, 1997, **12**, 449.
9. A. N. Savitsky, H. Paul and A. I. Shushin, *J. Phys. Chem. A*, 2000, **104**, 9091.
10. D. Goldfarb, In: *Electron Paramagnetic Resonance*, B. C. Gilbert, M. J. Davies, D. M. Murphy, Eds., Vol. **15**, Royal Society of Chemistry, Cambridge, 1996, pp. 182.
11. A. Schweiger and G. Jeschke, *Principles of Pulse Electron Paramagnetic Resonance*, Oxford University Press, Oxford, 2001.
12. H. van Willigen, P. R. Levstein and M. H. Ebersole, *Chem. Rev.*, 1993, **93**, 173.
13. G. A. Rinard, R. W. Quine, S. S. Eaton, G. R. Eaton and W. Froncisz, *J. Magn. Reson. A*, 1994, **108**, 71.
14. E. K. Zavoisky, PhD Thesis, Kazan State University, 1944.
15. E. Zavoisky, *J. Phys. USSR*, 1945, **9**, 211.
16. E. Zavoisky, *Zh. Eksper. Teor. Fiz.*, 1946, **16**, 603.
17. E. Zavoisky, *Zh. Eksper. Teor. Fiz.*, 1947, **17**, 155.
18. F. Bloch, W. W. Hansen and M. Packard, *Phys. Rev.*, 1946, **69**, 680.
19. F. Bloch, W. W. Hansen and M. Packard, *Phys. Rev.*, 1946, **70**, 474.
20. E. M. Purcell, H. C. Torrey and R. V. Pound, *Phys. Rev.*, 1946, **69**, 37.
21. H. C. Torrey, E. M. Purcell and R. V. Pound, *Phys. Rev.*, 1946, **69**, 680.
22. J. W. Emsley and J. Feeney, *Prog. Nucl. Magn. Reson. Spectrosc.*, 2007, **50**, 179.
23. J. S. Hyde, In: *Foundations of Modern EPR*, G. R. Eaton, S. S. Eaton, K. M. Salikhov, Eds., World Scientific, Singapore, 1998, pp. 695.
24. R. R. Ernst and W. A. Anderson, *Rev. Sci. Instrum.*, 1966, **37**, 93.
25. D. Schmalbein, In: *Foundations of Modern EPR*, G. R. Eaton, S. S. Eaton, K. M. Salikhov, Eds., World Scientific, Singapore, 1998, pp. 717.
26. J. B. Mock, *Rev. Sci. Instrum.*, 1960, **31**, 551.
27. M. Peter, *Phys. Rev.*, 1959, **113**, 801.
28. P. R. Elliston, G. J. Troup and D. R. Hutton, *J. Sci. Instrum.*, 1963, **40**, 586.
29. E. F. Slade and D. J. E. Ingram, *Proc. Phys. Soc. London, Sect. A*, 1969, **312**, 85.
30. D. J. E. Ingram, *Biological and Biochemical Applications of Electron Spin Resonance*, Adam Hilger, London, 1969.
31. H. van den Boom, *Rev. Sci. Instrum.*, 1969, **40**, 550.
32. I. Amity, *Rev. Sci. Instrum.*, 1970, **41**, 1492.

33. H. C. Box, H. G. Freund, K. T. Lilga and E. E. Budzinski, *J. Phys. Chem.*, 1970, **74**, 40.
34. Y. Alpert, Y. Couder, J. Tuchendler and H. Thome, *Biochim. Biophys. Acta*, 1973, **322**, 34.
35. O. Grinberg and A. A. Dubinskii, In: *Very High Frequency (VHF) ESR/EPR; Biological Magnetic Resonance*, Vol. **22**, O. Grinberg, L. J. Berliner, Eds., Kluwer/Plenum Publishers, New York, 2004, pp. 1.
36. Y. S. Lebedev, In: *Foundations of Modern EPR*, S. S. Eaton, K. M. Salikhov, Eds., World Scientific, Singapore, 1998, pp. 731.
37. Y. S. Lebedev, In: *Modern Pulsed and Continuous-Wave Electron Spin Resonance*, L. Kevan, M. K. Bowman, Eds., John Wiley, New York, 1990, pp. 365.
38. E. Haindl, K. Möbius and H. Oloff, *Z. Naturforsch. A*, 1985, **40**, 169.
39. O. Burghaus, M. Rohrer, T. Götzinger, M. Plato and K. Möbius, *Meas. Sci. Technol.*, 1992, **3**, 765.
40. R. T. Weber, J. Disselhorst, L. J. Prevo, J. Schmidt and W. T. Wenckebach, *J. Magn. Reson.*, 1989, **81**, 129.
41. J. Allgeier, J. A. J. M. Disselhorst, R. T. Weber, W. T. Wenckebach and J. Schmidt, In: *Modern Pulsed and Continuous-Wave Electron Spin Resonance*, L. Kevan, M. K. Bowman, Eds., John Wiley, New York, 1990, pp. 267–283.
42. O. Burghaus, A. Toth-Kischkat, R. Klette and K. Möbius, *J. Magn. Reson.*, 1988, **80**, 383.
43. T. F. Prisner, M. Rohrer and K. Möbius, *Appl. Magn. Reson.*, 1994, **7**, 167.
44. J. A. J. M. Disselhorst, H. van der Meer, O. G. Poluektov and J. Schmidt, *J. Magn. Reson. A*, 1995, **115**, 183.
45. W. B. Lynch, K. A. Earle and J. H. Freed, *Rev. Sci. Instrum.*, 1988, **59**, 1345.
46. F. Muller, M. A. Hopkins, N. Coron, M. Grynberg, L. C. Brunel and G. Martinez, *Rev. Sci. Instrum.*, 1989, **60**, 3681.
47. A. A. Galkin, O. Y. Grinberg, A. A. Dubinskii, N. N. Kabdin, V. N. Krymov, V. I. Kurochkin, Y. S. Lebedev, L. F. Oranskii and V. F. Shuvalov, *Instrum. Exp. Tech.*, 1977, **20**, 1229.
48. A. Y. Bresgunov, A. A. Dubinskii, V. N. Krimov, Y. G. Petrov, O. G. Poluektov and Y. S. Lebedev, *Appl. Magn. Reson.*, 1991, **2**, 715.
49. H. J. van der Meer, J. A. J. M. Disselhorst, J. Allgeier, J. Schmidt and W. T. Wenckebach, *Meas. Sci. Technol.*, 1990, **1**, 396.
50. D. Schmalbein, G. G. Maresch, A. Kamlowski and P. Höfer, *Appl. Magn. Reson.*, 1999, **16**, 185.
51. W. Wang, R. L. Belford, R. B. Clarkson, P. H. Davis, J. Forrer, M. J. Nilges, M. D. Timken, T. Walczak, M. C. Thurnauer, J. R. Norris, A. L. Morris and Y. Zhang, *Appl. Magn. Reson.*, 1994, **6**, 195.
52. M. J. Nilges, A. I. Smirnov, R. B. Clarkson and R. L. Belford, *Appl. Magn. Reson.*, 1999, **16**, 167.
53. I. Gromov, V. Krymov, P. Manikandan, D. Arieli and D. Goldfarb, *J. Magn. Reson.*, 1999, **139**, 8.

54. W. Hofbauer, K. A. Earle, C. R. Dunnam, J. K. Moscicki and J. H. Freed, *Rev. Sci. Instrum.*, 2004, **75**, 1194.
55. H. Brutlach, E. Bordignon, L. Urban, J. P. Klare, H. J. Reyher, M. Engelhard and H. J. Steinhoff, *Appl. Magn. Reson.*, 2006, **30**, 359.
56. J. S. Hyde, W. Froncisz, J. W. Sidabras, T. G. Camenisch, J. R. Anderson and R. A. Strangeway, *J. Magn. Reson.*, 2007, **185**, 259.
57. S. Hill, N. S. Dalal and J. S. Brooks, *Appl. Magn. Reson.*, 1999, **16**, 237.
58. P. Dorlet, S. A. Seibold, G. T. Babcock, G. J. Gerfen, W. L. Smith, A. L. Tsai and S. Un, *Biochemistry*, 2002, **41**, 6107.
59. A. Gunn, M. Brynda and R. D. Britt, *Abstracts of Papers of the American Chemical Society*, 2005, **229**, U751.
60. T. A. Stich, S. Lahiri, G. Yeagle, M. Dicus, M. Brynda, A. Gunn, C. Aznar, V. J. DeRose and R. D. Britt, *Appl. Magn. Reson.*, 2007, **31**, 321.
61. M. Bennati, C. T. Farrar, J. A. Bryant, S. J. Inati, V. Weis, G. J. Gerfen, P. Riggs-Gelasco, J. Stubbe and R. G. Griffin, *J. Magn. Reson.*, 1999, **138**, 232.
62. T. F. Prisner, S. Un and R. G. Griffin, *Israel J. Chem.*, 1992, **32**, 357.
63. L. R. Becerra, G. J. Gerfen, B. F. Bellew, J. A. Bryant, D. A. Hall, S. J. Inati, R. T. Weber, S. Un, T. F. Prisner, A. E. McDermott, K. W. Fishbein, K. E. Kreischer, R. J. Temkin, D. J. Singel and R. G. Griffin, *J. Magn. Reson. A*, 1995, **117**, 28.
64. M. Rohrer, O. Brugmann, B. Kinzer and T. F. Prisner, *Appl. Magn. Reson.*, 2001, **21**, 257.
65. J. T. Cardin, S. V. Kolaczkowski, J. R. Anderson and D. E. Budil, *Appl. Magn. Reson.*, 1999, **16**, 273.
66. E. Reijerse, P. P. Schmidt, G. Klihm and W. Lubitz, *Appl. Magn. Reson.*, 2007, **31**, 611.
67. J. van Tol, L. C. Brunel and R. J. Wylde, *Rev. Sci. Instrum.*, 2005, **76**, 07410.
68. H. Blok, J. Disselhorst, S. B. Orlinskii and J. Schmidt, *J. Magn. Reson.*, 2004, **166**, 92.
69. H. Blok, J. Disselhorst, H. van der Meer, S. B. Orlinskii and J. Schmidt, *J. Magn. Reson.*, 2005, **173**, 49.
70. A. Ivancich, T. A. Mattioli and S. Un, *J. Am. Chem. Soc.*, 1999, **121**, 5743.
71. S. Un, P. Dorlet and A. W. Rutherford, *Appl. Magn. Reson.*, 2001, **21**, 341.
72. M. R. Fuchs, T. F. Prisner and K. Möbius, *Rev. Sci. Instrum.*, 1999, **70**, 3681.
73. Y. A. Grishin, M. R. Fuchs, A. Schnegg, A. A. Dubinskii, B. S. Dumesh, F. S. Rusin, V. L. Bratman and K. Möbius, *Rev. Sci. Instrum.*, 2004, **75**, 2926.
74. G. Annino, M. Cassettari, M. Fittipaldi, I. Longo, M. Martinelli, C. A. Massa and L. A. Pardi, *J. Magn. Reson.*, 2000, **143**, 88.
75. A. K. Hassan, L. A. Pardi, J. Krzystek, A. Sienkiewicz, P. Goy, M. Rohrer and L. C. Brunel, *J. Magn. Reson.*, 2000, **142**, 300.

76. A. K. Hassan, A. L. Maniero, H. van Tol, C. Saylor and L. C. Brunel, *Appl. Magn. Reson.*, 1999, **16**, 299.
77. M. Rohrer, J. Krzystek, V. Williams and L. C. Brunel, *Meas. Sci. Technol.*, 1999, **10**, 275.
78. H. P. Moll, C. Kutter, J. van Tol, H. Zuckerman and P. Wyder, *J. Magn. Reson.*, 1999, **137**, 46.
79. K. A. Earle, D. S. Tipikin and J. H. Freed, *Rev. Sci. Instrum.*, 1996, **67**, 2502.
80. K. A. Earle and J. H. Freed, *Appl. Magn. Reson.*, 1999, **16**, 247.
81. H. Nojiri, M. Motokawa, K. Okuda, H. Kageyama, Y. Ueda and H. Tanaka, *J. Phys. Soc. Jpn.*, 2003, **72 Suppl. B**, 109.
82. T. Tatsukawa, T. Maeda, H. Sasai, T. Idehara, I. Mekata, T. Saito and T. Kanemaki, *Int. J. Infrared. Milli.*, 1995, **16**, 293.
83. T. Tatsukawa, T. Shirai, T. Imaizumi, T. Idehara, I. Ogawa and T. Kanemaki, *Int. J. Infrared. Milli.*, 1998, **19**, 859.
84. S. Mitsudo, T. Higuchi, K. Kanazawa, T. Idehara, I. Ogawa and M. Chiba, *J. Phys. Soc. Jpn.*, 2003, **72 Suppl. B**, 172.
85. R. J. Hulsebosch, I. V. Borovykh, S. V. Paschenko, P. Gast and A. J. Hoff, *J. Phys. Chem. B*, 1999, **103**, 6815.
86. K. V. Lakshmi, M. J. Reifler, G. W. Brudvig, O. G. Poluektov, A. M. Wagner and M. C. Thurnauer, *J. Phys. Chem. B*, 2000, **104**, 10445.
87. P. Höfer, A. Kamlowski, G. G. Maresch, D. Schmalbein and R. T. Weber, In: *Very High Frequency (VHF) ESR/EPR, Biological Magnetic Resonance*, Vol. **22**, O. Grinberg, L. J. Berliner, Eds., Kluwer/Plenum Publishers, New York, 2004, pp. 401.
88. P. Carl, R. Heilig, D. C. Maier, P. Höfer and D. Schmalbein, *Bruker Spinreport*, 2004, **154/155**, 35.
89. D. E. Budil, K. A. Earle, W. B. Lynch and J. H. Freed, In: *Advanced EPR: Applications in Biology and Biochemistry*, A. J. Hoff, Ed., Elsevier, Amsterdam, 1989, pp. 307.
90. A. A. Doubinski, In: *Electron Paramagnetic Resonance*, N .M. Atherton, M. J. Davies, B. C. Gilbert, Eds., Vol. **16**, Royal Society of Chemistry, Cambridge, 1998, pp. 211.
91. G. M. Smith and P. C. Riedi, In: *Electron Paramagnetic Resonance*, N. M. Atherton, M. J. Davies, B. C. Gilbert, Eds., Vol. **17**, Royal Society of Chemistry, Cambridge, 2000, pp. 164.
92. P. C. Riedi and G. M. Smith, In: *Electron Paramagnetic Resonance*, B. C. Gilbert, M. J. Davies, D. M. Murphy, Eds., Vol. **19**, Royal Society of Chemistry, Cambridge, 2004, pp. 338.
93. P. C. Riedi and G. M. Smith, In: *Electron Paramagnetic Resonance*, B. C. Gilbert, M. J. Davies, D. M. Murphy, Eds., Vol. **18**, Royal Society of Chemistry, Cambridge, 2002, pp. 254.
94. D. E. Budil and K. A. Earle, In: *Very High Frequency (VHF) ESR/EPR, Biological Magnetic Resonance*, Vol. **22**, O. Grinberg, L. J. Berliner, Eds., Kluwer/Plenum Publishers, New York, 2004, pp. 353.

95. L. C. Brunel, J. van Tol, A. Angerhofer, S. Hill, J. Krzystek and A. L. Maniero, In: *Very High Frequency (VHF) ESR/EPR, Biological Magnetic Resonance*, Vol. **22**, O. Grinberg, L. J. Berliner, Eds., Kluwer/Plenum Publishers, New York, 2004, pp. 466.
96. T. F. Prisner, In: *Advanced Magnetic and Optical Resonance*, Vol. **20**, W. Warren, Ed., Academic, New York, 1997, pp. 245.
97. J. E. Carlstrom and J. Zmuidzinas, In: *Reviews of Radio Science 1993–1995*, W.R. Stone, Ed., Oxford University Press Oxford, 1996.
98. J. L. Hesler, T. W. Crowe, "Responsivity and Noise Measurements of Zero-Bias Schottky Diode Detectors", 18th Intl. Symp. Space Terahertz Tech., 2007, Pasadena.
99. J. V. Grahn, P. Starski, J. Stake and T. Sergey, *IEICE Trans. Electron.*, 2006, **E89C**, 891.
100. V. Radisic, X. B. Mei, W. R. Deal, W. Yoshida, P. H. Liu, J. Uyeda, M. Barsky, L. Samoska, A. Fung, T. Gaier and R. Lai, *IEEE Microw. Wirel. Co.*, 2007, **17**, 223.
101. T. Gaier, L. Samoska, A. Fung, W. R. Deal, V. Radisic, X. B. Mei, W. Yoshida, P. H. Liu, J. Uyeda, M. Barsky and R. Lai, *IEEE Microw. Wirel. Co.*, 2007, **17**, 546.
102. R. Blundell and C. Y. E. Tong, *Proc. IEEE*, 1992, **80**, 1702.
103. B. R. Bennett, R. Magno, J. B. Boos, W. Kruppa and M. G. Ancona, *Solid State Electron.*, 2005, **49**, 1875.
104. E. R. Brown and C. D. Parker, *Philos. Trans. R. Soc. London, Ser. A*, 1996, **354**, 2365.
105. V. G. Bozhkov, *Radiophys. Quant. Electron.*, 2003, **46**, 631.
106. J. A. Dayton, V. O. Heinen, N. Stankiewicz and T. M. Wallett, *Int. J. Infrared. Milli.*, 1987, **8**, 1257.
107. G. Kantorowicz and P. Palluel, In: *Infrared and Millimeter Waves*, K. J. Button, Ed., Vol. **1**, Academic Press, New York, 1979.
108. N. Nakagawa, T. Yamada, K. Akioka, S. Okubo, S. Kimura and H. Ohta, *Int. J. Infrared. Milli.*, 1998, **19**, 167.
109. C. T. Farrar, D. A. Hall, G. J. Gerfen, S. J. Inati and R. G. Griffin, *J. Chem. Phys.*, 2001, **114**, 4922.
110. M. Bennati, J. Stubbe and R. G. Griffin, *Appl. Magn. Reson.*, 2001, **21**, 389.
111. V. Weis, M. Bennati, M. Rosay, J. A. Bryant and R. G. Griffin, *J. Magn. Reson.*, 1999, **140**, 293.
112. M. K. Hornstein, R. G. Griffin, J. Machuzak, M. A. Shapiro, R. J. Temkin and K. E. Kreischer, *Pulsed Power Plasma Science, 2001. IEEE Conference Record-Abstracts*, 2001, 516.
113. M. K. Hornstein, V. S. Bajaj, R. G. Griffin and R. J. Temkin, *IEEE Plasma Sci.*, 2006, **34**, 524.
114. A. L. Barra, *Appl. Magn. Reson.*, 2001, **21**, 619.
115. C. Kutter, H. P. Moll, J. van Tol, H. Zuckermann, J. C. Maan and P. Wyder, *Phys. Rev. Lett.*, 1995, **74**, 2925.

116. A. N. Savitsky, M. Galander and K. Möbius, *Chem. Phys. Lett.*, 2001, **340**, 458.
117. K. Möbius, A. Savitsky, A. Schnegg, M. Plato and M. Fuchs, *Phys. Chem. Chem. Phys.*, 2005, **7**, 19.
118. G. Annino, M. Cassettari and M. Martinelli, *Rev. Sci. Instrum.*, 2005, **76**, 084704.
119. N. Nandi, K. Bhattacharyya and B. Bagchi, *Chem. Rev.*, 2000, **100**, 2013.
120. H. Blok, PhD Thesis, University of Leiden, 2006.
121. H. Kogelnik and T. Li, *Proc. IEEE*, 1966, **54**, 1312.
122. H. Kogelnik and T. Li, *Appl. Opt.*, 1966, **5**, 1550.
123. G. D. Boyd and J. P. Gordon, *Bell System Tech. J.*, 1961, **40**, 489.
124. M. Fuchs, PhD Thesis, Freie Universtät Berlin, 1999.
125. T. A. Hargreaves, R. P. Fischer, R. B. McCowan and A. W. Fliflet, *Int. J. Infrared. Milli.*, 1991, **12**, 9.
126. R. N. Clarke and C. B. Rosenberg, *J. Phys. E*, 1982, **15**, 9.
127. O. Burghaus, PhD Thesis, Freie Universtät Berlin, 1991.
128. I. Tkach, U. Rogulis, S. Greulich-Weber and J. M. Spaeth, *Rev. Sci. Instrum.*, 2004, **75**, 4781.
129. R. Klette, PhD Thesis, Freie Universtät Berlin, 1994.
130. M. Fuhs, T. Prisner and K. Möbius, *J. Magn. Reson.*, 2001, **149**, 67.
131. J. S. Hyde and W. Froncisz, In: *Advanced EPR: Appications in Biology and Biochemistry*, A. J. Hoff, Ed., Elsevier, Amsterdam, 1989, pp. 277.
132. J. W. Sidabras, R. R. Mett, W. Froncisz, T. G. Camenisch, J. R. Anderson and J. S. Hyde, *Rev. Sci. Instrum.*, 2007, **78**, 034701.
133. G. Annino, M. Cassettari, I. Longo and M. Martinelli, *Appl. Magn. Reson.*, 1999, **16**, 45.
134. G. Annino, M. Cassettari, I. Longo, M. Martinelli, P. J. M. van Bentum and E. van der Horst, *Rev. Sci. Instrum.*, 1999, **70**, 1787.
135. G. Annino, M. Cassettari and M. Martinelli, *Rev. Sci. Instrum.*, 2005, **76**, 064702.
136. A. Okafuji, A. Schnegg, E. Schleicher, K. Möbius and S. Weber, *J. Phys. Chem. B*, 2008, **112**, 3568.
137. A. Schnegg, A. A. Dubinskii, M. R. Fuchs, Y. A. Grishin, E. P. Kirilina, W. Lubitz, M. Plato, A. Savitsky and K. Möbius, *Appl. Magn. Reson.*, 2007, **31**, 59.
138. J. P. Crenn, *Int. J. Infrared. Milli.*, 1993, **14**, 1947.
139. D. B. Montgomery, *Rep. Prog. Phys.*, 1963, **26**, 69.
140. G. Roth, *Bruker Spinreport*, 2003, **152/153**, 14.
141. O. Kirichek, P. Carr, C. Johnson and M. Atrey, *Rev. Sci. Instrum.*, 2005, **76**, 055104.
142. A. I. Smirnov, T. I. Smirnova, R. L. MacArthur, J. A. Good and R. Hall, *Rev. Sci. Instrum.*, 2006, **77**, 035108.
143. T. Maly, J. Bryant, D. Ruben and R. G. Griffin, *J. Magn. Reson.*, 2006, **183**, 303.
144. V. M. Borovikov, M. G. Fedurin, G. V. Karpov, D. A. Korshunov, E. A. Kuper, M. V. Kuzin, V. R. Mamkin, A. S. Medvedko, N. A. Mezentsev,

V. V. Repkov, V. A. Shkaruba, E. I. Shubin and V. F. Veremeenko, *Nucl. Instrum. Meth. A*, 2001, **467**, 198.
145. J. Haase, M. B. Kozlov, A. G. Webb, B. Buchner, H. Eschrig, K. H. Müller and H. Siegel, *Solid State Nucl. Magn. Reson.*, 2005, **27**, 206.
146. M. B. Kozlov, J. Haase, C. Baumann and A. G. Webb, *Solid State Nucl. Magn. Reson.*, 2005, **28**, 64.
147. M. Motokawa, *Rep. Prog. Phys.*, 2004, **67**, 1995.
148. O. Burghaus, Diploma Thesis, Freie Universtät Berlin, 1984.
149. O. Burghaus, E. Haindl, M. Plato and K. Möbius, *J. Phys. E*, 1985, **18**, 294.
150. T. Götzinger, Diploma Thesis, Freie Universtät Berlin, 1990.
151. M. Rohrer, Diploma Thesis, Freie Universtät Berlin, 1991.
152. M. Rohrer, PhD Thesis, Freie Universtät Berlin, 1995.
153. M. Rohrer, M. Plato, F. Macmillan, Y. Grishin, W. Lubitz and K. Möbius, *J. Magn. Reson. A*, 1995, **116**, 59.
154. R. Klette, J. T. Törring, M. Plato, K. Möbius, B. Bönigk and W. Lubitz, *J. Phys. Chem.*, 1993, **97**, 2015.
155. T. F. Prisner, A. van der Est, R. Bittl, W. Lubitz, D. Stehlik and K. Möbius, *Chem. Phys.*, 1995, **194**, 361.
156. A. Schnegg, Diploma Thesis, Freie Universtät Berlin, 1998.
157. M. Fuhs, PhD Thesis, Freie Universtät Berlin, 1999.
158. M. Fuhs, A. Schnegg, T. Prisner, I. Köhne, J. Hanley, A. W. Rutherford and K. Möbius, *Biochim. Biophys. Acta*, 2002, **1556**, 81.
159. A. Savitsky, M. Kühn, D. Duche, K. Möbius and H. J. Steinhoff, *J. Phys. Chem. B*, 2004, **108**, 9541.
160. A. Savitsky, A. A. Dubinskii, M. Flores, W. Lubitz and K. Möbius, *J. Phys. Chem. B*, 2007, **111**, 6245.
161. A. A. Dubinskii, Y. A. Grishin, A. N. Savitsky and K. Möbius, *Appl. Magn. Reson.*, 2002, **22**, 369.
162. A. W. Rutherford and P. Setif, *Biochim. Biophys. Acta*, 1990, **1019**, 128.
163. R. R. Ernst, *Bull. Magn. Reson.*, 1994, **16**, 5.
164. T. Prisner and K. P. Dinse, *J. Magn. Reson.*, 1989, **84**, 296.
165. D. T. Teaney, M. P. Klein and A. M. Portis, *Rev. Sci. Instrum.*, 1961, **32**, 721.
166. J. S. Hyde, J. C. W. Chen and J. H. Freed, *J. Chem. Phys.*, 1968, **48**, 4211.
167. M. R. Fuchs, A. Schnegg, M. Plato, C. Schulz, F. Müh, W. Lubitz and K. Möbius, *Chem. Phys.*, 2003, **294**, 371.
168. A. Schnegg, PhD Thesis, Freie Universtät Berlin, 2003.
169. P. F. Goldsmith, *Quasioptical Systems*, IEEE Press, New York, 1998.
170. J. C. G. Lesurf, *Milimetre-wave Optical, Devices and Systems*, Adam Hilger, New York, 1990.
171. A. F. Gulla and D. E. Budil, *Concept. Magn. Reson. B*, 2004, **22B**, 15.
172. K. A. Earle, R. Zeng and D. E. Budil, *Appl. Magn. Reson.*, 2001, **21**, 275.
173. J. L. Doane, In: *Infrared and Millimeter Waves*, K. J. Button, Ed., Vol. **13**, Academic Press, New York, 1985, pp. 123.

174. R. P. Leavitt, D. E. Wortman and C. A. Morrison, *Appl. Phys. Lett.*, 1979, **35**, 363.
175. F. S. Rusin and G. D. Bogomolov, *JEPT Letters-USSR*, 1966, **4**, 160.
176. K. Mizuno, S. Ono, In: *Infrared and Millimeter Waves*, Vol. **1**, K. J. Button, Ed., New York, 1979.
177. S. J. Smith and E. M. Purcell, *Phys. Rev.*, 1953, **92**, 1069.
178. D. E. Wortman, R. P. Leavitt, In: *Infrared and Millimeter Waves*, Vol. **7**, Part II, K. J. Button, Ed., Academic Press, New York, 1983.
179. V. L. Bratman, B. S. Dumesh, A. E. Fedotov, Y. A. Grishin and F. S. Rusin, *Int. J. Infrared. Milli.*, 2002, **23**, 1595.
180. G. A. Rinard, S. S. Eaton, G. R. Eaton, C. P. Poole and H. A. Farach, In: *Handbook of Electron Spin Resonance*, Vol. **2**, C. P. Poole, H. A. Farach, Eds., Springer, Berlin, 1999.

CHAPTER 4
Computational Methods for Data Interpretation

The development of high-resolution multifrequency EPR spectroscopy as a powerful tool to explore molecular structure and dynamics has prompted significant improvements both of molecular-dynamics and quantum-theoretical methods to translate EPR observables into information on molecular structure and dynamics of large (bio)systems. The obvious advantage of spectroscopic methods for structure determination over protein X-ray crystallography is that no single-crystal material is needed and that time-resolved measurements can be performed to detect short-lived intermediates. To obtain structural information, however, one must be able to interpret the spectral features in terms of molecular structure, greatly aided by quantum chemistry, foremost by density functional theory (DFT).[1]

DFT-based methods to calculate EPR spin-interaction parameters of biomolecules are state-of-the-art by now. For example, considerable emphasis is currently put on the calculation of environmental effects on the g-tensor and the other interaction tensors of the cofactors in their transient doublet or triplet states in photosynthetic bacterial reaction centers, *i.e.* specific electrostatic fields and hydrogen bonds from neighboring polar and protic amino acids in the binding sites. Another example is provided by molecular-dynamics (MD)-based simulation techniques for spin-label EPR spectra. Such methods allow the calculation of molecular-dynamics trajectories of spin-labelled proteins with water or lipid molecules in the vicinity of the spin-label side chains to learn about protein–membrane interactions. Although very successful in modern EPR on biomolecules, MD-based simulation methods will not be covered in this chapter for reasons of limited space. Rather, we will focus on DFT-based methods for calculating EPR interaction parameters. The aim of such calculations is, by comparing the theoretical results with the measured values, to learn about the electronic structure of the reactants of the protein in action.

Reliable computational methods have been developed recently to calculate the interaction parameters of the spin Hamiltonian, in particular electron–nuclear hyperfine-, quadrupole- and g-tensor components as well as zero-field splittings.[2] It is this combination of new technologies for measuring different molecular parameters and increased computing power that now allows biophysicists to describe quantitatively proteins while passing through various reaction stages, and to visualize complex biosystems in their three-dimensional structure and functional dynamics. But even today, the level of molecular theory chosen remains to be a compromise between reliability of the predictions and affordable computational time, and the compromise is dictated by the size of the molecular system. It is noted that, in comparison to hyperfine-tensors, the theory for g-tensor calculations of large molecular systems was rather neglected for a long time because experimental g-tensor data of sufficient precision were missing. This has changed now owing to the improved spectral resolution and information content of high-field EPR and ENDOR.

As a general statement one can say that the theory of g-tensor calculations is much more involved for organic systems with many carbon atoms and strongly delocalized electrons than for systems with dominant heavy-atom contributions, such as metallo-proteins, quinones or nitroxides.[3–5] Therefore, in this book we will refrain from describing in length the details of g-tensor theory for strongly delocalized systems, like the dimeric chlorophyll donor in photosynthetic systems, and rather present only the results of our computation (see Section 5.2.1.4). For nitroxide-labelled proteins, on the other hand, with strong localization of the unpaired electron on the N and O atoms, we will sketch some more details of the theory behind the computations when we describe the actual experiments in Section 5.2.2.3.

After earlier INDO- or AM1-type semiempirical quantum-chemical approaches to calculate g-tensors of bio-organic systems,[6–8] a couple of years ago our laboratory joined the still rather small club within the high-field EPR community that applies modern and quite pretentious DFT methodology to g-tensor calculations. One of the two first modern DFT approaches to the calculation of magnetic resonance parameters has been reported by van Lenthe and coworkers.[9] A serious limitation of their "zero-order regular approximation" (ZORA) is the use of a spin-restricted treatment that is inappropriate for the calculation of nuclear hyperfine couplings (hfcs), the evaluation of which along with the g-tensor is at least equally important.

The second modern DFT approach, which allows the calculation of hfcs and g-tensors at the same spin-unrestricted levels, has been developed by Schreckenbach and Ziegler[10] (the SZ method). More recently, two related implementations have been reported by Kaupp and coworkers[11] and by Neese.[12] These approaches all employ second-order perturbation theory based on the unrestricted Kohn–Sham equations.[13] They incorporate relativistic contributions to the g-tensor at different levels of approximation. The SZ method has the distinct advantage of using gauge-including atomic orbitals

(GIAOs) or "London orbitals".[14] These orbitals avoid the undesirable gauge dependence (choice of origin) in the calculation of g-tensors introduced by the orbital Zeeman operator.

In our recent study of mutation-induced g shifts of the primary donor $P^{\bullet+}$ in *Rb. sphaeroides* mutants (see Section 5.2.1.4), we used the SZ-method at the "relativistic scalar Pauli" level employing double-zeta atomic basis sets, the Becke exchange[15] and Lee–Yang–Parr correlation[16] options for the generalized gradient approximation (GGA) part of the exchange-correlation (XC) functional.[10]

The original idea behind DFT goes back to the Thomas–Fermi model of the electronic structure of atoms, which replaces the complicated N-electron wavefunction $\Psi(x_1, x_2, \ldots, x_N)$ and the associated Schrödinger equation by the much simpler electron density function $\rho(r)$ and its associated calculation scheme (here, x_i are the spatial coordinates of the N electrons, *i.e.* the dimensionality of the argument x_i is $3N$, whereas r is simply a 3-dimensional position vector). On the other hand, the DFT approach requires various approximations concerning electron correlation and exchange effects that limit the accuracy and universality of theoretical predictions of spectroscopic parameters. The so-called *ab-initio* methods, on the other hand, are largely free of such approximations although they suffer from large-sized atomic basis sets and excessive storage requirements for electron–electron interaction integrals. Additionally, *ab-initio* methods generally require extensive configuration interaction (CI) treatments for sufficient accuracy of ground and excited-state properties. The essential advantage of *ab-initio* treatments is that their accuracy in the prediction of spectroscopic properties may, in principle, be improved systematically towards the exact result for a given Hamiltonian. Generally, however, *ab-initio* approaches are presently only suitable for small molecules with up to at most 10 atoms, and they serve best as methods to benchmark more economic and approximate treatments.

Thus, for large bio-organic molecules and even for nitroxide spin labels in various environments (polarity effects, hydrogen bonding), DFT is the method of choice for the theoretical prediction of spectroscopic parameters, as it surpasses the accuracy and physical soundness of semiempirical methods. DFT implicitly includes electron-correlation effects at much lower cost than the *ab-initio* methods like restricted-open-shell Hartree–Fock (ROHF) and multireference configuration-interaction (MRCI) approaches.

A principal disadvantage of DFT is the lack of a tool for systematic improvement towards an exact limit, as the exact exchange-correlation functional is not known. Thus, DFT calculations of g-tensors and other observables require careful validation for a given functional. A further unsolved problem in DFT calculations of g-tensors (and chemical-shift tensors) is the dependence of the exchange-correlation potential on the paramagnetic currents induced by the magnetic field. Some open theoretical questions pertain also to the proper treatment of spin-orbit coupling within the DFT framework. On the other hand, one should keep in mind that a critical comparison between experimental and quantum-theoretical calculations of spectral observables (g-values, hyperfine couplings, quadrupole splittings, zero-field dipolar splittings) is indispensible to validate theoretical approaches and to substantiate the reliability of theoretical predictions of geometries, reaction energies and transition-state properties. This fruitful interplay between theory and experiment in determining molecular and electronic structures is indicated in Figure 4.1.

It is the task of ambitious molecular theory to predict the well-defined interaction parameters in the spin-Hamiltonian from first principles, *i.e.* from

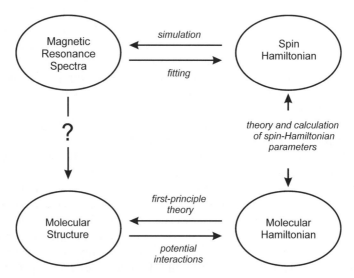

Figure 4.1 Relationship and interplay between experiment and theory in determining molecular structure. The objective is to decipher the experimental spectra (upper left) to obtain the encoded information on the molecular structure (lower left). First-principle quantum-mechanical calculations of the molecular structure requires the solution of the relativistic molecular Schrödinger equation with the full molecular Hamilton operator (lower right) that, for large molecules, is extremely complex and time consuming. Therefore, for biomolecules this direct path in Figure 4.1 is unrealistic and, instead, a bypass has to be created invoking experimental information from EPR spectroscopy: The molecular Hamiltonian is parameterized by a phenomenological spin-Hamiltonian (upper right) adapted to the specific spectroscopic observables. The spectra contain these parameters (g-values, hyperfine couplings, quadrupole splittings, zero-field dipolar splittings) which, in cases of well-resolved spectra, can be directly read off from line positions and lineshapes. In most cases the spectra of large biomolecules are badly resolved. The interaction parameters can then be obtained by simulating the spectra by means of an adequate spin-Hamiltonian and ensuing least-squares fitting of simulated and experimental spectra. The role of theory and subsequent quantum-chemical calculations is to predict the interaction parameters in the spin-Hamiltonian from the full molecular Hamiltonian.

the full molecular Hamiltonian. Because of the complexity of this task, EPR spectra are parameterized in terms of a phenomenological spin-Hamiltonian. These interaction parameters are determined by fitting calculated to experimental spectra. The energy-level differences probed by EPR are very small ($<1\,\mathrm{cm}^{-1}$, i.e. $\approx 10^{-4}\,\mathrm{eV}$) and even ≈ 1000 times smaller in NMR. They are much smaller than chemically relevant energy differences ($\approx 0.1\,\mathrm{eV}$). To interpret these small differences, one needs to take many small (mainly relativistic) interactions into account.

Thus, both spatial and electronic structures of the molecules are obtained from the interplay between theory and spectroscopic experiment. This is highly desired information because electronic structure and biochemical function of the reactants are intimately linked. It is noticed that only spectroscopy, preferably on the basis of prior knowledge of the ground-state molecular structure of the biological system from X-ray crystallography, provides the means of elucidating the electronic structure of transient reaction intermediates controlling biochemical function.

It is realized that by now DFT methods have become increasingly important for g-tensor calculations and the prediction of dipolar and quadrupolar hyperfine couplings. The apparent boom in the appreciation of DFT methods is mainly due to the overall high quality of the results combined with computational efficiency. As a suggestion for further reading, the anthology *"Calculation of NMR and EPR Parameters: Theory and Applications"*, edited by M. Kaupp, M. Bühl and V. G. Malkin in 2004,[2] can be considered as a standard reference for modern quantum-chemical calculation of magnetic-resonance parameters by DFT methods. Among the many enlightening contributions from experts in the fields of NMR and EPR, one finds essays on historical aspects of EPR parameter calculations (F. Neese, M. Munzarová), the concept of the effective spin-Hamiltonian from a quantum-chemical perspective (G. Lushington), the fundamentals of nonrelativistic and relativistic theory of NMR and ESR parameters (W. Kutzelnigg). Furthermore, methodological aspects and applications are presented, such as DFT calculations of EPR hyperfine coupling tensors (M. Munzarová), *ab initio* post-Hartree–Fock calculations of hyperfine coupling tensors and their comparison with DFT approaches (B. Engels), computation of hyperfine coupling tensors to complement EPR experiments (F. Ban, J. W. Gauld, R. J. Boyd), calculations of EPR g-tensors with DFT (S. Patchkovskii, G. Schreckenbach), *ab-initio* calculations of g-tensors (G. Lushington), critical assessments of calculations of zero-field splittings (F. Neese), DFT applications to EPR in bioinorganic chemistry (F. Neese).

Over the last years until very recently, Neese and coworkers have presented several indepth state-of-the-art reviews of the calculation and interpretation of spin-Hamiltonian parameters that we highly recommend.[17–20] In the following overview on modern theoretical methods used for interpreting EPR spectroscopical parameters, we strongly lean on a review article by F. Neese,[20] from which we borrow ideas and assessments.

Theoretical Approaches to g-values

Besides the constant intrinsic spin magnetic moment of the electron, there are two additional contributions to the g-value of an organic molecule: the contribution of orbital angular momentum of the unpaired electrons and the contribution from spin-orbit coupling (SOC). In second-order perturbation theory, the additional molecular magnetic dipole moment is proportional to the external magnetic field.[21] Hence, on a phenomenological level, the extra angular momentum is taken care of by modifying the g-value of the free electron, $g_e = 2.002319$, when writing the EPR resonance condition $h \cdot \nu = g \cdot \mu_B \cdot B_0$ in terms of the externally applied static magnetic B_0 field (μ_B is the Bohr magneton). Because the electronic interactions and induced molecular currents are anisotropic, the spin-Hamiltonian contains the modified g-value in terms of a 3×3 matrix ("g-tensor"), and the molecular information is encoded in the g-shift ($g-g_e$). For the calculation of the SOC-induced molecular currents and, thereby, for the prediction of theoretical g-values, basically two different philosophies exist within the DFT framework: The first strategy uses second-order perturbation theory to estimate SOC and magnetic-field effects.[10–12,22,23] The second strategy is to include the SOC in the self-consistent DFT calculation using a relativistic DFT method and then to use first-order perturbation theory to estimate the effect of the external magnetic field.[24,25]

From the results obtained so far in the field of DFT calculations of EPR parameters, one may draw several conclusions:[20]

1. For organic and inorganic radicals, all methods are able to predict g-values with good accuracy. They are typically within 500 ppm of the experimental values.
2. For transition-metal complexes, errors of a factor of about 2 in the g-shifts are common.
3. The performance of different density functionals is similar. Hybrid functionals such as B3LYP tend to be slightly more accurate for transition-metal complexes than GGA functionals such as BP86. This is attributed to a better description of metal–ligand covalency.
4. Standard basis sets of double-ζ plus polarization quality already give useful results.
5. The relativistic ZORA method[24] in its present form tends to be less accurate for organic radicals and first-row transition-metal complexes than the approaches based on perturbation theory.[12] In addition, the ZORA formalism is currently limited to nonhybrid functionals, a single unpaired electron and to a spin-restricted treatment.
6. The perturbation-based methods do not have these restrictions but tend to be less accurate for heavy elements beyond the first transition row. They may be less accurate for the molecules in which the ground state is nearly orbitally degenerate.

Some more statements can be made:

Good theoretical predictions for g-tensor components can also be obtained by the semiempirical INDO/S-CI method.[8,26-30]

Modern DFT calculations of g-values (and ^{1}H, ^{13}C and ^{17}O hyperfine couplings) in biochemistry have only recently started to emerge. They mainly deal with biological radicals for which DFT is generally accurate, for example quinone radicals in hydrogen-bond networks[3,31-34] or porphyrin radicals.[35,36]

Also for metal-proteins, pioneering work has been done by various groups, and up to 2003 their results have been reviewed.[20] Illustrative examples are the blue copper proteins such as the blue-copper site of azurin.[37] In this case, an elaborate CI study of the electronic structure of azurin was performed, and the authors were able to obtain good agreement with experimental results from optical and EPR spectra, including g-values and nitrogen hyperfine couplings. This demonstrates the potential and accuracy of such an approach.

Theoretical Approaches to Hyperfine Couplings

The magnetic nuclei in a molecule lead to hyperfine splittings in the EPR spectrum providing that these nuclei are interacting with the unpaired electron. In the spin-Hamiltonian this interaction of unpaired electron dipole moments and nearby magnetic nuclei is parameterized by a sum of hyperfine coupling (hfc) terms, the summation being done over all interacting nuclei. The hfc consists of three parts: (*i*), the nonclassical Fermi contact term that arises from the presence of unpaired s-spin density at the nucleus in question; (*ii*), the magnetic electron–nuclear dipole–dipole interaction; and (*iii*), the interaction of the electronic angular momentum with the magnetic dipole moment of the nucleus. Contribution (*iii*) is negligible for ligand nuclei but essential for calculating metal-nucleus hfcs.[20] Contribution (*ii*) is essentially the classical, *i.e.* nonquantum-mechanical interaction between two magnetic dipoles, and falls off with the inverse third power of the distance between the unpaired electron and the nucleus. It contains local parts arising from the spin density in non-spherical orbitals at the nucleus in question, but also distant contributions that are governed by interatomic distances.[38]

The isotropic Fermi contact and magnetic dipole contributions are both of first order and have to be calculated as expectation values over the ground-state wavefunction. This calculation can be done with most of the available standard quantum-chemistry programs. Hfc calculations require the use of basis sets that are more flexible in the core region than standard basis sets to insure convergence. Current DFT methods, for example B3LYP, usually give accurate predictions for the hfcs of light nuclei[20] However, for metal nuclei the DFT results are often not satisfactory.[39,40] Fortunately, using perturbation theory, a method was developed for calculating reliably metal and ligand hfcs of transition-metal complexes with the simultaneous inclusion of spin-polarization and SOC effects.[23,41,42]

Most DFT studies have dealt with ligands hfcs and biological radicals. In general, the results are fairly good when using, for example, the B3LYP functional. The number of prominent studies of hfcs in biological systems is far too large to be covered in this chapter, and we refer to recent reviews for further reading, for example refs. [19,20]

Theoretical Approaches to Zero-Field Splittings

For more than one unpaired electron in the molecular system ($S > 1/2$, referred to as "high-spin systems"), the additional dipole–dipole interaction between the unpaired electrons and the second-order effect of spin-orbit coupling lead to zero-field splitting (ZFS) energy contributions in the EPR spectrum. Effectively, this interaction takes the form $\hat{S} \cdot \tilde{D} \cdot \hat{S}$ in the spin-Hamiltonian and lifts the $(2S+1)$-fold degeneracy of the different M_S magnetic sublevels in zero external magnetic field.[43] In the spin-Hamiltonian, the 3×3 ZFS tensor, $\tilde{D}$, can be reduced, by a suitable choice of reference axes, to only two parameters $D = D_{zz} - (D_{xx} + D_{yy})/2$ (called the ZFS parameter) and $E = (D_{xx} - D_{yy})/2$ ($0 \leq E/D \leq 1/3$, called the rhombicity or asymmetry parameter). E measures the deviation from axial symmetry.

For triplet-state systems containing only light atoms, the spin-orbit contribution to the zero-field splitting can be neglected, and the dipole–dipole interaction between the unpaired electrons can be calculated with good success by available DFT methods, see for instance.[44] For transition-metal high-spin complexes, however, the situation is more complicated because of significant contributions from spin-orbit coupling. The general equations to compute the full D-tensor from nonrelativistic many-electron wavefunctions show that not only excited states with the same total spin as the ground state contribute to the ZFS, but also states with higher and lower multiplicity must be taken into account.[28] This complication is probably responsible for the fact that no consistent DFT theory for zero-field splittings of transition-metal complexes has been developed so far.[20] Nevertheless, in recent years considerable progress is noted in DFT calculations of zero-field splittings in transition-metal complexes.[45–49] Generally, however, a widely applicable and computationally affordable solution to this problem appears to be the INDO/S-CI treatment.[28]

Available Computer Program Packages Incorporating DFT Methods

The first modern DFT approach to the calculation of magnetic resonance parameters by van Lenthe and coworkers[24] is implemented in the widely used Amsterdam density function (ADF) code.[50] Appropriate computer programs with extensive documentation are commercially available as part of the ADF program package.[50] Two related implementations have been reported by Kaupp and coworkers[11] and by Neese.[12,17] These three approaches all employ second-order perturbation theory based on the unrestricted Kohn–Sham equations.[13] They incorporate relativistic contributions to the g-tensor at different levels of approximation.

The group of F. Neese (now at Bonn University) has created the very efficient and easy-to-use quantum-chemical program package ORCA. This general-purpose tool, which is being extensively used in our research group at FU Berlin (see Chapter 5), features a wide variety of quantum-chemical methods ranging from semiempirical to DFT methods (see below). Specific emphasis is put on spectroscopic properties of open-shell molecules, *i.e.* the objects of EPR studies.

A few of the most important features of ORCA are as follows:

Implemented Methods

1. Semiempirical AM1, PM3, ZINDO, NDDO, MNDO Hartree–Fock theory (RHF, UHF, ROHF and CASSCF).
2. DFT including a reasonably large number of exchange and correlation functionals, also including hybrid and double hybrid functionals.
3. Geometry optimization in internal coordinates using analytical gradient techniques for all SCF methods as well as MP2.
4. Scalar relativistic ZORA, IORA and Douglas–Kroll–Hess (DKH) approaches.
5. The COSMO solvation model is available throughout the package for continuum dielectric modelling of the environment of the solvated molecule.
6. A large number of user-defined point charges can be read by the program; this allows interfacing to existing quantum-mechanical/molecular-mechanics (QM/MM) methods.

Basis Sets

1. A large number of built-in Gaussian basis sets is available.
2. User-defined basis sets can be easily specified.

EPR-Spectroscopic Parameters

Zero-field splitting tensors, g-tensors, hyperfine-coupling tensors, quadrupole-tensors from Hartree–Fock, DFT and MR-CI. Scalar relativistic corrections at the ZORA level.

More detailed information on ORCA and downloads are available from F. Neese through: www.thch.uni-bonn.de/tc/orca

References

1. W. Koch and M. C. Holthausen, *A Chemist's Guide to Density Functional Theory*, Wiley-VCH, Weinheim, 2001.

2. *Calculation of NMR and EPR Parameters*, M. Kaupp, M. Bühl, V. G. Malkin, Eds., Wiley-VCH, Weinheim, 2004.
3. M. Kaupp, In: *EPR of Free Radicals in Solids: Trends in Methods and Applications*, A. Lund, M. Shiotani, Eds., Kluwer, Dordrecht, 2003, p. 267.
4. K. M. Neyman, D. I. Ganyushin, Z. Rinkevicius and N. Rösch, *Int. J. Quant. Chem.*, 2002, **90**, 1404.
5. F. Neese and M. L. Munzarova, In: *The Quantum Chemical Calculation of NMR and EPR Properties*, M. Kaupp, M. Bühl, V. Malkin, Eds., Wiley-VCH., Weinheim, 2004.
6. M. Plato and K. Möbius, *Chem. Phys.*, 1995, **197**, 289.
7. J.T. Törring, PhD Thesis, Freie Universität Berlin, 1995.
8. J. T. Törring, S. Un, M. Knüpling, M. Plato and K. Möbius, *J. Chem. Phys.*, 1997, **107**, 3905.
9. E. van Lenthe, E. T. Baerends and J. G. Snijders, *J. Chem. Phys.*, 1993, **99**, 4597.
10. G. Schreckenbach and T. Ziegler, *J. Phys. Chem. A*, 1997, **101**, 3388.
11. O. L. Malkina, J. Vaara, B. Schimmelpfennig, M. Munzarova, V. G. Malkin and M. Kaupp, *J. Am. Chem. Soc.*, 2000, **122**, 9206.
12. F. Neese, *J. Chem. Phys.*, 2001, **115**, 11080.
13. W. Kohn and L. T. Sham, *Phys. Rev.*, 1965, **140**, A1133.
14. R. Ditchfield, *Mol. Phys.*, 1974, **27**, 789.
15. A. D. Becke, *Phys. Rev. A*, 1988, **38**, 3098.
16. C. Lee, W. Yang and R. G. Parr, *Phys. Rev. B*, 1988, **37**, 785.
17. F. Neese and E. I. Solomon, In: *Magnetoscience–From Molecules to Materials*, J. S. Miller, M. Drillon, Eds., Vol. IV, Wiley-VCH, Weinheim, 2003, pp. 345.
18. F. Neese, *J. Biol. Inorg. Chem.*, 2006, **11**, 702.
19. F. Neese, In: *Electron Paramagnetic Resonance*, B. C. Gilbert, M. J. Davies, D. M. Murphy, Eds., Vol. **20**, The Royal Society of Chemistry, Cambridge, 2006, pp. 73.
20. F. Neese, *Curr. Opin. Chem. Biol.*, 2003, **7**, 125.
21. C. P. Slichter, *Principles of Magnetic Resonance*, Springer, Heidelberg, 1989.
22. M. Kaupp, R. Reviakine, O. L. Malkina, A. Arbuznikov, B. Schimmelpfennig and V. G. Malkin, *J. Comp. Chem.*, 2002, **23**, 794.
23. F. Neese, *J. Chem. Phys.*, 2005, **122**, 034107.
24. E. van Lenthe, P. E. S. Wormer and A. van der Avoird, *J. Chem. Phys.*, 1997, **107**, 2488.
25. K. M. Neyman, D. I. Ganyushin, A. V. Matveev and V. A. Nasluzov, *J Phys. Chem. A*, 2002, **106**, 5022.
26. G. Peng, J. Nichols, E. A. McCullough and J. T. Spence, *Inorg. Chem.*, 1994, **33**, 2857.
27. Y. W. Hsiao and M. C. Zerner, *Int. J. Quant. Chem.*, 1999, **75**, 577.
28. F. Neese and E. I. Solomon, *Inorg. Chem.*, 1998, **37**, 6568.

29. F. Neese, R. Kappl, J. Hüttermann, W. G. Zumft and P. M. H. Kroneck, *J. Biol. Inorg. Chem.*, 1998, **3**, 53.
30. F. Neese, *Int. J. Quant. Chem.*, 2001, **83**, 104.
31. S. Kacprzak, M. Kaupp and F. MacMillan, *J. Am. Chem. Soc.*, 2006, **128**, 5659.
32. M. Kaupp, *Biochemistry*, 2002, **41**, 2895.
33. M. Kaupp, C. Remenyi, J. Vaara, O. L. Malkina and V. G. Malkin, *J. Am. Chem. Soc.*, 2002, **124**, 2709.
34. S. Sinnecker, E. Reijerse, F. Neese and W. Lubitz, *J. Am. Chem. Soc.*, 2004, **126**, 3280.
35. S. Patchkovskii and T. Ziegler, *Inorg. Chem.*, 2000, **39**, 5354.
36. S. Patchkovskii and T. Ziegler, *J. Am. Chem. Soc.*, 2000, **122**, 3506.
37. M. van Gastel, J. W. A. Coremans, H. Sommerdijk, M. van C. Hemert and E. J. J. Groenen, *J. Am. Chem. Soc.*, 2002, **124**, 2035.
38. J. Hüttermann, In: *EMR of Paramagnetic Molecules, Biological Magnetic Resonance*, Vol. **13**, L. J. Berliner, J. Reuben, Eds., Plenum Press, New York, 1993, p. 219.
39. M. Munzarova and M. Kaupp, *J. Phys. Chem. A*, 1999, **103**, 9966.
40. M. L. Munzarova, P. Kubacek and M. Kaupp, *J. Am. Chem. Soc.*, 2000, **122**, 11900.
41. F. Neese, *J. Chem. Phys.*, 2003, **118**, 3939.
42. F. Neese, *Magn. Reson. Chem.*, 2004, **42**, S187.
43. N. M. Atherton, *Principles of Electron Spin Resonance*, Ellis Horwood, New York, 1993.
44. S. Sinnecker and F. Neese, *J. Phys. Chem. A*, 2006, **110**, 12267.
45. S. Kababya, J. Nelson, C. Calle, F. Neese and D. Goldfarb, *J. Am. Chem. Soc.*, 2006, **128**, 2017.
46. F. Neese, *J. Am. Chem. Soc.*, 2006, **128**, 10213.
47. C. Duboc, T. Phoeung, S. Zein, J. Pecaut, M. N. Collomb and F. Neese, *Inorg. Chem.*, 2007, **46**, 4905.
48. F. Neese, *J. Chem. Phys.*, 2007, **127**, 164112.
49. S. Zein, C. Duboc, W. Lubitz and F. Neese, *Inorg. Chem.*, 2008, **47**, 134.
50. G. Te Velde, F. M. Bickelhaupt, E. T. Baerends, C. F. Fonseca Guerra, S. J. A. van Gisbergen, J. G. Snijders and T. Ziegler, *J. Comp. Chem.*, 2001, **22**, 931.

CHAPTER 5

Applications of High-Field EPR on Selected Proteins and their Model Systems

5.1 Introduction

In this overview of high-field EPR applications we focus on a selection of protein systems, and some of their biomimetic model systems, that were previously crystallized and for which X-ray structures of high or, at least medium resolution are now available: bacterial reaction centers (RCs) from non-oxygenic as well as photosystems I and II (PS I and PS II) from oxygenic photosynthetic organisms executing light-induced electron transfer across the membrane; bacteriorhodopsin (BR), the light-driven transmembrane proton pump; DNA photolyases accomplishing DNA repair after UV damage by light-induced electron-transfer mechanisms; colicin A, the transmembrane ion-channel forming bacterial toxin. These proteins have been characterized in detail over recent years also by powerful spectroscopic techniques including ultrafast laser spectroscopy,[1] FT-IR,[2–5] solid-state NMR[6–8] and, last but not least, multifrequency EPR.[9–34]

In addition, sophisticated theoretical studies had been performed to elucidate their light-induced electron and proton transfer characteristics[35–37] emphasizing the importance of electrostatic energies in proteins[38–42] as well as of the electronic coupling.[43–53] Hence, the chosen proteins represent paradigm systems of general interest. They are well suited for new multifrequency high-field EPR experiments to study functionally important transient states at time windows relevant for biological action. These states are either directly detectable by EPR via their transient paramagnetism as, for instance, the radical-ion and radical-pair states of cofactors involved in photosynthetic electron transfer in RCs. Or, they are detected more indirectly via site-specific nitroxide spin labelling of the protein as, for instance, in site-directed mutants of BR and colicin A. It will be shown that high-field EPR experiments allow specific protein regions to be identified and characterized as molecular switches for vectorial transfer processes across membranes. Such switches are envisaged to function by controlling the electronic structure of cofactors, for example by invoking weak cofactor–protein interactions via H-bonds, by polarity gradients or even by substantial cofactor and/or helix displacements. RCs and BR are examples of involving molecular switches of this kind to control their vectorial transmembrane electron and proton transport. Colicin A toxin, on the other hand, requires massive protein refolding for transmembrane channel formation with subsequent ion depletion and cell death.

We realize that the examples chosen as attractive applications of high-field EPR on proteins and their model systems are dominated by studies of light-induced electron-transfer processes in photosynthetic reaction centers. Hence, it seems appropriate to give, as an introduction to this chapter, two short summaries, *Prologue 1* and *Prologue 2*, of the main aspects of (*1*) electron-transfer theory and (*2*) photosynthesis, as far as they apply to EPR studies.

Prologue 1

According to the famous Marcus electron-transfer theory[54,55] (Nobel Prize in Chemistry to R. Marcus, 1992), intramolecular electron transfer (ET) in a molecular donor(D)–spacer(s)–acceptor(A) complex, D–s–A, can be described as the motion of the excited donor reactant in the complex on the free-energy surface of the initial state, *D–s–A, intersecting with another energy surface of the final state, D^+–s–A^-, that corresponds to the charge-separated product. ET occurs when the reactant crosses from one surface to the other. Once the crossing has occurred, the charge distribution in the reacting molecule is no longer the same and, thus, the solvent equilibrium will be disturbed. Therefore, a dielectric relaxation mechanism is required to realign the solvent molecules with respect to the new charge distribution. This relaxation is carried out by a reorientation of the solvent dipoles around the solute molecules. It is characterized by the Debye relaxation time, τ_D, which is a function of the solvent properties.[35,56–58] Thus, the motion of the reactant on the free-energy surface is coupled to the reorganization energy, λ, which is the energy of the final charge-separated state in the equilibrium solvent configuration of the initial state (see Figure 5.1). For fast solvent-relaxation dynamics, where the vibrational motion of the nuclei is fast on the ET time scale (nonadiabatic ET), the ET rate k_{ET} is determined, in first-order time-dependent perturbation theory, by the product of the Franck–Condon factor, F, and the electronic coupling element, V, between the initial state and the final state:

$$k_{ET}(\text{na}) = (2\pi/\hbar) \cdot V^2 \cdot F \quad (5.1)$$

The electronic coupling V is the Hamiltonian matrix element between the electronic wavefunctions of the initial and final states. Hence, its magnitude

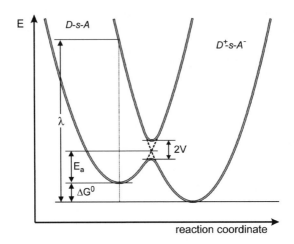

Figure 5.1 A schematic representation of the electronic potential energy surfaces (initial and final states of donor–spacer–acceptor complex) as a function of reaction coordinate, the intersection point split by $2V$, the reorganization energy (λ), the free energy of reaction (ΔG^0), and the activation energy (E_a).

depends critically on the overlap integral of their wavefunctions, *i.e.* on the distance and relative orientation of the two molecules, D and A. An important role is played by the medium in which the donor and acceptor molecules are embedded, for example the protein microenvironment of the D and A binding sites. Effects of conjugation of D and A may be strongly amplified by hydrogen-bonded networks and pathway tubes between D and A,[59] superexchange[60,61] or structured water molecules near the redox factors.[62]

The ET reaction proceeds predominantly at the intersection region between the initial and final potential energy functions. This intersection (anticrossing) is split by $2V$, and the electronic coupling energy, as well as the activation energy E_a, become the most important parameters for controlling k_{ET}. The Marcus expression for the averaged Franck–Condon factor is given by:

$$F = (4\pi \cdot \lambda \cdot k \cdot T)^{-1/2} \cdot \exp[-E_a/kT] \quad (5.2)$$

where the activation energy is given as

$$E_a = (\Delta G^0 + \lambda)^2/4\lambda \quad (5.3)$$

and ΔG^0 is the thermodynamic standard free energy of the ET process.[54,55] All the energetic quantities entering the Marcus expression for k_{ET} are visualized in Figure 5.1.

The critical quantity that describes whether the solvent relaxation (approximated by one time constant, τ_S) is fast on the ET time scale, $(k_{ET})^{-1}$, is the adiabaticity factor κ in the Marcus theory

$$\kappa = (4\pi \cdot V^2 \cdot \tau_S)/\hbar\lambda \qquad (5.4)$$

In ET processes, two limiting cases are generally considered, which are represented also in the examples reviewed in this chapter, *i.e.* in donor–acceptor complexes in RCs and their biomimetic model systems: the nonadiabatic limit ($\kappa \ll 1$),[54,63–65] and the adiabatic limit ($\kappa > 1$).[66,67] The respective ET rate constants are related to each other by a correction term $(1+\kappa)$:

$$k_{ET}(a) = k_{ET}(na)/(1+\kappa) \qquad (5.5)$$

The case of adiabaticity is given for long relaxation time and large electronic coupling. Then, the rate constant becomes independent of the electronic coupling and the medium dynamics cannot be neglected. The ET rate, $k_{ET}(a)$, is governed by the solvent's longitudinal dielectric relaxation time, τ_L. The relation between the relaxation times given by:

$$\tau_L = (\varepsilon_\infty/\varepsilon_S) \cdot \tau_D \qquad (5.6)$$

where ε_S and ε_∞ are the static and optical dielectric constants of the solvent matrix. The temperature dependence of τ_L is governed by that of the Debye relaxation time, τ_D.

A transition from the nonadiabatic limit (*e.g.*, in RCs) to the solvent-controlled adiabatic limit (*e.g.*, in model systems) will occur upon an increase of τ_L, that can be controlled by lowering the temperature and/or increasing the solvent viscosity.[67–72] Since rotational diffusion is a function of the solvent viscosity, dielectric relaxation will be absent at very low temperatures and the ET process will effectively stop. In isotropic solvents, the attenuation of ET processes from a fast nonadiabatic regime into the adiabatic one may be carried out either by lowering the temperature to the soft-glass region[72] or by using a highly viscous solvent. However, to use liquid crystals in their anisotropic mesophase as solvent for donor–acceptor complexes is often the better option for doing EPR studies on light-induced ET processes,[73] as will be exemplified in Section 5.4.

Prologue 2
Photosynthetic processes span a huge range of time scales of 20 orders of magnitude, ranging from less than 100 femtoseconds to more than 100 megaseconds. Consequently, photosynthesis is the research target of vastly different fields of science, ranging from ultrashort radiation physics, condensed-matter physics and chemistry, photochemistry, biophysics and biochemistry, physiology and botany.[74] Consequently, there exists a huge literature on the diverse subjects subsumed under the heading photosynthesis, and no attempt is made here to make a suggestion for further reading. With two exceptions: A cross

section from structure to electron-transfer dynamics in photosynthesis, discussing current problems, speculations, controversies and future trends, has been provided in a 1992 special issue of the *Israel Journal of Chemistry* edited by H. Levanon (Jerusalem),[57] and an exhaustive coverage of the photophysics of photosynthesis with restriction to structure and spectroscopy of reaction centers of purple bacteria has been presented in a 1997 special issue of *Physics Reports* by A.J. Hoff (Leiden) and J. Deisenhofer (Dallas).[75]

The photosynthetic process converts the energy of sunlight into electrochemical energy, thereby enabling life on Earth. Photosynthesis can be performed by certain archaea, bacteria, algae and green plants and provides the energetic basis needed by all higher organisms for synthesis, growth and replication. All these organisms utilize sunlight to power cellular processes and ultimately derive their biomass through secondary, *i.e.* dark chemical reactions driven by the light-initiated primary reactions of energy and electron transfer between pigment molecules in a protein scaffold. Energy is first collected by an "antenna" system consisting of highly organized protein–pigment complexes, and is rapidly transferred, most likely via efficient Förster-type processes, to a "reaction center" complex where primary electron transfer takes place. Secondary reactions lead to reduction of carbon dioxide (CO_2) with or without oxidation of water, depending on the class of organisms. Antennas permit an organism to increase greatly the absorption cross section for light needed to drive photosynthetic electron transfer. Hence, antenna systems are also termed "light-harvesting complexes". Antenna systems are regulating aggregates optimized by evolution to maximize energy collection efficiency while avoiding photodamage of the pigments. Such a pigment can be (bacterio)chlorophyll, carotenoid or bilin (open chain tetrapyrrole) molecules depending on the type of organism. A wide variety of different antenna complexes are found in different photosynthetic systems.

The so-called primary processes in the protein reaction-center complexes of photosynthetic organisms are the light-induced electron-transfer (ET) reactions between protein-bound donor (D) and acceptor (A) pigments across the cytoplasmic membrane. The successive charge-separating ET steps between the various redox partners in the transmembrane reaction center protein complexes have very different reaction rates, k_{ET}. The lifetimes, $t_{1/2}=(k_{ET})^{-1}$, of the transient charge-separated states range from less than 1 ps for neighboring D–A pigments to more than 1 ms for large D–A separations on opposite sides of the membrane (about 40 Å). The cascade of charge-separating ET steps of primary photosynthesis competes extremely favorably with wasteful charge-recombination ET steps, thereby providing almost 100% quantum yield.

The largest impact of photosynthesis on life on Earth is due to green plants, algae and cyanobacteria, in whose two interconnected reaction centers, photosystems I and II (PS I and PS II) a reversible ET photocycle occurs, for which water serves as an electron donor. CO_2 is fixed in the form of carbohydrates, and oxygen gas (O_2) is released as a byproduct, thereby stabilizing the CO_2/O_2 composition of the Earth's atmosphere.

Thus, the net reaction of oxygenic photosynthesis can be summarized to

$$n \cdot CO_2 + n \cdot H_2O \xrightarrow{h\nu, Chl} C_n(H_2O)_n + n \cdot O_2 \tag{5.7}$$

For this photoreaction to proceed, chlorophylls (and other cofactors) are needed as biocatalysts. To set the stage for an overview on what we learn from high-field EPR in photosynthesis research, some additional remarks may be appropriate concerning photosynthetic RC protein complexes.

There exist two different, but related types of RCs that differ in their ability to reduce the terminal electron acceptors, either iron–sulfur complexes (type I) or quinones (type II). The purple nonsulfur photosynthetic bacteria contain only type-II RCs, whereas cyanobacteria, algae, and green plants contain both type-I and type-II RCs. They are connected in series to split water and to produce the transmembrane electrochemical potential for NADPH and ATP synthesis as well as for CO_2 assimilation.

Organisms capable of oxygenic photosynthesis use two light reactions associated with the PS I and PS II reaction centers in a sequential electron-transport scheme, known as the Z-scheme[76] (see Figure 5.2), in which PS I and PS II operate in tandem. After light excitation of the primary donors (chlorophyll dimers known as P_{700} and P_{680} that absorb predominantly at 700 and 680 nm, respectively), ET proceeds from the water-splitting Mn complex (*left*) via the cytochrome b_6f complex and PS I to $NADP^+$ (*right*). There the energetic coupling of the light reactions to the dark reactions for CO_2 fixation occurs. The ellipses in Figure 5.2 represent the various protein–cofactor complexes. The vertical position of each cofactor indicates its redox potential. Light excitation of the primary donors to their first excited singlet states, P*, is symbolized by wavy arrows $h\nu$. The other arrows signify the ET pathways.

The series of light-dependent ET reactions culminate with the reduction of $NADP^+$ to NADPH and the generation of a transmembrane proton gradient whose energy is tapped by the membrane-bound proton-translocation ATP-synthase protein, *i.e.* the proton gradient powers the enzymatic synthesis of ATP from $ADP + P_i$ (adenosindiphosphate + inorganic phosphate). Hence, the final products of the chloroplasts' light reactions are the energy-rich molecules NADPH and ATP. They provide reducing power and free energy, respectively, for the subsequent light-independent reactions to drive the Calvin cycle, in which CO_2 is reduced and incorporated into carbohydrates (M. Calvin, Nobel Prize in Chemistry 1961). If, for example, the CO_2 molecules are incorporated into hexose sugars, such as glucose or fructose, the overall reaction of the Calvin cycle can be formulated as:

$$6CO_2 + 18ATP + 12NADPH + 12H_2O \longrightarrow \\ C_6(H_2O)_6 + 18ADP + 18P_i + 12NADP^+ + 6H^+ \tag{5.8}$$

i.e. the fixation of one CO_2 to form one hexose molecule needs three ATP and two NADPH molecules. From this net reaction the energy efficiency of photosynthesis can be estimated as 30%.[77]

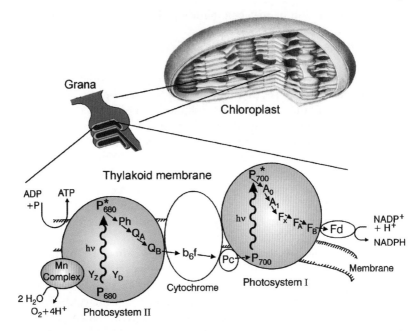

Figure 5.2 The thylakoid phospholipid membrane in the chloroplast organelles of plant cells, where photosynthesis is conducted by absorbing light energy and storing it in the form of sugars. The chloroplast contains stacks of thylakoid membranes. Each stack of thylakoid disks represents one granum. The light reactions of photosynthesis occur in the grana. The area between the grana is called the stroma. This is where the dark reactions of photosynthesis occur. The light reactions follow the "Z-scheme" according to which electrons from excited chlorophyll molecules flow through a cytochrome transport system along the thylakoid membranes. During this electron-transport process, ATP and NADPH are generated. They are needed to fuel the dark reactions in the stroma, by which CO_2 is converted into glucose through a series of reactions called the Calvin cycle. In the Z-scheme of light-induced electron transfer in oxygenic photosynthetic systems the photosystems I and II (PS I, PS II) are interconnected by the cytochrome b_6f and plastocyanin (Pc) complexes. Abbreviations: P_{680} and P_{700}, primary electron donors; Y_Z and Y_D, tyrosines; Mn complex, oxygen-evolving complex; Ph, pheophytin; Q_A and Q_B, quinones; A_0, monomeric chlorophyll; A_1, phylloquinone; F_X, F_B, F_A, iron-sulfur centers; Fd, ferredoxin/flavodoxin complexes.

It is stressed again that a detailed analysis of the primary ET processes on the molecular level requires the three-dimensional structures of the RC protein complexes. Only recently, as another milestone of protein crystallography, the X-ray structures of PS I and PS II from the oxygenic photosynthetic cyanobacterium *Thermosynechococcus (T.) elongatus* were resolved, by the groups of H.T. Witt (TU Berlin) and W. Saenger (FU Berlin), to atomic resolution: for PS I to 2.5 Å,[78] for PS II to 3.8 Å.[79] Very recently, for PS II improved crystallization protocols and new X-ray diffraction data led to electron-density maps calculated and interpreted to 3.2 Å resolution, and even 3.0 Å resolution is in reach now.[80–83] It has been the holy grail of photosynthesis research to reveal the organization of the Mn_4Ca^{2+} cluster and describe its protein environment with the view to elucidate fully the molecular mechanisms of the water-splitting reaction.[84] It is not surprising that, in the scientific race to this holy grail, different and in some respect conflicting interpretations of the electron-density maps at resolutions in the range from 3.2 to 3.8 Å have been published.[79,80,85–87] The electron density of such a large protein complex is well defined in some parts of the protein but poorly defined in others, for example in loop segments and side chains of cofactors. For an unambiguous assignment of amino-acid side chains in the electron-density map, a resolution of at least 2.8 Å is required and, hence, a conservative rather than a speculative interpretation is preferable.[88] Anyway, the solution of the X-ray structures of PS I and PS II is, indeed, an exciting scientific achievement. It was felt to be overdue by the photosynthesis community, since it lagged behind the high-resolution X-ray structures of the RCs from purple photosynthetic bacteria for a long time (H. Michel, J. Deisenhofer, R. Huber, Nobel Prize in Chemistry 1988). This is mainly because type-I reaction centers are much larger protein complexes than type-II RCs and bind a considerable number of light-harvesting antenna pigments. For several purple bacterial RCs, the X-ray crystallographic models are now refined to 2.3 Å resolution for wild-type RCs[89–94] and even to 1.8 Å resolution for site-specific mutant RCs.[95]

Recently, EPR in photosynthesis has been reviewed comprehensively by W. Lubitz.[15] High-field EPR applications on transient intermediates of non-oxygenic photosynthesis will be discussed in Section 5.2, followed by oxygenic photosynthesis in Section 5.3.

5.2 Nonoxygenic Photosynthesis

Light-Induced Electron Transfer
Three billion years before green plants evolved, photosynthetic energy conversion could already be achieved by certain bacteria, for instance the purple bacteria *Rhodobacter (Rb.) sphaeroides* and *Rhodopseudomonas (R.) viridis*. These early photosynthetic organisms are simple, one-cellular protein-bound donor–acceptor complexes that contain only one RC for light-induced charge separation. They cannot split water, but rather use hydrogen sulfide or organic compounds as electron donors to reduce CO_2 to carbohydrates with the help of sunlight and bacteriochlorophylls and quinones as biocatalysts. The net reaction for nonoxygenic bacterial photosynthesis is

$$2H_2S + CO_2 \xrightarrow{h\nu, BChl} C(H_2O) + H_2O + 2S \tag{5.9}$$

In primary photosynthesis the incoming light quanta are harvested by "antenna" pigment/protein complexes and channeled to the actual "reaction center" (RC) complexes by ultrafast energy transfer to excite the primary donor, a "special pair" of bacteriochlorophylls, to its excited electronic singlet state. This initiates a series of ET steps energetically downhill from the protein-bound donor to the quinone acceptors at the opposite side of the cytoplasmic membrane. A quantum yield of 100% represents a unique feature of the RC when converting sun energy into electrochemical potential.

In Figure 5.3 models of the light-harvesting complexes, LH1 and LH2, from photosynthetic bacteria are shown; they are based on crystal structures and molecular modelling calculations by various research groups.[97–103] A nice collection of images of light-harvesting antenna systems has been generated by R.J. Cogdell (Glasgow) and can be found on the webpages: www.gla.ac.uk/ibls/BMB/rjc/lh2galery.htm.

The antenna complexes LH1 and LH2 of photosynthetic purple bacteria, e.g., R. acidophila or R. palustris, contain circular arrays of bacteriochlorophyll (BChl) and carotenoid molecules as pigments. The crystal structure of LH2 from R. acidophila[97] exhibits two rings of molecules in the "B800" and "B850" units with 9 and 18 BChls, respectively. The LH2 subunit is not in direct contact with the RC, whereas the larger LH1 domain encloses the RC complex. The arrays of LH2 complexes transfer the absorbed light energy as singlet excitons to the RC through LH1 mostly by the incoherent Förster dipolar mechanism, although it is sometimes argued that the coherent Dexter exchange mechanism is also involved.[104] The LH1 subunit has been cocrystallized with the RC in its center from R. palustris[98] and shows a ring of pigment molecules, the "B880" unit, with 30 BChls arranged in pairs. Owing to the favorable spacing and relative orientation of the pigment molecules in the ring structures, the excitation energy spreads extremely rapidly (<1 ps) in the antenna system, jumping between the LH2 units and between LH2 and LH1, which serves as an energy trap absorbing at longer wavelengths than LH1. Ultimately, LH1 delivers the trapped excitation energy to the RC where the primary charge separation occurs.

Since the antenna complexes LH1 and LH2 operate via singlet exciton transfer between the pigment molecules, no paramagnetic species are created during light harvesting. This means that EPR spectroscopy cannot be used *a priori* for characterizing transient intermediates of the energy-transfer processes. There are, however, means to turn antenna pigments into paramagnetic doublet or triplet states by manipulating the antenna systems, for example by chemical oxidation of LH1 subcomplexes or intact LH1 domains[105–109] or by 532 nm pulsed laser excitation of carotinoidless preparations of LH1 and LH2 complexes.[110] X-band EPR experiments on such manipulated antenna complexes and their results in comparison to the cation radical of the primary donor in the electron-transfer cascade of illuminated RCs have been thoroughly reviewed by W. Lubitz up to 2003.[15]

The question of how the high quantum yield of primary photosynthesis is achieved by Nature has stimulated worldwide activities in fundamental and applied research with the goal to understand the structure–function relationship on the molecular level. Moreover, it is hoped to make practical use of such an understanding in the challenging construction of tailor-made artificial

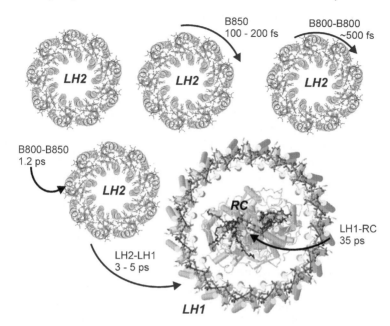

Figure 5.3 Light-harvesting complexes, LH1 and LH2, of the antenna domain of purple bacteria.[96] LH1 and LH2 are integral membrane proteins that form ring-like structures, oligomers of alpha beta-heterodimers, in the photosynthetic membranes of purple bacteria. They contain a large number of chromophores organized in ring structures that are optimal for light absorption and rapid light-energy migration. One ring comprises sixteen B850 BChl molecules perpendicular to the membrane plane, the other ring eight B800 BChl molecules that are nearly parallel to the membrane plane. The transition dipole moments of neighbouring B850 and B800 BChl molecules are nearly parallel to each other and, thus, are optimally aligned for Förster exciton transfer. The time constants of the energy-transfer processes between the various BChl domains are written next to the arrows in the figure. Specifically, LH2 consists of a circular array of heterodimers, each comprising an α- and a β-apoprotein. Together they bind three BChl molecules and one carotenoid molecule. The full structure is an $\alpha_9\beta_9$ nonamer. The pigment molecules are enclosed by two concentric rings of transmembrane α-helices. The BChl molecules are arranged in two distinct groups absorbing light at somewhat different wavelengths. Nine monomeric BChl molecules absorb at 800 nm (B800) and 18 tightly coupled BChl molecules absorb at 850 nm (B850). The RC containing LH1 complex has a very similar, but larger ring structure, in the LH1 case it is a larger $\alpha_{16}\beta_{16}$ oligomer. Thereby, a central pocket is provided that is large enough to accommodate the RC. The LH1 complex only contains a single ring of 32 tightly coupled BChl molecules that absorb at 875 nm (B875). Thus, the protein cage provides, in addition to an optimal cofactor orientation, a subtle cofactor–protein interaction mechanism for tuning the absorption wavelength of the BChl aggregates involved in energy transfer into energetically low-lying exciton states. Thereby an energy funnel is created that is needed for initiating the primary electron-transfer processes in the RC. For details and references, see refs. [97,98].

photovoltaic cells with high quantum yield from organic biomimetic donor–acceptor complexes. In the last decade, remarkable progress has been made in understanding primary photosynthesis, mostly from biophysical and biochemical investigations of specifically engineered protein preparations from wild-type and site-specific mutant organisms, applying X-ray crystallography, ultrafast laser spectroscopy in the visible and IR regions and, last but not least, modern EPR and NMR techniques.[75,111–113] It should be pointed out that, in parallel to the experimental accomplishments, theory has contributed much during the last years for a better understanding of primary photosynthesis.[35] Just to give some examples: ET routes and kinetic data could be interpreted in depth on the basis of the Marcus theory.[114] Next to covalent bonds and van der Waals contacts, hydrogen bonds between cofactors and amino-acid residues turned out to be of utmost importance in ET-proteins, as has been elucidated by extended pathway analysis of H-bond networks.[115]

In addition to *static* donor–acceptor and cofactor–protein interactions, the protein *dynamics* can have pronounced effects on biological ET, in particular when molecular motion occurs on a time scale comparable to that of the respective ET step[116] providing dynamic amplification of the ET rate constant k_{ET}. The dynamics of proteins in their solvent shell ("slaved" protein motion) deserves some extra attention: It is well established by now (see, *e.g.*, refs. [117–124]) that the intrinsic flexibility of proteins, which are thermally fluctuating between different conformations of cofactors in their binding sites ("energy landscape") at wide time scales including ns–ms, can strongly affect functional processes with corresponding time constants. Since this ns–ms time window is open for EPR experiments, both cw and pulsed high-field EPR can be used to orientationally resolve T_2 anisotropies. They are due to "slaved" anisotropic librational fluctuations of donor and acceptor radical ions in the H-bonding network of their binding sites. Such dynamics have been studied by pulsed high-field EPR RCs from wild-type and mutant organisms, see Section 5.2.1.5.

Light-Induced Proton Transfer
It is intriguing to realize that, in addition to the bacterial photosynthetic RC protein for light-induced *electron* transfer between donor and acceptor cofactors with subsequent transmembrane proton transfer in the opposite direction, Nature has invented a second photosynthetic machinery, the protein bacteriorhodopsin (BR), for light-induced *proton* transfer between amino-acid residues, mediated by conformational changes of the only cofactor, a retinal. BR is found in certain archaea, most notably the *Halobacterium (H.) salinarium* flourishing in hypersaline lakes. Like photosynthetic purple bacteria, halobacteria are also able to use sunlight as an energy source to synthesize ATP. Different from the RC protein, the BR protein acts as a light-driven proton pump, moving protons across the membrane out of the cell. In both protein systems, RC and BR, the ultimately resulting proton gradient is subsequently converted into chemical energy. Another essential difference in the RC and BR photosynthetic systems is that the bacteriochlorophylls in RCs are aided in capturing light energy by the "antenna" protein complexes, whereas

in BR-based systems there are no antennas. Lastly, chlorophyll-based photosynthesis is coupled to CO_2 fixation by synthesizing carbohydrates, whereas this not the case for BR-based systems. It is, thus, likely that early photosynthesis independently evolved at least twice, once in bacteria and once in archaea. The three-dimensional tertiary structure of BR resembles that of vertebrate rhodopsins, the light-sensing pigment complex in the retina. Rhodopsins also contain retinal as cofactor, however, there is no homology of their amino acid sequences. The light-driven chloride pump halorhodopsin, however, has homology to BR.

Among the techniques useful for investigating photosynthetic electron-transfer systems, EPR plays generally a prominent role. This is because the light-induced single-electron transfer reactions in oxygenic and nonoxygenic photosynthesis or in DNA photolyase repair processes involve paramagnetic intermediates that can be trapped long enough for EPR detection. Furthermore, metallo-proteins like the oxygen-evolving complex in PS II from cyanobacteria and plants, pass through some light-initiated high-spin valence states of manganese that are paramagnetic. Hence, for characterizing the transient states of metal ions or of radical-ion and radical-pair state intermediates of the photosynthetic ET chain concerning their electronic structure, EPR is often the method of choice. But, even when the transient states of the photocycle are not paramagnetic at all, like in the photocycle of the proton-transfer protein BR, EPR remains at a top position among the methods of choice. This is because nowadays nitroxide spin labels can be introduced at almost any specific protein positions of interest.[125] Our pledge in favor of EPR holds for multifrequency experiments in general, and for the application of high magnetic fields in particular. To compensate, at least partially, for the particular choice of high-field EPR examples in this chapter, the reader is referred to several overviews that treat the subject in more breadth, for example, refs. [13,15,17,18,57,58,75,111,126–134].

Representative examples of high-field EPR applications in biology, chemistry and physics have been collected in two special issues dedicated to this subject: *Applied Magnetic Resonance* Vol. 21, No. 3–4 (2001), Guest Editors M. Huber, K. Möbius, and *Magnetic Resonance in Chemistry* Vol. 43, Special Issue November 2005, Guest Editors W. Lubitz, K. Möbius and K. P. Dinse.

5.2.1 Multifrequency EPR on Bacterial Photosynthetic Reaction Centers (RCs)

We start this section with the three-dimensional picture of the bacterial photosynthetic reaction center (RC). In Figure 5.4(a) the structural arrangement of the RC of *Rb. sphaeroides* is shown according to high-resolution X-ray crystallography.[92,93,135,136] The cofactors are embedded in the L, M, H protein domains forming two ET branches, A and B. The RC of the carotenoid-less strain R26 of *Rb. sphaeroides* contains nine cofactors: the primary donor P_{865} "special pair" (a bacteriochlorophyll *a* (BChl) dimer), two accessory BChls ($BChl_A$, $BChl_B$), two bacteriopheophytins *a* ($BPhe_A$, $BPhe_B$), two ubiquinones

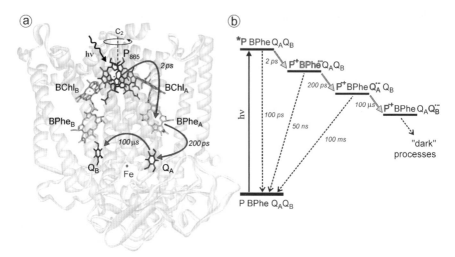

Figure 5.4 (a) X-ray structural model of the RC from *Rb. sphaeroides*[135] with the protein subunits and the cofactors P_{865}, BChl, BPhe, Q and Fe. Light-induced electron-transfer proceeds predominantly along the A branch of the cofactors ("unidirectionality" enigma) despite the approximate C_2 symmetry of the cofactor arrangement. (b) The ET time constants range from 2 ps to 100 μs in the cascade of transmembrane charge-separation steps. The abbreviations for the donor and acceptor molecules are explained in the text.

(Q_A, Q_B), one nonheme iron (Fe^{2+}). In Figure 5.4(b) the energy-level scheme of the redox partners in the cascade of the electron-transfer steps between the chlorophyll- and quinone-type donor and acceptor molecules is shown. The redox potentials of the cofactors span a range of about 1.5 V between the primary-donor ground state, P_{865}, and its first excited singlet state, P^*_{865}.

As a dominant *motif* in the evolution of photosynthetic bacteria, in the RC an approximate C_2 symmetry of the cofactor arrangement prevails. It is intriguing that, despite the apparent twofold local symmetry of the cofactor arrangement, the primary ET pathway is one-sided along the A branch, as indicated by the arrows in Figure 5.4(a). The origin of this "unidirectionality" of bacterial ET is not fully understood despite the numerous elaborate studies, both experimentally and theoretically, performed over the last decades. It is, however, clear by now that the ET pathway is largely due to the finely tuned energetics and electronic couplings of the primary donor and the intermediary acceptors. The fine tuning is the result of weak, though highly specific interactions between the cofactors and the amino-acid environment of their binding sites (for further reading, see the overviews in refs. [57,75,137]).

According to the nonadiabatic electron-transfer theory (see Section 5.1), the rate of the ET reaction is proportional to the product of two independent terms: the Franck–Condon factor, F, and the square of the electronic matrix transfer element, V. The Franck–Condon factor is associated with the energetics of the ET reaction, *e.g.*, the redox potentials of the donor and acceptor cofactors. The electronic matrix element measures the coupling between the electron wavefunctions of the donor and the acceptor. It determines the probability of an electronic transition occurring at a favorable medium conformation.

Theoretical and experimental studies have concentrated mostly on the F term, assuming that the unidirectional electron flow relies on differences in the energies of states P*BChlBPhe, P$^{\bullet+}$BChl$^{\bullet-}$BPhe, and P$^{\bullet+}$BChlBPhe$^{\bullet-}$ in the two ET branches, A and B.

In the last 20 years, the role of V in controlling the ET asymmetry has also received considerable theoretical and experimental attention, assuming that differences between the matrix elements V_A and V_B could significantly contribute to the observed asymmetry of primary ET in photosynthetic reaction centers along branches A and B, see refs. [43–45].

5.2.1.1 X-Band EPR and ENDOR Experiments

Primary photosynthesis provides a "Garden of Eden" for the EPR spectroscopists[138] because in each electron-transfer step a transient paramagnetic intermediate is formed. Our knowledge of the molecular structure of donor and acceptor cofactors can be classified as originating from "BC" or "AC" times (*i.e.* before and after crystallization of the RC.[139]). In BC times, much of what emerged from structural investigations came from EPR and ENDOR spectroscopy, but also in AC times important information about spatial and electronic structures was (and still is) obtained by EPR and ENDOR studies. In this overview, a brief account of such studies is presented that have led to the identification and characterization of the primary ion radicals of P_{865} and Q_A.

Optical spectra of RC preparations had disclosed a characteristic absorption at 865 nm that was ascribed to light-induced oxidation of some form of BChl, acting as the primary donor, P_{865}. The comparison of the frozen-solution EPR spectra of monomeric BChl$^{\bullet+}$ in an organic solvent and $P_{865}^{\bullet+}$ in the RC revealed a striking difference in the linewidth of both EPR spectra: for $P_{865}^{\bullet+}$ it is $1/\sqrt{2}$ narrower than for BChl$^{\bullet+}$, see Figure 5.5. Norris, Katz and coworkers[140] explained this observation by the "special pair" hypothesis, *i.e.* the unpaired electron on $P_{865}^{\bullet+}$ is equally shared between a (BChl)$_2$ dimer cation radical, which they assumed to be a symmetric dimer. Feher, Hoff and coworkers[141,142] confirmed the dimeric nature of $P_{865}^{\bullet+}$ by comparing the ENDOR spectra of frozen-solution samples of BChl$^{\bullet+}$ and $P_{865}^{\bullet+}$. Similar ENDOR experiments were performed independently around the same time by Norris, Katz and coworkers.[143,144] The ENDOR spectra showed that the few resolved hyperfine couplings (marked in Figure 5.5(c)) are indeed approximately halved in $P_{865}^{\bullet+}$, at least within the limits of the broad solid-state ENDOR lines of the experiments.

To resolve more hyperfine couplings for a thorough comparison of BChl$^{\bullet+}$ and $P_{865}^{\bullet+}$, Möbius and coworkers[146] applied ENDOR and electron–nuclear TRIPLE resonance to fluid-solution samples with intrinsically narrower lines. At physiological temperatures the RCs are still tumbling fast enough to average

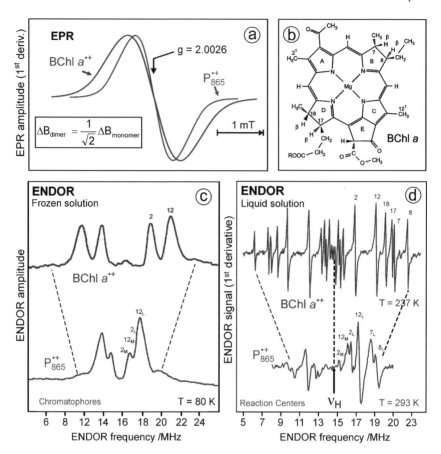

Figure 5.5 (a) EPR spectra (X-band) at ambient temperature of the radical cations of BChl a organic solution and of P in the RC. Both have the same g-value and saturation behavior but the linewidth of $P_{865}^{\bullet +}$ is approx. $1/\sqrt{2}$ narrower than that of BChl $a^{\bullet +}$ indicating a dimeric species. (b) Molecular structure of BChl a with numbering scheme. (c) Comparison of the ^{1}H ENDOR spectra of BChl $a^{\bullet +}$ and $P_{865}^{\bullet +}$ in frozen solution (80 K). Two strong resonances of the methyl group protons (pos. 2 and 12) are visible; the respective hyperfine couplings are reduced in $P_{865}^{\bullet +}$ indicating spin delocalization in the BChl a dimer. (d) Comparison of high-resolution ^{1}H ENDOR spectra of BChl $a^{\bullet +}$ and $P_{865}^{\bullet +}$ in isotropic liquid solution. From the assigned individual isotropic hyperfine couplings a detailed picture of the spin-density distribution in the monomeric BChl $a^{\bullet +}$ and in the dimer $P_{865}^{\bullet +}$ is obtained. Final assignments result from specific deuterations and investigations of $P_{865}^{\bullet +}$ in RC single crystals (see numbers on the respective resonance lines. The analysis shows an asymmetric spin distribution in favor of P_A, the BChl dimer half-bound to the L subunit ($\rho_L : \rho_M \cong 2:1$). For details, see ref. [145].

out the anisotropic hyperfine contributions resulting in highly resolved ENDOR-in-solution spectra (see Figure 5.5(b)).

In order to contribute to a solution of the "unidirectionality enigma", the electronic structure of the primary donor cation radicals, the dimeric $P^{\bullet+}_{865}$ in *Rb. sphaeroides* and $P^{\bullet+}_{960}$ in *Rps. viridis*, and their monomeric constituents, BChl $a^{\bullet+}$ and BChl $b^{\bullet+}$, respectively, have been studied in great detail by EPR, ENDOR and TRIPLE. This was done in liquid and frozen solutions as well as in single crystals of RCs (for a chronological account, see ref. [74], for reviews, see refs. [147,148]). Lendzian et al.,[146,149,150] for example, performed ENDOR/TRIPLE investigations on the cation radicals in fluid solution under physiological conditions. From the highly resolved hyperfine spectra of the monomers and dimers and their analysis by all-valence electron MO methods (RHF-INDO/SP, see ref. [148]) it was concluded *(i)* that for both organisms the primary donor dimer has to be viewed as a supermolecule with the wavefunction extending over both dimer halves, *(ii)* that the symmetry in the electron spin-density distribution over the two dimer halves is broken favoring the L half, on average, by 2:1, thereby manifesting that the "special pair" is an asymmetric dimer. This finding is corroborated by comparing the experimental spin densities of $P^{\bullet+}_{865}$ with theoretical predictions based on state-of-the-art quantum-chemical calculations[148] (see Figure 5.6). It is speculated that such an

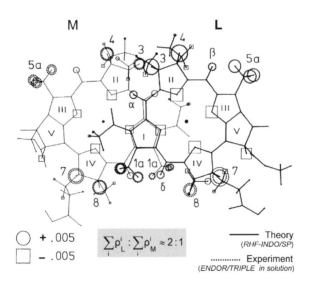

Figure 5.6 Comparison of experimental (dotted lines) and calculated (RHF-INDO/SP[148]) s-spin densities $\rho_H(1s)$ (solid lines) of $P^{\bullet+}_{865}$ in *Rb. sphaeroides* R26. Experimental values from isotropic proton hfcs using $A_{iso} = Q_H \cdot \rho_H(1s)$ with $Q_H = 1420$ MHz.[145] Geometry from X-ray structure analysis.[92] The s-spin densities are proportional to the areas of the respective squares ($\rho < 0$) and circles ($\rho > 0$). For rotating methyl protons, the average value of the three proton s-spin densities is shown. For details of the experiments and calculations, see refs. [145,148].

asymmetry in the electronic distribution of P_{865} is interrelated with the unidirectional electron transfer in the RC.

This asymmetry ratio 2:1 is primarily caused by subtle details of the dimer structure with some "fine-tuning" from neighboring amino-acid residues. These results have been fully confirmed by ENDOR/TRIPLE experiments on $P_{865}^{\bullet+}$ in RC single crystals of *Rb. sphaeroides* near room temperature.[145] The single-crystal work at physiological temperatures, performed independently by three groups, Feher at UC San Diego, Lubitz at TU Berlin, Möbius at FU Berlin, but ultimately published jointly, represents a culmination of two decades of ENDOR work on the primary donor in bacterial RCs. For the first time it was possible to assign ENDOR lines unambiguously to the individual dimer halves. An important – and comforting – aspect was the result that the experimental isotropic hfcs, as determined from the traces of the single-crystal hyperfine data, agree well the hfcs from liquid-solution ENDOR. This shows that there is no significant change of the spatial structure of $P_{865}^{\bullet+}$ and its immediate protein environment upon crystallization of the RC. The ENDOR-in-solution experiments described above have been extended to other BChl *a* containing organisms,[151] to various mutants with specifically changed amino acids in the vicinity of P,[152,153] and to RCs of *Rb. sphaeroides* reconstituted with chemically modified bacteriochlorophylls.[154]

As we see, the good news of fluid-solution EPR and ENDOR is the narrow linewidth with concomitant increase of spectral resolution. The bad news, however, is that the information of anisotropic hyperfine interactions is lost, *i.e.* of the molecular orientation in the laboratory coordinate system. This information is retained by studying RC single crystals. Single-crystal ENDOR/TRIPLE can be used to assign individual hyperfine lines to which half of the "special pair" dimer they belong to. On the other hand, such assignment procedures of hyperfine couplings to individual molecular positions is a time-consuming endeavor often requiring selective isotope labelling by ^{2}H and ^{13}C. An easier procedure would be desirable to determine the symmetry of the electronic structure of $P_{865}^{\bullet+}$, not relying on hyperfine couplings as local probes for the spin-density distribution, but rather on global probes such as the *g*-tensor. For nonmetallo-proteins like the bacterial RC, the *g*-anisotropy is very small, *i.e.* less than 10^{-3}. Hence, to resolve the *g*-tensor components of $P_{865}^{\bullet+}$ definitely requires high-field EPR on RC single crystals (by 95-GHz EPR) or on RC frozen solutions (by 360-GHz EPR).

Next, a few introductory remarks are made concerning the quinone acceptors in bacterial photosynthesis, Q_A and Q_B. In the light-driven ET processes of *Rb. sphaeroides* the primary and secondary quinones, Q_A and Q_B, are the same ubiquinones-10. They act as one- and two-electron gates, respectively. Obviously, their different function in the ET processes is induced by different interactions with the amino-acid environment in their binding sites. To learn about these interactions within the binding pocket, for example the specific H-bonding patterns, EPR and ENDOR on quinone anion radicals in bacterial RCs (with Fe^{2+} replaced by Zn^{2+} to avoid fast spin relaxation) and in organic solvents have been performed at several mw frequencies by various groups, both in fluid and frozen solution. Such multifrequency experiments on the primary donor and acceptor radical ions will be reviewed in the next sections.

5.2.1.2 95-GHz EPR on Primary Donor Cations $P^{\bullet +}$ in Single-Crystal RCs

One of the intriguing puzzles in bacterial photosynthesis is the so-called unidirectionality of the photoinduced ET, *i.e.* the electron is transferred preferentially along the L protein subunit, despite the approximate C_2 symmetry of the cofactor location in the L and M proteins (see previous section). Hence, it is anticipated that within the RC a molecular switch is operating to ensure vectorial electron and proton transfer across the cytoplasmic membrane. It may function due to involving weak, but subtle cofactor–protein interactions, for example hydrogen bonds between cofactors and amino-acid residues of the local protein environment, or by conformational changes of the cofactor binding sites during the photocycle, *e.g.*, cofactor displacements and reorientations, or helix tilting, or by dynamic amplification of the electronic coupling between donor and acceptor cofactors by librational motion at time scales adequate for the respective transfer steps. Most probably, it is a combination thereof to achieve the necessary robustness of the biological transfer process against external stress and perturbations.

In an attempt to determine, by means of the global *g*-tensor probe, the characteristic symmetry properties of the electronic structure of the primary donor, 95-GHz high-field EPR on illuminated single-crystal RCs of *Rb. sphaeroides* was performed at 12EC, *i.e.* at physiological temperature.[155] At a Zeeman field of 3.4 T even the magnetically inequivalent sites in the unit cell of the RC crystal could be resolved, and the angular dependence of their *g*-factors in the three symmetry planes of the crystal has been measured and analyzed.

As microwave resonance structure, we used a Fabry–Perot resonator with tunable frequency and coupling, which can incorporate a precision 2-axes goniometer allowing for crystal rotation around three orthogonal axes (see Figure 3.9 in Section 3.4.1.3). The RC crystal is placed in a sealed capillary containing a drop of the aqueous mother liquor out of which it was grown. The crystal was cw photoexcited *in situ* by a quartz light fiber connected to a filtered 50-W halogen lamp. A 5-cm water filter and subsequent edge filters restricted the excitation wavelengths to a range between 830 and 900 nm. Thus, the RCs were excited predominantly in the BChl dimer band at 865 nm.

The space group of the studied crystals of the orthorhombic RC from *Rb. sphaeroides* is $P2_12_12_1$ and, thus, a complication is encountered in that there are four RCs ("sites") per unit cell, pairwise related by a twofold symmetry axis.[156] In an arbitrary direction of the Zeeman field B_0 in the crystal axes system $\{a, b, c\}$ there is a fourfold site-splitting of the $P^{\bullet +}_{865}$ EPR signal. Fortunately, for B_0 lying in the symmetry planes *ab, ac, bc* there only two magnetically inequivalent sites, A and B, each of them consisting of two magnetically equivalent RCs; and for $B_0 \parallel a$, $B_0 \parallel b$, $B_0 \parallel c$ all four RCs are magnetically equivalent.[157]

Figure 5.7 shows the rotation pattern $g(\theta)$ of the g-values of $P_{865}^{\bullet+}$ in the three crystallographic symmetry planes. The site splitting (A, B) is clearly resolved in the *ac* and *ab* planes, but not in the *bc* plane.[155] Since $g=2+\delta g$ with δg of the order of 10^{-3}, the function $g(\theta)$ can be linearized to the form (see Section 2.1.1)

$$g(\theta) = g_{ii} \cdot \cos^2 \theta + g_{jj} \cdot \sin^2 \theta + 2g_{ij} \cdot \sin\theta \cdot \cos\theta \qquad (5.10)$$

where the subscripts i and j denote the crystal axes a, b and c. The solid lines in Figure 5.7 are the least-squares fits of the experimental g-values (error $\Delta g = 1 \times 10^{-5}$) with eqn (5.7), from which the g_{ii}, g_{jj} and g_{ij} components in the crystal axes system can be directly determined. Diagonalization yields the principal g-values ($g_{\alpha\alpha}=2.00329$, $g_{\beta\beta}=2.00239$, $g_{\gamma\gamma}=2.00203$) in the g-tensor axes system (α, β, γ) together with the principal axes directions of the g-tensor.

There remains a fourfold ambiguity in the g-tensor orientation with respect to the crystal (or molecular) axes system, because the signs of the off-diagonal elements can principally not be determined by EPR.[155,158] Only two physically reasonable solutions prevail, however, the tensors $\tilde{g}_1$ and $\tilde{g}_2$, between which to decide is not possible by cw EPR, not even at high fields (but needs addition pulse 95-GHz experiments on transient spin-polarized radical pairs (see ref. [159]). Their orientation with respect to the molecular dimer axes system (x, y, z) is given in Table 5.1 and depicted in Figure 5.8. A suggestive construction of the molecular dimer axes system happens in such a way that the z-axis is the average dimer normal (defining the average monomer planes as having the least sum of squares of distances to the four nitrogen atoms on the respective BChl dimer half), the x-axis is the projection of the Mg–Mg direction onto the plane normal to the z-axis, and the y-axis then describes the approximate local C_2 axis of the "special-pair" P_{865} dimer. As is seen from Figure 5.8, the principal directions (α, β, γ) of the g-tensor in the molecular axes system (x, y, z) reveal a breaking of the local C_2 symmetry in the electronic structure of $P_{865}^{\bullet+}$. Coinciding axes systems would reflect local C_2 symmetry also for the electronic structure. This finding is consistent with earlier ENDOR results for the hyperfine structure (see above). It is comforting to realize that the g-tensor of large molecular systems, such as the special pair in its anisotropic protein microenvironment, is sensitive to the breaking of C_2 symmetry, provided the g-tensor components can be resolved and measured with sufficient accuracy. Thus, this application example of high-field EPR has substantiated the claim that the g-tensor – complementary to the hyperfine-tensor – can also be used as a probe for the symmetry properties of a molecular complex.

It remains to be seen whether the breaking of C_2 symmetry in $P^{\bullet+}$ is a relevant factor in controlling the unidirectional charge-separating ET in primary bacterial photosynthesis. Other factors could contribute, such as cofactor- or protein-mediated superexchange to modify the electronic coupling in the long-distance ET along the L protein subunit and/or dielectric asymmetry of the RC protein complex along the two potential ET pathways. This asymmetry could

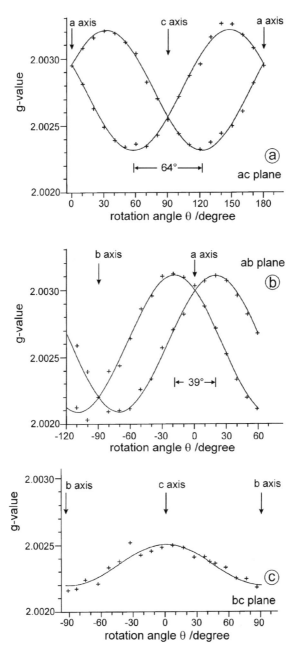

Figure 5.7 W-band EPR rotation patterns of the g-factor of $P_{865}^{\cdot+}$ in the three symmetry planes of an RC single crystal of *Rb. sphaeroides* R26. The angle θ is taken between the respective crystallographic axis and the B_0 direction. The two solid lines in each pattern are the least-squares fits. They belong to the pairwise magnetically inequivalent RC sites, A, B. The crossings of the curves correspond to orientations, where B_0 lies parallel to one of the crystallographic axes: (a) *ac* plane; (b) *ab* plane; (c) *bc* plane. For details, see ref. [155].

Table 5.1 Angles (degrees, error ±5°) between the principal axes (α, β, γ) of the g-tensors $\tilde{g}_1$ and $\tilde{g}_2$ and the dimer axes (x, y, z), see ref. [155].

Tensor	Angle between axes	x	y	z
$\tilde{g}_1$	α	45	69	128
($g_{ab} < 0$, $g_{ac} > 0$)	β	121	32	96
	γ	61	67	38
$\tilde{g}_2$	α	20	90	70
($g_{ab} > 0$, $g_{ac} < 0$)	β	86	12	101
	γ	109	78	23

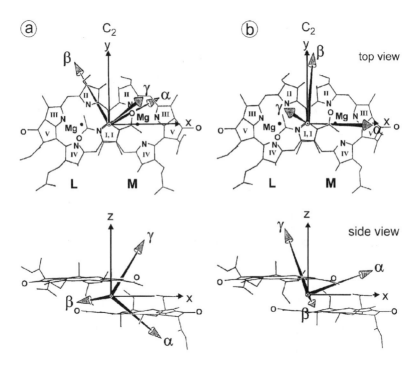

Figure 5.8 Tensor principal axes system of (a): $\tilde{g}_1$ and (b): $\tilde{g}_2$ in $P_{865}^{\cdot +}$: (top) view from the average dimer normal (z-direction) onto the monomer planes; (bottom) view into the direction of the local C_2 symmetry axis.[155]

originate in the different amino-acid environment of the redox cofactors along the two branches, A and B, and also an asymmetric proticity of this environment, *i.e.* asymmetric hydrogen-bond formation may play a role.

In this respect, it is important to note that most spectroscopic investigations of the photosynthetic electron-transfer processes by optical and EPR techniques are being done on detergent-solubilized complexes like RCs because the same resolution and sensitivity cannot be achieved with chromatophores in membrane fragments or even with whole cells. It has been demonstrated by

X-band ENDOR/TRIPLE spectroscopy on RCs from *Rb. sphaeroides*[160,161] that commonly used detergents can, indeed, have pronounced effects on the electronic structure of the special pair dimer, affecting the asymmetry ratio of the spin densities on the L and M dimer half. The shifts of the spin-density ratio depend on the detergent properties, the detergent/RC ratio, and the temperature. The largest effect is observed for ionic and zwitterionic detergents, even leading to two distinct conformations of P$^{\bullet+}$ with different absorption maxima of the Q$_y$ band of the neutral state, λ_1=866 nm and λ_2=850 nm at room temperature. P$^{\bullet+}_{866}$ is found in chromatophores and in RCs solubilized with nonionic detergents, whereas P$^{\bullet+}_{850}$ is induced in ionic and zwitterionic detergents. It appears that nonionic detergents should be used when comparing the electronic structure of RCs from various mutants and organisms to avoid artefacts caused by the solubilization procedure – as has been done in our studies on RCs from *Rb. sphaeroides* reviewed in this chapter.

Many more organisms, both wild types and mutants, will have to be studied in detail before a definite conclusion can be reached regarding other factors potentially operative to secure unidirectional charge-separation ET in bacterial photosynthesis. The results for the doublet state, P$^{\bullet+}$, should also be compared with those for the photoexcited triplet state, ^{T}P. This is because the *orientation* of the zero-field-splitting tensor is a sensitive probe for the symmetry properties of the triplet-state wavefunction extending over the dimer halves[111] (see Section 5.2.1.8).

High-field EPR (95 GHz) on RC single crystals was also used to characterize the g-tensor of P$^{\bullet+}_{865}$ in the heterodimer mutant HL(M202) of *Rb. sphaeroides*.[162] Since in this mutant the unpaired electron of P$^{\bullet+}_{865}$ is localized on the bacteriochlorophyll, the g-tensor reflects the monomer properties. The directions of the principal axes of the g-tensor were found to be similar for the mutant and the wild type.

With the availability of precise g-tensor data, there is an increasing need for a reliable quantum-chemical analysis of g-tensors of biomolecules. This is a demanding endeavor,[163] and an overview of modern computational methods for the spin-interaction parameters is given in Chapter 4.

The starting point for the calculations of the g-tensor components is the theory of Stone.[164] The dominant contribution to the inplane g-tensor components of interest, g_{xx} and g_{yy}, arises from the spin-orbit interaction. It couples the *s*ingly *o*ccupied *m*olecular *o*rbital SOMO of the ground state (given the index 0) to an excited-state SOMO (index i) and modifies the free-electron electron g_e-value. The principal g-tensor components are given by

$$g_{rt} = g_e \delta_{rt} - 2 \sum_{i \neq 0} \frac{\langle \Psi_0 | \zeta(r) \hat{L}_r | \Psi_i \rangle \langle \Psi_i | \hat{L}_t | \Psi_0 \rangle}{E_i - E_0} \qquad (5.11)$$

where g_e=2.002322; Ψ_0, Ψ_i are the ground and excited SOMO states, respectively; E_0, E_i are the respective state energies; $\zeta(r)$ is the spin-orbit coupling function, and $\hat{L}_r$, $\hat{L}_t$ (r,t=x,y,z) are the orbital angular momentum operators.

Very promising g-tensor calculations for cofactor radical ions in photosynthesis have been performed recently on the basis of DFT methods (see Section 5.2.1.4).

5.2.1.3 360-GHz EPR on Primary Donor Cations $P^{\bullet+}$ in Mutant RCs

The unidirectional nature of the primary ET route is probably not determined by a single structural feature, but rather by the concerted effects of small contributions of several different optimized factors, for example the redox energetics of the various intermediate states (affecting the Franck–Condon factor) and the coupling scheme of the cofactor wavefunctions (affecting the electronic matrix element), and controversially discussed theoretical concepts invoke either "overlap" or "superexchange" coupling mechanisms or a combination thereof.[35] Both the wavefunctions and the energetics of the cofactors involved can be systematically changed by selectively exchanging neighbouring amino-acid residues of the protein environment by means of site-specific mutation. This can be accomplished, for instance, by introducing or disrupting H-bonds of the cofactors or by changing the ligation of the Mg in the chlorophyll macrocycles of P. In this context, it is particularly interesting to systematically study the influence of the environment of the primary donor P, since P is generally considered to play a key role in the origin of the unidirectionality of the primary charge-separation steps.

When constructing site-specific mutants of RCs one often modulates simultaneously the Franck–Condon factor, F, and the electronic coupling matrix element, V. This is, of course, unfavorable when trying to pin down the dominant contribution to the complex network of ET pathways. Hence, it is the strategy of several molecular biology groups cooperating with spectroscopists to select their mutations in such a way that either F or V are predominantly modulated. Such a research strategy is certainly a very difficult task. For example, those mutants with the axial histidine ligand of the central Mg in one of the bacteriochlorophyll (BChl a) molecules of the special-pair dimer P_{865} removed to form a heterodimer of a BChl a and a bacteriopheophytin (BPhe a) molecule, are extreme cases of modulating the inherent asymmetry of the electronic structure of P_{856} due to the large difference in the redox potentials of BChl a and BPhe a (BPhe is harder to oxidize than BChl by ≈ 300 meV). Hence, heterodimer RCs are good candidates for mutants to study effects of the energetics on the ET characteristics.

The effect on the electronic structure caused by such mutations can be measured via the oxidation-potential changes of the primary donor.[165–167] An alternative is to measure the mutation effect via characteristic shifts of g-tensor or hyperfine-tensor components, as can be resolved by high-field EPR[155,162,168] and ENDOR,[169–173] respectively. As was pointed out above, in the course of X-band ENDOR and TRIPLE experiments on the R26 mutant and wild-type RCs,[74,75,145,174] a pronounced asymmetry of 2:1 of the spin-density distribution over the two dimer halves in $P_{865}^{\bullet+}$ was determined, favoring the BChl axially ligated by the L-subunit (see Section 5.2.1.1). The electronic structure of $P_{865}^{\bullet+}$ in the R26 mutant RC turned out to be similar to the carotenoid containing wild-type RC.

To gain further insight into the origins and consequences of this asymmetry in the electronic structure, a program was started in our laboratory to investigate various tailor-made site-directed mutants of the RC from *Rb. sphaeroides* by 360 GHz/12.9 T high-field EPR. We started with the heterodimer mutants in which the ligands to the magnesium of the bacteriochlorophylls were altered.[14,168] Figure 5.9 shows the exchange of the histidine His(M202) by leucine (L) or glutamic acid (E) to generate the mutants HL(M202) and HE(M202), respectively. Since the characteristic optical absorption band of $P^{\bullet+}$ in wild-type and R26 at $\lambda = 865$ nm is absent in these mutants,[165,175,176] we will omit the index of $P^{\bullet+}$ in the following. In the symmetry-related heterodimer mutants HL(M202) and HL(L173),[170] the His that ligates the Mg in the BChl is exchanged for a Leu that does not ligate to the metal center, resulting in a BChl:BPhe heterodimer.[135] In HL(M202), the unpaired electron in $P^{\bullet+}$ was shown to reside on the BChl(L),[169,170] in HL(L173) on the BChl(M) dimer half.[170] It was, however, also shown that in both mutants the main ET pathway is still provided by the A-branch.[175] This finding emphasizes that the asymmetry in the electron spin-density (and charge) distribution in $P^{\bullet+}$ is not *a priori* the major contributor to the unidirectionality of the ET in bacterial RCs, but its effect is enhanced by specific cofactor–protein interactions in addition to those affecting predominantly the spin distribution over the $P^{\bullet+}$ supermolecule.

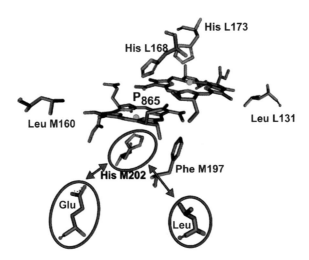

Figure 5.9 X-ray structure of the primary-donor special pair P_{865} and its immediate amino-acid environment in RCs from *Rb. sphaeroides*. In the site-directed mutants HL (M202) and HE (M202) the histidine His M202 is replaced by a leucine Leu and by a glutamic acid Glu, respectively.

As discussed above, ENDOR/TRIPLE experiments provide hfcs of the various magnetic nuclei in the $P^{\bullet+}$ molecule that represent *local* probes of the electronic wavefunction. The assignment of the individual hfcs to molecular positions is a difficult task. It involved selective isotope labelling and additional measurements on RC single crystals.[145] A less time-consuming approach would be desirable to obtain information on the symmetry properties of the electronic structure. This is offered by g-tensor measurements. In contrast to the *local* hfc probes, the g-tensor represents a more *global* probe of the electronic wavefunction of the unpaired electron. For many organic biosystems the g-tensor anisotropies Δg are very small and cannot be resolved by standard EPR at 9.5 GHz (X-band) or 35 GHz (Q-band), because the Zeeman-field shifts of the g-tensor components are still smaller than the inhomogeneous linewidth. $P^{\bullet+}$ in photosynthetic RCs exhibits extremely small g-tensor anisotropies of $\approx 10^{-3}$. It is, therefore, a prime example of a situation where even high-field EPR at 3.4 T/ 95 GHz (W band) can only partially resolve the powder pattern of the isotropically disordered samples.[155,162,177–179] To move these samples into the "true" high-field regime already at W-band frequencies (see Section 2.1), one has to reduce the linewidth by elaborate sample-preparation techniques like deuteration or single-crystal growth. In an attempt to determine, by means of the global g-tensor probe, the characteristic symmetry properties of the electronic structure of the primary donor, 95-GHz high-field EPR on illuminated single-crystal RCs of *Rb. sphaeroides* was performed at 12°C, *i.e.* at physiological temperatures.[155] The single-crystal EPR linewidth is so small that at a field of 3.3 T even the magnetically inequivalent sites in the unit cell of the RC crystal could be resolved. The angular dependences of their g-factors in the three symmetry planes of the crystal have been measured and analyzed. The remarkable result is that the principal directions of the g-tensor are tilted in the molecular axes system and, thereby, reveal a breaking of the local C_2 symmetry of the electronic structure of $P_{865}^{\bullet+}$. This finding is consistent with the ENDOR/ TRIPLE results for the hyperfine structure.

The expected *shifts* of the g-tensor components upon mutation will be even smaller, of the order of 10^{-4}, as has been shown by our previous W-band measurements on similar mutants[162,179] The further increase of EPR frequency and field by a factor of 4 to 360 GHz and 12.9 T, however, provides the spectral resolution necessary to both fully resolve all three principal g-tensor components of $P^{\bullet+}$ randomly oriented in frozen solution,[180] and to measure their mutation-induced shifts with high precision.[168]

Figure 5.10 shows the 360-GHz EPR spectra of $P^{\bullet+}$ of R26 in comparison with those of the HL(M202) and HE(M202) mutants.[168] The mutant R26 can be considered as standing for the wild-type *Rb. sphaeroides* since both hyperfine- and g-tensor values are indistinguishable for both organisms within experimental accuracy. Minimum least-squares fits to a model spin Hamiltonian, including only the Zeeman interaction of radicals with an isotropic orientation distribution, are overlaid as dashed lines over the measured spectra. The g-values obtained from these fits[168] are summarized in Table 5.2. They are in accord with previous measurements within error limits.[155,162,177–181]

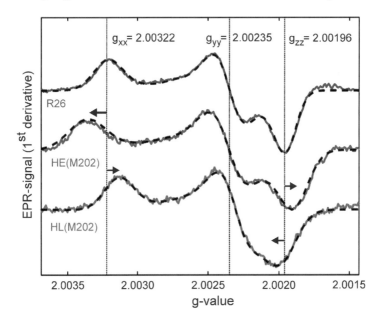

Figure 5.10 First-derivative 360-GHz cw EPR signal of P•+ in RCs from *Rb. sphaeroides* mutants at 160 K. The most prominent shifts of the *g*-tensor components g_{xx} and g_{zz} due to the mutation at M202 are indicated with solid arrows. Minimum least-squared fits for each spectrum are overlaid with dashed lines.[168]

Table 5.2 360-GHz EPR fit data of the *g*-tensor components for P•+ of RCs from *Rb. sphaeroides* mutants.[168]

Sample	g_{xx}	g_{yy}	g_{zz}	g_{iso}
HE(M202)[a]	2.00333(3)	2.00236	2.00191	2.00253(1)
HE(M202)[b]	2.00337	2.00236	2.00190	2.00254(1)
HL(M202)[a,b]	2.00313	2.00231	2.00198	2.00247(1)
R26[a]	2.00324	2.00236	2.00195	2.00252(1)
R26[b]	2.00322	2.00235	2.00196	2.00251(1)

Relative error 2×10^{-5}, except where noted otherwise. [a]Cation generation by light excitation. [b]Cation generation by chemical oxidation.

In discussing differences of observables, such as $\Delta g = (g_{xx} - g_{zz})$, the error in the presented 360-GHz EPR measurements is only 2×10^{-5}, and the small shifts of the order of 10^{-4}, that are induced by the mutations of the protein environment of P•+, can therefore be detected with high significance. A small *g*-strain induced increase of the intrinsic linewidth compared to W-band EPR measurements[162] from 0.89(6) mT to 1.4(2) mT is noticed[168] but, owing to the concurrent increase in spectral resolution by a factor of 4, this does not pose any resolution problems for bacterial RCs.

5.2.1.4 Results of g-tensor Computations of $P^{\bullet+}$

All previous experimental results on the M202 mutants from X-ray structural analyzes, ENDOR/TRIPLE and optical and FT-IR measurements strongly indicate that the primary donors of both the HL(M202) and the HE(M202) mutants are heterodimers of bacteriochlorophyll a and bacteriopheophytin a, abbreviated BChl:BPhe, in contrast to the homodimeric special-pair species $(BChl)_2$ of wild-type and R26 RCs. Both EPR studies and energetic considerations have lead to the conclusion that this structural modification of the primary donor in its oxidized paramagnetic state is characterized by an almost complete localization of the spin density on the BChl(L) half-bound to the L protein subunit.[165,170] From this it seems straightforward to expect a shift of the g-tensor components of the mutants towards those of an isolated $BChl^{\bullet+}$ monomer, as was also concluded earlier from a comparison of HL(M202) with R26.[162] The g-component shifts observed in our 360-GHz EPR experiments, however, are very surprising in this respect: While for $P^{\bullet+}$ of HL(M202), as noted above, the overall g-tensor anisotropy $\Delta g = (g_{xx} - g_{zz})$ becomes smaller (more like that of the monomer), for HE(M202) Δg increases to a value considerably larger than for R26 (see Table 5.2).

This unexpected behavior of g-tensor strongly suggests a local structural change of BChl(L) as a consequence of the M202 ligand mutations. This conclusion is also supported by the observation of considerable rearrangements in the spin-density distributions on BChl(L) in the two mutants. For example, previous ENDOR/TRIPLE studies[165] yielded quite different ratios $R_L = a(12_L^1)/a(2_L^1)$ of the isotropic methyl proton hfcs at positions 12^1 and 2^1 on BChl(L) (see Figure 5.5):

$$R_L(\text{HL(M202)}) = 1.29; \quad R_L(\text{R26}) = 1.44; \quad R_L(\text{HE(M202)}) = 1.66.$$

Earlier semiempirical MO studies[154] and experimental ENDOR/TRIPLE results on $P^{\bullet+}$ in RCs[169,172,182] suggest that such a variation of R_L might originate in different torsional angles $\theta_{ac}(L)$ of the acetyl group at position 3 with respect to the plane of the adjacent aromatic ring A (see Figure 5.5). Different values of $\theta_{ac}(L)$ for R26, HL(M202) and HE(M202) are, in fact, expected since the acetyl-methyl groups point towards the region of the molecular centers of the dimeric counter parts. This is BChl(M) in the case of R26, and BPhe in the case of HL(M202) and HE(M202), where the changes in the chemical structures due to the mutations occur. Obviously, the experimental ratios R_L, as given above, correlate with the corresponding values of Δg (see Table 5.2). This observation urged us to perform theoretical calculations of the g-tensors of the three $P^{\bullet+}$ species as a function of $\theta_{ac}(L)$. Since earlier semiempirical MO calculations based on AM1 or ZINDO/S wavefunctions[183] failed to reveal any changes in Δg as a function of θ_{ac} (but rather revealed changes in g-tensor axes orientation), we have applied more advanced relativistic DFT methods employing the ADF package,[184] and the Schreckenbach–Ziegler (SZ) approach[185] for the calculation of magnetic resonance parameters. Details of

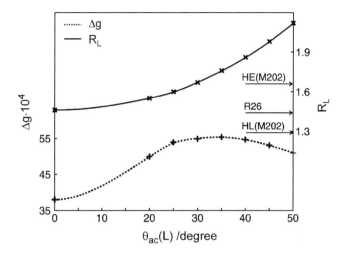

Figure 5.11 The results of DFT calculations for $\Delta g = g_{xx} - g_{zz}$ (dashed line) and R_L (solid line) as a function of the acetyl-group torsional angle $\theta_{ac}(L)$. Horizontal arrows point to the experimental values of R_L for HL(M202), R26 and HE(M202). Both R_L and Δg show increasing DFT values in the range $0° \leq \theta_{ac}(L) \leq 35°$ in accordance with the observed ordering $R_L(\text{HL(M202)}) < R_L(\text{R26}) < R_L(\text{HE(M202)})$ and $\Delta g(\text{HL(M202)}) < \Delta g(\text{R26}) < \Delta g(\text{HE(M202)})$. For details, see ref. [168].

the employed packages and approximations to reduce computation times can be found in our previous publication;[14,168] in brief the results are as follows:

The computational results for R_L and Δg as function of $\theta_{ac}(L)$ are presented in Figure 5.11 and summarized in Table 5.3. The curves shown in Figure 5.11 were obtained by a numerical spline interpolating through eight calculated points between $\theta_{ac}(L)=0°$ and $50°$; for symmetry reasons, their slopes at $\theta_{ac}(L)=0°$ were constrained to zero. The calculated values of R_L show a marked monotonic increase with increasing values of $\theta_{ac}(L)$. Although quite close to the experimental values of R_L given above for the three species HL(M202), R26 and HE(M202), the range of calculated R_L values does not fully enclose the range of the observed ones. From a few test calculations we conclude that structural details are responsible for this discrepancy and they have not been considered in the calculations. Since, at present, we only aim at a qualitative interpretation of the observed ordering of the three mutant species with respect to R_L and Δg, we refrained from considering such structural details any further.

The calculated principal components g_{xx}, g_{yy} and g_{zz} show only moderate *quantitative* agreement with the experimental values. Such a deficiency is frequently observed for DFT-based calculations of g-tensor components.[168,185–187] However, it is expected that a correct *qualitative* prediction of trends of g_{ii} as a function of $\theta_{ac}(L)$ should be provided by the present DFT results. The essential conclusion from the current computational results is that our model predicts a range of increasing angles $\theta_{ac}(L)$ between $0°$ and 30–$35°$ in which both R_L and Δg increase parallel to each other, in accordance with our experimental results.

Table 5.3 Results of DFT calculations of the g-tensor components and the R_L ratio of methyl hfcs of $P^{•+}$ in *Rb. sphaeroides* mutants as function of the acetyl-group torsional angle θ_{ac}. For details, see ref. [14].

θ_{ac}	g_{xx}	g_{yy}	g_{zz}	R_L	$10^4 \Delta g^b$
0°	2.00566	2.00378	2.00185	1.46	38
20°	2.00539	2.00376	2.00036	1.55	50
25°	2.00536	2.00355	2.00003	1.60	53
30°	2.00534	2.00331	1.99985	1.67	55
35°	2.00534	2.00308	1.99980	1.76	55
40°	2.00533	2.00288	1.99987	1.86	55
45°	2.00534	2.00273	2.00003	1.98	53
50°	2.00534	2.00263	2.00024	2.12	51
30°^{a}	2.00645	2.00282	2.00198	1.36	45

^{a}with H-bond.
$^b \Delta g = g_{xx} - g_{zz}$.

The increase in Δg in this series is entirely due to a decrease of g_{zz}. This is reasonable, since the rotation of the acetyl group perturbs the planarity of the electronic π-system and, therefore, primarily leads to changes in g_{zz}. In the same angular region, the value of $g_{yy}-g_{zz}$ also increases parallel to R_L and Δg, in concert with the experimental results, see Table 5.2.

To model the remaining experimentally observed strong shift of the g_{xx} component, one needs to take into account H-bonding interactions with the environment. An H-bond of the acetyl-group oxygen atom to the neighbouring histidine shifts g_{xx} to larger values, and we conclude that the strength of this H-bond, *i.e.* the distance $r(H \ldots O_{ac})$, varies in the three mutant species according to $r(HE) < r(R26) < r(HL)$. This conclusion cannot be tested directly by means of the X-ray structure because of the insufficient precision of atomic positions from X-ray crystallography. Based on the results described in ref. [168], further DFT calculations are in progress with the aim to gain a deeper insight into the general influence of polar side groups, superimposed electric fields and H-bond effects on g-tensor components in photosynthetic systems.

Also, further 360-GHz EPR experiments with RC mutants, in which the H-bond network around the L-half of the special pair has been altered, are in preparation. The high sensitivity of very high field EPR to subtle changes in the structure and microenvironment of the ET chromophores should provide valuable insight into the influence of structural features and energetics on the effectiveness and directionality of the electron-transfer steps.

5.2.1.5 95-GHz EPR and ENDOR on the Acceptors $Q_A^{•-}$ and $Q_B^{•-}$

Quinones play an important role in many biological systems, prominent examples are the light-driven ET processes of photosynthesis. In the photosynthetic bacterium *Rb. sphaeroides*, for example, the primary and secondary

Applications of High-Field EPR on Selected Proteins and their Model Systems 235

quinones, Q_A and Q_B, act as one- and two-electron gates, respectively: $Q_A^{\bullet -}$ just passes the extra electron to Q_B that, in a second photoinitiated ET step, gets doubly reduced, binds two protons, dissociates from the RC, and releases the protons on the periplasmic side of the membrane. In *Rb. sphaeroides*, Q_A and Q_B are the same ubiquinones-10. Apparently, their different functions in the ET processes are induced by different interactions with the protein environment at their binding sites.

In Figure 5.12 the orientation of the principal axes of the *g*-tensor and of the methyl hyperfine-tensor of $Q_A^{\bullet -}$ or $Q_B^{\bullet -}$ with respect to the molecular frame are shown. The g_{xx} component lies along the line connecting the two oxygen atoms carrying most of the spin density, g_{zz} is perpendicular to the molecular plane and g_{yy} is perpendicular to both.[188]

To learn about such cofactor–protein interactions within the quinone $Q_A^{\bullet -}$ or $Q_B^{\bullet -}$ binding pockets, for example the specific H-bonding networks, EPR and ENDOR on quinone anion radicals in bacterial RCs (with Fe^{2+} replaced by Zn^{2+} to avoid fast spin relaxation) and in organic solvents have been performed at several mw frequencies by various groups, both in fluid and frozen solution. For example, in the Möbius group W-band high-field EPR and ENDOR experiments on a series of quinones related to photosynthesis were performed and their intramolecular and intermolecular proton hyperfine interactions were discerned.[177,189] As another example, in the Feher group ^{2}H ENDOR at Q-band frequencies was used to detect both the electron–nuclear hyperfine and the nuclear quadrupole couplings of the H-bonded deuterons.[15,190]

The aim of the W-band high-field EPR and ENDOR experiments on quinone radical anions in frozen solutions was to measure their anisotropic interactions with the organic solvent matrix ("*in vitro*") or protein microenvironment

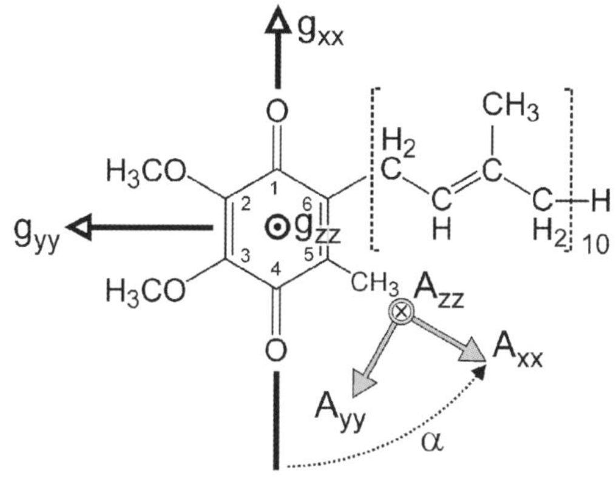

Figure 5.12 Molecular structure and *g*- and methyl hyperfine-tensor axes systems of the ubiquinone-10 anion radical. The arrows indicate the canonical components of the molecular *g*-tensor directions and the methyl *A*-tensor, respectively, and α indicates the angle between g_{xx} and A_{xx}. The axes g_{zz} and A_{zz} are assumed to be collinear and oriented perpendicular to the molecular plane.

("*in vivo*") by means of the cofactors' *g*- and hyperfine-tensor components as well as T_2 relaxation times. It was envisaged to learn from these measurements about the anisotropic hydrogen bonding of the quinones to specific amino-acid residues, and about the motional dynamics of the quinones at their binding sites. From more than a dozen quinone anion radicals, both natural and model systems, powder high-field EPR spectra were recorded. Owing to the high Zeeman magnetoselection capability of W-band EPR, a high degree of orientational selectivity is achieved that is inaccessible by X-band EPR (compare Figures 5.13(a) and 5.13(b)).[189,191] The measured *g*-tensor components follow the sequence $g_{xx} > g_{yy} > g_{zz}$, where *x* is along the $>$C=O bond direction and *z* is perpendicular to the quinone plane.[177] The *g*-tensor values of anion radicals of Q_A in RCs from *Rb. sphaeroides* (*in vivo*) and of ubiquinones-10 in the organic solvent isopropyl alcohol (*in vitro*) are collected in Table 5.4.

The inspection of this table reveals the close correspondence between the *g*-tensor values of the ubiquinones-10 (*in vitro*) and Q_A (*in vivo*) anion radicals, which indicates that hydrogen bonds of similar strength are present in both cases.

Exploiting the Zeeman magnetoselection even further, W-band pulsed ENDOR was performed at the rather well separated canonical peaks of the powder EPR spectrum. Figure 5.13(c) shows the Davies-type ENDOR spectra of the radical anion of ubiquinone-10 in frozen perdeuterated isopropanol (*in vitro*) at *T*=115 K. (For the Davies-ENDOR pulse sequence, see Figure 1.7.) At least for the g_{xx} and g_{zz} canonical field positions, the ENDOR spectra are single-crystal-like and, accordingly, the representations of the orientational selections of molecules show narrow distributions (Figure 5.13(c)). These representations follow from the simulations of the spectra on the basis of the spin Hamiltonian containing Zeeman and hyperfine interaction terms. The orientational distribution of molecules is considerably broader for the g_{yy} canonical value, which reflects its still rather poor resolution by W-band EPR. When varying the solvent (protic and aprotic, with and without perdeuteration), characteristic changes of hyperfine couplings (predominantly along the *y*-direction) and *g*-tensor components (predominantly along the *x*-direction) could be discerned. They are attributed to hydrogen-bond formation at the lone-pair orbitals on the oxygens: Dipolar hyperfine interactions with the solvent protons will result in line broadening along the oxygen lone-pair direction, *i.e.* broadening of the g_{yy} part of the EPR spectrum, while changes in the lone-pair excitation energy $\Delta E_{n\pi^*}$ and/or spin density ρ_π^O at the oxygen due to H-bonding will predominantly shift the g_{xx} component of the *g*-tensor.[189]

This reasoning adheres to the simplified approach to *g*-factor theory, as suggested by Stone,[164] to approximate the state energies E_0, E_i in the energy denominator of the *g*-tensor components by the corresponding molecular orbital energies. Since the spin-orbit coupling parameter for p-electrons at the oxygen in the C=O bond is much larger than that of the carbon atoms (ζ(C)=28 cm^{-1}, ζ(O)=151 cm^{-1})[192] the dominant contribution to the *g*-tensor components comes from the π-spin density at the oxygen and, in first-order

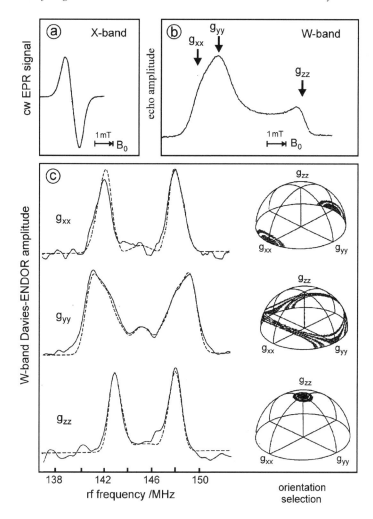

Figure 5.13 (a) X-band cw-detected and (b) W-band ESE-detected EPR spectra of ubiquinone-10 anion radicals in frozen perdeuterated isopropanol solution ($T = 115$ K). (c) Davies-ENDOR spectra taken at the three B_0 positions marked by the principal g-tensor components g_{xx}, g_{yy}, g_{zz}. Dotted lines show the simulated ENDOR spectra from which the degrees of orientation selection of contributing molecules (*right*) were derived.[177,189]

Table 5.4 Experimental g-tensor components of the primary acceptor radical anion $Q_A^{\bullet-}$ in *Rb. sphaeroides* reaction centers and of the ubiquinone radical anion (in organic solvent).[189]

radical	g_{xx}	g_{yy}	g_{zz}	g_{iso}
$Q_A^{\bullet-}$ (Zn-RC)	2.0066(1)	2.0054(1)	2.0022(1)	2.0047(1)
UQ – 10$^{\bullet-}$	2.00646(5)	2.00542(5)	2.00222(5)	2.00470(5)

approximation, g_{rt} is given by:[177]

$$g_{xx} \approx g_e + 2 \cdot \zeta(O) \cdot \rho_\pi^O \cdot c_{ny}^2 / \Delta E_{n \to \pi}$$
$$g_{yy} \approx g_e + 2 \cdot \zeta(O) \cdot \rho_\pi^O \cdot c_{nx}^2 / \Delta E_{n \to \pi} \qquad (5.12)$$
$$g_{zz} \approx g_e$$
$$g_{rt} = 0 \quad for \ r \neq t, \ r,t = x,y,z,$$

where ρ_π^O is the π-spin density c_{SOMO}^2 on the oxygen $2p_z$ atomic orbital; c_{nx}, c_{ny} are the MO coefficients of the $2p_x$ and $2p_y$ atomic orbitals contributing to the oxygen lone-pair orbital Ψ_n (briefly n), and $\Delta E_{n \to \pi}$ is the $n \to \pi$ excitation energy. Equations (5.12) can be justified by the fact that the lone-pair orbital n lies energetically very close to the lowest half-filled π orbital (ground state SOMO).

Similar W-band pulsed Davies-type ENDOR experiments were performed on the secondary quinone $Q_B^{\bullet-}$ in Zn-substituted RCs from *Rb. sphaeroides*.[191,193,194] The orientation-selection benefit of EPR at high B_0 field is highlighted in Figure 5.14. It shows the echo-detected W-band EPR spectrum of $Q_B^{\bullet-}$ in RCs in frozen solution at $T=120$ K. The lineshape exhibits the typical powder pattern of a well-resolved anisotropic *g*-tensor. The respective resonance positions of the canonical *g*-tensor values are indicated by arrows. In this case of resolved *g*-tensor components ENDOR measurements on defined spectral positions yield single-crystal-like information about the relative orientation of the hyperfine-tensors with respect to the *g*-tensor. Thereby, the assignment of the hyperfine couplings is simplified and, even more important, valuable information about the orientation of the hyperfine-tensor axes with respect to the molecular frame can be extracted.

This capability is demonstrated in Figure 5.14. It depicts W-band Davies-ENDOR spectra recorded at the canonical *g*-tensor positions of $Q_B^{\bullet-}$. Pronounced orientation selectivity of the *g*-tensor, *i.e.* single-crystal-like ENDOR spectra, are expected at the g_{xx} and g_{zz} positions of the EPR spectrum, while in the center of the spectrum at g_{yy}, a broadened ENDOR spectrum is anticipated, as is indeed observed. According to the extensive W-band ENDOR work on quinone radical anions,[191,194,195] the outer couplings in the range between 4 and 8 MHz can be assigned to the methyl group at position 5 (see Figure 5.12).

At a temperature of 120 K methyl groups rotate almost freely about their C–C bond. Hence, the three hyperfine-tensors of the individual protons of the methyl group in $Q_B^{\bullet-}$ collapse to only one hyperfine-tensor, with the largest component A_{xx} oriented along the C–C bond.[191,196] In the ENDOR spectrum taken at the position g_{zz}, only the A_{zz} component appears in the ENDOR spectrum, while at the resonance position of g_{yy}, both A_{xx} and A_{yy} contribute. A precise determination of the principal values and orientation of the methyl hyperfine-tensor with respect to the *g*-tensor and, thereby, to the molecular frame can be obtained by simulating the contribution of the methyl hyperfine couplings to the ENDOR spectra obtained at g_{xx}, g_{yy} and g_{zz}, respectively. The dashed lines in Figure 5.14 show the results of the spectral simulations

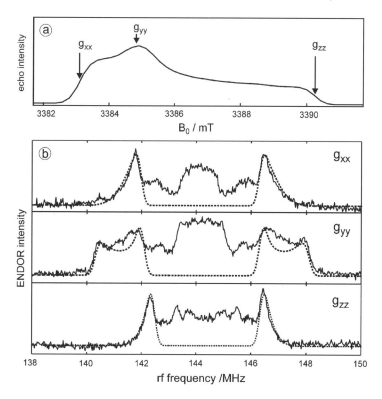

Figure 5.14 (a) Echo-detected W-band powder EPR spectrum of $Q_B^{\bullet-}$ in frozen-solution RCs from *Rb. sphaeroides* ($T = 120$ K). The resonance positions of the principal g-tensor values g_{xx}, g_{yy} and g_{zz} are indicated by arrows. (b) W-band Davies-ENDOR spectra (solid lines) recorded at the g_{xx}, g_{yy} and g_{zz} positions of $Q_B^{\bullet-}$ in RCs from *Rb. sphaeroides* ($T = 120$ K). Dashed lines indicate the respective contribution of the methyl hfc to the ENDOR spectra. Note that the Larmor frequency and, thereby, the center of the ENDOR spectra increases proportional to the external magnetic field from g_{xx} to g_{zz}. For details, see ref. [193].

Table 5.5 EPR and ENDOR parameters obtained from the simulations of the W-band Davies ENDOR spectra of $Q_B^{\bullet-}$ in ZnRCs from *Rb. sphaeroides* R26.[193]

g_{yy}	g_{yy}	g_{zz}	α	A_{xx}/MHz	A_{yy}/MHz	A_{zz}/MHz
2.00626(5)	2.00527(5)	2.00210(5)	$(66 \pm 2)°$	7.8(2)	4.4(2)	3.9(2)

providing the parameter set given in Table 5.5 (the hyperfine couplings contributing to the central part of the ENDOR spectra have been excluded from the simulations). The angle α describes the inplane deviation of the g- and hyperfine-tensor axes systems. The coupling values obtained for the methyl hyperfine-tensor of $Q_B^{\bullet-}$ by W-band Davies-ENDOR match nicely the results of cw Q-band measurements on $Q_B^{\bullet-}$.[196]

The most elaborate EPR and ENDOR studies on the trapped radical anion states $Q_A^{\bullet-}$ and $Q_B^{\bullet-}$ of the ubiquinone acceptors in Zn-substituted RCs from *Rb. sphaeroides* were performed by the collaborating groups of Feher (San Diego) and Lubitz (Mülheim). Their work, as published until 2004, has been described in detailed review articles.[13,15,128,196] More recently, by means of 35-GHz (Q-band) EPR and ENDOR, the hydrogen-bonding network of $Q_A^{\bullet-}$[197] and that of $Q_B^{\bullet-}$[198] were studied. From frozen solutions of deuterated Q_A in H_2O buffer and protonated Q_A in D_2O buffer, the proton hyperfine and deuteron quadrupole coupling tensors of $Q_A^{\bullet-}$ in its binding site were obtained. This was feasible by taking ENDOR spectra at several magnetic-field values. These Q-band experiments represent an illustrative example for fulfilling the "high-field EPR" condition, *i.e.* $\Delta g \cdot B_0/g_{iso} > \Delta B_{1/2}^{hf}$, already with Q-band EPR by reducing the inhomogeneous linewidth $\Delta B_{1/2}^{hf}$ via deuteration. The results provide a precise picture of the three-dimensional geometry of the Q_A binding site in the charge-separated state of the RC. Nonequivalent hydrogen bonds from the carbonyl oxygens of the quinone to close-by histidine and alanine residues dominate the bonding network. This is important information for an understanding of the competition between charge-separation and -recombination kinetics in bacterial photosynthesis.

To conclude this discourse on the *static* spin interactions of the ET cofactors: The multifrequency EPR and ENDOR results show that both $Q_A^{\bullet-}$ and $Q_B^{\bullet-}$ have two different H-bonds to the two carbonyl oxygens, their strengths, however, are different in $Q_A^{\bullet-}$ and $Q_B^{\bullet-}$. A detailed analysis of the 35-GHz and 95-GHz ENDOR spectra revealed the geometry of the H-bonds; this information is not directly obtainable from X-ray crystallography. DFT calculations[199] show that the H-bond distances are changing upon reduction of the quinones in the ET process. This leads to the unusual electronic properties and function of the quinones and their radical anions in the binding pocket of the RC. The EPR and ENDOR results of the primary donor and acceptor ion radicals nicely amend the geometric structure information from X-ray crystallography of RCs. Moreover, they provide detailed information of the electronic structure of the transient intermediate states of primary photosynthesis. This information is not directly accessible by X-ray crystallography, but is crucial for understanding the light-induced reactions on the molecular level.

Magnetic resonance cannot only probe the *static* structural details of a molecule but also details of its *dynamic* properties.[200–202] If the motion is on the time scale of the EPR experiment, spin relaxation and, thereby, line broadening can be observed in the cw EPR spectrum. In many cases the analysis of this effect is obscured by static ("inhomogeneous") broadening effects from unresolved hyperfine interactions or g-strain. Therefore, pulsed spin-echo techniques, which can separate dynamic and static contributions to the spectrum, are the methods of choice to study molecular motion.[203,204] The molecular fluctuations that contribute to the spin relaxation in biomolecules may be vibrational and rotational motion, intra- and intermolecular conformational dynamics, spin flips of surrounding nuclei or fluctuating coupling to solvent phonons. By choosing proper pulse sequences and resonance conditions, these effects may be discriminated against each other.

It is well known that in photosynthetic RCs molecular dynamics of the cofactors and the protein matrix play an important role in the kinetics of the electron transfer.[205,206] To investigate the molecular motion of the ubiquinone-10 (UQ-10) cofactor anions, $Q_A^{\bullet-}$ and $Q_B^{\bullet-}$, in RCs of *Rb. sphaeroides* R26, we performed high-field/high-frequency (3.4 T/95 GHz, W-band) two-pulse echo experiments. In the investigated RCs the paramagnetic Fe^{2+} was replaced by diamagnetic Zn^{2+} to reduce the linewidth of the EPR spectra.[207] Thereby, the g-tensor anisotropy is left as the linewidth-determining relaxation contribution to the W-band spectrum. Because the g-tensor anisotropies of quinone radical anions are about 4×10^{-3} and the inhomogeneous linewidth is about 0.5 mT, W-band EPR is required to spread the spectrum sufficiently (over almost 10 mT) and to separate the x-, y- and z-components of the g-tensor.[177] Therefore, one can spectroscopically select those molecules in the sample that are oriented with one of these principal g-tensor axes along B_0 to measure their phase-memory time T_{mem}.[194] T_{mem} is the time constant of the echo decay and, since this is monoexponential, T_{mem} can be identified as the transverse relaxation time T_2. At high magnetic fields, the dominant anisotropic T_2 relaxation is induced by molecular motion modulating the effective g-value of the tumbling radical. At W-band Zeeman fields this leads to orientation-dependent modulation of the resonance condition. The anisotropic T_2 contributions have minima (longest T_2) along the canonical orientations of the g-tensor. Thus, the determination of T_2 as a function of the resonance position in the high-field EPR spectrum provides information about the directions and amplitudes of molecular motions and their correlation times.

The dynamics of the cofactor–protein complexes become particularly important for controlling the primary processes in photosynthesis when the time constant of a characteristic molecular motion becomes comparable to one of the time constants, k_{ET}^{-1}, in the ET chain for charge separation (see Figure 5.3(b)) or recombination (which are orders of magnitude slower). For this reason, there is an increasing interest in the slow motional modes of thermal fluctuations of protein–cofactor complexes. They may affect the electron-tunnelling mechanism and thereby the biological function. In order to learn more about slow motion of the quinone cofactors in photosynthesis, the anisotropic stochastic oscillatory motion of $Q_A^{\bullet-}$ in frozen RC solutions of *Rb. sphaeroides* were studied by pulsed ESE detected EPR at high field.[194] The two-dimensional field-swept electron spin-echo technique was chosen because it directly reveals the homogeneous linewidth parameter T_2 and, due to the high Zeeman field, resolves its variation over the powder spectrum. Figure 5.15(a)[208] shows the two-dimensional W-band ESE spectrum of $Q_A^{\bullet-}$ at 115 K in frozen-solution RCs of the *Rb. sphaeroides* mutant HC(M266), in which Fe^{2+} is replaced by Zn^{2+}. The canonical orientations of the g-tensor are rather well resolved at this field. The monoexponential echo decay curves $S(2\tau, B_0)$ at the

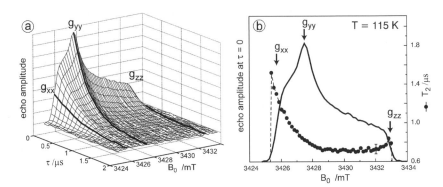

Figure 5.15 (a) Field-swept two-pulse echo-decay spectrum of $Q_A^{\bullet-}$ in frozen-solution RCs ($T=115$ K) of the $Fe^{2+} \rightarrow Zn^{2+}$ mutant MC(M266) of *Rb. sphaeroides*. (b) Extrapolated $\tau = 0$ slice of the ESE-detected EPR spectrum and field-dependent T_2 relaxation times, as extracted from $S(2\tau, B_0) = S_0 \cdot \exp(-2\tau/T_2(B_0))$, the echo decays.[194,208]

g_{xx}, g_{yy} and g_{zz} orientations are highlighted in Figure 5.15(a):

$$S(2\tau, B_0) = S_0 \cdot \exp[-2\tau/T_2(B_0)] \qquad (5.13)$$

Obviously, the decays have different time constants T_2 in different directions with respect to the B_0 field. Since T_2 relates solely to the dynamic process, the resolved anisotropy of T_2 directly provides information about the axes of torsional fluctuations of $Q_A^{\bullet-}$ at a given temperature. As is shown in Figure 5.15(b)[208] at $T=115$ K, the magnitude of T_2 varies over the powder EPR spectrum and clearly peaks at the g_{xx} orientation.

At high B_0 fields, the dominant contribution to anisotropic T_2 relaxation stems from the wobbling motion of $Q_A^{\bullet-}$, and depends on the orientation of the *g*-tensor with respect to the direction of B_0 (see Figure 5.16). The magnitudes of the T_2 contributions are determined by a random walk on the surface of the *g*-tensor ellipsoid. This leads to time- and angular-dependent fluctuations δg that translate to fluctuations of the Larmor frequency of the electron spins.[194] As is obvious from Figure 5.16, the Larmor frequency fluctuations will be minimal for $Q_A^{\bullet-}$ oscillations around the principal axes x, y, z. In other words: The T_2 values will be the largest along the directions of the respective axes of oscillation. Thus, the maximum of T_2 in the g_{xx} region (see Figure 5.15(b)) tells us that, at 115 K, the wobbling motion of $Q_A^{\bullet-}$ predominantly occurs around the *x*-direction, *i.e.* along the C=O bond direction.

For $Q_A^{\bullet-}$, this result can be understood from the X-ray structure of the RC in its ground state Q_A.[209] The *x*-axis of the quinone, which is along the C=O bonds, points to the nearby histidine residue, His M219, and to the more distant alanine, Ala M260, allowing a strong and a weak H-bond to be formed between the imidazole and peptide nitrogens and the two respective carbonyl oxygens. This asymmetric H-bond pattern is obviously preserved in the anion

Applications of High-Field EPR on Selected Proteins and their Model Systems 243

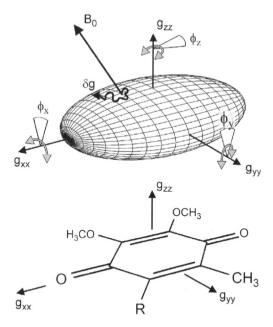

Figure 5.16 Schematic representation of the g-tensor connected with $Q_A^{\bullet-}$ (R stands for the isoprenoid chain). When $Q_A^{\bullet-}$ undergoes stochastic thermal fluctuations with angular amplitudes ϕ_x, ϕ_y, ϕ_z, the g-tensor orientation fluctuates with respect to B_0. This causes g-factor fluctuations δg of varying magnitude and, thereby, anisotropic T_2 relaxation. For details, see ref. [194].

state $Q_A^{\bullet-}$, which is probed by EPR. It is indistinguishable within the accuracy limits from that of the ground state Q_A. This conclusion is in accordance with the H-bond situation observed for ^{13}C-labelled Q_A by NMR[210,211] and FT-IR[212] and for ^{13}C-labelled $Q_A^{\bullet-}$ by X- and Q-band EPR,[196,213–215] and contrasts with the $Q_B^{\bullet-}$ situation (see below). In other words: The Q_A binding site does not noticeably change its structure upon light-induced charge separation within an estimated error margin of about ± 0.5 Å and the resolution limits of this single-frequency ESE-detected EPR experiment (see, however, the results of our W-band two-frequency pulsed ELDOR (PELDOR) experiment[216] reviewed in Section 5.2.1.7).

Obviously, the next step in this molecular dynamics study is to go to the secondary quinone, Q_B, with its pronounced differences in amino-acid environment as compared to the Q_A site. The 95-GHz ESE-detected EPR work on $Q_B^{\bullet-}$ is reviewed in the following paragraph:[14]

Figure 5.17 shows how the orientation dependence of T_2 changes the lineshape of the echo-detected EPR spectrum of $Q_B^{\bullet-}$ in the RC of *Rb. sphaeroides* R26.[217]

In Figure 5.18, T_2 of $Q_A^{\bullet-}$ and $Q_B^{\bullet-}$ is plotted *vs.* the magnetic field for different temperatures. The relaxation times of $Q_A^{\bullet-}$ and $Q_B^{\bullet-}$ at the resonance positions

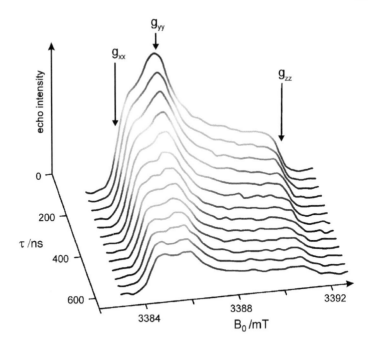

Figure 5.17 W-band echo-detected EPR spectra of $Q_A^{\bullet-}$ in RCs (Fe^{2+} substituted by Zn^{2+}) from *Rb. sphaeroides* at 120 K. The spectra are measured with the pulse sequence shown below for different values of τ. For longer τ, the echo amplitudes decrease and the spectral shape changes because of the orientation dependence of the phase-memory time. For details, see refs. [14,217].

for g_{xx}, g_{yy} and g_{zz} are clearly different, but their orientation dependence is strikingly similar: At 120 K, T_2 reaches its maximum at the g_{xx} position of the spectrum, while the shortest relaxation times are found at the intermediate position between g_{yy} and g_{zz}. At the positions g_{yy} and g_{zz}, the T_2 values are nearly the same and only slightly above the minimum value.

From experiments on one quinone alone it is not possible to discriminate between fluctuations of the quinone in its binding pocket and fluctuations of the quinone's protein microenvironment. However, this discrimination could be achieved by comparing the temperature dependence of the T_2 anisotropy of the echo-detected spectra of both $Q_A^{\bullet-}$ and $Q_B^{\bullet-}$ (see Figure 5.18): Despite their different binding pockets in terms of amino-acid composition and spatial constraints (see Figure 5.19), the observed field dependence of the relaxation times of $Q_A^{\bullet-}$ and $Q_B^{\bullet-}$ is very similar, and at 120 K exhibits a pronounced T_2 maximum along the g_{xx} axis. Such an anisotropic mode of the motion of the whole protein, which effects $Q_A^{\bullet-}$ and $Q_B^{\bullet-}$ in the same way, is highly improbable. Hence, we conclude that at 120 K the whole protein does not fluctuate on the EPR time scale, but the cofactors librate individually in random fashion around an H-bond axis in the *x*-direction.[194,217]

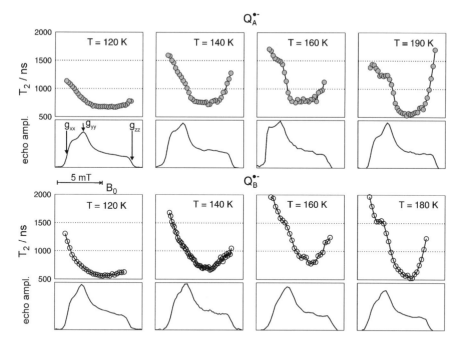

Figure 5.18 Field dependence of T_2 for $Q_A^{\bullet-}$ (upper panels) and $Q_B^{\bullet-}$ (lower panels) at various temperatures. In the upper part of the panels, T_2, as extracted from the echo decays by means of monoexponential functions, are presented. In the lower parts of the panels, the W-band echo-detected EPR spectra ($\tau = 100$ ns) are shown as a function of B_0. The absolute errors of the relaxation times differ at different spectral positions. The biggest errors (± 100 ns) are found at the edges of the spectrum, while the smallest errors are observed at the maximum of the spectrum (± 20 ns). For details, see refs. [14,217].

The T_2 values in the various directions mirror the local restrictions to the motion of the quinone in its binding environment. In their protein pockets both $Q_A^{\bullet-}$ and $Q_B^{\bullet-}$ fluctuate uniaxially at 120 K about a dominant binding axis, which lies along a strong H-bond to an amino acid of the protein pocket. For $Q_A^{\bullet-}$ and $Q_B^{\bullet-}$ the H-bond directions are approximately parallel to g_{xx}, as is reflected by their similar relaxation pattern.

But what about the Q_B binding site with its pronounced differences in the amino-acid environment as compared with the Q_A site? In the ground-state X-ray structure[209] the C=O bonds of Q_B do not point to a nearby H-bonding amino-acid candidate! Hence, our high-field spin-echo result, that also $Q_B^{\bullet-}$ undergoes uniaxial librational motion at 120 K around the quinone x-axis, can only be understood by assuming structural changes of the Q_B binding site upon charge separation, shifting an H-bond candidate like a histidine close to the C=O bond. Such structural changes of the Q_B site have, indeed, been unveiled in a more recent high-resolution (1.9 Å) X-ray crystallography experiment by

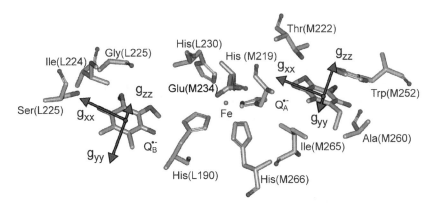

Figure 5.19 Structure of the $Q_A^{\bullet -}$ and $Q_B^{\bullet -}$ binding sites in the RC from *Rb. sphaeroides* as obtained from X-ray structure analysis on the charge-separated state $P^{\bullet +}Q_B^{\bullet -}$.[93] For clarity, the isoprenoid chains of the two ubiquinones are truncated. X-ray and ENDOR studies deduced possible H-bonds to His(M219) and to Ala(M260) for $Q_A^{\bullet -}$ and to His(L190), Ser(L223), Gly(L225) and Ile(L224) for $Q_B^{\bullet -}$.[196] The angles that relate the *g*-tensors of $Q_A^{\bullet -}$ and $Q_B^{\bullet -}$ to each other in the state $Q_A^{\bullet -}Q_B^{\bullet -}$ are obtained from cw-EPR measurements.[218] Note that for EPR work the Fe^{2+} was replaced by Zn^{2+}, the *z*-direction of the *g*-tensor axis system is perpendicular to the quinone plane. For details, see ref. [14].

Stowell and coworkers.[93] These authors claim to have detected dramatic displacements of $Q_B^{\bullet -}$ by almost 5 Å in the light-adapted X-ray structure in comparison to the dark-adapted one, *i.e.* the ground-state structure. This light-induced movement of the negatively charged quinone in the Q_B binding site brings the histidine, His L190, into the proximity of a carbonyl oxygen of $Q_B^{\bullet -}$ thereby favoring the formation of an H-bond in the *x*-direction (see Figure 5.19). The authors[93] postulate that this structural change is a necessary prerequisite for electron transfer to proceed from $Q_A^{\bullet -}Q_B$ to $Q_AQ_B^{\bullet -}$. Moreover, it offers an explanation of the observation by Kleinfeld and coworkers[205] in the Feher group that at 90 K the ET from $Q_A^{\bullet -}Q_B$ to $Q_AQ_B^{\bullet -}$ is blocked in RCs frozen in the dark (ground state), but effective in RCs frozen under illumination (charge-separated state). In other words: The structural change of the Q_B site under illumination provides a molecular switch for vectorial ET steps in bacterial photosynthesis. The present results show that low-temperature high-field spin-echo measurements of the relaxation times at different spectral positions can be used to reveal important structural information such as H-bond directions in functional proteins at work.

When going to higher temperatures, the orientation dependence of the relaxation times of $Q_A^{\bullet -}$ and $Q_B^{\bullet -}$ changes. The differences in T_2 values at canonical positions and intermediate positions increase, while the T_2 values at the canonical positions become equal. This shows that at temperatures higher than 150 K, the motional modes lose their directional preference, probably because of additional fluctuations between conformational substrates of the protein[206] superimposing the fluctuations of the quinones around their H-bond axis.

At this point, we refer to molecular-dynamics simulations carried out to investigate the dynamics of ubiquinone (UQ) with respect to its hydrogen bond structure in water as a solvent.[219] The H-bonds in water do not necessarily resemble the H-bonds in the RC protein of *Rb. sphaeroides*, but rather suggest possible hydrogen bonds of UQ in the binding site. Interestingly, it was found that only weak H-bonds to the C=O carbonyl oxygens are formed between the neutral form of UQ and water molecules, while the anionic form of UQ shows a distinct solute–solvent hydrogen bond structure. This finding is in agreement with the results of an NMR study on neutral Q_A in the RC from *Rb. sphaeroides*.[220] Moreover, in the case of the anionic UQ, the details of the H-bond orientations to O1 and O4 of the quinone ring are in agreement with an ENDOR study of the radical anion of UQ-10 in 2-propanol as a solvent.[191] The results indicate that the directions of the H-bonds are dependent on the adjacent groups: methyl or methoxy groups force the H-bond towards the symmetry axis through O1–O4, while an adjacent ethyl group forces the H-bonds out of the ring plane of the quinone.

We refer to another important work related to quinone radical anions embedded in hydrogen-bond networks:[199] In a combined Q-band pulse EPR/ENDOR and DFT study of quinone–solvent interactions the impact of hydrogen bonding on the *p*-benzosemiquinone radical anion $BQ^{\bullet -}$ in coordination with heavy water, D_2O, or alcohol molecules was investigated. After complete geometry optimizations, ^1H, ^{13}C and ^{17}O hyperfine as well as ^2H nuclear quadrupole coupling constants and the *g*-tensor were computed. Different model systems with one, two, four, and 20 water molecules were tested; best agreement between theory and experiment could be obtained for the largest model system with 20 water molecules. The ^2H ENDOR spectra could be very well simulated with interaction parameters calculated from DFT. For $BQ^{\bullet -}$ in coordination with four water or alcohol molecules, rather similar hydrogen-bond lengths between 1.75 and 1.78 Å were calculated. The distance dependence of the EPR parameters (dipolar and quadrupolar hyperfine coupling constants, g_{xx} component of the *g*-tensor) on the hydrogen-bond length shows interesting characteristics: The nuclear quadrupole and the dipolar hyperfine coupling constants of the bridging hydrogens show the expected $1/R^3_{O\ldots H}$ dependencies on the H-bond length $R_{O\ldots H}$. For the *g*-tensor component g_{xx} a linear correlation with $1/R^3_{O\ldots H}$ was theoretically predicted, the slope strongly depending on the number of water molecules assumed in the solvation model.[199] The point-dipole model is suitable for estimating hydrogen-bond lengths from the anisotropic hyperfine coupling constants of the bridging ^1H nuclei provided that the H-bonds are longer than ≈1.7 Å. Earlier observations,[221–224] that there exist simple empirical relations between ^2H nuclear quadrupole coupling constants (NQCC) of bridging deuterium nuclei and H-bond lengths, $e^2qQ/h = a - b/R^3_{O\ldots D}$, have been reproduced by the DFT calculations. Here, the parameter a corresponds to an extrapolated ($R_{O\ldots D}$ to infinity) value of the ^2H NQCC in the isolated hydrogen donor molecule, while b describes the effect of the O $\cdots$ D bond formation on the ^{2}H NQCC. The confirmation of the empirical relations between e^2qQ values and O–D distances by semiempirical MO calculations[225] and DFT[199] calculations is an important result encouraging practical applications. The estimation of hydrogen-bond lengths from nuclear quadrupole coupling constants is definitely a very attractive approach since no other parameters besides the ^{2}H NQCC are required. A challenge is, however, to obtain the NQCC from the experiments with sufficient accuracy. Modern ENDOR and ESEEM techniques are judged to be prone to meet this challenge (see Chapter 2). It is apt to comment on the DFT result that there exists a linear correlation between $1/R^3_{O \cdots H}$ and the g_{xx} tensor component: The strong dependence of the g-shifts on the detailed structure and polarity of the surroundings of the radicals makes it difficult, if not impossible to derive a quantitative relationship between the g-shift and the H-bond length that is sufficiently general to be used in practice.

5.2.1.6 95-GHz ESE-Detected EPR on the Spin-Correlated Radical Pair $P^{\bullet +} Q_A^{\bullet -}$

In the following, we review pulsed W-band high-field electron-spin-echo (ESE) experiments on the laser-pulse generated short-lived radical pair in frozen RC solution of *Rb. sphaeroides*.[159] These experiments were performed with the aim to determine, via spin-polarization effects, the three-dimensional structure of the charge-separated donor–acceptor system. Principally, this excited-state structure may differ from the ground-state structure and, indeed, as was discussed above, upon illumination of RC crystals of *Rb. sphaeroides* drastic changes have been observed in the X-ray structure of the secondary quinone (Q_B) binding site in comparison with the dark-adapted X-ray structure.[93] The high-field EPR spectra were recorded using the field-swept two-pulse ESE technique. To avoid fast spin relaxation of the $Q_A^{\bullet -}$, the nonheme Fe^{2+} ion was replaced by Zn^{2+}. The charge-separated radical pairs (RPs) were generated by 10-ns laser flashes. Their time-resolved EPR spectrum is strongly electron-spin polarized because the transient RPs are suddenly born in a spin-correlated noneigenstate of the spin Hamiltonian with pure singlet character. Such spin-polarized spectra with lines in enhanced absorption and emission (see Figure 5.20) originate from the CCRP mechanism described in Section 2.2.1.2. They contain important structural information of magnitude and orientation of the g-tensors of the two radical partners, $P^{\bullet +}$ and $Q_A^{\bullet -}$ with respect to each other and to the dipolar axis r_{QP} connecting the two radicals (see Figure 5.20). Several parameters critically determine the lineshape of the CCRP polarization pattern, such as the principal values and orientations of the g- and electron-dipolar-coupling tensors, the exchange coupling J, and the inhomogeneous linewidths of both radicals.[159]

From earlier time-resolved EPR measurements on at X-band (9.5 GHz), K-band (24 GHz) and Q-band (35 GHz), g-tensor orientations could not be extracted unambiguously from spectra simulations (for references, see ref. [126]). This was mainly because of strongly overlapping lines, even when

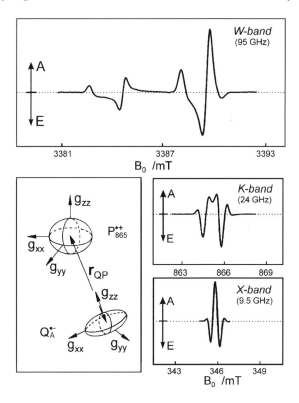

Figure 5.20 Schematic representation of the relative orientation of the g-tensors and dipolar axis r_{QP} of the transient radical pair $P^{\bullet+}_{865}Q^{\bullet-}_A$ in deuterated frozen RC solution from *Rb. sphaeroides*. The spin-polarized EPR spectra are recorded at various settings of mw frequency and Zeeman field. A and E stand for absorption and emission, respectively. For details, see ref. [159].

deuterated samples were used to reduce hyperfine contributions. In the pulsed W-band ESE experiments, however, the Zeeman field is strong enough to largely separate the spectral contributions from $P^{\bullet+}$ and $Q^{\bullet-}_A$. Thus, the overall spectrum is dominated by the characteristics of the two g-tensors, and its interpretation is simplified. It allows for an unambiguous analysis of the tensor orientations. The most important result of this high-field ESE study is that, within an estimated error margin of ±0.3 Å, no detectable light-induced structural changes of the quinone site occur, as compared to the ground-state configuration $P_{865}Q_A$. This finding is in accordance with recent results from various other studies, including X-ray crystallography,[93] and contrasts with the $Q^{\bullet-}_B$ situation. However, as will be discussed in Section 5.2.1.7, there exist small, but significant conformational changes of $Q^{\bullet-}_A$ with respect to Q_A, but it needs the enhanced orientational resolution of high-field PELDOR to detect them.[216]

At the end of this section we mention recent studies of spin-correlated donor–acceptor ion radical pairs in the bacterial reaction center from *Rb. sphaeroides* and photosystem I from *S. lividus* by high-frequency/high-field time-resolved

ENDOR and quantum beat oscillations.[226] In these experiments the dominant nuclear hyperfine structure of the spin-correlated pairs is probed. The analysis of the spin-polarized transient spectra requires consideration of three interacting spins: two correlated electron spins together with one nuclear spin. The results illustrate the importance of resolving the hyperfine structure for obtaining details of structure–function relationships in photosynthetic electron transfer.

We conclude by emphasizing that by time-resolved high-field EPR, for example pulsed ESE-detected EPR experiments, on spin-correlated coupled radical pairs a detailed picture of the electronic structure and spin dynamics of the ET partners can be obtained. Moreover, ET-induced structural changes in the relative orientation of donor and acceptor can be detected, even for disordered samples. Such information is important for a deeper understanding of the ET characteristics of charge-separation and charge-recombination processes on the molecular level. The charge-separated RP state represents the initial state for ET recombination, *i.e.* it is one of the important working states of the photocycle in the photosynthetic RC.

5.2.1.7 95-GHz RIDME and PELDOR on the Spin-Correlated Radical Pair $P^{\bullet+}Q_A^{\bullet-}$

Light-induced transient radical pairs in photosynthetic reactions centers of plants and bacteria have been in the focus of numerous spectroscopic investigations over the past decades.[13,227] In particular, the charge-separated radical pairs $P^{\bullet+}Q^{\bullet-}$ of the primary electron donor and quinone acceptors have been characterized by a variety of time-resolved EPR methods in bacterial RCs, as well as in plant photosystem I and photosystem II. After pulsed laser excitation of the primary donor, $P^{\bullet+}Q^{\bullet-}$ appears in the spin-correlated coupled radical-pair (CCRP) state, which is characterized by a weak electron spin–spin coupling in a fixed geometry of the radicals in the pair and an initial singlet state of the system. The EPR responses of the CCRP state display a number of interesting and useful spectroscopic features: spin polarization, quantum beats, transient nutations, as well as echo-envelope modulation and out-of-phase echo effects. From their analysis the magnetic interaction parameters and the geometry of the system can be obtained. Three-dimensional geometrical information about the radical pair is of particular importance. It allows extraction of structural information about the transient charge-separated states in photosynthetic RCs for which only rarely[93] detailed X-ray data are available. Moreover, it allows us to recognize and characterize the structural changes occurring in RCs upon charge-separation and charge-recombination processes. Although the time-resolved EPR signals of the CCRP state are very sensitive to the relative orientation of the radicals in the pair, the number of parameters on which the CCRP lineshape depends is large, and for unique solutions of the spectra simulations additional independent information is needed. As a result, it is often difficult to judge the accuracy of the resulting pair geometry. Thus, an additional EPR methodology has to be developed, which allows the orientation information to be obtained directly, and this with predictable accuracy. This methodology is pulsed electron–electron double resonance (PELDOR or DEER) dipolar spectroscopy[228,229] applying two microwave frequencies, which is favorably performed in conjunction with one-frequency RIDME dipolar spectroscopy (relaxation-induced dipolar modulation enhancement[230]) on the CCRP state at high magnetic fields. In the following, we review our recent work on orientation-resolving dipolar spectroscopy on $P^{\bullet+}_{865}Q^{\bullet-}_A$ via PELDOR and RIDME at 95 GHz/3.4 T [193,216] summarizing the principles of the PELDOR and RIDME spectroscopy on the CCPR state in RCs in a frozen solution of Zn-substituted *Rb. sphaeroides* R26 as an example. We briefly describe the analysis of the CCRP spectra and, subsequently, the high-field PELDOR experiment. Finally, we discuss the extracted three-dimensional structure of the radical pair. A more extended description of high-field PELDOR experiments including the complete theoretical model and graphical analysis of the experimental data is given elsewhere.[216] It is noted that independent work of orientation-selective high-field PELDOR (or DEER) on large two-spin systems has been published very recently.[231–233]

For large interspin distances of well-localized electron spins in the radical partners, $P^{\bullet+}_{865}$ and $Q^{\bullet-}_A$, of a spin pair, for which the point-dipole approximation holds (as is often the case for the weakly coupled donor–acceptor radical pairs in photosynthesis) the electron–electron dipolar coupling frequency, ν_{QP}, is given by (see Section 2.2.2):

$$\nu_{QP}(\theta) = \frac{\mu_0 \cdot \mu_B^2 \cdot g_P \cdot g_Q}{4\pi \cdot h} \cdot \frac{(3 \cdot \cos^2 \theta - 1)}{r_{QP}^3} \qquad (5.14)$$

Here, r_{QP} is the interspin distance in the radical pair and θ the angle between the Zeeman field B_0 and the distance (dipolar) vector $\mathbf{r}_{QP}$, see Figure 5.21(a). The g-values $g_Q(\theta_Q, \varphi_Q)$ and $g_P(\theta_P, \varphi_P)$ refer to the radical pair selected by the external field B_0, as is explained for $Q^{\bullet-}_A$ radical in the insert of Figure 5.21(a).

Figure 5.21(b) shows the EPR spectra of the individual radicals $P^{\bullet+}_{865}$ and $Q^{\bullet-}_A$ in frozen solution as calculated for X-band (9.5 GHz) and W-band (95 GHz) using the previously measured principal g-values $g_Q=[g_{xx}; g_{yy}; g_{zz}]=[2.00647;$ 2.00532; 2.00215], see ref. [188], and $g_P=[2.00325; 2.00234; 2.00196]$, see ref. [168]. At X-band the shape of the EPR spectra is governed by the inhomogeneous linewidth masking the canonical orientations. Thus, an isotropic distribution of angles θ is observed when probing the dipolar coupling between the radicals in the pair. Consequently, only a typical powder-type dipolar spectrum (Pake pattern) is obtained, from which the interradical distance r_{QP} can be determined, the orientational information, however, is averaged out. In principle, it is possible to retrieve the orientational information by investigating the single crystals of RCs. In this case, certain pair orientations can be selected by rotating the single crystal in its symmetry planes with respect to the direction of B_0.[234] Unfortunately, such single-crystal experiments and the subsequent data

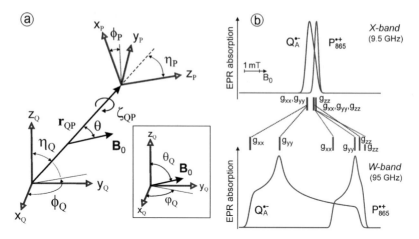

Figure 5.21 (a) Geometrical representation of the respective g-tensor frames of $R_Q(x_Q,y_Q,z_Q)$ and $R_P(x_P,y_P,z_P)$ in terms of the polar angles η_Q, ϕ_Q and η_P, ϕ_P and the dipolar vector $\mathbf{r}_{QP}$. The angle ζ_{QP} defines the relative tilt of the Q and P frames around the dipolar axis. The insert shows the g-value selection in Q-frame by the external magnetic field B_0. (b) Individual absorption EPR spectra of $P_{865}^{\bullet+}$ and $Q_A^{\bullet-}$ radicals as calculated for X- and W-band. The corresponding canonical g-values of both radicals are indicated. The individual spectra are normalized to the maximum amplitude. For details, see ref. [216].

analysis of the rotation patterns are very time consuming. Thus, orientational information from disordered solids would be desirable. It is, of course, only available at a sufficient degree of orientation selectivity in the EPR powder spectrum by means of the anisotropy of a dominating spin-interaction parameter. For the radical pair $P_{865}^{\bullet+}Q_A^{\bullet-}$ this is the electronic g-tensor, which at sufficiently high magnetic fields provides the desired Zeeman magnetoselection. At W-band the condition for "true" high-field EPR is fulfilled for $P_{865}^{\bullet+}Q_A^{\bullet-}$. The spectra appear with resolved canonical positions of the g-tensors of the respective radicals, Figure 5.21(b). Thus, if one can determine the dipolar frequencies at all resolved field positions within the EPR spectra of pair partners, i.e. is selectively probing the electron dipolar coupling frequencies within the EPR spectrum of $P_{865}^{\bullet+}$ and $Q_A^{\bullet-}$, the full geometry of the radical pair can be reconstructed: the five angles η_Q, ϕ_Q, η_P, ϕ_P, ζ_{QP} and the interradical distance r_{QP}, see Figure 5.21(a).

For CCRP systems in frozen isotropic solution, the dipolar coupling frequencies at the canonical orientations of the g-tensors can be extracted from the spin-polarized EPR powder spectrum.[235–237] Photoexcitation of the primary RC donor leads to its excited singlet state, and this spin state is conserved during the electron transfer because of the short lifetime of all intermediate states compared to the time scale of all magnetic interactions. The radical pair $P_{865}^{\bullet+}Q_A^{\bullet-}$ is, thus, generated in a pure singlet state, $|S\rangle$. Because only two of the four eigenstates of the system contain admixtures of singlet character, the

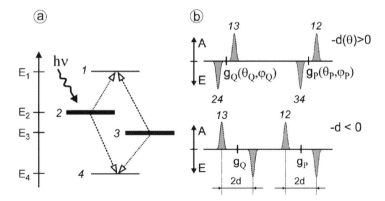

Figure 5.22 (a) Energy-level scheme for a coupled two-spin system formed in singlet state after light excitation. The arrows denote the single-quantum transitions between the energy levels. For the energy values and wavefunctions, see ref. [235]. (b) EPR absorption spectra of the spin-correlated radical pair $P^{\bullet+}_{865}Q^{\bullet-}_A$ at a selected orientation with respect to the external magnetic field. The number on the EPR lines denotes the transitions between the respective energy levels. The separation between the two line pairs is governed by the g-difference between the radicals. The separation within the line pair is given by $-2d(\theta)$, see eqn (5.15), because the exchange interaction J is neglected. The top spectrum is observed for an orientation with negative $d(\theta)$, i.e. for $\theta < 54.7°$ (magic angle), the bottom one for a different orientation $\theta > 54.7°$ (positive $d(\theta)$). For details, see ref. [216].

population of the energy levels of the CCRP is far from thermal equilibrium. In the high-field limit, when the dipolar and the exchange interactions are small compared to the Zeeman interactions, the eigenstates of the 4-level system of the radical pair can be approximated by the unperturbed triplet states $|T_+\rangle$, $|T_-\rangle$ and the mixtures of $|S\rangle$ and $|T_0\rangle$ states, as indicated in Figure 5.22(a) by the energy levels 1, 4 and 2, 3, respectively. In each orientation of the radical pair with respect to the Zeeman field one observes two pairs of EPR lines (antiphase doublets), corresponding to the four single-quantum transitions between the energy levels, Figure 5.22(a). They are equal in intensity, but one line appears in absorption (A), the other in emission (E).[235–237] Both antiphase doublets are symmetrically positioned around the g-factors of the individual partner radicals, $g_Q(\theta_Q, \varphi_Q)$ and $g_P(\theta_P, \varphi_P)$, orientation selected by the B_0 field, see Figure 5.22(b). The separation of each pair of lines is given by the dipolar coupling $-2d(\theta)$, if exchange interaction is neglected. Hence, the polarization pattern depends on the sign of the dipolar interaction. Inspection of eqn (5.14) shows that at $\theta=54.7°$ (magic angle) the dipolar coupling changes sign. Thus, E/A/E/A and A/E/A/E polarization patterns are observed for radical-pair orientations with $\theta<54.7°$ and $>54.7°$, respectively, see Figure 5.22(b). Consequently, it would be possible to construct the complete architecture of the radical pair from the rotation patterns of single-crystal high-field CCRP EPR spectra with sufficient spectral selectivity, because the dipolar frequency could be selectively determined for any field position within the EPR spectrum of the pair. This single-crystal approach, however, would be very demanding in terms of experimental efforts and measuring time. In contrast, CCRP EPR spectra of frozen-solution samples are much easier and faster to record, but have the disadvantage that they reflect only the powder average over all possible pair orientations. This averaging leads to a loss of information about the relative orientations of the partner radicals of the pair,[238] *i.e.* the angle ζ_{QP} (see Figure 5.21(a)) cannot be obtained from frozen-solution spin-polarized SCRP (Spin-Correlated Radical Pair) EPR spectra alone. This is illustrated by Figure 5.23(a), which serves, in conjunction with a short excursion of CCRP theory, to explain the limited angular information content of EPR spectra of CCRPs in frozen solution, and how to overcome this limitation by additional PELDOR experiments on the same sample.

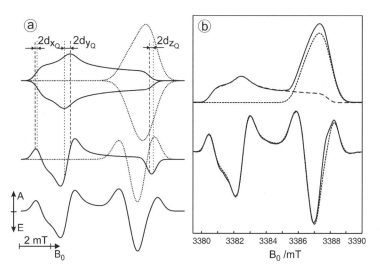

Figure 5.23 (a) Calculated SCRP W-band EPR spectrum of randomly distributed $P_{865}^{\bullet+}Q_A^{\bullet-}$: top spectra: the distribution of the four EPR lines in absorption and emission (solid lines for $Q_A^{\bullet-}$, dotted lines for $P_{865}^{\bullet+}$) as determined by the anisotropy of the g-tensors (see Figure 5.22(a)); middle spectra: summation over corresponding absorptive and emissive contributions to the spectra of the individual radicals $P_{865}^{\bullet+}$ and $Q_A^{\bullet-}$ of the pair; bottom spectrum: resulting EPR spectrum of the spin-correlated coupled radical pair. For g-values and linewidth, see ref. [216]. A dipolar field of $d_0 = 0.15$ mT was used for this demonstration. (b) Field-swept two-pulse echo-detected W-band EPR spectra of $P_{865}^{\bullet+}Q_A^{\bullet-}$ in the Zn-substituted RCs from Rb. sphaeroides R26 at 150 K for thermally equilibrated (top) and SCRP (bottom) states of the radical pair, taken at a delay-after-laser-flash time, T_{DAF}, of 500 μs and 200 ns, respectively. The signals from residual Fe^{2+}-containing RCs and the background steady-state signals are removed. The dashed lines give the calculated spectra, as described in ref. [216]. For the thermally equilibrated state (top), the calculated spectra of $Q_A^{\bullet-}$ and $P_{865}^{\bullet+}$ are shown separately, by long and short dashes, respectively.

Figure 5.23(a) shows the formation principle of the EPR spectrum of CCRPs in randomly oriented RCs. At the top of the figure, the distribution of the four absorptive and emissive EPR lines from Figure 5.22(b) are depicted, as determined by the respective anisotropy of the g-tensors of the $P^{\bullet+}_{865}$ and $Q^{\bullet-}_A$ radicals. The middle spectra are obtained by summation over corresponding absorptive and emissive contributions to the spectra of the individual radicals. At the bottom of Figure 5.23(a), the resulting spectrum of the spin-correlated radical pair is shown. In the resulting CCRP spectrum the information about the orientation of corresponding g-frames relative to each other is averaged out. This means that after the powder averaging the contributions of CCRP spectra of the individual radicals can be considered to be independent. The spectral shape of the individual CCRP spectra depends only on the polar angles defining the distance vector (dipolar axis) in the corresponding g-tensor frame, i.e. η_Q, ϕ_Q in the $Q^{\bullet-}_A$ frame and η_P, ϕ_P in the $P^{\bullet+}_{865}$ frame. Characteristic derivative-like features are observed at the canonical positions of the g-tensor, see Figure 5.23(a). The dipolar fields projected onto the canonical orientations of the g-tensor of $Q^{\bullet-}_A$, for example, are given by

$$d_{x_Q} = d_0(1 - 3 \cdot \sin^2 \eta_Q \cos^2 \phi_Q);$$
$$d_{y_Q} = d_0(1 - 3 \cdot \sin^2 \eta_Q \sin^2 \phi_Q); \quad (5.15)$$
$$d_{z_Q} = - d_{x_Q} - d_{y_Q} = d_0(1 - 3 \cdot \cos^2 \eta_Q),$$

where $d_0 = h \cdot v_{QP}(\theta = 90°)/\mu_B \cdot g_e$ is the canonical dipolar field when approximating $g_Q \cong g_P \cong g_e = 2.0023$. Thus, the angle ζ_{QP}, which defines the relative tilt of the Q- and P-frames around the dipolar axis, cannot be obtained from the powder-averaged SCRP spectrum. Moreover, the d_z values for the $P^{\bullet+}_{865}$ and $Q^{\bullet-}_A$ radicals cannot be measured precisely from the SCRP spectrum due to the close proximity of the g_{zz} values. Thus, the full equation system, combining eqn (5.15) for $P^{\bullet+}_{865}$ and $Q^{\bullet-}_A$, cannot be solved because there are only four equations for five independent parameters. Additionally, the experimental SCRP spectrum contains an unknown scaling factor of the spectral intensities, which does not allow to use the dipolar parameter d_0 obtained from independent experiments.[239] This problem can be solved by scaling the CCRP spectrum to the EPR spectrum of the radical pairs in the same sample after their thermal equilibration,[238] thereby providing the needed correlations for the polar angles. Figure 5.23(b) depicts the experimental EPR spectra of $P^{\bullet+}_{865}Q^{\bullet-}_A$ radical pairs in their spin-correlated and thermally equilibrated states, i.e. when recording the EPR spectra 200 ns and 500 μs, respectively, after the laser flash. The simultaneous fit of both spectra allows determination of the following correlations for the polar angles η_Q, ϕ_Q for $Q^{\bullet-}_A$:

$$d_{y_Q}/d_{x_Q} = \frac{1 - 3 \cdot \sin^2 \eta_Q \sin^2 \phi_Q}{1 - 3 \cdot \sin^2 \eta_Q \cos^2 \phi_Q} = -2.90,$$
$$d_{x_Q}/d_0 = 1 - 3 \cdot \sin^2 \eta_Q \cos^2 \phi_Q = 0.44 \quad (5.16)$$

and the corresponding values of -1.25 and 0.92 for $P_{865}^{\bullet+}$. The solution of these correlations yields several sets of possible polar angles.

Figure 5.24(a) shows the multiple solutions that were obtained using a graphical method. Eight possible directions of the dipolar axis in the P coordinate system were found (designated I, II, ..., VIII in Figure 5.24(a)). These ambiguities occur because the observed g-value for an orientation defined by a vector **h** is given by $(\mathbf{h} \cdot \tilde{g} \cdot \tilde{g} \cdot \mathbf{h})^{1/2}$ where $\tilde{g}$ is the g-tensor and the components of the vector **h** are the direction cosines of the external magnetic field B_0 in the reference axis.[158] Moreover, the observed dipolar interaction follows the same rule. A similar situation is found for the dipolar axis in the Q coordinate system. Thus, the total number of solutions for the radical-pair geometry is 64 (8×8). These solutions are mathematically equivalent since they are symmetry related, but they differ in the resulting geometry of the radical pair. The resulting geometries can be grouped by the transformations: rotation (4×4) and mirroring (2×2). Hence, at this stage of data analysis we follow convention and choose as the basic pair configuration the one where the dipolar axis points towards the first octants of the molecular frames of the radicals: $\eta_Q = 76.5°$, $\phi_Q = 64°$, $\eta_P = 59°$ and $\phi_P = 81°$. Other solutions can be evaluated according to the corresponding symmetry transformations. It is important to emphasize that the accuracy of this set of angles is difficult to estimate, because the simulation of the EPR spectra requires also additional input data, such as exact g-tensor values and orientation-dependent linewidth parameters.[216]

The angular parameter set can be improved by RIDME experiment. The single-frequency pulse dipolar EPR experiment of the RIDME technique is based on measuring the three-pulse stimulated spin-echo signal as a function of the preparation time τ under the following condition: The fixed mixing time T should be long enough to allow the longitudinal spin relaxation to flip the partner spins in the pair.[240,241] The relaxational flip of the interaction partner spin causes the change of the dipolar frequency sign. This leads to the distinct modulation of the spin-echo decay. When performed consecutively at stepped field positions within the HF-EPR spectra of the paired radicals, RIDME is potentially capable of revealing the specific field positions where the dipolar modulation occurs at the principal dipolar frequencies. Orientational selection at these field positions, *i.e.* $g_Q(\eta_Q, \phi_Q)$ and $g_P(\eta_P, \phi_P)$, establishes correlations between the polar angles that define possible orientations of the dipolar axis with respect to the radical frames. RIDME spectra of CCRPs show spectral maxima corresponding to the perpendicular and parallel orientation of the dipolar axis with respect to the field direction.[216] Particular positions along the B-axis where the "parallel" dipolar peaks are observed could, in principle, serve for selecting the relevant pair orientations, *i.e.* for finding the selected traces that define the allowed orientations of the pair axis in the g-frames of the radicals in the pair, analogous to those orientations derived from the lineshape analysis above. Because the parallel peaks are too shallow for such quantitative treatment, we used RIDME-derived parallel positions as prerequisite information for further PELDOR measurements. They significantly increase selectivity and sensitivity to dipolar modulations. Importantly, only the PELDOR

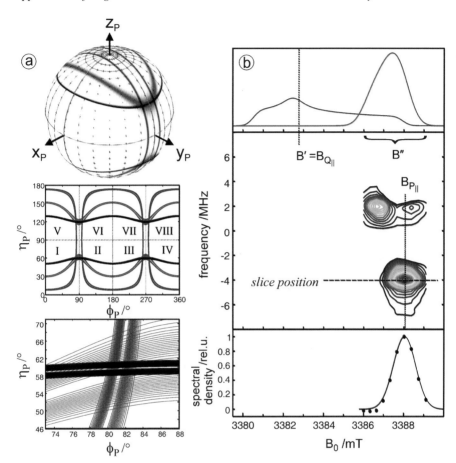

Figure 5.24 (a) Graphical solution for the angles η_P and ϕ_P in the $P_{865}^{\bullet+}$ molecular frame. In the upper part the selection traces obtained from the analysis of CCRP EPR spectra, eqn (5.16) are plotted on the P-frame sphere in black and blue. The red trace shows selections corresponding to orientations responding at $B_{P\|}$ in the PELDOR experiment in (b). The middle part shows (η_P, ϕ_P) projections of the selection traces on the spheres. The lower part shows a close-up of the projections around the crossing points of the selection traces in the first octant. For details, see ref. [216]. (b) Example of dipolar PELDOR spectra of the SC radical pair $P_{865}^{\bullet+}Q_A^{\bullet-}$ at 150 K. The observer mw frequency is fixed at a value corresponding to the resonance field B', while the pump microwave is swept through a region around the resonance frequency determined by the chosen field value B''. The middle part shows contour plots of the positive Fourier amplitudes of PELDOR echo-decay traces detected out-of-phase. The positive Fourier amplitudes as the function of the magnetic field (slice curves) at the slice position $v_{d\|} = -4.08$ MHz are shown by dots. Additionally, Gaussian lines are shown with widths that correspond to the inhomogeneously broadened EPR linewidth of the $P_{865}^{\bullet+}$, see ref. [216].

technique gives additional angular correlations that complete the description of the radical-pair geometry. A RIDME experiment does not yield reliable information about the tilt angle ζ_{QP}. This is because the RIDME responses of two interacting partners have to be considered independently from each other. When probing a certain spectral position within the spectrum of one radical, the dipolar contributions from all coupled orientations of the second radical are obtained because they are induced by orientation-independent relaxation of the coupled spin.

Thus, only four angular parameters can be extracted from the analysis of the CCRP spectra of $P_{865}^{\bullet+}Q_A^{\bullet-}$ in frozen-solution RCs, because the fifth angle, ζ_{QP}, is averaged out in the powder spectrum of the interacting radicals. This tilt angle and an independent proof of the preliminary polar angles extracted from the CCRP spectra, cannot be determined by single-frequency high-field techniques like ESE or RIDME,[216] but need spectrally selective two-frequency PELDOR measurements.[242] The PELDOR technique simultaneously uses two microwave fields with different frequencies employing selective pulsed mw excitation within the EPR spectra of both interacting radicals (see Section 2.2.2) and, thereby, can measure directly the dipolar coupling frequency, eqn (5.14).

Figure 5.25 shows the mw pulse scheme for dual-frequency PELDOR on pulsed laser-generated CCRPs. Within the mixing period between the three-pulse stimulated echo pulses, which are frequency-adjusted (v_1) to an EPR transition of the observer spins (either $Q_A^{\bullet-}$ or $P_{865}^{\bullet+}$), an additional mw π-pulse in resonance (at v_2) with the partner spins is applied to flip them. The two frequencies v_1 and v_2 for the observer spins and partner spins, respectively, correspond to the two Zeeman resonance fields B'' and B' (see Figure 5.25), which are interrelated by $B'' = B' + h \cdot (v_1 - v_2)/(g_e \cdot \mu_B)$. This pulse scheme allows reversal of the sign of the dipolar interaction field during the preparation (time

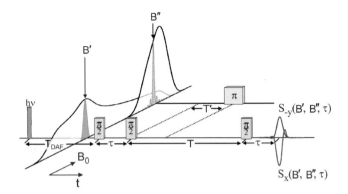

Figure 5.25 Pulse sequence of the dual-frequency W-band PELDOR experiment on laser-pulse-generated spin-correlated radical pairs with in-phase (S_{-y}) and out-of-phase (S_x) echo detection. The thermally equilibrated EPR spectra of both radicals, $P_{865}^{\bullet+}$ and $Q_A^{\bullet-}$, as well as the microwave excitation bandwidths for typical mw pulse-length settings are shown for clarity. For details, see ref. [216].

τ between the first and the second $\pi/2$ mw pulse of the observer sequence) and for detection (time τ between third $\pi/2$ pulsed and spin-echo formation). As the result, the echo responses in dependence on τ become amplitude modulated with the frequency of the dipolar interaction, eqn (5.14). For the CCRP state two echo responses are formed, the in-phase signal S_{-y} and the out-of-phase signal S_x.[239,242–248] It is often advantageous to use the out-of-phase signal for CCRP measurements. They are amplitude enhanced as compared to in-phase echoes, especially in the spectral regions where the EPR absorption of the CCRP state approaches zero, see Figure 5.23. Additionally, the out-of-phase echoes are amplitude modulated by $\sin(2\pi \cdot v_{QP} \cdot \tau)$, whereas the modulation of in-phase echo signals shows cosine function behavior. Figure 5.25 pertains to one specific data point of the PELDOR experiment. The complete PELDOR data surfaces are obtained by taking the echo signals at different positions of B'' within the EPR spectrum of the partner radical, *i.e.* by stepping the mw frequency v_2 in typically 5.6 MHz steps (corresponding to 0.2 mT field steps). Thereby, the PELDOR spectra of only those radicals $P^{\bullet+}_{865}$ and $Q^{\bullet-}_A$ are detected that are conjugated by dipolar interaction. In principle, these spectra contain information about the full 3D geometry of the radical pair, which is encoded in the characteristic modulations of the spin-echo amplitude.

The specific conditions to be fulfilled for such a double magnetoselection provide the needed correlations between those angular parameters that complete the determination of the full pair structure. As stated above, this is not possible from the analysis of single magnetoselection of CCRP spectra by RIDME alone. The experimental procedure for determining the full pair structure was as follows:

In a first step, the magnetic field B' was fixed at the value $B_{Q\|}$ (Figure 5.24(b)), where the preceding RIDME measurements had indicated the parallel peak within the $Q^{\bullet-}_A$ domain, while the magnetic field B'' was swept through the $P^{\bullet+}_{865}$ spectrum. The recorded out-of-phase PELDOR decays were extended to negative τ values, extrapolated to the deadtime interval and Fourier transformed. The contour plot of the resulting dipolar spectra, as shown in Figure 5.24(b), reveals the parallel peak at $B_{P\|}$. In a second step, the magnetic field B' was fixed at the found value $B_{P\|}$, while the magnetic field B'' was swept through the spectral range around the value previously kept fixed at $B_{Q\|}$. The resulting spectrum also shows the well-developed parallel dipolar peak, now in the $Q^{\bullet-}_A$ domain. The "parallel" orientation selection traces $g_P(\eta_P, \phi_P)$ and $g_Q(\eta_Q, \phi_Q)$ can be plotted on the selection spheres of P- and Q-frames. The PELDOR selection trace for $P^{\bullet+}_{865}$ plotted jointly with correlations from CCRP spectra analysis is depicted in Figure 5.24(a). All three types of correlations have a common crossing point at angles $\eta_P = 59 \pm 2°$; $\phi_Q = 82 \pm 2°$. Analogously, the angles $\eta_Q = 70.5 \pm 2°$; $\phi_Q = 62 \pm 2°$ are obtained, see ref. [216].

As mentioned above, the analysis reveals a fundamental ambiguity: a manifold of eight orientations for every radical pair. Solutions that direct the dipolar axis either to the 1st, 3rd, 6th or 8th octants (denoted by I, III, VI, and VIII in Figure 5.24(a)) of both coordinate frames replace each other by π-rotations around the principal g-axes of the radicals $Q^{\bullet-}_A$ or $P^{\bullet+}_{865}$. Because any g-frame is attached to its radical with the same sign ambiguity, there is a choice to present these subgroups of orientations by the basic configurations, $Con_Q(I)$ and $Con_P(I)$, where the dipolar axis has positive projections and points to the 1st octant at the Q- and P-frames. Analogously, orientation solutions that direct the dipolar axis either to the 2nd, 4th, 5th or 7th octants replace each other by the same π-rotations and can be represented by the basic configurations, $Con_Q(VII)$ and $Con_P(VII)$, where the dipolar axis has negative projections and points to the 7th octant in the Q- and P-frames, *i.e.* by mirroring. Importantly, the π-rotations do not transpose orientations between $Con_R(I)$ and $Con_R(VII)$; such transposition does occur by inversion of the dipolar axis in the Q-frame or in the P-frame. Because the dipole director ***d*** is defined in the paired frames, it can be inverted only in both of them simultaneously. Therefore, the manifold of eight configurations derived above can be reduced to the four basic group configurations of the radical pair: $Con_Q(I)$-$Con_P(I)$, $Con_Q(VII)$-$Con_P(VII)$, $Con_Q(I)$-$Con_P(VII)$ and $Con_Q(VII)$-$Con_P(I)$. The polar angles characterizing the alignment of the dipolar axis in the configurations $Con_Q(I)$ and $Con_P(I)$ were evaluated above: $\eta_Q(I) = 70.5°$, $\phi_Q(I) = 62°$, $\eta_P(I) = 59°$ and $\phi_P(I) = 82°$. Other solutions can be evaluated according to the corresponding symmetry transformations, for instance: $\eta_Q(VII) = 180° - \eta_Q(I)$, $\phi_Q(VII) = 180° + \phi_Q(I)$, *etc.* The complete set of the polar angles derived for the basic group configurations is presented in Table 5.6. A choice between the four groups of configurations is principally outside the possibility of an EPR study, but requires information provided by other methods, for instance X-ray crystallography.

After the orientation of the dipolar vector in each radical frame is known, specific PELDOR experiments were performed to find the spectral field positions at which only the "perpendicular" dipolar frequency, $v_{QP}(\theta = 90°)$, can be detected. The analysis of these positions yields, as possible options, four allowed ζ_{QP} tilt angles of the pair: 54°, 124°, 234° or 304° for $Con_Q(I)$-$Con_P(I)$. To choose between them, in a final step the analysis of the PELDOR responses in noncanonical dipolar positions was performed, which fixes the tilt angle to $\zeta_{QP} = 54 \pm 7°$. Additionally, an interradical distance of 2.89 ± 0.02 nm was calculated from the "parallel" and "perpendicular" dipolar frequencies (2.05 and 4.08 MHz, respectively). The exchange interaction between the radical-pair

Table 5.6 Angular parameters of the pair geometry.[216]

Configuration[a]	ϕ_Q[b]	η_Q[b]	ϕ_P[b]	η_P[b]	ζ_{QP}[c]
$Con_Q(I)$-$Con_P(I)$	62°	70.5°	82°	59°	54°
$Con_Q(I)$-$Con_P(VII)$	62°	70.5°	262°	121°	310°
$Con_Q(VII)$-$Con_P(I)$	242°	109.5°	82°	59°	50°
$Con_Q(VII)$-$Con_P(VII)$	242°	109.5°	262°	121°	306°

[a]The four basic group configurations are defined in the text.
[b]Directions of the dipolar axis defined by angles (ϕ_R,η_R) are evaluated with an accuracy of $\pm 2°$ for deviations in any direction.
[c]Evaluation accuracy $\pm 7°$.

partners was found to be negligible, as were contributions from the pseudo-secular term in the dipolar Hamiltonian.[216]

The solution of the geometry of the $P_{865}^{\bullet+}Q_A^{\bullet-}$ radical pair found by high-field PELDOR ought to be compared with the X-ray structure of a model reference pair $P_{865}Q_A$ in its ground state. The cofactor radical ions $P_{865}^{\bullet+}$ and $Q_A^{\bullet-}$ are assumed to remain in the same binding pocket as their precursor cofactors, P_{865} and Q_A, in the dark-state RC.

These positions were determined previously[93] by high-resolution X-ray crystallography. To describe the positions of the radical ions within the reference pair in the same terms of distance and angle parameters, the respective Q_{ref}- and P_{ref}-frames were attached to the radicals to be coaxial to their g-tensors. For $Q_A^{\bullet-}$, the principal x-axis of the g-tensor was aligned to be collinear with the averaged director of the C=O bonds of ubiquinone-10. The y-axis was aligned to be perpendicular to the x-axis in the averaged plane of the quinone ring. The z-axis was taken to be perpendicular to both x- and y-axes, i.e. normal to the quinone plane. The origin of the Q_{ref}-frame was set at the center of the quinone ring that approximates the electron spin-density center of the radical. This attachment is conventional for quinone molecules and follows from their symmetry. Alignment of the $P_{865}^{\bullet+}$ g-tensor (P_{ref}-frame) within the primary donor was calculated using the results of the previous W-band EPR study[155] on single crystals of *Rb. sphaeroides* R26. The origin of the P_{ref}-frame was positioned at the electron spin-density center of $P_{865}^{\bullet+}$, as determined from DFT calculations of the primary-donor special pair (82-atom $BChl_L$, 82-atom $BChl_M$).

Such a reconstructed geometry of the paired g-frames still suffers from the symmetry ambiguity discussed above: The π-rotations around the principal g-axes generate four possibilities to place the right-handed g-tensor coordinate system within each of the cofactor molecules. However, the basic configuration, for which the dipole director has either all positive or all negative projections on the coordinate axes, is unique: The reference pair has the configuration Con_Q (VII)–Con_P(I), as shown in Figure 5.26. For this basic configuration, the parameters characterizing the relative positions of the Q_{ref}- and P_{ref}-frames, are listed in Table 5.7.

To begin with the comparison of the two distance evaluations, it is noted that their difference, $\Delta r = r_{QP} - r_{ref} = 0.05$ nm, is close to the upper limit of their joint error estimates. Therefore, a definite increase of the $Q_A - P_{865}$ distance to occur after the transition from the ground state to the charge-separated state of the RC cannot be concluded. Nevertheless, this result may indicate a small translational displacement of the $Q_A^{\bullet-}$ position (such a displacement is less probable for the donor special pair, which is much larger than the quinone and more confined by the surrounding protein scaffold).

In an attempt to visualize the rotational realignment indicated by the observed differences between the corresponding angles, the EPR-derived dipolar axis is attached to the reference frame P_{ref}. This axis, d_P, is defined by the polar angles (ϕ_P, η_P) and is shown as dotted line in Figure 5.26. This axis was found to be tilted by 7.5° from the reference-pair axis (solid line).

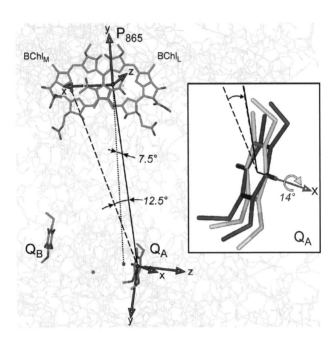

Figure 5.26 Reference structure of the $P_{865}Q_A$ pair overlaid with the $P_{865}^{•+}Q_A^{•-}$ radical pair derived by the pulsed high-field EPR methods of this work. Dotted line: the pair axis in the P_{ref}-frame at the EPR-derived direction (ϕ_P, η_P); dashed line: the pair axis aligned in the Q_{ref}-frame at the EPR-derived direction (ϕ_Q, η_Q); solid line: the reference-pair axis. In the insert the figure visualizes the turning around the x-axis of the quinone that matches the direction of the pair axis, while the lightly colored quinone corresponds to the structure obtained by X-ray crystallography. For details, see ref. [216].

Table 5.7 Comparison of the geometry parameters characterizing the relative positioning of the paired radicals $Q_A^{•-}$ and $P_{865}^{•+}$, as determined by high-field dipolar EPR methods,[216] with those derived for the model pair from the X-ray structures of RCs from *Rb. sphaeroides*.

Structure	r_{QP} /nm	ϕ_P /°	η_P /°	ϕ_Q /°	η_Q /°	ζ_{QP} /°
Dipolar EPR	2.89(2)	82[a]	59[a]	242[a]	110[a]	50[a]
X-ray model	2.84(2)	79[b]	66[b]	246[c]	97[c]	49[b]
Model with turned Q_A	2.84(3)	79[b]	66[b]	245[c]	110[c]	44[b]

[a]Directions of the dipolar axis defined by angles (ϕ_R, η_R) are evaluated with an accuracy of ±2° for deviations in any direction. ζ_{QP} is evaluated with an accuracy of ±7°.
[b]The pair of angles (ϕ_P, η_P) defines the direction of the dipolar axis in the P-frame. For the model pair a general accuracy ±5° is estimated due to uncertainties of the attachment of principal g-axes to the donor special pair. The same accuracy is specified for the turning angle ζ_{QP}.
[c]The pair of angles (ϕ_Q, η_Q) defines the direction of the dipolar axis in the Q-frame. The general deviation accuracy is better than for the P-frame.

Following this deviation, the radical $Q_A^{\bullet-}$ should be shifted from its reference position by about 0.39 nm, which looks unrealistically large. Actually, locating the dipolar axis in the g-frame of $P_{865}^{\bullet+}$ by EPR was performed with an estimated error of 2°. This would allow tilting of the dipolar axis within the error margin and to bring it slightly nearer to the reference-pair axis. The remaining part of the observed realignment may be explained by accounting for an inexact attachment of the g-axes at the donor site. Indeed, within the error limits given in ref. [155], the P_{ref}-frame may deviate from the orientation calculated above by about 5°. This would allow tilting of the P_{ref}-frame together with the vector d_P in a way that matches d_P to the reference-pair axis. Therefore, the observed realignment still lies within the error limit, which is relatively large, as long as the donor site is chosen as the basis for angular triangulation.

The situation is different when the g-axes are assigned more accurately to the quinone acceptor, and a consideration of the dipolar axis in the Q_{ref}-frame leads to more distinct conclusions. This axis, d_Q, is defined by the polar angles (ϕ_Q, η_Q) and is shown as dashed line in Figure 5.26. Again, as above, this axis was found to be tilted from the reference-pair axis, now by the even larger angle of 12.5°. In this case, however, such a considerable deviation is clearly outside the error limits, and can be compensated only by a turning of the quinone together with its precisely attached g-frame. Indeed, if we allow the quinone molecule to leave the orientation derived from the dark-state X-ray data and turn it around the x-axis by about 14°, the dipolar axis turns together with this Q_{ref}-frame and closely approaches the direction of the pair axis. This model pair structure (characterized in the 3rd line of Table 5.7) matches the structure derived in our study (1st line) within certified error margins. In the last step of our comparison, we have to exclude possible errors in the positioning of Q_A of the X-ray structure of choice. For this aim we first examined 31 available X-ray structures of RCs from *Rb. sphaeroides* in terms of η_Q, ϕ_Q and r_{QP}.[216] The Q-frame and the center of the $P_{865}^{\bullet+}$ electron spin density were defined as described above. The polar angles were determined using the director interconnecting the center of the Q-frame and the density center as the dipolar vector. Finally, the distance between these two centers was calculated. The differences between the X-ray structures were found to be small. On average, excluding the X-ray structures of the mutants and the four wild-type structures that show unexpected deviations, the angles $\phi_Q = 247 \pm 2°$ and $\eta_Q = 101 \pm 3°$, as well as the distance $r_{QP} = 2.84 \pm 0.02$ nm were obtained. Thus, the X-ray structure we used for the construction of the reference cofactor configuration lies within the dispersion of the polar angles (ϕ_Q, η_Q), as obtained from various X-ray structures. Therefore, the observed structural change of $Q_A^{\bullet-}$ is outside the error margin of analysis, and a distinct turning of the quinone cofactor $Q_A^{\bullet-}$ around its molecular x-axis occurs as a consequence of the charge-separation process.

This conclusion is supported by earlier 95-GHz pulsed EPR investigations of librational dynamics of $Q_A^{\bullet-}$.[194,217] The analysis of the T_2 relaxation anisotropy of $Q_A^{\bullet-}$ undergoing librational dynamics in its binding site, as observed by

W-band ESE-detected EPR, shows that above 120 K $Q_A^{\bullet-}$ performs rotational fluctuations predominantly around its x-axis. This axis is along the H-bonds between the carbonyl oxygens of the quinone and neighboring histidine (M219) and alanine (M260) amino acids of the Q_A binding site.[194,217] At temperatures around 150 K the rms amplitude of these fluctuations exceeds 20°, while the aperture of the nutation cone, which the libration axis describes around the g-tensor x-axis, was determined to be below 12°.[217] In the present study we conclude that a static distortion in the alignment of $Q_A^{\bullet-}$ is also present, which leads to a distribution of quinone orientations. These observations show that $Q_A^{\bullet-}$ does not experience severe constraints in its protein binding pocket, otherwise they would prohibit such pronounced excursions of the quinone molecule. Hence, the observed conformational change that significantly shifts the equilibrium orientation of Q_A upon light-induced reduction appears reasonable, albeit it had escaped previous detection by other methods. As expected from the X-ray structures of the quinone binding sites, the observed light-induced reorientation of Q_A is rather small compared with the dramatic changes of the Q_B binding site under light excitation.[93] Nevertheless, even small reorientations of Q_A under illumination may be of functional importance for optimizing transmembrane electron-transfer rates: Such reorientations affect the electronic coupling matrix elements between the transient donors and acceptors in the charge-separation and charge-recombination processes of bacterial photosynthesis.

We conclude this digression on the 3D structure of the light-induced transient radical pair $P_{865}^{\bullet+}Q_A^{\bullet-}$ by emphasizing that such information is essential for a detailed understanding of the ET characteristics of charge-recombination processes on the molecular level. The charge-separated radical-pair state represents the initial state for ET recombination, *i.e.* it is one of the working states of the photocycle in the RC, whose slow decay guarantees the high quantum yield of bacterial photosynthesis. RCs from *Rb. sphaeroides* R26 exhibit drastic changes in the recombination kinetics of the charge-separated radical-pair state, $P_{865}^{\bullet+}Q_A^{\bullet-}$, depending on whether the RCs are cooled to cryogenic temperatures in the dark or under continuous illumination.[205] For Q_A light-induced structural changes were suggested to explain the observed difference in charge-recombination kinetics in RCs cooled in the dark and cooled in the charge-separated state. The authors explained the slower recombination kinetics in the charge-separated state in comparison with the ground state by changes in the P_{865}...Q_A donor–acceptor average distance and its distribution. This hypothesis has been tested using distance information obtained from pulsed X-band EPR studies,[249,250] however, without a conclusive result. The X-band experiments were insensitive to the relative *orientation* of the radicals.

W-band EPR and PELDOR had to be applied to obtain the orientation selectivity and spectral resolution of the radical-pair spectra[216,251] necessary to explore the nature of the proposed structural changes. Using deuterated RCs for optimum resolution[251] we addressed the following question: Is the light-induced charge separation accompanied by changes in the orientation and/or in the orientation distribution of the $P_{865}^{\bullet+}Q_A^{\bullet-}$ electron-transfer cofactors? Field-swept ESE-detected W-band EPR was used to monitor the charge separation in the frozen state. From the deuterated RC sample frozen in the dark, no EPR signal was observed prior to illumination. Upon continuous illumination, a reversible radical-pair $P_{865}^{\bullet+}Q_A^{\bullet-}$ EPR signal was formed from 100% of the sample, see Figure 5.27(a). From the sample frozen under continuous illumination, charge separation was also observed in 100% of the RCs (solid line, Figure 5.27(b)). However, only about 30% of RCs shows cyclic ET, whereas 70% of the RCs remained in the $P_{865}^{\bullet+}Q_A^{\bullet-}$ state (stable for days), dotted line in Figure 5.27(b).

Transient W-band EPR with laser-flash illumination was used to monitor the kinetics of charge recombination (see Figure 5.27(c)). The respective rates were

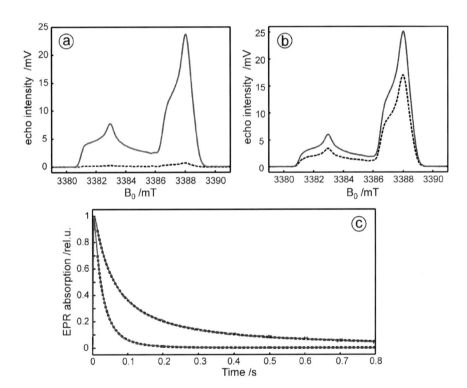

Figure 5.27 Field-swept ESE-detected W-band EPR spectra of $P_{865}^{\bullet+}Q_A^{\bullet-}$ in the perdeuterated Zn-substituted RCs from *Rb. sphaeroides* R26 at 90 K, (a) in RCs frozen in the dark and (b) frozen under continuous illumination. Light-off and light-on spectra are shown by dashed and solid lines, respectively. Samples were illuminated with a cw laser (690 nm). (c) Charge-recombination kinetics ($P_{865}^{\bullet+}Q_A^{\bullet-} \rightarrow P_{865}Q_A$) in RCs frozen in the dark (green line) and frozen under illumination (blue line) are measured by time-resolved W-band EPR with pulsed laser (532 nm) excitation with the corresponding best fits to a power-law function. For details, see ref. [251].

Table 5.8 The average $P_{865}^{\bullet+}Q_A^{\bullet-} \to P_{865}Q_A$ charge-recombination rate constant (k) and its distribution (σ) in RCs frozen in the dark and frozen under illumination (see Figure 5.27(c)) at 90 K. Numbers in parenthesis are the values given by Kleinfeld et al.[205]

Sample	k/s^{-1}	σ/s^{-1}
Frozen in the dark	46.3 ± 4.3 (45.4)	22.7 ± 3.0 (18.4)
Frozen in the light	19.1 ± 1.4 (15.4)	16.7 ± 1.9 (14.3)

determined by fitting the kinetic decays to the commonly used function.[205]

$$S(t) = S_0 + S_1 \cdot (1 + \lambda \cdot \tau)^{-n} \tag{5.17}$$

The fitting parameters λ and n are related to the average charge-recombination rate, $k = n \cdot \lambda$, and to the width of the rate distribution, $\sigma^2 = n \cdot \lambda^2$. The results are shown in Table 5.8. The difference in recombination kinetics observed in this work by EPR is in good agreement with that reported earlier[205] using optical spectroscopy.

Orientation-selective W-band PELDOR spectroscopy[216] was used to monitor the relative orientation of the cofactor ions, the orientational distribution of $Q_A^{\bullet-}$ and the distance between the donor and acceptor ions in the pair $P_{865}^{\bullet+}Q_A^{\bullet-}$ of deuterated RCs frozen in the dark or frozen under illumination.

Figure 5.28 (middle part) shows PELDOR spectra of the spin-correlated radical-pair (CCRP) corresponding to the sample frozen in the dark. The electron–electron dipolar coupling along the direction parallel to the donor–acceptor axis ($v_{||}$) is 4.11 ± 0.02 MHz. It is positioned at a magnetic field ($B_{Q||}$) value of 3383.08 ± 0.02 mT (EPR of $Q_A^{\bullet-}$). The dipolar coupling frequency represents a direct measure of the donor–acceptor distance. The $B_{Q||}$ value yields information about the orientation of $Q_A^{\bullet-}$ with respect to $P_{865}^{\bullet+}$. The broadening of the PELDOR response at $v_{||}$ is a direct measure of angular distribution. The spectral density (dots in Figure 5.28, bottom) was modelled using a pseudo-Voigt function (90% Lorentzian and 10% Gaussian) (dashed line in Figure 5.28, bottom) with a linewidth ($\Delta B_{1/2}$) of 1.00 ± 0.07 mT. This considerably exceeds the intrinsic inhomogeneous $Q_A^{\bullet-}$ EPR linewidth of 0.29 mT. Thus, the geometry of $P_{865}^{\bullet+}Q_A^{\bullet-}$ is characterized by a significant distribution of $Q_A^{\bullet-}$ orientations. Similar results were obtained for $P_{865}^{\bullet+}Q_A^{\bullet-}$ in the recombining fraction (approx. 30%) of the sample frozen under illumination. In this case the following values were obtained from the PELDOR spectra: $v_{||} = 4.10 \pm 0.02$ MHz, $B_{Q||} = 3383.09 \pm 0.03$ mT and $\Delta B_{1/2} = 1.06 \pm 0.09$ mT.

Thus, all the structural parameters of the $P_{865}^{\bullet+}Q_A^{\bullet-}$ radical pair are the same for both RC states! These results indicate the existence of three kinetically different conformations associated with the charge-separated state $P_{865}^{\bullet+}Q_A^{\bullet-}$. Furthermore, they show that the difference in charge-recombination kinetics

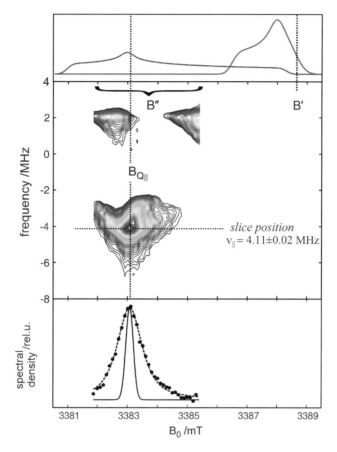

Figure 5.28 W-band PELDOR spectra of the spin-correlated radical pair $P_{865}^{\cdot+}Q_A^{\cdot-}$ in the deuterated Zn-substituted RCs from *Rb. sphaeroides* R26 frozen in the dark at 90 K. The observer microwave (mw) frequency is fixed at the value corresponding to the field B', while the pump mw is swept through a frequency region determined by the field region B''. The middle part shows a contour plot of the positive Fourier intensities of the echo decays. Its amplitude (at the slice position) *vs.* magnetic field (dots) and the best fit to a pseudo-Voigt function (dashed line) are shown at the bottom. The intrinsic EPR linewidth (solid line) is 0.29 mT. For details, see ref. [251].

between the fast and slow conformations, as reported first by Kleinfeld and coworkers[205] is *not* related to changes in the distance or in the relative orientation of the donor and acceptor ions in $P_{865}^{\cdot+}Q_A^{\cdot-}$. Our results[251] rather suggest that the difference in recombination kinetics of $P_{865}^{\cdot+}Q_A^{\cdot-}$ in the two RC states is due to structural changes involving other cofactors of the ET pathway, *e.g.*, the intermediate electron-acceptor bacteriopheophytin (BPhe$_A$), potentially involving structured water molecules in the protein surrounding. Further experiments are needed to clarify this point.

At this point we refer to recent 180-GHz PELDOR experiments on the tyrosyl radicals in R2 from mouse ribonucleotide reductase.[252] The pronounced orientational selectivity was employed to determine the biradical structure. This high-field PELDOR study shows once more that the relative orientation of radical sites of a protein complex can be elucidated. This is an important step towards full structure determination of proteins by EPR methodologies, provided suitable paramagnetic probes are available. For the frequently occurring case of proteins with no paramagnetic states involved in their biological action, the construction of site-directed mutants for selective nitroxide spin labelling has developed to a widely used technique in molecular biology (see Section 5.2.2.1). There is growing interest in doubly spin-labelled proteins and use of high-field RIDME and PELDOR for measuring distance and orientation of two nitroxides placed at well-separated molecular moieties with weak interspin interactions.

We conclude this section by drawing attention to the strategy of combining site-directed spin labelling with natural cofactor ion radicals to generate two-spin systems in bacterial photosynthetic reaction centers.[253,254] So far, X-band EPR techniques were applied, but this strategy is certainly promising also for high-field EPR investigations. The RC of *Rb. sphaeroides* contains five native cysteines necessary for site-directed spin labelling. The EPR experiments in conjunction with molecular dynamics (MD) simulations show that only one cysteine, C156 located on the H protein subunit, is accessible for spin labelling. Using the two-frequency pulsed double-electron–electron spin resonance (DEER) method, a distance of 3.05 nm between the paramagnetic semiquinone anion state of the primary acceptor (Q_A) and the spin label at the native cysteine at position 156 in the H-subunit is found. MD simulations are performed to interpret the distance. The average distance found by the MD simulation is smaller than the distance obtained by DEER by at least 0.2 nm. For a discussion of possible reasons for this discrepancy and for the relatively large width of the distance distribution, see refs. [253,254].

5.2.1.8 Multifrequency EPR on Primary Donor Triplet States in RCs

Most high-field EPR and ENDOR studies at various microwave frequencies have focused on doublet state ($S=1/2$) cofactor ions occuring as paramagnetic intermediates in the electron-transfer cascade initiated by light excitation of the primary donor, P, in the photosynthetic reaction center. This is different from conventional X-band EPR and ENDOR, where the primary donor excited triplet state, $^3P^*$, has attracted much attention, albeit not involved in natural photosynthesis, as a rich source of information about the electronic structure of excited states involved in the photodynamics of the RC (see, for example, refs. [13,15,75,128,255–258]. Pulsed EPR at Q-band (34 GHz) on the triplet state of the primary donor P_{865} in wild-type and mutant RCs from *Rb. sphaeroides* at

10 K have been performed recently[259] to investigate the relative activities of the A- and B-branch for charge separation in the RC protein complex. Only rather few studies on ^{3}P* employing high-field EPR have appeared so far, for example.[260–263] There exist excellent review articles dealing with triplet states in photosynthesis, for instance[166] focusing on the primary donor in bacterial photosynthesis,[137] in photosystem II of oxygenic photosynthesis, and[15] in both bacterial and oxygenic photosynthesis, including photosystems I and II. The most interesting excited state involved in primary photosynthesis is, of course, the short-lived singlet excited state, ^{1}P*, which is not paramagnetic and most of our information about its electronic structure stems from fast laser spectroscopy. Hence, ^{1}P* is not accessible to EPR techniques, but the excited triplet state, ^{3}P*, is. ^{3}P* is formed by radical-pair recombination reaction of the primary donor cation and the bacteriopheophytin anion when the quinone acceptor is blocked by prereduction or even removed. In the triplet state, the unpaired electrons probe the highest occupied molecular orbital (HOMO) *and* the lowest unoccupied molecular orbital (LUMO), which are both singly occupied in ^{3}P*. Hence, the triplet state reflects properties of both the HOMO and the LUMO and, therefore, is very informative about the electronic structure of the excited state.

An interesting question of triplet-state EPR spectroscopy is the extent to which the triplet wavefunction is delocalized over the special pair primary donor in its excited state. This has to be seen in comparison to what has been found for the doublet-state wavefunction of P$^{\bullet +}$, which is delocalized over the dimer halves in an asymmetric fashion (see Section 5.2.1.1). In principle, all interaction parameters of the triplet-state Hamiltonian, *i.e.* the zero-field splitting tensor parameters D and E, the g-tensor and the hyperfine- and quadrupole-tensors, are expected to be sensitive to the details of the electronic and spatial structures and dynamics of ^{3}P* and its interaction with the protein microenvironment. Principal values and orientation of all these interaction tensors are the target of EPR and ENDOR experiments on ^{3}P*. At X-band this full information can be extracted only from single-crystal spectra, as was achieved by J.R. Norris and coworkers[255] in a classic paper comparing the symmetry of the triplet-state electronic structure of the primary donor with its X-ray spatial structure in RCs from the photosynthetic bacteria *Rb. sphaeroides* R26 and *R. viridis*. The orientations of the principal axes of the zero-field splitting tensor, as measured in single-crystal RCs from both organisms, differ significantly from each other. The measured directions of the axes indicate that the triplet excitation is almost completely localized on the L-subunit half of the dimer in *R. viridis*, but is more symmetrically, though not equally, distributed over the L- and M-halves of the dimer in *Rb. sphaeroides* R26. From the sizeable reduction of the zero-field splitting parameters relative to monomeric triplet-state BChl in an organic solvent (the ratio is D(RC)/D(BChl) = 0.83 for *Rb. sphaeroides* R26 and 0.73 for *R. viridis*[264]) it is concluded that asymmetrical charge-transfer configurations significantly participate in the electronic configuration of the special pair triplet state of both organisms, more in *R. viridis* than in *Rb. sphaeroides* R26. Descriptively, this means that the triplet-state wavefunction contains components that place each one of the two electrons on different BChl macrocycles of the conjugated dimer.

An interesting application of 240-GHz high-field EPR on the primary donor triplet state in randomly oriented RCs from *Rb. sphaeroides* R26, embedded in dried plastic films, was reported by Budil and coworkers[263] who studied the temperature dependence of the ^{3}P* g-tensor. Thanks to the high Zeeman resolution of 240-GHz EPR, the measured g-tensor data allow the determination of both the principal values and the principal axes directions with respect to the axes of the zero-field splitting tensor. In contrast to the doublet-state cation-state P$^{\bullet+}$, the triplet-state ^{3}P* exhibits a significant temperature dependence of its g-tensor, particularly in the directions of the principal axes. Over the temperature range studied (10–230 K), the g-tensor principal axes system rotates by about 30° around the x-direction. The only previous observation of temperature dependent principal g-values of ^{3}P* was by X-band EPR.[265] Quantitatively, the data differ considerably from the 240-GHz data. Apparently, the much lower resolution available at 9 GHz greatly limits the reliability of the reported temperature effects. Among discussed reasons for the observed temperature dependence of the g-tensor, the most probable ones include *(i)* temperature-dependent conformational changes of the P molecule,[266,267] *(ii)* temperature-dependent delocalization of triplet excitation onto an accessory bacteriochlorophyll,[265,268,269] and *(iii)* spin-orbit coupling (SOC) between ^{3}P and an electronic state with temperature-dependent energy, such as the first excited state of P. It is suggested that SOC of the acetyl oxygen atom may be quite significant in determining the rotation of the g-tensor in ^{3}P as a function of temperature.

The measurement of the g-tensor of ^{3}P* offers a unique opportunity to compare the g-tensors of two different states of the same molecule. For ^{3}P* the principal g-values, particularly g_{xx} and g_{yy}, are considerably larger than the corresponding values of P$^{\bullet+}$. The g-axis orientations in the two states are also significantly different well beyond experimental uncertainty. The observed differences between the g-tensors of ^{3}P* and P$^{\bullet+}$ may be rationalized in terms of the electronic structures of the two states. Whereas the unpaired spin in P$^{\bullet+}$ occupies only the HOMO of P, the second unpaired electron in ^{3}P* occupies the LUMO; thus the g-tensor of ^{3}P probes the properties of both HOMO and LUMO. Because the LUMO is singly occupied in the radical anion states, one might expect, to a first approximation, the g-tensor of ^{3}P* to reflect the "average" properties of the radical cation and anion states, P$^{\bullet+}$ and P$^{\bullet-}$. This arguing rests on the simplifying assumption that the BChl macrocycles obey, to first approximation, the pairing theorem of alternant hydrocarbons (AHs).[270,271] This theorem predicts pairing of the molecular orbital energy levels of AHs, as calculated within the Hückel molecular orbital (HMO) model. Specifically it predicts *(i)* the bonding and antibonding molecular orbitals (MOs) of an even-AH to occur in pairs with energies $\alpha \pm |m_j| \cdot \beta$, where α is the Coulomb integral for a $2p_z$ electron of carbon, $\beta < 0$ the bond integral, and m_j the MO energy parameter with $m_j > 0$ for bonding MOs, $m_j < 0$ for antibonding MOs and $m_j = 0$ for nonbonding MOs; *(ii)* the coefficients of atomic orbitals (AOs) of the paired

MOs to be numerically the same, but to differ in sign. In other words, according to the pairing theorem, the spin density of the unpaired electron at a specific atomic position in the molecule, defined as the squared AO coefficient of the singly occupied MO at that position, is the same for the cation and anion radical of an even-AH. More than 65 years later, when large-scale DFT computations have become fashionable and the HMO model is hardly known any longer, experience still teaches us that qualitative rationalization of measured electronic structure parameters is often supported by arguments deduced from HMO theory.

That said, it is envisaged that the spin densities of the triplet state MO should be close to the average of the spin densities of the singly occupied HOMO and LUMO of the cation and anion radicals, *i.e.* they should be the same in all three electronic states, at least as long one is dealing with alternant hydrocarbons. Concerning the isotropic g-values of AH radicals: According to A.J. Stone's g-factor theory[164] the isotropic g-values of hydrocarbon radicals should be linearly dependent on the Hückel energy coefficient m_0 of the odd electron MO, *i.e.* the deviation Δg from the free electron g-value ($g_e = 2.002319$) is given by $\Delta g = A + B \cdot m_0$, with the constants $A > 0$, $B < 0$. Both predictions from the pairing theorem, concerning the orbital energies and the coefficients of bonding and antibonding level pairs symmetrically displayed around the nonbonding level $m = 0$, have been experimentally verified for a series of even-AHs by early EPR measurements of the hyperfine couplings and g-factors of the cation and anion radicals in isotropic solution.[272] Hence, for AHs the isotropic g-value for $^3P^*$ is expected to be close to the average of the g-values of $P^{\bullet +}$ and $P^{\bullet -}$, *i.e.* $\Delta g = A$, the g-value for an odd-AH neutral radical with the unpaired electron occupying the nonbonding orbital $m_0 = 0$.

But what about BChl and the special-pair donor P with their conjugated macrocycles containing heteroatoms? In the general case, the pairing theorem does not apply to nonalternant hydrocarbons or alternant conjugated systems containing heteroatoms. However, it was demonstrated[273] that in certain heteroconjugated molecules (both alternant and nonalternant) some HMO energy levels remain paired irrespective of the nature of the heteroatoms. When one compares the isotropic g-factors, $g_{iso} = 1/3 \cdot (g_{xx} + g_{yy} + g_{zz})$ measured for $^3P^*$,[263] and an odd-alternant hydrocarbon radical such as triphenylmethyl (TPM)[272] one observes that the predictions of the pairing theorem are surprisingly well fulfilled ($g_{iso}(^3P^*) = 2.00279$, $g_{iso}(TPM) = 2.00257$). And also the second prediction of Stone's g-factor theory in the HMO approximation of the pairing theorem, $g_{iso}(^3P^*) > g_{iso}(P^{\bullet +})$, is reflected by the high-field EPR measurements $g_{iso}(P^{\bullet +}) = 2.00264$.[155] Advanced semiempirical calculations[45,274,275] suggest that the HOMO and LUMO of P should have quite different symmetries. Thus, it is reasonable to expect the g-tensors of $^3P^*$ and $P^{\bullet +}$ to differ substantially in both their principal values and principal axis orientations.[263]

Another prediction of the pairing theorem, that $g_{iso}(^3P^*) = 1/2 \cdot [g_{iso}(P^{\bullet +}) + g_{iso}(P^{\bullet -})]$ should hold, cannot be tested directly, because the anion state $P^{\bullet -}$ is not a viable state in primary photosynthetic electron transfer. But monomeric $BChl^{\bullet -}$ is, as a transient accessory ("voyeur") BChl acceptor state. Hence, it would be interesting to compare the g-factors of BChl in its doublet states of cation and anion radicals with that of the excited triplet state. Although the hyperfine structure of all three states of BChl has been studied in detail by X-and W-band EPR and ENDOR (for references, see refs. [15,147,257,276], so far there are only very few precision measurements of the g-value of BChl$^{\bullet+}$ by high-field EPR available.[177] Future high-field EPR experiments on the anion radical BChl$^{\bullet-}$ and the triplet-state ^{3}BChl* diluted in diamagnetic hosts should provide the additional g-factor data needed for checking the approximate validity of the pairing theorem for large conjugated macrocycles containing heteroatoms, for which the BChl molecule is a typical representative in nonoxygenic photosynthesis.

5.2.2 Multifrequency EPR on Bacteriorhodopsin (BR)

The strategy to use sunlight energy to synthesize ATP, the multifunctional energy source of life for metabolism and cellular processes including biosynthetic reactions and cell divisions, has been invented by Nature twice: In the photosynthetic reaction center (RC) protein complex of purple bacteria, the ATP synthesis is initiated by light-induced *electron* transfer between bacteriochlorophyll and quinone cofactors. In the bacteriorhodopsin (BR) protein complex of halobacteria, the ATP synthesis is set going by light-initiated *proton* transfer between amino-acid residues, mediated by conformational changes of the only cofactor, the retinal.

BR is an integral membrane protein complex of 26 kDa located in the cell membrane of halophilic archaea such as *Halobacterium salinarium*. High-resolution X-ray crystallography coordinates are available for the ground-state structure[278] (see Figure 5.29(b)). Seven transmembrane helices (A–G) enclose the chromophore retinal that is covalently attached to the amino acid lysine, K216 on helix G, via a protonated Schiff base. Absorption of 570-nm photons initiates the all-*trans* to 13-*cis* photoisomerization of the retinal. The Schiff base then releases a proton to the extracellular medium and is subsequently reprotonated from the cytoplasm. Transient intermediates of this catalytic photocycle can be distinguished by the different absorption properties of the retinal, and a sequence of intermediates, J, K, L, M, N and O, has been characterized by time-resolved absorption spectroscopy.[279] Double-flash experiments revealed that the M intermediate is divided into two substrates, M_1 and M_2.[280,281] During this photocycle conformational changes of the protein (and the retinal) occur, as has been detected by a variety of experimental techniques (for a review, see, *e.g.*, ref. [282]). Such changes ensure that release and uptake of protons do not occur from the same side of the membrane, but rather enable BR to work as a vectorial transmembrane proton pump. To this end, in wild-type BR conformational changes associated with the M_1 to M_2 transition are suggested to function as a "reprotonation switch" required for the vectorial proton transport. It is believed that during the lifetime of the M state the accessibility of the Schiff base for protons is switched from the extracellular to the cytoplasmic side of the membrane. Detailed analyzes of the nature of the

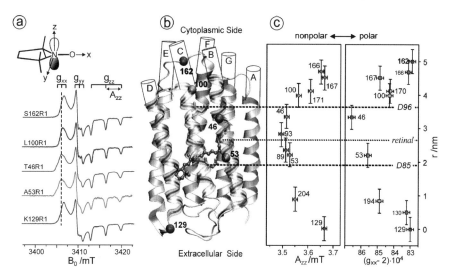

Figure 5.29 (a) Experimental W-band cw EPR spectra for a set of bacteriorhodopsin (BR) mutants spin labelled with the MTSSL nitroxide side chain (R1). (b) Structural model of BR. The C_α atom of the spin-labelled residues, seven α-helices A to G, the chromophore retinal and D96 and D85 participating in the H^+ transfer are indicated. (c) The magnitude of the tensor elements A_{zz} and g_{xx} of the spin labels are plotted as function of the nitroxide location in the protein with respect to position 129. For details, see ref. [277].

conformational changes include neutron diffraction, electron microscopy, X-ray diffraction, solid-state NMR or EPR spectroscopy. The studies reveal major changes to occur at the cytoplasmic moieties of helices F and G. In wild-type BR such helix movements have been shown to provide an "opening" of the protein to protons on the cytoplasmic end of the transmembrane proton channel.[283] Thereby, proton transfer can occur from the internal aspartic-acid proton donor, D96, to the Schiff base during the M to N transition. The reprotonation of D96 from the cytoplasm occurs during the recovery of the BR initial state. A detailed inspection of the structure of the unilluminated state of the protein reveals that certain amino-acid side chains block the proton pathway from the cytoplasm to D96.[283] The region between D96 and the Schiff base is largely nonpolar, packed with bulky amino-acid residues. Hence, in this unilluminated state the Schiff base is effectively inaccessible to protons from the cytoplasm. In the light-driven M_1 to M_2 transition, this region is opened for access of protons to the Schiff-base nitrogen atom. However, the question remains whether the relatively large conformational changes observed in the photocycle of the wild-type BR and many BR mutants are a prerequisite for vectorial proton transport, *i.e.* if they really represent the proposed reprotonation switch.

Obviously, additional spectroscopic experiments are needed on wild-type BR and strategically constructed BR mutants to follow even small, but significant conformational changes of protein and cofactor during the photocycle. In this respect, the combination of site-directed mutagenesis for spin labelling and high-field EPR for resolving structural changes of proteins at work is particularly powerful.

5.2.2.1 Site-Directed Nitroxide Spin Labelling

During the photocycle of BR no paramagnetic intermediates occur, *i.e.* neither radicals, radical pairs nor triplet states. To enable the application of EPR, doublet-state spin labels ($S = 1/2$) have to be introduced to the protein (in contrast to NMR that probes $S = 0$ (singlet states)). The site-directed spin-labelling (SDSL) technique in combination with X-band EPR spectroscopy was pioneered by W.L. Hubbell[125,284–287] (UC Los Angeles). The recent extension to SDSL/high-field EPR has opened new perspectives for studying structure and dynamics of large nitroxide-labelled proteins during biological action.[11,12,34,129,277,288–291] Another biologically important application of multifrequency EPR, particularly at high fields, is the investigation of the complex modes of motion of spin labels in membranes[292,293] and their orientation and alignment relative to the magnetic field.[294] Figure 5.30 demonstrates the remarkable gain in resolution of nitroxide-radical spectra, *i.e.* the separation of the g_{xx}, g_{yy}, g_{zz} components in relation to the A_{zz} hyperfine-tensor component, when increasing the Zeeman field from X-band EPR to 95-GHz and 360-GHz EPR. It has been shown[34,277,289,295–300] that the isotropic and anisotropic components of the g- and hyperfine-tensors of the nitroxide spin label can be used as sensitive probes for the polarity and proticity properties of the

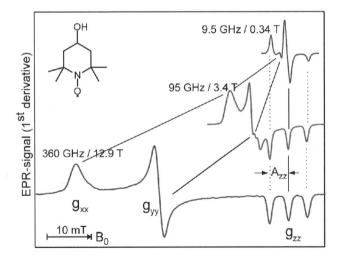

Figure 5.30 First-derivative cw EPR spectra of a nitroxide radical (OH-TEMPO) in frozen water solution at different microwave frequency/B_0 settings. The spectra are plotted relative to the fixed g_{zz} value. For details, see ref. [14].

immediate environment of the reporter group. This is also true for the principal components of the ^{14}N quadrupole-tensor, see Section 2.2.1.4.

The SDSL technique requires selective cysteine-substitution mutagenesis of the protein with subsequent modification of the unique sulfhydryl group of cysteine with a nitroxide reagent, for example (1-oxyl-2,2,5,5-tetramethylpyrroline-3-methyl)methanethiosulfonate, commonly abbreviated as an MTS spin label. For systematic studies, a set of SDSL mutants is constructed, each containing a single nitroxide-containing amino-acid side chain, differing by position in the protein sequence. It is self-evident that the photocycle of all spin-labelled mutants has to be checked to ensure that the overall function of the protein is retained.[34,277,288] This was approved for all spin-labelled BR mutants discussed in this chapter.

In conclusion, it is emphasized that, owing to the achievements of molecular biology in site-directed mutagenesis techniques, SDSL of one or two specific positions has gained an enormous impact in protein research, with EPR as the most specific tool for measuring structure and dynamics of protein domains. Moreover, the accessibility of the spin-labelled sites to collisions with paramagnetic reagents in solution, *e.g.*, polar or apolar spin relaxers such as hydrophilic chromium(III)oxalate or hydrophobic O_2 molecules, respectively, can be used to determine the protein topography, *i.e.* its secondary structure as well as its orientation.[301] This is achieved by measuring, for a series of spin-labelled point mutants, the different mw power-saturation behavior of the nitroxide EPR spectrum with and without the polar or apolar spin relaxers. It is important to point out that the EPR data of this type of study have contributed to the general notion of SDSL investigations that the perturbation by introducing a nitroxide spin label is sufficiently small to avoid disruption of the protein structure at the level of the backbone fold.

5.2.2.2 *Hydrophobic Barrier of the BR Proton-Transfer Channel*

By 95-GHz (W-band) high-field EPR details of the polarity profile along the putative proton channel were probed by *g*- and hyperfine-tensor components from a series of 10 site-specifically nitroxide spin-labelled BR mutants, with MTS spin label as the reporter side chain R1.[277] Previous studies of a large number of spin-labelled proteins have shown that the A_{zz} component of the hyperfine-tensor and the g_{xx} component of the *g*-tensor are particularly sensitive probes of the microenvironment of the nitroxide side chain R1. They allow measurement of changes in polarity and proticity, *i.e.* g_{xx} and A_{zz} probe the local electric fields and the availability of H-bond forming partners of nearby amino-acid residues or water molecules.[277,288,289] Moreover, the dynamic properties of the nitroxide side chain and, thus, the EPR spectral lineshape have been shown to contain direct information about constraints to motion that are introduced by the secondary and tertiary structures of the protein in the vicinity of the nitroxide binding site.[277,288]

For measuring the polarity changes, W-band EPR spectra were recorded at temperatures below 200 K to avoid motional averaging of the anisotropic magnetic tensors. At these temperatures, R1 can be considered as immobilized on the EPR time scale. The spectra of selected mutants are shown in Figure 5.29(a). They exhibit the typical nitroxide powder-pattern lineshape expected for an isotropic distribution of diluted radicals. The spectra are clearly resolved into three separate regions corresponding to the components g_{xx}, g_{yy} and g_{zz}, the latter with resolved A_{zz} splitting. The variations of g_{xx} and A_{zz} with the nitroxide binding site can be measured with high precision. The plots of g_{xx} and A_{zz} vs. R1 position r along the proton channel (Figure 5.29(b)) demonstrate distinct variations in the polarity and proticity of the nitroxide microenvironment. According to the structure model,[302] residue S162R1 is located in the E–F loop at the cytoplasmic surface, whereas residue K129R1 is positioned in the D–E loop on the extracellular surface. The high polarity in the environment of these residues is clear evidence that the nitroxides are accessible to water, which is in agreement with the structure. The environmental polarity of the nitroxide at positions 100, 167 and 170 is significantly smaller and reaches its minimum at position 46 between the proton donor D96 and the retinal. The plots directly reflect the hydrophobic barrier that the proton has to overcome on its way though the protein channel.

We point out that there exist situations when the polarity or proticity differences between selected protein sites are so small that even high-field EPR at 95 GHz cannot resolve the correspondingly small differences in their g_{xx} components. For such situations it has been demonstrated that by EPR at considerably higher microwave frequencies the g_{xx} differences can be resolved, for example by 360-GHz EPR[303,304] in the case of the spin label in the BR mutant V167C in comparison to the free MTS spin label at the membrane surface, or by 275-GHz EPR[297] for the spin labels attached to the surface sites 12, 27, 42 and 118 of azurin mutants.

The analysis of both tensor components, g_{xx} and A_{zz}, allows characterization of the R1 environment in terms of *protic* and *aprotic* surroundings. Theoretically, both g_{xx} and A_{zz} are expected to be linearly dependent on the π-spin density ρ_π^O at the oxygen atom of the nitroxide group (see below). For g_{xx}, however, apart from a direct proportionality to ρ_π^O, there is an additional dependence on specific properties of the oxygen lone-pair orbitals. The lone-pair orbital energy E_n affects g_{xx} via the excitation energy $\Delta E_{n\pi^*} = E_{\pi^*} - E_n$ and is known to be sensitive to the polarity of the environment. It is particularly sensitive to H-bonding of the lone pairs to water or to polar amino-acid residues. Thus, the plot of g_{xx} vs. A_{zz} should indicate the presence or absence of H-bonds in the spin-label environment, *i.e.* its proticity (see below). This dependence is plotted in Figure 5.31 for various spin-label positions in BR.[277] Obviously, two straight-line correlations can be deduced. The points corresponding to positions 46, 171 and 167 belong to a line whose slope is different from that for the remaining points. These three positions can be classified to be exposed to an aprotic environment,[299,300] the other ones to a protic environment. This allows the hydrophobic barrier of the BR proton channel to be characterized in terms of different accessibilities of the respective protein regions to water molecules.

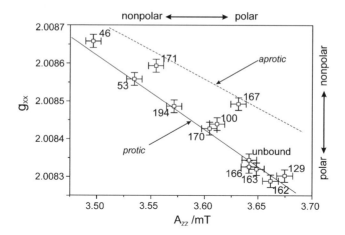

Figure 5.31 Experimental plot of g_{xx} vs. A_{zz} of the nitroxide side chains for various spin-label positions in bacteriorhodopsin (see text). The "protic" and "aprotic" limiting cases are placed with reference to theory, see ref. [289] and Figure 5.32.

Another point has to be mentioned that is of particular interest for answering the question whether a protein domain has a regular secondary structure of α-helical or β-sheet character. From a periodical polarity profile along a specific sequence of spin-labelled positions one can infer the torsion angles of the polypeptide backbone to be consistent with either a α-helix or a β-sheet. This strategy was recently used in a W-band high-field EPR study of the phototransducer NpSRII in complex with sensory rhodopsin, NpSRII.[305] The observed periodicity of the polarity properties of the respective spin-label microenvironment agrees with an α-helical secondary structure of the chosen section of the protein complex.

A theoretical interpretation of the observed g_{xx} vs. A_{zz} dependence in site-directed nitroxide spin labelled BR was attempted on the basis of semiempirical MO theory and by using simple models for varying solute–solvent interactions.[289] The ultimate goal was to extract a measure for the different magnitudes of polar and protic contributions from the spin-label environment to g_{xx} and A_{zz} and to find a way to discriminate between these two effects. The calculation of the g-tensor components g_{ii}, $i = x, y, z$, is based on the MO treatment of Stone[164,306] for organic radicals. The evaluation of the resulting formulae requires the input of the ground and excited states, molecular wavefunctions in their linear-combination-of-atomic-orbitals (LCAO) representation, and their respective state energies in addition to the spin-orbit coupling parameters of the various contributing atoms. All the required MO quantities were computed at the ZINDO/S[307] level from restricted Hartree–Fock (RHF) ground-state MOs. The required spin-orbit coupling parameters were taken from the literature.[192] The variations of g_{xx} due to solute–solvent interactions can be understood qualitatively by analyzing the dominant constituents of this component according to the theory of Stone.[164,306] These are approximately given by

$$g_{xx} = g_e + \frac{2 \cdot \zeta(O) \cdot \rho_\pi^O \cdot c_{ny}^2}{\Delta E_{n\pi}} \qquad (5.18)$$

where $g_e = 2.0023$ is the free-electron g-value; $\zeta(O)$ is the oxygen spin-orbit coupling parameter; ρ_π^O is the π-spin density on the oxygen $2p_z$ atomic orbital; c_{ny} is the LCAO coefficient of the $2p_y$ atomic orbital contributing to the oxygen lone-pair MO n; and $\Delta E_{n\pi^*}$ is the $n \to \pi^*$ excitation energy.

Solute–solvent interactions can affect on all three quantities ρ_π^O, c_{ny}^2, $E_{n\pi^*}$, on the first two of these by electron-density rearrangements and on the latter by energy level shifts either by superimposed electric fields or by H-bond formation. Thus, any change in g_{xx} can be broken down into three contributions:

$$\frac{\delta \Delta g_{xx}}{\Delta g_{xx}} = -\frac{\rho_\pi^N}{\rho_\pi^O} \cdot \frac{\delta A_{zz}}{A_{zz}} - \frac{\delta \Delta E_{n\pi}}{\Delta E_{n\pi}} + \frac{\delta c_{ny}^2}{c_{ny}^2} \qquad (5.19)$$

where $\Delta g_{xx} = g_{xx} - g_e$ and where $\delta \rho_\pi^O$ has been substituted by the corresponding change in A_{zz}, making use of the sum condition $\rho_\pi^O + \rho_\pi^N \cong 1$ and of the proportionality between A_{zz} and ρ_π^N, the latter standing for the π-spin density at the nitrogen atom (see below). Equation (5.19) correctly predicts the observed negative slope of g_{xx} vs. A_{zz}. The last two terms are responsible for additional vertical deviations. The nitrogen hyperfine-tensor component A_{zz} is given by[289]

$$A_{zz} = Q_\pi^N \cdot \rho_\pi^N \qquad (5.20)$$

for a strictly planar structure. The spin density ρ_π^N is supplied directly by the calculated ZINDO/S MOs. The value of $Q_\pi^N = 7.3$ mT was adjusted to the measured value of $A_{zz} = 3.36$ mT observed for the MTS spin label in a nonpolar environment where $\rho_\pi^N = 0.46$ ($\rho_\pi^O \cong 0.54$).

5.2.2.3 Modelling of Solute–Solvent Interactions

Polarity effects from the various intermolecular electric fields originating in dispersion forces, permanent electric dipole interactions, induced dipole interactions, *etc.*, are most conveniently described by a single collective parameter, the average local electric field **E** in the NO bond region. This approach was first taken by Griffith and coworkers.[299] For example, these authors show that the π spin density at the nitrogen atom, ρ_π^N, changes to first order as

$$\delta \rho_\pi^N = C_1 \cdot E_x \qquad (5.21)$$

where E_x is the electric-field component along the NO bond and $C_1 \cong 2 \times 10^{-9}$ V^{-1} cm. Regarding the observed maximum change of A_{zz} of 0.4 mT and applying eqn (5.20), eqn (5.21) establishes an effective "polarity range" $0 < E_x \leq 2 \times 10^7$ V/cm for the MTS spin label at the different sites in BR. In our calculations we thus imposed a uniform electric field E_x in the nitroxide spin label model system as effective polarity parameter, applying values of E_x within the range defined above. This was achieved by using the electric-field setup options within the molecular modelling package Hyperchem (Hypercube Inc., Gainesville FL, USA).

The effects of H-bonding (hb) on the g_{xx} vs. A_{zz} dependence have been studied theoretically on a nitroxide model structure (see inset of Figure 5.32) with one water H-bonded to the O atom. Our approach was similar to that of a theoretical study on tyrosyl radicals.[308] For the quantum-chemical method (ZINDO/S) used for calculating the various quantities determining g_{xx} and A_{zz}, and for further details, see ref. [289].

Figure 5.32 shows the calculated g_{xx} vs. A_{zz} dependence for this nitroxide spin-label model. The upper dashed line, defined as "aprotic", is based on the calculated g_{xx} values *without* hydrogen-bond (hb) formation, the lower dashed line *with* hb formation. The lower line is shifted against the aprotic line by a constant value of -4×10^{-4} over the whole range of E_x values between 0 and 0.02 a.u. (1 a.u. = 5×10^9 V cm^{-1}). The solid line, defined as "protic" in analogy with earlier work,[300] is obtained by linear interpolation between points A and B, which are characterized by the nonpolar/aprotic limit with the experimental value $A_{zz} = 3.36$ mT, and by the highly polar/protic situation in water for which a value $A_{zz} = 3.64$ mT was measured.[289] Both limiting values of A_{zz} could be theoretically reproduced,[289] and the molecular parameters obtained from these

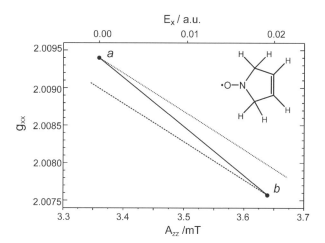

Figure 5.32 Theoretical plot of g_{xx} vs. A_{zz} for a planar nitroxide model radical as calculated for the "aprotic" case (without hydrogen bonding, short dashes) and for the case with additional hydrogen bonding (long dashes). The "protic" line (solid line) is obtained by linear interpolation between the nonpolar limit (point *a*, $A_{zz} = 3.36$ mT) and the water limit (point *b*, $A_{zz} = 3.64$ mT). For details, see ref. [289].

calculations were, in turn, used in the calculations of the limiting values of g_{xx} to provide points a and b in Figure 5.32.

Any point between the two limiting lines with and without H-bonding can be assigned a fractional H-bonding parameter q_{prot} between 0 (point a) and 100 % (point b) to serve as a measure of the protic interaction (proticity). In terms of the g_{xx} values at a particular polarity E_x (*i.e.* for $A_{zz}=\text{const.}$) we thus define

$$q_{\text{prot}} = \frac{g_{xx} - g_{xx}^{\text{aprotic}}}{\Delta g_{xx}^{\text{hb}}} \quad (5.22)$$

where $\Delta g_{xx}^{\text{hb}}$ is assigned an average value of $-(4\pm 1)\times 10^{-4}$.

The calculated average slope $|dg_{xx}/dE_x| \cong 1.5 \times 10^{-11}\,\text{cm V}^{-1}$ is close to the corresponding experimental value $(2.0\pm 0.3)\times 10^{-11}\,\text{cm V}^{-1}$ found by Gullá and Budil.[309] This value was also ascertained theoretically in the same laboratory by a subsequent *ab-initio* calculation[310] on the structurally similar nitroxide radical 2,2,5,5-tetramethyl-3,4-dehydropyrrolidine-1-oxyl (TMDP).

For comparison, experimental results for spin-labelled BR are given in Figure 5.31. It shows the g_{xx} vs. A_{zz} dependence for various spin-label positions in BR.[277] Obviously, two straight-line correlations can be deduced, in accordance with the theoretical predictions. The points corresponding to positions 46, 171 and 167 belong to a line whose slope is different from that for the remaining points. These three positions can be classified to be exposed to an aprotic environment,[299,300] the others to a protic environment. Admittedly, there remains a discrepancy between calculated and experimental slopes of the g_{xx} vs. A_{zz} plots.[289] As in earlier studies,[311] it is anticipated that sterically induced strains on the attached nitroxide spin label may influence the slopes dg_{xx}/dA_{zz} and explain the discrepancy.

Hence, structural changes of the nitroxide label by superimposed local electric fields were theoretically analyzed as a possible reason for the striking differences between the calculated and measured slopes.[289] Indeed, the analysis shows that such local electric fields exert "polarity-induced steric strain" on the nitroxide spin label, resulting in deviations from planarity of the nitroxide atomic skeleton. They, in turn, reduce the calculated slope dg_{xx}/dA_{zz}, via combined action of energy and spin-density changes, towards the experimental slope value (for details of the calculations, see ref. [289]).

To summarize this section:

By means of 10 site-specifically spin-labelled BR mutants with the NO˙ side chain located in the cytoplasmic loop region and in the protein interior along the putative proton channel, the polarity gradient, *i.e.* the hydrophobic barrier for proton migration, could be determined from g_{xx} and A_{zz} shifts. Moreover, from a plot of g_{xx} vs. A_{zz}, protic and aprotic regions of the ion channel could be revealed. Much of the detailed information about polarity and proticity of protein sites in the vicinity of NO side chains can be deduced from the different slopes of the linear g_{xx} vs. A_{zz} plots. To put these observations on a theoretical footing, we have calculated polarity and H-bonding effects on the g- and hyperfine-tensors of an NO spin-label model by semiempirical molecular-orbital as well as by DFT methods. The results of this analysis suggests that polarity effects can be described by an electric field oriented along the NO bond, whereas H-bonding effects can be reproduced by an energy-minimized NO/H$_2$O-pair model.

We make the point that the experimental g_{xx} vs. A_{zz} plots enables the hydrophobic barrier of the BR proton channel to be characterized in terms of different accessibilities of the respective protein regions to water molecules. No wonder that the discussed sensitivity of the resolved g_{xx} and A_{zz} tensor components of nitroxide radicals to the polarity of their microenvironment was recognized also by other high-field EPR research groups. For instance, more than 20 years ago it was utilized for characterizing a large variety of nitroxide radicals in different frozen solvents[312] or, more recently, it was applied for investigating spin-labelled phospholipid membranes.[11,311,313] Up to 2001, the subject "spin-labelling in high-field EPR" was rigorously reviewed by A.I. Smirnov.[11] Here, we allude to a particularly detailed study along these lines: The influence of solvent polarity and hydrogen bonding on the g- and ^{14}N hyperfine-tensor components of the nitroxide spin label MTS was investigated by X- and W-band EPR spectroscopy in combination with DFT calculations.[296] Again, a linear correlation was found for the isotropic (g_{iso}, A_{iso}) and anisotropic (g_{xx}, A_{zz}) parameters. DFT calculations of these parameters were performed for a model spin label with varying dielectric constants (ε) of the medium and different number of hydrogen bonds formed with the nitroxide oxygen. In the "apolar region" ($\varepsilon < 25$), the sensitivity of A_{iso} and A_{zz} to ε is large. However, in the "polar region" ($\varepsilon > 25$), the sensitivity to ε is small, and the shifts in A_{iso} and A_{zz} are mainly determined by the proticity of the solvent. Methanol was found to form, on the average, one hydrogen bond, whereas water forms two hydrogen bonds to the nitroxide oxygen.

5.2.2.4 Conformational changes during the BR photocycle

From "dark minus light" W-band EPR on side-directed BR mutants conformational changes of specific helices in the M-intermediate state of the photocycle could be detected.[277,288] These results provide important contributions to a better understanding of the mechanisms for vectorial ion transfer in membrane proteins invoking, for example, specific helix movement as "molecular switches".

The discussion about conformational changes of distinct helices in BR, to activate a molecular switch for vectorial proton transfer, has been revived recently by the investigation of the tailor-made triple mutant D96G/F171C/F219L that modifies the cytoplasmic proton entrance region, see Figure 5.33(a). This BR triple mutant reveals a remarkable conformation of its dark state: It was shown to resemble that of the late M intermediate (preceding N in the photocycle) in wild-type BR, but with a conformation that is retained upon illumination (*pseudo* M state).[283,288,315,316] In our work, the triple mutant was

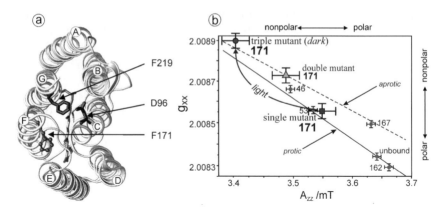

Figure 5.33 (a) Top view of the X-ray structural model of bacteriorhodopsin (BR).[283] The seven α-helices (A–G) and the interconnecting loops are visible as well as positions D96, F171 and F219 in their nonmutated forms are shown. (b) Polarity plot of g_{xx} vs. A_{zz} of the nitroxide side chains R1 for various spin-label positions in BR mutants with emphasis on the R1 position 171 in the single mutant F171R1 (square), the double mutant D96G/F171R1 (triangle) and the triple mutant D96G/F171R1/F219L (circle). The dark state of the triple mutant reveals a very nonpolar protein microenvironment of R1 in the vicinity of the cytoplasmic entrance of the proton channel. Light excitation of the single mutant switches the microenvironment of R1 to the same nonpolar properties as experienced by the triple mutant in the dark. For details, see refs. [288,314].

MTS labelled and studied by W-band high-field EPR without and with light irradiation.[288] The goal was to test the sensitivity of selectively spin-labelled helix segments in singly, doubly and triply mutated BR towards changes of the microenvironment in the cytoplasmic region during the photocycle. In these studies, we chose position 171 at the cytoplasmic end of helix F, see Figure 5.33(a), in the single mutant F171C, in the double mutant D96G/F171C, and in the triple mutant D96G/F171C/F219L for attaching an MTS spin label to the unique cysteine, thus forming the side chain R1. The R1 probe is used to measure the polarity changes in this region via light-induced shifts of the g_{xx} and A_{zz} tensor components. The results are depicted in Figure 5.33(b). It convincingly shows that upon light excitation of the single mutant to its M state, the NO• residue at position 171 experiences the same nonpolar microenvironment as the triple mutant with its *pseudo* M state open for proton uptake already in the dark.[288]

Pronounced conformational changes of wild-type BR during the photocycle were also observed in the high-field EPR spectra of selectively spin-labelled double mutants D96N/V167R1 and D96N/V101R1 in their ground states recorded in the dark and under light illumination. The dark-minus-light spectra clearly show an *increased* reorientation mobility of the nitroxide side chain R1 in the M intermediate of the D96N/V161R1 mutant, but a *decreased* mobility for D96N/V101R1. Obviously, the residual anisotropy of the nitroxide motion changes during the photocycle owing to changing space restrictions within the R1 binding site. Label V167R1 is located at the cytoplasmic moiety of helix F and oriented towards helix C. Thus, an increase of the interhelical distance, caused by displacement of helix F or helix C, would account for the experimental data – in accordance with neutron-diffraction and FT-IR results and with EPR interspin distance measurements of doubly labelled BR variants[314] (for details, see ref. [277]).

At the end of this section, we want to emphasize that multifrequency EPR on nitroxide spin-labelled biopolymers, such as proteins or phospholipid membranes, has proven to be extremely useful not only in determining molecular structure, but also in elucidating molecular motion over wide time scales, for example slow tumbling motion of the whole protein or fast internal dynamics of cofactors in their binding sites. For the study of molecular dynamics, high-frequency/high-field EPR extends the "snapshot" window for sensitive detection of the motional modes of nitroxide spin labels to the order of 1 ps.[130,317–320] Pulsed EPR techniques, such as ESE-detected EPR, are particularly sensitive to anisotropic changes of the transverse relaxation time T_2 due to anisotropic motion. They are blind to static linewidth contributions like inhomogeneous broadening due to unresolved hyperfine interactions. In a combined effort of four laboratories, the molecular dynamics of nitroxides in glasses was studied at resonance frequencies of 3, 9.5, 95 and 180 GHz and corresponding magnetic-field settings.[202] The investigation focused on orientational motion of the nitroxide spin probe (Fremy's salt) in glycerol glass near the glass transition temperature. By measuring echo-detected multifrequency EPR spectra at different pulse separation times, different relaxation mechanisms could be discriminated and characterized by the correlation times of molecular reorientations (10^{-7}–10^{-10} s). Near the glass transition temperature, the orientation-dependent transverse relaxation T_2 is dominated by fast reorientational fluctuations overlapping with fast modulations of the canonical values of the g-tensor. The data was interpreted by using a new spectra simulation program for the orientation-dependent relaxation rate $1/T_2$ of nitroxides that is based on different models for the molecular motion. The validity of these models was assessed by comparing least-squares fits of the simulated relaxation behavior to the experimental data.

5.3 Oxygenic Photosynthesis

Our present understanding of the structure–function relationship for the primary processes in oxygenic photosynthesis of green plants is still considerably less than in bacterial photosynthesis owing to the much higher complexity of the charge-transfer photomachine of the interconnected photosystems I and II (PS I, PS II). Moreover, until recently the three-dimensional structures of PS I and PS II were not known with sufficient accuracy. This situation was in contrast to that of RCs from purple photosynthetic bacteria, but has dramatically improved now thanks to the recent progress in high-resolution X-ray crystallography of PS I and PS II (see Section 5.1). Detailed reviews on doublet-state radicals occurring in PS I and PS II during light-induced electron-transfer reactions as characterized by EPR in general[15] and high-field EPR in particular[17] have been presented. The authors provide precise g-tensor data of tyrosine-, quinone-, pheophytin- and chlorophyll-based radicals from which valuable information about radical-protein interactions can be derived. Moreover, EPR results concerning spin pairs in PS I and PS II are thoroughly reviewed.

5.3.1 Multifrequency EPR on Doublet States in Photosystem I (PS I)

Taking PS I as an example, the 2001 X-ray structure has reached 2.5 Å resolution,[78] while until then only 4 Å resolution could be achieved,[321] and it was impossible to even locate the quinone acceptor A_1, a phylloquinone. It was, therefore, a challenge to apply the whole arsenal of modern multifrequency time-resolved EPR and ENDOR techniques to PS I with the goal to determine the $A_1^{\bullet-}$ location and orientation (for overviews, see refs. [15,137,322–324]). For instance, time-resolved EPR studies, either with direct detection or with ESE detection, at X-, K- and W-band frequency/field settings have been performed on the transient spin-correlated radical pair $P_{700}^{\bullet+}A_1^{\bullet-}$ of PS I, and the results were compared with those of $P_{865}^{\bullet+}Q_A^{\bullet-}$ in RCs from *Rb. sphaeroides*.[325] The spin-polarized EPR spectra of the radical pair $P_{700}^{\bullet+}A_1^{\bullet-}$ in highly purified PS I particles are analyzed to obtain both the magnetic parameters of the radical pair and the relative orientation of the two radical species. From the analysis, the g-tensor of $A_1^{\bullet-}$ is determined to be $g_{xx} = 2.0062$, $g_{yy} = 2.0051$, and $g_{zz} = 2.0022$. It is shown that $A_1^{\bullet-}$ is oriented such that the carbonyl bonds are parallel to the vector joining the centers of $P_{700}^{\bullet+}$ and $A_1^{\bullet-}$. The anisotropy of the g-tensor is considerably larger than that obtained for chemically reduced phylloquinone in frozen 2-propanol solution. Reasons for this difference are specific cofactor–protein interactions in the $A_1^{\bullet-}$ binding site. The relative orientation of $P_{700}^{\bullet+}$ and $A_1^{\bullet-}$ is compared with that measured earlier, by W-band ESE-detected EPR, for the radical pair $P_{865}^{\bullet+}Q_A^{\bullet-}$ in Zn-substituted bacterial RCs from *Rb. sphaeroides* R26 in which the nonheme iron has been replaced by zinc,[159] see Figure 5.34. This allows the structural and magnetic properties of the charge-separated state in the two systems to be compared. From the similarity of the two W-band spectra in the region around the free electron g-value it is clear that the dipolar vector, $\mathbf{r}_{AP}$, between $P^{\bullet+}$ and $Q_A^{\bullet-}/A_1^{\bullet-}$ has a similar orientation relative to P_{700} in PS I and P_{865} in bacterial RCs. This is compatible with the similar overall structural arrangement of the primary donors, P_{700} and P_{865}. In contrast, the low-field parts of the two W-band spectra (see Figure 5.34) are very different as a result of differences in the orientation of A_1 and Q_A with respect to $\mathbf{r}_{AP}$ as a consequence of their different binding sites. This finding has been fully confirmed by the later 2.5 Å high-resolution X-ray structure of PS I.[78]

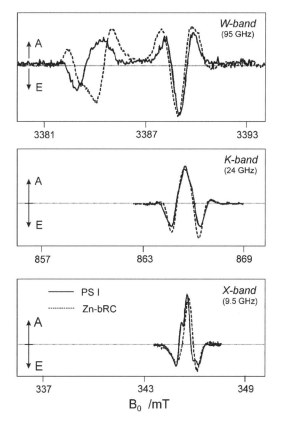

Figure 5.34 Comparison of spin-polarized EPR spectra of $P_{700}^{\bullet+}A_1^{\bullet-}$ in PSI and $P_{865}^{\bullet+}Q_A^{\bullet-}$ in Zn-bRCs at three different microwave frequencies. For details, see ref. [159].

Additional variations on the PS I theme have been published with special emphasis on the advantages of high-field EPR: W-band EPR was used to resolve the g-tensor components of the primary chlorophyll donor cation $P_{700}^{\bullet+}$ and quinone acceptor $A_1^{\bullet-}$ in frozen-solution and single-crystal preparations of PS I.[324] The measured principal values of the g-tensor of $P_{700}^{\bullet+}$ were compared with those of the isolated pigment radicals in organic solvents. From the observed differences information about the cofactor–protein interactions in the binding sites was obtained. The measured g-tensor principal axes of $P_{700}^{\bullet+}$ were assigned to the molecular structure of the primary donor by means of an analysis of the spin-polarized spectra of the photoinduced radical pair $P_{700}^{\bullet+}A_1^{\bullet-}$. DFT calculations on a structural model of the A_1 binding pocket, as derived from the 2.5-Å X-ray structure,[78] were used to correlate the EPR parameters with structural elements of the protein. Another detailed g-tensor analysis has been made for the cation radicals and triplet states of the primary donor P_{700} in PS I and chlorophyll *a in vitro*.[326]

To summarize: Combining all the pieces of information from the various state-of-the-art EPR and ENDOR experiments, finally the goal of localizing A_1 in PS I was achieved in times before crystallization ("BC times"). It is, of course, satisfying that the results obtained by EPR are in agreement with the structures appearing from high-resolution X-ray crystallography.[78] Hence, multifrequency time-resolved EPR proved again to be particularly suited for the characterization of the radical and radical-pair intermediates in the reaction centers of different photosynthetic organisms, both wild-type and mutants. Aspects of mutation-induced changes of the PS I complex and their characterization motivated a number of pulsed high-field EPR and ENDOR studies.[327,328] For example, specific biosynthetic pathways were developed to recruit foreign quinones into the A_1 site of PS I. From multifrequency pulsed EPR and ENDOR experiments the structural and dynamic characteristics of the reconstituted A_1 site were determined.

The advantage of orientation selection by high-field EPR was exploited when studying the transient radical pair $P_{700}^{\bullet+}A_1^{\bullet-}$ in photosynthetic PS I multilayers applying time-resolved W-band EPR (TREPR) with direct detection.[329] For these experiments novel variants of plane–concave Fabry–Perot resonators have been developed (see Section 3.3.4.2) to measure the orientation dependence of spin-polarized spectra from frozen-solution PS I in oriented multilayer samples. To obtain good orientation selection with the multilayers on plane surfaces, the two plane–concave resonators were exchanged to align the external magnetic field either parallel or perpendicular to the plane mirror. The goal of this work is to provide the basis for future studies, in which the improved resolution of such multilayers may be used to detect mutation-induced structural changes of PS I in membrane preparations. This approach is particularly interesting for protein systems that cannot be prepared as single crystals suitable for X-ray crystallography. However, in order to use such multilayers for structural investigations of protein complexes, it is necessary to know their orientation distribution. PS I was chosen as a test example because the wild type was crystallized and its X-ray structure determined to 2.5 Å resolution.[78] From the W-band TREPR data, the width of the orientation distribution of the transient radical pairs $P_{700}^{\bullet+}A_1^{\bullet-}$ was determined as $(30 \pm 10)°$. Based on this orientation distribution width, the simulations of the 1D oriented spectra show that subtle structural changes in the radical-pair complex in membrane fragments, for example by point mutations, can be detected more easily in oriented multilayer than in disordered samples of PS I in frozen solution. This is particularly promising for future studies on mutant PS I complexes, for which no single crystals are available. Apparently, the combination of multifrequency TREPR on disordered samples and on oriented multilayers is a very appealing strategy for structural analysis of transient radical-pair systems, and their mutation-induced changes, in photoinduced electron-transfer processes.

A significant step forward in complexity towards electron-transport processes in whole cells was taken recently by combining X-band and W-band experiments on PS I complexes of intact cyanobacterial cells *Synechocystis* sp. PCC 6803.[330] The energy of light quanta absorbed by the light-harvesting complexes is delivered to the pigment–protein complexes, the photosystems PS I and PS II and then converted via an electron-transfer chain (ETC) into the energy of macroenergetic chemical compounds, NADPH and ATP, see Figure 5.2. The ETC contains three large protein complexes located in the thylakoid membrane: PS I, PS II and the cytochrome b_6f–complex (bf-complex). Light-induced electron transport from the water-splitting complex of PS II to the terminal electron acceptor of PS I ($NADP^+$) is mediated by the bf-complex and three mobile electron carriers, plastoquinone (Q), plastocyanine (Pc) and ferredoxin (Fd), that provide the electron transport between PS II, bf-complex, PS I and the ferredoxin-NADP-reductase complex. Plastoquinone is an essential component of the photosynthetic ETC that couples electron transfer to proton translocation across the thylakoid membrane (for further details of the ETC partners in cyanobacterial cells, see ref. [330] and references therein). Currently, the elucidation of the regulatory mechanisms of electron-transport processes in native photosynthetic systems and their adaptation to variable environmental conditions are central problems in bioenergetics of photosynthesis. The cyanobacterial cell serves as an excellent model for studying the regulation of electron-transport processes in intact photosynthetic systems of oxygenic type. Cyanobacteria contain both oxygenic photosynthetic and respiratory electron-transfer chains incorporated into the same membrane. This makes cyanobacteria a convenient model system for the investigation of interactions between the photosynthetic and respiratory ETCs.

The light-induced redox transients of the primary electron donor P_{700} and the secondary acceptor A_1 were analyzed in both their electronic structure and kinetic behavior.[330] The kinetics of the light-induced redox transients of P_{700} is a rather good indicator of the cell's metabolic state. It was found that the kinetic behavior of the cation-radical $P_{700}^{\bullet+}$, generated by illumination with continuous light, and the EPR intensity of the radical pair $P_{700}^{\bullet+}A_1^{\bullet-}$, generated by laser pulse illumination, strongly depend on the illumination prehistory (either the sample was frozen in the dark or during illumination). Both processes were sensitive to the presence of electron-transport inhibitors, which block the electron flow between the two photosystems, PS I and PS II, in the cell. In line with the X-band EPR data on the kinetics of light-induced redox transients of P_{700}, the high-field W-band EPR study of the spin-correlated radical-pair state $P_{700}^{\bullet+}A_1^{\bullet-}$ shows that photosynthetic electron flow through PS I is controlled both on the donor and on the acceptor sides of PS I, *i.e.* it is controlled at two check-points: (*i*) the plastoquinone segment of the electron-transport chain, and (*ii*) the acceptor side of PS I. In conclusion, it is emphasized that the high spectral and time resolution as well as excellent detection sensitivity of high-field W-band EPR allow elucidation of the details of the complex electron-transfer pathways in the interconnected ETCs of both intact and chemically treated cells of photosynthetic organisms.

5.3.2 Multifrequency EPR on Doublet States in Photosystem II (PS II)

Taking PS II as an example, one notices many similarities to the simpler RC of purple bacteria concerning the primary electron-transfer cofactors embedded in their protein subunits, also exhibiting a pseudo-C_2 symmetry axis running through the nonheme iron and the dimeric primary donor sitting in the symmetry-related D1 and D2 protein subunits. Concerning the complete PS II protein complex, however, the increased complexity is attributed to the evolutionary invention of photosynthetic water splitting by the oxygen evolving complex (OEC) with its protein-bound tetranuclear manganese cluster, and to protect the fragile photosynthetic machinery against the destructive interaction with molecular oxygen produced in the OEC of PS II.[331] Comparing the central parts of PS II and bacterial RC, one realizes a highly conserved arrangement of the cofactors: They are arranged in two branches bound to two protein subunits, D1/D2 and L/M in PS II and RC, respectively. On the donor side a pair of interacting chlorophyll or bacteriochlorophyll molecules is located ("special pair"), the acceptor chains contain monomeric (bacterio)chlorophylls, (bacterio)pheophytins and two quinones, Q_A and Q_B. In PS II, Q_A and Q_B are plastoquinones, in bacterial RCs they are either both ubiquininones (*Rb. sphaeroides*) or a ubiquinone and a menaquinone (*R. viridis*). Both quinones are coupled to a high-spin iron (Fe^{2+} with $S=2$). The OEC in PS II is stabilized by extrinsic proteins. Between the Mn cluster of the OEC and chlorophyll dimer a redox-active tyrosine (Tyr 161 of D1, Y_Z) is located. Related to this position by the pseudo-C_2 axis, a second tyrosine (Tyr 161 of D2, Y_D) is located whose role in the electron-transfer process is not clear yet. Moreover, PS II contains intrinsic antenna subunits (CP43 and CP47) with additional chlorophylls and carotenoids, as well as a number of other subunits, for example the Cyt b_{559} and Cyt c_{550} cytochromes.[79]

In the following we focus on a small selection of multifrequency EPR studies on PS II cofactors and refer to the comprehensive reviews mentioned above[15,17] for a broader coverage of the field. There is a growing interest in advanced EPR studies of the tyrosyl radicals in the D1 and D2 polypeptides. In particular, the determination of individual g-tensor components in frozen PS II preparations by high-field/high-frequency EPR proved to be very informative in probing the different hydrogen-bonding interactions of the Y_Z and Y_D tyrosyl and the Q_A and Q_B quinone radicals. Their frozen-solution[126,308] and single-crystal[332] spectra exhibit resolved Zeeman and, in some cases, hyperfine structure. Similar cw high-field EPR studies have been performed to characterize the pheophytin anion radical in the wild-type and D1-E130 mutants of PS II from *Chlamydomonas (C.) reinhardtii*.[333] The mutants were constructed site specifically to weaken or remove the hydrogen bond of the carbonyl group at ring V of the pheophytin molecule to the D1 polypeptide. A strong mutation-induced shift of the g_{xx} component of the g-tensor was observed. DFT calculations were used to rationalize the observed trend in the variations of the g_{xx} component. The results again demonstrate the sensitivity of g-values in probing the protein microenvironment of cofactor radicals.

Next, we refer to recent [55]Mn pulse Davies-ENDOR experiments at Q-band (34 GHz) of the S_0 and S_2 states of the oxygen-evolving complex (OEC).[334,335] During the catalytic cycle of water splitting (Kok cycle)[336] the OEC passes through five different redox states (S_n states, S_0 to S_4). The advancement of the OEC to the next higher S_n state occurs after the absorption of a suitable light quantum by a chlorophyll molecule of the PSII antenna, which transfers the excitation energy into the reaction center of PSII, where the primary charge separation occurs. Knowledge of the spatial and electronic structures of the OEC in its different S_n states is highly important for understanding its reactivity. No wonder, therefore, that numerous efforts to this end are found in the literature using different techniques, see, for example, a special issue of *Biochimica Biophysica Acta*, 2001, dedicated to photosynthetic water oxidation. It contains a tribute by R. D. Britt, K. Sauer and V. K. Yachandra in remembrance of M.P. Klein (Berkeley, 1921–2000)[337] whose major contributions to photosynthesis research included the structure of the Mn complex involved in photosynthetic oxygen evolution. EPR and X-ray absorption spectroscopies, and their complementary information content, have provided much of what is now known about the structure of the Mn complex.

Recently, the combined efforts of laboratories using X-ray diffraction, EXAFS, XANES and multifrequency EPR and ENDOR techniques have led to a significant improvement in spatial resolution. For the Mn_4O_xCa complex, more than 10 different structural models are suggested containing two or three 2.7 Å Mn–Mn distances and one or two 3.3 Å Mn–Mn distances.[334–335,338–350]

However, up to now, the resolution of all existing PSII crystal structures is still not high enough to assign precise positions for the four manganese ions and the one calcium ion within the OEC protein matrix and to determine the details of the bridging between these metal ions. Moreover, it has to be pointed out that the current X-ray diffraction data-collection conditions lead to a severe impairment of the integrity of the Mn_4O_xCa complex due to radiation damage.[351] In this respect, EPR studies are, of course, much less prone to experiment-inherent damages of the protein.

Normally, Q-band EPR is not considered to be a "true" high-field experiment (as determined by the condition that Zeeman splitting is larger than the inhomogeneous linewidth, see Chapter 2). However, despite the rather small Zeeman magnetoselection of Q-band EPR on PS II, the [55]Mn-ENDOR spectrum allowed the determination of the principal values of the hyperfine-tensors for all four Mn ions. The hyperfine components were found to be slightly different in the S_0 and S_2 states. The ultimate goal of such experiments is to derive the electronic and spatial structure of the OEC. Although this goal was not yet achieved by these experiments (see, however, below), their results already safely exclude all models in the literature proposing two magnetically uncoupled dimers as structural elements of the OEC in the S_2 state.

In principle, high-field EPR at 95 GHz and above should be suitable for studying the Mn cluster that constitutes the OEC. To our knowledge and surprise, however, such high-field studies on the Mn cluster in PS II have not been reported so far, although multifrequency (9, 95, and 285 GHz) high-field

EPR studies of binuclear Mn(III)Mn(IV) complexes as relevant model systems were conducted already ten years ago.[352]

We conclude this excursion to the hidden secrets of the electronic structure of the Mn_4O_xCa cluster in the OEC of PS II by reviewing a very recent breakthrough in the theoretical interpretation of pulse Q-band EPR and ^{55}Mn-ENDOR results.[335] In this study, the recently obtained ^{55}Mn hyperfine coupling constants of the S_0 and S_2 states of the OEC were analyzed on the basis of Y-shaped spin-coupling schemes with up to four nonzero exchange coupling constants, J. This analysis rules out the presence of one or more Mn(II) ions in S_0 and establishes that the oxidation states of the manganese ions in S_0 and S_2 are, at 4 K, Mn_4(III, III, III, IV) and Mn_4(III, IV, IV, IV), respectively. The result of the analysis favors a new structural model that is fully consistent with the EPR and ^{55}Mn-ENDOR data. Furthermore, the Mn oxidation states were assigned to the individual Mn ions. It is proposed that the known structural changes of the Mn_4O_xCa cluster when passing through the Kok cycle, namely shortening of one 2.85 Å Mn–Mn distance in S_0 to 2.75 Å in S_1, correspond to a deprotonation of a μ-hydroxo bridge between MnA and MnB, *i.e.* between the outer Mn and its neighboring Mn of the μ$_3$-oxo bridged moiety of the cluster. The exchange coupling J_{AB} for the suggested models of the Mn_4O_xCa cluster is significantly smaller in the S_0 state as compared to the S_2 state, while the other J couplings between the four Mn ions in the possible coupling schemes (dimer of dimers, trimer-monomers, tetramers) hardly change. This allows, for the first time, the assignment of a structural change within the Mn_4O_xCa cluster to a specific Mn–Mn bridge, namely that between Mn_A and Mn_B. The presented analysis of ENDOR and EXAFS data[335] is a first step towards a molecular understanding of the mechanism of water oxidation. The Mn_4O_xCa cluster actively takes part in the photosynthetic water splitting chemistry via structural changes within the cluster. It is hoped that these results will be useful for the synthesis of artificial catalysts for solar water splitting, certainly a most challenging project for renewable energy resources.

5.4 Photoinduced Electron Transfer in Biomimetic Donor–Acceptor Model Systems

5.4.1 Introduction

Understanding the structure–dynamics–function relationship of light-induced electron transfer (ET) in the natural photosynthetic apparatus at the atomic level is a fascinating, but extremely difficult task. Much of our current understanding of the structure–dynamics–function relationships of the photosynthetic apparatus stems from studies of "biomimetic" donor–acceptor model systems in homogeneous solvents (isotropic and liquid crystalline) that might mimic some of the important events in primary photosynthesis. Such model studies can provide complementary information to *in-vivo* studies – and *vice versa*. To solve the task of constructing such biomimetics is not only challenging in its own right, but also in view of possible implications for the design of artificial solar-energy conversion systems with high quantum efficiency. Long-lived charge-separated radical pairs (RPs) occur as secondary ET intermediates of the light-driven cascade of transmembrane ET steps in the photosynthetic reaction center (RC) protein complex. Because of the complexity of the natural RC – which during evolution was optimized by following strategies not only for high-yield photoinduced charge separation, but also for efficient protection against internal and external stress factors – supramolecular organic donor (D)–acceptor (A) model complexes have been synthesized in many laboratories to mimic essential features of photosynthetic ET and, hopefully, to simplify the task of understanding its mechanism (for more recent reviews, see refs. [353–355,58]. It is established by now that multicomponent biomimetic model complexes (D–s–A), consisting of donor and acceptor moities linked by special spacer (s) groups, can be designed – and synthesized – to produce, with high quantum yield, long-lived charge-separated RPs in homogeneous media like isotropic or liquid-crystalline solvents.[58,353,356–368] Such supramolecular biomimetics consist of a chromophore, that absorbs in the visible spectral region, and several tailor-made substituents linked to it by covalent bonds or hydrogen-bond networks, thus establishing a spacer bridge to provide electronic coupling between the donor and acceptor moieties. The synthesis of biomimetic D–s–A complexes generally adheres to strategic principles: The redox potentials of the substituents are chosen in a way to be operative either as electron acceptors or donors. In other words: The substituents will either accept electrons from the photoexcited chromophore or donate electrons to the chromophore after it became oxidized by ejecting an electron from its excited state. By varying the number of substituents as well as the bonding characteristics and spatial arrangement of the spacer bridge, the factors governing long-range ET can be selectively controlled to ultimately stabilize the charge-separated donor–acceptor RP state against energy-wasting charge-recombination reactions. An extended π-electron system is normally chosen as chromophore, *e.g.*, a porphyrin moiety, but also for acceptors extended π-electron systems, such as free-base porphyrins or fullerenes, are favorably chosen for stabilizing the charge-separated state against recombination. This stabilization can be further increased by adding secondary acceptors to the D–s–A dyads, thereby creating supramolecular triads, tetrads, *etc.*[353–356] The characteristics of the linkage between D and A, *e.g.*, σ-bonds, H-bonds, ligation, or electrostatic interaction, will affect the electronic coupling matrix element, V, between donors and acceptors, ranging from covalent through-bond, via direct through-space to superexchange coupling situations. This is similar to what occurs in natural photosynthetic systems along the various ET pathways in the protein.

Thus, the electronic and spatial properties of the spacer bridge decide about the magnitude of the electronic coupling matrix element, V, between donor cation and acceptor anion within the charge-separated RP state. Controlling charge recombination (CR) via V_{CR} is essential for retarding wasteful ET recombination dynamics also in biomimetic devices, in analogy to what is operative in natural photosynthesis. Unfortunately, determining reliable values of V_{CR} is difficult, both experimentally and theoretically. There exist only a few spectroscopic methods by which V_{CR}^2 can be measured indirectly through the exchange coupling parameter, J, between the electron spins of the radicals within a radical pair. Among these methods are ns transient absorption measurements of magnetic-field effects on the reaction yield[369,370] and ns time-resolved EPR detection of transient spin-polarized radical-pair spectra (in this section we will use the same short-hand notation "TREPR" for time-resolved EPR, be it the direct-detection method with cw microwave irradiation or the pulsed method with spin-echo pulse trains, see above and refs. [218,356,371–374].

The TREPR method excels by providing reliable estimates of the exchange interaction J from the spectral simulation routines of the spin-polarized spectra of charge-separated spin-correlated RPs in their triplet state. The theoretical analysis of the relation between the singlet–triplet splitting within the RP, $2J = E_S - E_T$, and the electronic coupling, V_{CR}^2, goes back to P.W. Anderson,[375] and was later adapted to charge-recombination ET reactions of the RP.[126,218,369,370,376] According to theory, $2J$ is given by

$$2J = E_S - E_T = \left[\sum_n V_n^2 / \Delta E_n \right]_S - \left[\sum_n V_n^2 / \Delta E_n \right]_T \quad (5.23)$$

where V_n^2 are the squared electronic coupling matrix elements between the ground state and nearby excited states, weighted by the energy separation, ΔE_n, between the RP state and those excited states n to which it is coupled at the nuclear coordinate of the relaxed equilibrium nuclear configuration of the charge-separated RP. The energy gap is given by $\Delta E_n = \Delta E + \lambda$, i.e. by the free-energy difference ΔE for the reaction and the total reorganization energy λ for the electron transfer. The individual terms in the equation for $2J$ may have positive or negative signs, depending on the sign of the energy denominator, thereby determining the coupling to be ferromagnetic ($J < 0$) or antiferromagnetic ($J > 0$). Thus, from combining TREPR and cyclic voltammetry experiments, the desired quantity V_{CR}^2 can be determined after estimating the solvent reorganization energy $\lambda(r)$ from the seize of the donor and acceptor molecules and the dielectric properties of the solvent[54] according to:

$$\lambda(r) = \frac{e^2}{4\pi \cdot \varepsilon_0} \cdot \left(\frac{d_A}{2} + \frac{d_D}{2} - \frac{1}{r} \right) \cdot \left(\frac{1}{n_S^2} - \frac{1}{\varepsilon} \right) \quad (5.24)$$

where r is the center-to-center separation of the D and A molecules, and d_A and d_D are the radii of the D and A molecules (assumed to be spherical). The vacuum permittivity constant, the refractive index and dielectric constant of the solvent are represented by ε_0, n_S, and ε, respectively. We see that from TREPR experiments crucial information about the D–s–A coupling schemes, for instance direct coupling versus superexchange coupling, can be revealed.

An important aspect of biomimetic long-range ET is the strategic selection of the solvent matrix in which the redox partners are embedded. Intramolecular ET can occur when the photoexcited reactant D, which is linked to A by a spacer s, *D–s–A, moves on a free-energy surface until it reaches a crossover point with another intersecting surface that corresponds to the final state, *i.e.* to the charge-separated product $D^{\bullet+} - s - A^{\bullet-}$. Once curve crossing occurs, the charge distribution in the molecular complex changes, and the solvation equilibrium will be perturbed. Dielectric relaxation will realign the solvent molecular electric dipoles until – with a characteristic time constant, the Debye relaxation time τ_D of the solvent – a new equilibrium with respect to the new charge distribution is obtained.[377] According to the Marcus theory of electron transfer[54] (see *Prologue 1* of Chapter 5), the controlling factors of the ET rate, k_{ET}, are both the energetics of the redox couple, *i.e.* the Franck–Condon factor, F, and the squared electronic coupling matrix element, V^2, between donor and acceptor. F is governed by the free-energy change ΔE of the reactants and the reorganization energy λ of the final charge-separated state in the equilibrium solvent configuration of the initial state: $F = \exp[-E_a/kT]/\sqrt{4\pi \cdot \lambda \cdot kT}$ with the activation energy $E_a = (\Delta E + \lambda)^2/4 \cdot \lambda$.

For "fast" solvent relaxation dynamics, where the nuclear motion is fast on the ET time scale (nonadiabatic ET), the ET rate is given by $k_{ET}(\text{na}) = (2\pi/\hbar) \cdot V^2 \cdot F$,[57] see above, where (na) stands for nonadiabatic. The critical quantity describing whether the solvent relaxation (approximated by one relaxation time τ_L) is fast on the ET time scale, is the adiabaticity factor κ in the Marcus theory: $\kappa = 4\pi \cdot V^2 \cdot \tau_L/\hbar \cdot \lambda$. In ET processes, two limiting cases are generally considered, the nonadiabatic case with $\kappa \ll 1$, and the adiabatic case with $\kappa > 1$. In the adiabatic case, the solvent dynamics cannot be neglected, and the ET rate becomes solvent-controlled, *i.e.* k_{ET} (a) is governed by the solvent's longitudinal dielectric relaxation time, τ_L. Consequently, the basic nonadiabatic ET rate, as is prevailing for photosynthetic ET in protein RCs at ambient temperatures, has to be corrected for describing the adiabatic case: $k_{ET}(\text{a}) = k_{ET}(\text{na})/(1 + \kappa)$. The temperature dependence of τ_L is determined by the temperature dependence of the Debye relaxation time, τ_D, since $\tau_L = (\varepsilon_\infty/\varepsilon_s) \cdot \tau_D$. Here, ε_s and ε_∞ are the static and high-frequency (optical) dielectric constants of the solvent medium, respectively.

An experimental handle to control the transition from the nonadiabatic to the solvent-controlled adiabatic limit, is provided by the temperature and by the choice of the solvent matrix.[58] To monitor the transients of light-induced ET processes by time-resolved EPR, such a solvent control is a prerequisite. It must slow down the normally sub-ns fast photoinduced ET near room temperature in biomimetic DA systems dissolved in isotropic solution into the time window of transient EPR, which is restricted to, say, 10 ns upwards. Generally, in isotropic solvents the attenuation of ET processes from a fast nonadiabatic regime into the slower adiabatic regime can be achieved by lowering the temperature to the soft-glass region,[72] where the ET process has not yet been frozen out, or by using highly viscous solvents. It is intriguing, however, that anisotropic liquid-crystal (LC) solvents are often the better choice, allowing study of the ET transient intermediates over the wide temperature range of the nematic mesophase of the LC, where molecular motion still prevails resulting in well-resolved EPR spectra.[73] The inherent properties of LCs "dial" the relevant ET rates into the readily monitored $10^6 \, s^{-1}$ range.

The clue of explaining this ability of LCs to retard dramatically ET rates, is the change of the internal electric field, which accompanies the intramolecular ET. It produces a nematic potential that hinders molecular reorientation, thereby prolonging τ_D in the nematic mesophase of the LC, $\tau_D(\text{nem})$, as compared to the Debye relaxation time in isotropic media, $\tau_D(\text{iso})$. Both time constants are related by $\tau_D(\text{nem}) = G \cdot \tau_D(\text{iso})$ with the retardation factor G.[378] Here, $G = (\exp[q/kT] - 1) \cdot q/kT$, where q is the barrier height of the nematic potential. The power of LCs as solvents for studying the transient ET reaction intermediates in artificial photosynthetic DA complexes by time-resolved EPR has been widely recognized.[58,353,356,368,371,379–381] These studies show how by employing LCs with different polarity, viscosity, structure and diamagnetic susceptibility anisotropy, the ET and spin dynamics can be affected, allowing for a better understanding of the controlling factors of intra- and intermolecular ET reactions in artificial photosynthesis model complexes.

It is clear by now that such studies of model systems provide complementary information also to *in-vivo* studies – and *vice versa*. In most of our multi-frequency time-resolved EPR studies we had concentrated on biomimetic DA complexes that are covalently linked by selected spacer groups. These studies mainly deal with dyad and triad porphyrin–spacer–quinone systems to elucidate their ET and spin dynamics during light-induced charge-separation and recombination processes involving transient triplet and radical-pair states. The dyad and triad, and even tetrad porphyrin–spacer–quinone systems were synthesized in the group of H. Kurreck (FU Berlin). We also included DA models that are not covalently linked, but rather linked by H-bonding networks, thus mimicking more realistically the natural RC situation. They involve pre-organized supramolecular porphyrin–quinone or porphyrin–nitrobenzene aggregates that are functionalized by appropriate guanine and cytosine residues to allow for molecular recognition via Watson–Crick base pairing. They were synthesized in the group of J.L. Sessler (University of Texas, Austin). In the following two sections, we will review a few of these experiments taking prominent examples from both types of supramolecular biomimetics.

5.4.2 Covalently Linked Porphyrin–Quinone Dyad and Triad Model Systems

Employing 9.5-GHz and 95-GHz TREPR spectroscopy, four different porphyrin (P)-quinone (Q) systems covalently linked by a cyclohexylene bridge (see Figure 5.35) have been investigated to study the effect of molecular dynamics on electron-spin polarization and light-driven electron-transfer characteristics.[379] The reaction scheme for P-*trans*CH-BQ for light-initiated ET is depicted in Figure 5.36. By absorbing a photon $h\nu$, the zinc-tetraphenyl-porphyrin (ZnTPP)

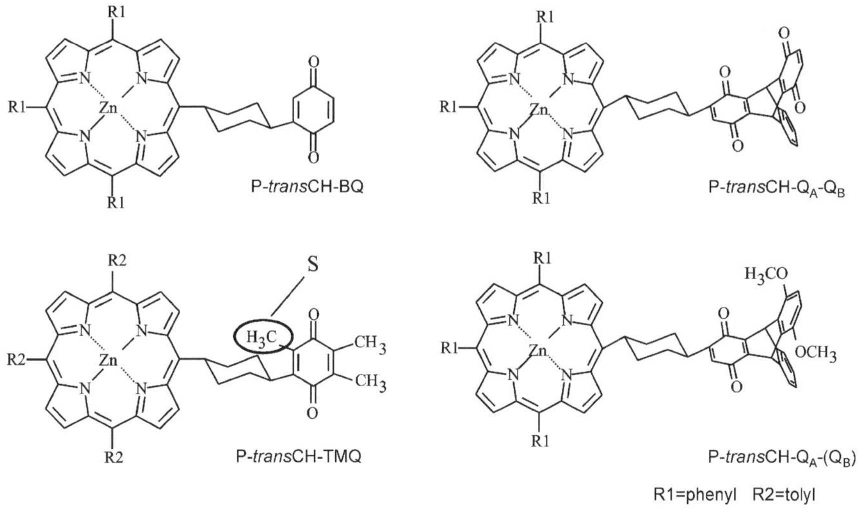

Figure 5.35 Molecular structure of investigated porphyrin-quinone (PQ) dyad and triad model systems. The methyl substitution that restricts the motion about the bond between cyclohexylene and quinone is indicated by S ('bulky' PQ).

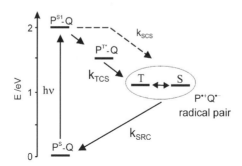

Figure 5.36 Reaction scheme for light-induced electron transfer in the porphyrin-quinone dyad P-*trans*CH-BQ. The rate k_{SCS} denotes the singlet charge-separation rate, k_{TCS} the triplet charge-separation rate, and k_{SRC} the singlet charge-recombination rate.

is excited to its first excited singlet state P^{S1}. At room temperature, fast singlet-ET produces the singlet radical-pair (RP) state, $(P^{\bullet+}Q^{\bullet-})^S$. Since the singlet charge-recombination (SRC) rate, k_{SRC}, is larger than $10^8 \, s^{-1}$ for most porphyrin–quinones, the singlet-RP has decayed before singlet–triplet mixing can generate any detectable EPR signal intensity. At lower temperatures, however, singlet-ET is slowed down and, hence, spin-orbit intersystem crossing in the porphyrin moiety to the excited triplet state P^{T*} can compete. Subsequently, triplet-ET to the triplet-RP state, $(P^{\bullet+}Q^{\bullet-})^T$, can occur.

This RP state is detectable by TREPR in polar solvents in a temperature window of about 30 degrees below the melting point of the solvent. Alternatively, the triplet-RP can be observed over a wider temperature range in the soft-glass phase of the "room-temperature" liquid crystal E7 (*Merck*).[382] At considerably lower temperatures, ET is frozen out, and the EPR spectrum of P^{T*} is observed. In the reviewed work,[379] it was shown that the investigated $(P^{\bullet+}Q^{\bullet-})^T$ triplet-RPs are all strongly coupled, *i.e.* the exchange interaction $(J) \gg (\Delta\omega)$ with $\Delta\omega = (\mu_B/2\hbar) \cdot B_0 \cdot [g_1(\varphi,\theta) - g_2(\varphi,\theta)]$. Here, φ and θ denote the polar angles of the magnetic B_0 field with respect to the molecule in the principal axes system of the zero-field tensor. Moreover, in the chosen solvent matrix the overall tumbling motion of the complexes is slowed down such that the anisotropic contributions to the zero-field and *g*-tensors are resolved in the EPR spectra. Nevertheless, small-angle fluctuations inside the solvent cage prevail and lead to spin-lattice relaxation that is anisotropic, *i.e.* the reaction rates and chosen transfer pathways depend on the orientation of the molecule with respect to the external magnetic field. These fluctuations also lead to a pronounced modulation of the isotropic exchange interaction *J* affecting the singlet-triplet (ST_i, $i = 0, -1, +1$) mixing efficiency. Hence, the TREPR spectra are sensitive not only to molecular structure but also to molecular motion, thereby probing the flexibility and microenvironment of the molecule.

When comparing the present PQ systems with the bacterial photosynthetic reaction center, several points should be mentioned: First, the radical pairs of the PQ systems resemble the secondary RP in RCs rather well because in both cases the individual radicals are nondiffusing on the EPR time scale and their anisotropic interactions are resolved by TREPR. Secondly, in the RC charge recombination is slow, the RP is only weakly coupled, and the TREPR spectrum can be described by using the correlated-coupled-radical-pair (CCRP, see above) model.[235,244] For example, by analyzing 95-GHz TREPR signals on the basis of the CCRP model, relative orientations of donor and acceptor sites in the RC could be revealed with high accuracy.[159] For strongly coupled RPs, TREPR was observed to be sensitive to motion and relaxation of freely diffusing biradical chains.[383–385] In this case, the exchange interaction is modulated when the chain is folding. In contrast to these cases, here we describe the dramatic effects of spin relaxation and modulation of *J* on TREPR lineshapes in highly viscous solvents where anisotropic contributions of the spectral parameters are resolved. The results of our multifrequency EPR study indicate that caution has to be exercised with the interpretation of spin and ET dynamics that is based on TREPR experiments at one microwave frequency only.

In Figure 5.37 the X- and W-band spectra of two different PQ systems are shown. They are measured in ethanol and an ethanol/toluene mixture, about 10 degrees below the melting point, 1 μs and 4 μs after the laser flash. For delay times longer than 1 μs the lineshapes remain unchanged and the signal decays within 10 μs. From inspection of the time evolution of the X-band spectrum of P–*trans*CH-TMQ it turns out that the PQ systems can be divided into two

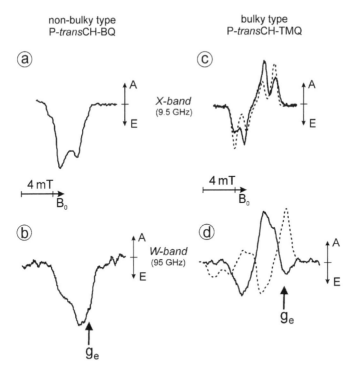

Figure 5.37 TREPR spectra (solid lines) of P-*trans*CH-BQ and P-*trans*CH-TMQ in ethanol at 150 K (a) [386] and ethanol/toluene at 166 K (b, c,[387] d) in X- and W-band, 1 μs (c) and 5 μs (a, b, d) after laser excitation. The integration time was 1 μs. The letters A and E stand for absorption and emission, respectively. The resonant position for the *g*-value of the free electron, g_e, is indicated in the W-band spectrum. The spectra simulations in c and d (broken lines) are calculated for a weakly coupled radical pair using the CCRP model, for triplet electron transfer, for $J = -1.1$ mT, for $D = -2.3$ mT and are based on the X-ray structure.[388]

categories that are related to molecular flexibility: Systems in which the quinone carries no methyl substituent next to the cyclohexylene bridge have totally emissive TREPR spectra. We term them "nonbulky" PQs. Systems with such an additional methyl group (indicated by S in Figure 5.35) exhibit TREPR spectra that have emissive (E) and absorptive (A) parts. We refer to them as "bulky" PQs. On cooling the nonbulky systems down to 140 K in ethanol, the spectra become equal to the EEAA spectra of the bulky systems. This was observed before,[387] both at X-band and at W-band, and a polarization mechanism for the nonbulky PQs was suggested on the basis of a dynamic exchange interaction due to the higher flexibility of the nonbulky quinine.[389] For the bulky PQs it was suggested that the X-band spectra may be understood in terms of the CCRP model of weakly coupled radical pairs. A simulation based on that model with $J = -1.15$ mT and $D = -2.3$ mT is shown as the dotted line in Figure 5.37(c). However, for W-band EPR the CCRP model predicts, on the low-field side, resolved spectral contributions from the quinone anion radical due to its relatively large g-value. However, these contributions are not observed at W-band, and the predicted lineshape is totally different from the experimental result (Figure 5.37(d)). Additionally, it was shown by X-band saturation recovery experiments that the polarization is generated within the radical pair, and is neither due to polarization transfer from the triplet precursor (denoted P^{T*} or $^3P^*$) nor due to the CCRP mechanism.[390] It was suggested that the bulky PQs form strongly coupled radical pairs, and the polarization is caused by relaxation phenomena. This suggestion is now supported by the T_1 measurements on the dimethylbenzoquinone anion radical at W-band.

In the following we discuss a reaction model on the basis of a strongly coupled RP that includes relaxation. It leads to the convincing simulations of the time evolution of the spectra. All the TREPR results, both at X- and W-band, on bulky PQs can be understood in terms of this model.[379]

Reaction Model for "Bulky" PQs

Several mechanisms exist that can produce spin polarization in a strongly coupled radical pair, and population/depopulation pathways involved are depicted in Figure 5.38. We shall discuss them in the following.

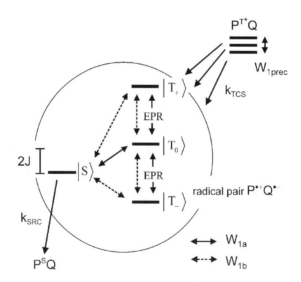

Figure 5.38 States and population/depopulation pathways for a strongly coupled radical pair. Only the pathways for uncorrelated spin-lattice relaxation W_{1a} and W_{1b} are shown. k_{SRC}: singlet recombination rate; k_{TCS}: triplet charge separation rate; J: exchange coupling; W_{1prec}: precursor spin-lattice relaxation rate; P^SQ: PQ ground state; $P^{T*}Q$: excited triplet state of P. For details, see ref. [379].

(1) **Triplet mechanism:** The P^{T*} (zinc porphyrin) precursor triplet states $|T_-^p\rangle$, $|T_0^p\rangle$ and $|T_+^p\rangle$ are selectively populated by intersystem crossing according to the contribution of $|T_Z^p\rangle$ to these states. This spin polarization is transferred to the RP if the ET rate is large compared to the spin-lattice relaxation rate W_1 of the precursor, which is between 6×10^6 and $2 \times 10^7\,s^{-1}$. The EEAA TREPR spectra cannot be due to the triplet mechanism for the following reasons: (*i*) their rise time is larger than the decay time of the precursor triplet signal, (*ii*) the saturation recovery measurements show that polarization is produced within the RP [390] and (*iii*) the triplet mechanism produces spin polarization symmetric to the porphyrin Z-axis, which is the PQ dyad's Y-axis. The spectra simulations show that, instead, the AE spectra, that are observed shortly after the laser pulse, are due to the triplet mechanism.

(2) **Singlet–triplet (ST) mixing:** The state $|T_0\rangle$ in Figure 5.38 is not an eigenstate, but is depopulated according to the contribution of $|S\rangle$ to the proper eigenstates and according to the singlet recombination rate k_{SRC}. This mechanism cannot depopulate $|T_-\rangle$ and $|T_+\rangle$ simultaneously and, moreover, is strongly field dependent. Therefore it cannot cause the observed spin polarization.

(3) **Spin-orbit coupling:** According to the nonvanishing electronic matrix element of the spin-orbit operator between the RP triplet and the RP singlet ground state, it is possible that electron charge recombination takes place also directly from the RP triplet states.[391] This mechanism may act selectively on the three triplet states and, thus, might cause spin polarization. It is strongly dependent on the D–A distance and on the environment. It is not only necessary for there to be a large overlap of the wavefunctions involved, but there must be also a heavy nucleus nearby in order to produce noticeable spin-orbit coupling. Since this is not the case for our systems, this mechanism probably does not operate efficiently in the nearly planar PQ systems with interradical distances as large as 1–1.5 nm.

(4) **Spin relaxation:** In a stable spin system stochastic mixing of states drives the system to thermal equilibrium by spin-lattice relaxation. In a transient spin system, such as a light-generated RP whose lifetime is restricted by rapid charge recombination, the same mechanisms might produce spin polarization.[392] Stochastic modulation of the dipolar coupling leads to correlated relaxation, *i.e.* it connects the triplet states. Modulation of the exchange interaction J leads to transitions between $|S\rangle$ and $|T_0\rangle$. On the other hand, modulation of local interactions that are different for the two electron spins, leads to uncorrelated relaxation, *i.e.* to singlet–triplet transitions indicated as W_{1a} and W_{1b} in Figure 5.38.[392] Basically, these two rates are different and, because of the rapid electron recombination from $|S\rangle$, spin polarization in the RP is generated. Both the transition energies and the spin-lattice relaxation may depend on the orientation of the molecule with respect to the magnetic field. This is either because the magnetic interaction of the

relaxation mechanism is orientation dependent or because parts of the molecule undergo anisotropic small-angle fluctuations. This would result in the characteristic orientation-dependent spin polarization that is shown in Figure 5.38. Hence, we conclude that the observed spin polarization is mainly due to anisotropic spin-lattice relaxation. Indeed, the simulations calculated by using orientation dependent relaxation rates give very satisfying agreement. For the parameters used in the simulations, see ref. [379].

The value of the exchange interaction, J, is crucial for the interpretation of the spectra. The semiempirical predictions for the distance dependence of $J(r)$ are not very helpful in our case: In the through-space model, an exponential dependence $J(r) = J_0 \cdot \exp[-\delta \cdot (r - r_1 - r_2)]$ is assumed (r_i are the radii of the molecules). P and Q in our systems have distances of 0.98 nm to 1.5 nm. Thus, from the distance dependences published, $|J|$ might be between 0 and 10^4 mT! (see references in ref. [389]). On the other hand, it was shown that in the case of a weakly interacting radical pair one expects an EPR lineshape with different spectral contributions from the pair partners and, moreover, a significant change between X- and W-band spectra. This is not observed, while for a strongly interacting radical pair one can easily understand the experimental results described. To reach the limit of strong interaction, singlet and triplet states of the radical pair must be separated by at least 50 mT. The simulations of the EPR signals reproduce the transition from the early AE spectrum to the EEAA spectrum at later times.

We conclude that, with the model and parameters presented, lineshape and time development of the TREPR spectra can be well simulated. The value of the exchange interaction is not critical as long as $|S\rangle$ is energetically separated from $|T_0\rangle$ and $|T_-\rangle$ by at least 50 mT, and as long as the radical pair is strongly coupled. The orientation selection of the PQ systems in the liquid crystal E7 (Merck, UK) was used to directly measure the anisotropy of the spin polarization. The TREPR spectra obtained in a liquid crystal give further evidence for the spin-polarization scheme presented. For details, see ref. [379].

Orientation-Dependent Spin-Rotation Relaxation

In the previous section it was discussed why in the radical pair of the 'bulky' PQs spin relaxation is most probably responsible for the observed electron-spin polarization. TREPR spectra of RPs could be obtained both in the entire nematic phase of the liquid crystal E7 as well as in organic solvents in a small temperature range below their melting points. Earle and coworkers[393] have shown that in this situation molecular motion can be described as a slow diffusion of the solvent cage and a fast diffusion of the molecules in the potential of the solvent cage. Furthermore, they have shown that the Einstein–Stokes–Debye relation between temperature, viscosity and reorientational correlation time is valid down to temperatures that are about 20% higher than the glass transition temperature. In highly viscous solvents the main contribution to spin-lattice relaxation is caused by stochastic modulation of the spin-rotation interaction.[394,395] The spin-rotation interaction Hamiltonian for the spin S_2 localized on the quinone site is given by

$$\hat{H}_{SR} = \boldsymbol{J} \cdot \tilde{C} \cdot \hat{S}_2 \qquad (5.25)$$

with the angular momentum $\boldsymbol{J}$. The tensor $\tilde{C}$ is related to the g-tensor by $\tilde{C} = -2 \cdot \tilde{A} \cdot (\tilde{g} - g_e \cdot \tilde{1}) \cdot \hbar^2$, where $\tilde{A}$ denotes the rotational tensor. The spin S_1 localized on the porphyrin site is not subject to spin-rotational interaction, because the g-tensor of $P^{\bullet+}$ is very close to that of the free electron. Generally, the correlation time τ_J of the angular momentum is much smaller than the reorientation correlation time τ_R and we can neglect the modulation of $\tilde{C}$. In ref. [379] it is shown that the observed spin relaxation is caused by spin-rotation relaxation and anisotropic motion of the molecular fragments.

Model for "Nonbulky" PQs

It was rather surprising when it was observed that both W-band and X-band TREPR spectra of P-transCH-BQ in frozen polar solvents are totally emissive (Figures 5.37(a) and (b)), *i.e.* for most molecular orientations of strongly coupled two-spin systems both transitions, $|T_+\rangle \leftrightarrow |T_0\rangle$ and $|T_0\rangle \leftrightarrow |T_-\rangle$, are emissive. Generally, this may be caused by ST_- mixing and fast singlet electron recombination. ST_- mixing is most effective when $|S\rangle$ and $|T_-\rangle$ are degenerate or at least energetically close. Therefore, because the spectra are emissive in both frequency bands and corresponding magnetic fields, the exchange interaction must be modulated over a wide range. Whether this mechanism leads to emissive spectra or to EEAA spectra depends on the amplitude of the motion. It is restricted by forces within the molecule and between the quinone and the solvent cage. The potential well set up by these forces is increased upon cooling or by substitution with the methyl group in the "bulky" PQ (*e.g.*, P-*trans*CH-TMQ, see Figure 5.35. In this situation, ST_- mixing is no longer active. In ref. [379], a description of this diffusion model is presented that leads to a consistent interpretation of the TREPR spectra of nonbulky PQs both at X- and W-band. The modulation of J must take place on an intermediate time scale within two limits: It must be fast enough that every PQ molecule could adopt a conformation for which $|S\rangle$ and $|T_-\rangle$ are energetically close in both frequency bands before the TREPR signal is observed. On the other hand, the rotational diffusion must be slow enough that the ST_- mixing has enough time to evolve.

Molecular Dynamics of PQ Systems

From the TREPR study described, the following qualitative conclusions can be drawn: The observed spin polarization is caused by two different mechanisms for bulky and nonbulky PQs. They are related to the two different molecular motions shown in Figure 5.39:

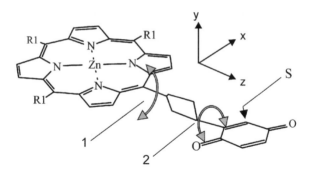

Figure 5.39 Stochastic motions of PQ dyads: Motion of type 1 is active in bulky PQs. Motion of type 2 is active in nonbulky PQs. Substitution of position S with a methyl group renders the system bulky.

Type-1 motion: The PQs have enough flexibility to allow, in the solvent cage, for small-angle fluctuations about an axis parallel to the molecular X-axis. The excursion is very small, *i.e.* the potential well is rather deep. An estimate of the correlation time yields values between 3×10^{-11} s and 2×10^{-10} s. This is similar to results from measurements on triplet states of porphyrins with their planar structure being distorted by substituents. The X-ray structures of different PQs gives additional evidence for the suggested motional effect: The structures of PQs that only differ at the quinone sites show slightly different twist angles of the bond between porphyrin and the cyclohexylene spacer (*e.g.*, P-*trans*CH-BQ, P-*cis*CH-BQ,[388] P-*trans*CH-MQ[396]).

Therefore, this bond is probably flexible enough to allow for the molecular motion in question. The temperature dependence of the suggested effect is much less pronounced than one would expect from the Einstein–Stokes–Debye relation. The effect is visible over the whole temperature range in which the RP state is observed, and as long as motion of *Type 2* (see below) is not active. This is the case for the bulky PQs which have a methyl substituent at point S in Figure 5.39. This also holds for the nonbulky systems, both dyads and triads, in polar solvent at the lower limit of the temperature range and in liquid crystals.

Type-2 motion: The rotation of the quinone plane with respect to the porphyrin plane is the only motion that can modulate J over the necessary range of values and that would be affected by the substitution at point S. Therefore, it is most probable that rotational diffusion about the bond axis between quinone and cyclohexylene takes place with correlation times $\tau < 10^{-6}$ s for nonbulky PQs at the upper limit of the temperature range for which the RP state is observed. This motion has a strong temperature dependence. On cooling down by 10–20 degrees, the potential barrier for the rotation increases by at least a factor of ten. The same effect has substitution at position S of the quinone or using a liquid crystal as solvent environment. Hence, for this situation, the behavior of all PQ systems can be understood in terms of the motion of *Type 1*.

In the systems discussed here, the porphyrin is linked to the quinone by a cyclohexylene bridge. PQ systems that are linked by a phenylene bridge behave differently.[387] In this case TREPR measurements were only possible in liquid crystals, and the observed lineshapes were inversely polarized. Therefore, it was suggested that the RP polarizations are formed through singlet electron transfer and anisotropic relaxation pathways. In the cyclohexylene-bridged systems this polarization mechanism depopulates the triplet states, whereas in the phenylene-bridged systems it populates the triplet states. This is a plausible mechanism, but it requires that singlet recombination takes place on the μs rather than on the ns time scale.

On the other hand, it is also plausible to assume that molecular flexibility is different for the different bridges. Hence, the TREPR results on the phenylene-bridged systems may be explained also by triplet electron transfers, and assuming rotational diffusion about the molecular Z-axis, *i.e.* as is the case for motion of *Type 2*. The correlation times, however, must be in a range that permits spin-rotational relaxation to occur, as is the case in the bulky cyclohexylene-bridged systems. Obviously, by means of TREPR measurements alone, the different mechanisms cannot be distinguished and an ambiguity in the analysis remains.

The lifetime of the charge-separated RP state of about 5 μs is rather long for systems in which the electrons are separated by only 1–1.5 nm. When comparing this with reaction centers of natural photosynthesis, one should keep in mind that the mechanism that leads to this long lifetime in model PQs is very different from that operating in RCs. Also, in photosynthesis, the primary charge-separated radical pair, formed in the first ET step after light irradiation, is electronically strongly coupled, but it is born in the singlet state. A long lifetime of a charge-separated state is achieved by the second ET step to the Q_A quinone acceptor by which the distance between the two unpaired electrons is strongly increased. The $P^{\bullet+}Q_A^{\bullet-}$ radical pair, therefore, is electronically weakly coupled. In contrast to that, the charge-separated states of the porphyrin-quinone dyad and triad model systems are trapped as triplet states of a strongly coupled RP from which charge recombination to the ground state is spin-forbidden.

Nevertheless, we believe that the covalently linked triad systems represent a good starting point for creating long-lived RP states in model tetrad systems of covalently linked porphyrins and quinines.[397] In such 4-component model systems an additional ET step is possible resulting in an even larger separation of the two electrons. The longer the lifetimes of the dyad and triad RPs, the higher the probability for the additional electron-transfer step. Since the lifetime in the dyad and triad systems is limited by the motion-induced triplet–singlet transitions described above, the important strategy for long lifetimes is to restrict molecular flexibility. On the other hand, a certain degree of molecular flexibility is necessary for the first electron-transfer step to take place at all. The challenge for constructing suitable tetrad systems is to allow for enough flexibility to initiate electron transfer, but simultaneously to provide enough rigidity to restrict the lifetime limiting triplet–singlet transitions.

A completely different approach for a covalently linked donor–acceptor model system is shown in Figure 5.40.[398] This linear structure is abbreviated

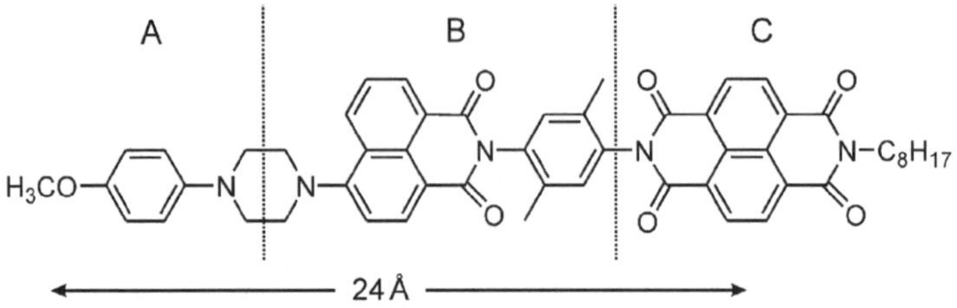

Figure 5.40 Structure of donor–acceptor linear model system ABC.

ABC, it consists of a 4-aminonaphthalene-1,8-dicarboxyimide (B) chromophore, positioned at a fixed distance between a p-methoxyaniline (A) donor and a 1,4,5,8-naphthalene-tetracarboxydiimide (C) acceptor, with an A-to-C separation of ∼24 Å.

The ABC complex, embedded in different LCs, turns out to be a highly efficient photosynthetic model system.[398–400] This is especially evident from the participation of a spin-polarized triplet state that has been previously observed only in photosynthetic RCs, *i.e.* bacterial[401] and PS I,[402] which is attributed to radical-pair intersystem crossing (RP-ISC)[403,404] and occurs when the light-induced ET to the quinone acceptor has been blocked chemically. The ABC-LC system is the first example that exhibits this unique triplet state, as well as other features that fulfill the essential requirements for an efficient photosynthetic model system. The ABC system in its photoexcited state A-B-^{3}C* exhibits characteristic electron-spin polarization (ESP) patterns depending on the orientation of the director of the nematic LC mesophase with respect to the external magnetic field direction, from which the unique multistep intramolecular ET processes in the LC solvent, with singlet and triplet genesis, could be extracted. Unlike ordinary triplet states, which are formed via spin-orbit intersystem crossing (SO-ISC), the triplet here develops from a RP precursor via the RP-ISC mechanism. These two mechanisms can be differentiated via the polarization pattern of the six canonical transitions of the triplet spectrum.[403,404]

A remarkable step forward to considerable quantum yield for light-induced charge separation in an artificial photosynthetic electron-transfer system was achieved recently with a fullerene–porphyrin-linked triad ZnP-H2P-C60.[405] Here, ZnP stands for Zn-porphyrin, H2P for free-base porphyrin. Photoinduced primary charge separation (CS) and charge recombination (CR) were characterized by time-resolved optical and EPR measurements. The electronic coupling element V for the energy-wasting charge recombination is found to be smaller by ≈40% than that for the primary charge separation. This inhibition of the electronic interaction for the charge recombination to excited triplet state, a phenomenon artificial photosynthetic complexes are normally suffering from, largely results from a symmetry-broken electronic structure of the porphyrin moiety that is modulated by configuration interaction between relevant triplet states of the free-base porphyrin. These findings demonstrate the importance of the excited-state electronic character in donor and acceptor components for designing light-generated molecular wires with high quantum yield. A strong dominance of V for CS processes over the competing CR processes, as in natural photosynthesis, can apparently be induced by state mixing in the excited triplet states, leading to a loss in electronic-structure symmetry. This strategy should add to that of controlling the energetic and electronic parameters needed for optimized Franck–Condon factor and electronic coupling matrix element for efficient organic photovoltaic cells. Accordingly, it is hoped that such an approach may lead to the development of chemical strategies for synthesizing high quantum yield donor–acceptor systems.

Next, we will review a TREPR study[356] of a D–s–A complex (Figure 5.41) formed by a pyridino-fullerene (PyrF) ligand of zinc tetraphenylporphyrin (ZnP), where ZnP is the donor (D) and PyrF is the acceptor attached to the spacer (s–A). TREPR both at X-band (9.5 GHz) and W-band (95 GHz) was employed utilizing the corresponding increase of the external magnetic field to discriminate between field-dependent and field-independent spin-polarization mechanisms in photoinduced ET (electron transfer) and EnT (energy transfer) processes. This enabled us to unambiguously characterize the radical-pair (RP) formation.

Among biomimetic systems, porphyrin–fullerene complexes are most attractive as electron donor–acceptor (D–A) systems, due to the celebrated electron-donor properties of porphyrins (D) and the unique electron-accepting properties of fullerenes (A). An important requirement for an effective ET

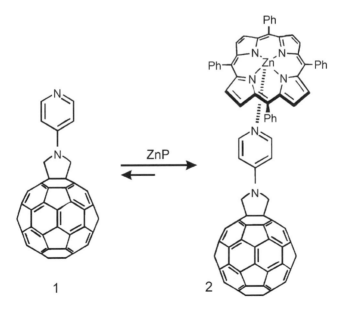

Figure 5.41 Complexation of pyridino-fullerene (1) with zinc tetraphenylporphyrin (ZnP).

process is to minimize undesirable energy-wasting reactions such as charge recombination and EnT from the donor to the acceptor, *i.e.* processes that may compete with efficient charge separation. Increasing the solvent polarity results in stabilization of the charge-separated state, thus favoring ET over EnT. To overcome fast back electron transfer that is, unfortunately, characteristic of many covalently linked D–A systems, employing organic systems linked via hydrogen bonds, van der Waals forces, electrostatic interactions, π–π stacking or metal–ligand coordination has attracted attention. In such noncovalent complexes, fast and efficient charge separation takes place within their intramolecular configuration generated via molecular recognition. In the case of metal–ligand complexation, dissociation of the charge-separated state may occur, in the limit turning charge recombination into an intermolecular process. With respect to porphyrin–fullerene D–A systems, the utilization of fullerene derivatives with an attached pyridine spacer enables complexation with the kinetically labile metal center of zinc porphyrin. Association by such an interaction is reversible, as an equilibrium between bound and unbound states is established. The complex dissociation rate depends on the strength of the coordination bond, thus permitting stabilization of the charge-separated states.

The studies over a wide temperature range were carried out in media of different polarity, including isotropic toluene and tetrahydrofuran, and anisotropic nematic liquid crystals (LCs). At low temperatures (frozen solutions), photoexcitation of the ZnP donor results mainly in singlet–singlet EnT to the pyridine-appended fullerene acceptor. In fluid phases, ET is the dominant process. Specifically, in isotropic solvents the generated RPs are long-lived, with lifetime exceeding that observed for covalently linked donor–acceptor systems. It is concluded that in liquid phases of both polar and nonpolar solvents the separation of the tightly bound complex into the more loosely bound structure slows down the back ET process. Photoexcitation of the donor in fluid phases of LCs does not result in the creation of long-lived RPs, since the ordered LC matrix hinders the separation of the complex constituents. As a result, fast intramolecular back ET takes place in the tightly bound complex. Contrarily to the behavior of covalently linked donor–acceptor systems in different LCs, the polarity of the LC matrix affects the ET process. Moreover, in contrast to covalently linked D–s–A systems, utilization of LCs for the coordinatively linked D–s–A complexes does not reduce the ET rates significantly.

As a last topic of this section, we will briefly introduce multifrequency time-resolved EPR experiments, performed primarily by the groups of C. Corvaja (Padova) and S. Yamauchi (Sendai), on excited states with high spin multiplicity, such as quartet states ($S = 3/2$) and quintet states ($S = 2$). Photoinduced switching of molecular magnetism in organic systems is considered as an important process for future applications of molecular-based magnetism. Again, TREPR turned out to be a powerful tool for investigating short-lived paramagnetic intermediates, formed after laser excitation of organic molecular complexes containing stable radical adducts. Excited states of high spin multiplicity such as quartet and quintet states have been observed for organic complexes consisting of nitroxide and nitronyl nitroxide spin-labelled chromophores. Up to 1999, the subject of excited multiplet states has been thoroughly reviewed.[406] Prominent examples of excited radical-labelled chromophores include porphyrins[406–414] and fullerenes.[415–424] Their excited states with doublet ($S = 1/2$, D), triplet ($S = 1$, T), quartet ($S = 3/2$, Qa) and quintet ($S = 2$, Qi) spin multiplicity arises from the coupling of the excited triplet spins of the chromophore with the spins of the radical labels. The resulting spin-multiplicity scheme of exchange-coupled spin angular momentum operators has been derived in a classic textbook[425] for different limits of exchange in comparison to the other interactions in the spin Hamiltonian. In the strong exchange limit, the observed EPR spectra are the superposition of the spectra for the different total spin states. In thermal equilibrium, each total spin state is Boltzmann populated according to the temperature of the cw experiment. In transient TREPR spectra taken at short delays after the light excitation pulse, strong deviations from the Boltzmann-type EPR intensities are normally observed due to specific spin-polarization mechanisms.[410,424]

In solution, the EPR signal contributions from the individual spin-state multiplicity are characterized by isotropic Zeeman and hyperfine parameters, g and a_k, which are linear combinations of the respective values in the contributing multiplet state,[425] for example[406]

$$g(D) = -1/3g(R) + 4/3g(T);$$
$$g(Qa) = 1/3g(R) + 2/3g(T);$$
$$g(Qi) = 1/4g(R_1) + 1/4g(R_2) + 1/2g(T)$$

with analogous expressions for the isotropic hyperfine constant a_k of nucleus k. Here, D, T, Qa and Qi denote the excited doublet, triplet, quartet and quintet states, respectively, and R_1, R_2 stand for the spin labels.

In solids, the effective zero-field splitting parameters D of the interacting spins in the triplet and spin pairs of radicals R are given by[406]

$$D(Qa) = 1/3[D(T) + D(RT)]; \quad D(Qi) = 1/6[D(T) + D(R_1T) + D(R_2T)]$$

To extract these parameters and determine the multiplicity of the contributing total spin states from EPR experiments, the spectra are simulated using appropriate spin Hamiltonians and trying suitable models for the spin-polarization mechanisms.

In some cases, the contributions can only be disentangled by high-field EPR to resolve even small differences in the g-values of the different total spin states. For a radical complex of Mg-tetraphenylporphyrin (MgTPP) and an axially ligating p-pyridylnitronylnitroxide radical (nit-p-py), time-resolved W-band high-field EPR enabled the first observation of the excited doublet state of a radical–triplet pair in isotropic liquid solution.[411] The measured g-value of the lowest excited doublet (D_1) state is in good agreement with the one calculated in the strong exchange limit from $g(D_1) = -1/3\,g(R) + 4/3\,g(T)$ with the known

g-values of the radical R (nit-p-py) and excited triplet T (MgTPP*). The spin-polarization patterns of D_1, Q_1 and D_0 varied from emission to absorption with time delay after the laser flash. The spectra simulation results show that the spin polarizations are produced by a radical–triplet mechanism with a ferromagnetic exchange coupling ($J > 0$) between the triplet and the radical.

Taking the fullerene–bisnitroxide system as an example,[424] the hyperfine structure of two equivalent ^{14}N nuclei of an excited quintet state in liquid solution could be observed. The quintet state multiplicity was determined by the transient nutation frequency of the TREPR spectrum recorded in rigid glass matrices. The TREPR spectrum in liquid solution is complex and asymmetric. It could be simulated by the superposition of the EPR lines of the biradical in its ground state and of an additional five hyperfine lines with intensity ratios 1 : 2 : 3 : 2 : 1, centered at $g = 2.0039 \pm 0.0005$ and separated by $a = 0.385$ mT. The anomalous intensity of the ground-state lines, even showing emissive contributions, can be explained by a selective intersystem crossing from the excited states. The additional lines in the TREPR spectrum are assigned to the hyperfine-resolved lines of the excited quintet state.

Quartet state species have been observed before[415] not only in rigid glass matrices but also in liquid solutions. The latter is not surprising because the $-1/2 \leftrightarrow +1/2$ EPR transition of a quartet state does not depend on the molecular orientation. That these species have, indeed, quartet-state multiplicity could be deduced from the values of the g-factor and of the ^{14}N hyperfine coupling constants. On the other hand, for triplet and quintet states all EPR transitions depend strongly on the molecular orientation in the magnetic field, and $a\ priori$ narrow EPR lines are not expected in liquids. In liquid solution the EPR linewidth ΔB is determined by the rotational diffusion correlation time τ and the anisotropy of the EPR transition that, in turn, is determined by the electron–electron spin dipolar interaction parameter D: $\Delta B \propto \tau \cdot D^2$. For a quintet state, D is smaller by a factor of 3 than for a triplet state and, hence, only a quintet state multiplicity can explain the observed narrow EPR linewidth of $\Delta B = 0.1$ mT. This is an additional argument for excluding the triplet state to be responsible for the five-line spectrum observed in liquid solution.

Next, we return to biomimetic electron-transfer systems, but to those without covalent bonds between donor and acceptor. It is an interesting observation that, in contrast to the covalently bridged systems, Watson–Crick base-paired donor–acceptor systems, although electronically only weakly coupled, can nevertheless show fast charge-separation rates in conjunction with substantially slower charge-recombination rates, thereby providing rather long-lived charge-separated RP states. Such systems are discussed in the following section.

5.4.3 Base-Paired Porphyrin–Quinone and Porphyrin–Dinitrobenzene Complexes

Among the hotly debated issues with respect to biomimetic modeling of cofactor–protein interactions in the ET pathways of H-bond networks are supramolecular D–A complexes that self-organize by H-bonds. These complexes attract considerable interest as being biologically more realistic model system than the commonly used DA systems covalently linked via a spacer bridge.[58,426,427]

Since the discovery of photoinduced electron transfer between bacteriochlorophyll and cytochrome,[428] many of the topical issues in the realm of biological ET have involved questions of how long-range ET proceeds through various noncovalently linked protein pathways.[76,114,115,429] To answer such questions, among several different spectroscopies also TREPR has been applied to study biomimetic model systems wherein the donor and acceptor molecules are tethered together noncovalently via hydrogen bonds.[368,380,430] These studies, in conjunction with work on covalently linked models[58,382,390,431] have broadened our knowledge of structure–dynamics–function relationships associated with ET processes. Still, many facets of ET processes remain poorly understood, in particular the role of noncovalent interactions. Work on proteins served to establish that specific hydrogen-bonding interactions are critical in achieving long-range D–A electronic coupling.[432–434] Furthermore, studies on model systems revealed that the electronic coupling for ET through hydrogen bonds may be larger than that for ET processes mediated by either σ- or π-bonding networks.[435]

An interesting example of biomimetic modeling primary photosynthesis involves preorganized supramolecular porphyrin–quinone aggregates that are not covalently linked, but molecular recognition rather proceeds via base pairing.[430] The Watson–Crick base-paired complex is shown in Figure 5.42(a). While fluorescence measurements of 1:1 mixtures of guanine-functionalized Zn-porphyrin (ZnG) and cytosine-functionalized quinone (QC) could only suggest that a critical recognition process is established via Watson–Crick base-pairing interactions,[436] TREPR experiments could demonstrate unambiguously, that a charge-separated radical-pair state, $MG^{\bullet+} \cdots QC^{\bullet-}$ (M=Zn, H_2) is formed as a result of a light-induced intraensemble ET. Due to the ultrafast rates at room temperature, attempts to monitor by EPR directly the ET processes using isotropic solvents failed. However, as was true for the D–s–A systems, the inherent difficulty of TREPR, namely relatively low time resolution compared to optical spectroscopy, could be overcome by using LCs as solvents that retard the kinetic processes. Indeed, the results unambiguously confirm the formation of photoinduced charge-separated states, i.e. $ZnG^{\bullet+} \cdots QC^{\bullet-}$ and $H_2G^{\bullet+} \cdots QC^{\bullet-}$, with lifetimes of a few microseconds. Furthermore, unique spin-polarized EPR spectra, which depend on the reorganization energy, the temperature and the specific LC properties, allowed to determine the genesis of the ET route, i.e. whether it goes via the porphyrin precursor in its photoexcited singlet or triplet state.

Next, we review the first study that combines time-resolved EPR at X-band and W-band of a photoexcited base-paired D–A complex.[368] It demonstrates how the critical supramolecular geometry and the ET pathways are stabilized by multiple hydrogen-bonding interactions. TREPR experiments at high Zeeman fields, with significantly improved spectral and time resolution, allowed to

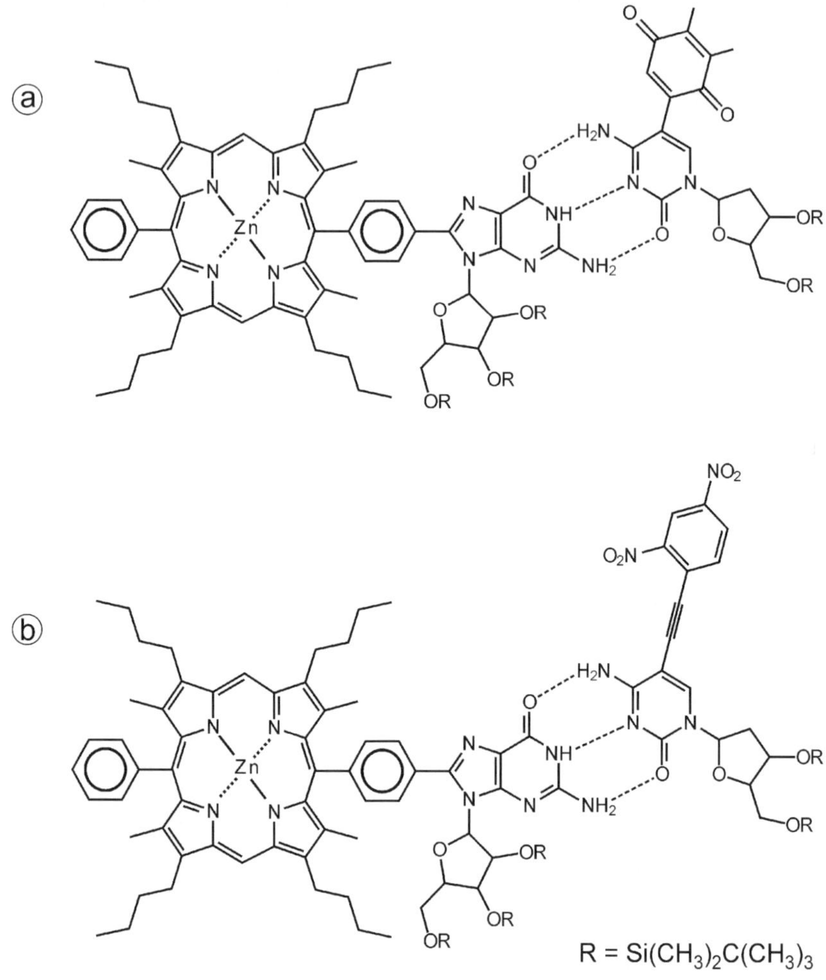

Figure 5.42 Watson–Crick-type base-paired (a) porphyrin–quinone; (b) porphyrin–dinitrobenzene complexes.

directly identify the partners of the charge-separated radical pair generated by selective light excitation and to determine unambiguously the genesis of the spin-correlated coupled radical pair in the ET reaction. In the reviewed study, the focus is on X- and W-band TREPR experiments on the base-paired porphyrin–dinitrobenzene complex shown in Figure 5.42(b). It consists[380] of a guanine-functionalized zinc-porphyrin (ZnP) linked to a cytosine-functionalized dinitrobenzene (DN) via noncovalent base-pairing interactions [ZnP···DN]. A mixture of ZnP and DN was dissolved (5×10^{-4} M) in the liquid crystal E7 (*Merck*), transferred into a thin-walled quartz capillary and flushed with argon. Selective pulsed laser excitation (532 nm, 2 mJ, 5 ns) of the porphyrin part in [ZnP···DN] results in long-range, triplet-initiated ET to the dinitrobenzene, approximately 18 Å apart, in the nematic phase of the LC^{380,430}

$$[\text{ZnP}\cdots\text{DN}] \xrightarrow{h\nu} [^{1*}\text{ZnP}\cdots\text{DN}] \xrightarrow{\text{ISC}} [^{3*}\text{ZnP}\cdots\text{DN}]$$
$$\xrightarrow{\text{ET}} {}^{3}[\text{ZnP}^{\bullet+}\cdots\text{DN}^{\bullet-}]$$
(5.26)

Here, the transient paramagnetic species are TREPR detected in their absorption (A) or emission (E) modes shortly (ns to μs) after the laser pulse in order to detect them in their spin-polarized states.

Figure 5.43(a) shows the X-band TREPR broad and narrow spin-polarized signals, which are ascribed to ^{3*}(ZnP) and to the RP $^{3}[\text{ZnP}^{\bullet+}\cdots\text{DN}^{\bullet-}]$, respectively.[380] The narrow derivative-like AE signal (Figure 5.43(b)) is assigned to a weakly coupled, but still spin-correlated RP, originating from photoinduced triplet-initiated ET, eqn (5.26). This assignment is based on CCRP theory of a weakly coupled RP,[237,437] see above. For small values of zero-field splitting and exchange parameters, D and J, as compared to the RP's EPR linewidth, a derivative-like signal is predicted for X-band EPR whose phase pattern, absorption or emission, should depend on the molecular orientation of the RP with respect to the magnetic field. Indeed, when changing the direction of the LC axis and, thereby, the RP molecular orientation by $\pi/2$, it is observed that the phase pattern changes from AE to EA, which is typical for the CCRP case. Such findings are in contrast to the strongly coupled triplet RP case with relatively large $|D|$ and $|J|$ values.[379,437]

The X-band EPR method for differentiating between weak and strong exchange coupling is, admittedly, rather ambiguous and asks for an independent, more direct high-field TREPR confirmation. Also, the question remained open whether the derivative-like line consists of contributions from both radicals of the pair or from only one. To identify the partners of the RP and their interactions and clarify the origin of the narrow spin-polarized signal, W-band high-field EPR experiments (10 ns time resolution) were carried out.[368] Figure 5.43(c) shows the spin-polarized RP W-band spectrum with two resolved AE features. In fact, for a weakly coupled RP in disordered samples, CCRP theory[237,437] predicts a pair of derivative-like lines. The line separation is field dependent and, at a given B_0 field, is determined by the difference in g-factors of the two radicals in the CCRP, while the splitting of the antiphase components of the individual derivative-like lines is defined by the dipolar and exchange interactions. The isotropic g-factors of $\text{ZnP}^{\bullet+}$ and $\text{DN}^{\bullet-}$ are 2.0025 ± 0.0003[438] and 2.0049 ± 0.0003,[439] respectively. At X-band, their Δg corresponds to a line separation of 0.42 mT that remains unresolved because of the larger dipolar coupling ($|D| = 0.47$ mT for $[\text{ZnP}^{\bullet+}\cdots\text{DN}^{\bullet-}]$ and larger inhomogeneous linewidths.[380] At W-band, however, for a weakly coupled CCRP such as the present one, it is expected that two well-resolved AE lines, separated by 4.2 mT, would be observed. For a strongly coupled triplet pair, on the other hand, only one line would be expected, even at high field. As can be

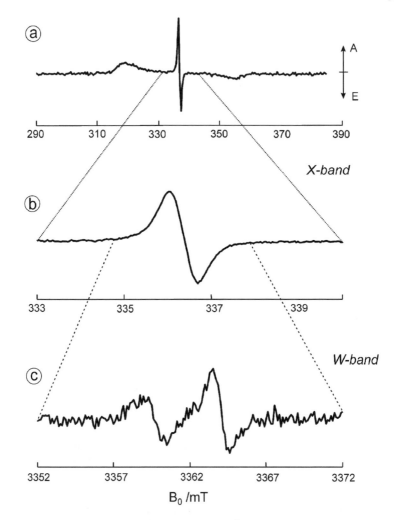

Figure 5.43 (a) X-band AE TREPR spectra of ^{3}[ZnP···DN] and the superimposed RP ^{3}[ZnP$^{•+}$···DN$^{•-}$] (narrow signal), taken in the nematic phase of the liquid crystal E7 at 298 K, 450 ns after the laser pulse.[380] (b) Expanded X-band TREPR spectrum of the RP.[380] (c) W-band TREPR spectrum of the RP, taken 250 ns after the laser pulse at 280 K in E7.[368]

seen from Fig 5.43(c), a weakly coupled RP spectrum with two separated AE features of both radicals is indeed observed for the base-paired ensemble studied. This confirms that light-induced ET proceeds in the steps postulated by eqn (5.26).

Figure 5.44 depicts the time evolution of the W-band TREPR spectra. At early times after the laser pulse, two derivative-like AE lines are observed centered around 2.0052 ± 0.0001 and 2.0025 ± 0.0001. This is in close agreement with the literature values cited above, and exactly what CCRP theory

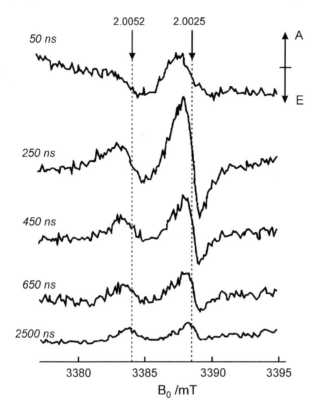

Figure 5.44 W-band EPR time evolution of the spin-polarized radical pair spectra at different delay times after the laser pulse (280 K, in E7). The different g-factors of ZnP•+ and DN•− are indicated.[368]

predicts for a weakly coupled [ZnP•+ ··· DN•−] radical pair: The low-field AE line corresponds to the spin-polarized spectrum of DN•−, and the high-field AE line to the spin-polarized spectrum of ZnP•+. The different signal amplitudes are due to different linewidths of the two radicals. The larger linewidth of DN•− is caused by its larger g-anisotropy, reflecting the orientational distribution of this guest molecule in the liquid crystal. At sufficiently long times after the laser pulse, the AE lines turn into features characteristic of an absorption spectrum (Figure 5.44). This time evolution is ascribed to the CCRP undergoing relaxation to a state of thermal equilibrium, with a spin-lattice relaxation time T_1 of approximately 100 ns. This relaxation, therefore, competes with the building-up of the TREPR spectrum with an ET time constant of approximately 50 ns. This means that back ET must be slower than several µs.

To summarize: The covalently linked ABC system[398] clearly demonstrates how a combination of novel synthesis, advanced TREPR spectroscopy and variations in microenvironment effects can provide a detailed structure–function relationship for light-induced ET processes that cannot be obtained readily using other approaches. The reviewed study on base-paired donor–acceptor complexes[368] has demonstrated that high-field time-resolved EPR opens a new direction in the elucidation of complex photochemical ET reactions, where different paramagnetic states and species are involved. This conclusion applies not only to base-paired and covalently linked donor–acceptor supramolecular ensembles as described here, but also to large ET proteins, such as photosynthetic reaction centers[159,325] and DNA photolyases, as will be discussed in the next section.

5.5 DNA Repair Photolyases

5.5.1 Introduction

The photolyase enzymes are probably Nature's earliest solution to repair lethal and carcinogenic DNA damages in the genome of many organisms caused by life under sunlight. Intriguingly, photolyases utilize light to repair light-induced DNA damages employing light of longer wavelengths than the damaging UV light. Thus, DNA photolyases are among the rare kinds of enzymes that are driven by light. The repair of UV-damaged DNA by photolyases, which belong to the blue-light sensitive flavoenzymes, is a light-induced electron-transfer photoreaction that starts from the fully reduced flavin-adenine dinucleotide (FAD) chromophore, $FADH^-$, see recent review articles.[29,440,441] The leading cause of light-induced skin cancer is epidermal absorption of UV-B or UV-C radiation, and DNA is believed to be the primary chromophore. Photolyases are monomeric proteins operative in many organisms, though not in mammals to directly repair, via a cyclic electron-transfer mechanism, the DNA damages. The most prevalent lesions formed by UV irradiation of DNA are the results of photoreactions of pyrimidine bases. In particular, thymine (T) dimers (T <> T) can form to generate cyclobutane pyrimidine dimer (CPD) lesions following light absorption by a thymine monomer. Photolyases are found in present-day prokaryotes, plants, and a variety of animals including fish, frogs, snakes and kangaroos. They use blue or near-UV sunlight to drive the cleavage of the cyclobutane ring of CPD lesions. Humans use a different set of enzymes to repair CPD lesions. Another type of DNA damage is the formation of (6–4) photoproducts. Accordingly, the photolyase enzymes are categorized as CPD photolyase (also called DNA photolyase) and (6–4) photolyase. DNA photolyases catalyze the repair of UV-damage to cellular DNA by splitting the cyclobutane ring of the major UV photoproduct, the *cis-syn* or the *trans-syn* CPD. The (6–4) photolyase enzyme specifically repairs the pyrimidine-pyrimidone (6–4) photoproduct. Here, we restrict our discussion mainly to the DNA photolyases.

The DNA repair is initiated by visible light, and proceeds through energy transfer, ET and catalysis by a radical-pair mechanism;[442,443] it results in splitting of two C–C bonds. The accepted model for the individual steps of the repair reaction splitting the CPD lesion is shown in Figure 5.45.

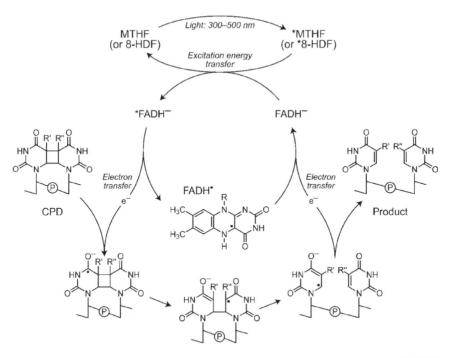

Figure 5.45 The putative reaction mechanism of CPDs by DNA photolyase.[20,442,443] R is defined in Figure 5.46. R' and R" are either CH_3 in thymine or H in uracil.

Photolyase contains a single polypeptide chain and two noncovalently attached prosthetic groups: a flavin adenine dinucleotide (FAD) as the essential catalytic, redox-active cofactor, and a second chromophore, an antenna complex, which absorbs light and transfers excitation energy to the FAD. The function of this antenna complex is to enhance the blue-light quantum yield. For this purpose, all photolyases have a second cofactor, either 5,10-methenyl-tetrahydrofolyl-polyglutamate or 8-hydroxy-5-deazaflavin. So far, high-resolution three-dimensional X-ray structures are available only for DNA photolyase[444–447] but not for the (6-4) photolyase enzyme. Nevertheless, from an alignment of the amino-acid sequences a high degree of similarity between the FAD-binding sites in both photolyases is expected.[448,449]

Two types of photoreactions are involved in the reaction mechanism of photolyases, (1) photorepair of the CPD lesion by the catalytically active enzyme with its FAD cofactor in the fully reduced redox state, $FADH^-$, and (2) photoactivation of the catalytically inactive enzyme. Reaction (2) proceeds when the FAD cofactor is in a redox state different from $FADH^-$, either in the semiquinone state, $FADH^{\bullet}$, or in the fully oxidized state, FAD^{ox}. The biologically relevant redox states of the FAD cofactor are given in Figure 5.46.

To repair DNA lesions, photolyases form a complex with the substrate in a light-independent step. Following blue-light illumination, the excitation energy

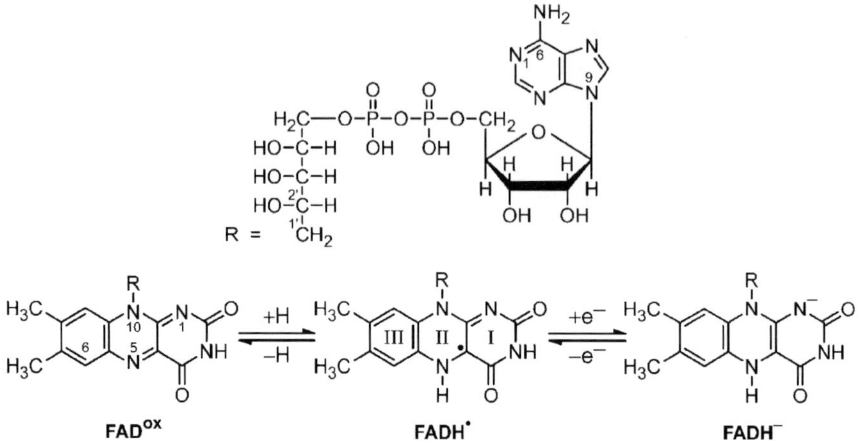

Figure 5.46 The biologically relevant redox states of flavin adenine dinucleotide. The IUPAC numbering scheme is shown for FAD^{ox}.[20]

is transferred from the antenna pigment via Förster dipole–dipole interaction to the fully reduced flavin cofactor, which in turn transfers an electron to the DNA lesion.[440,444] After cleavage of the dimer, an electron is back transferred from the DNA strand to the neutral flavin radical, FADH•, which is restored to its catalytically active FADH⁻ form.[450–453]

CPD photolyases are classified into two classes, I and II, based on their amino-acid sequence similarity. Class-I photolyases are found in many micro-organisms, whereas most of class-II photolyases are found in higher eukaryotes.[442] The class-I photolyases are further categorized according to their light-harvesting chromophore into either a folate- or a deazaflavin-type, with 5,10-methenyl-tetrahydrofolyl-polyglutamate (MTHF) or 8-hydroxy-5-deazaflavin (8-HDF) as antenna cofactor, respectively. To date, the three-dimensional structures of the class-I CPD photolyases from *Escherichia (E.) coli* (MTHF-type) and *Anacystis (A.) nidulans* (8-HDF-type) have been determined by X-ray crystallography at atomic resolution.[444,445] They are the first structures for each type of photolyase, in which the geometry of the two cofactors became clear and an analysis of the energy-transfer process was made. Recently, the crystal structure of the thermostable class-I CPD photolyase from *Thermus (T.) thermophilus* has also been reported.[446] Its protein backbone structure shows a high similarity in overall folding and FAD cofactor binding when compared with *E. coli* and *A. nidulans* photolyases; 8-HDF is thought to be the antenna chromophore.

In photolyases, binding of DNA substrate and repair catalysis occur in separate steps in view of type and mechanism: Binding of DNA is light independent, whereas catalysis is light initiated. The mechanism of substrate recognition remained largely unclear for many years because of the lack of structural information regarding the photolyase–substrate complex. This

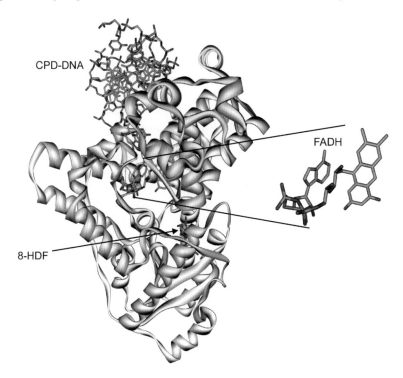

Figure 5.47 X-ray-crystallography-based ribbon model of the recognition mode of CPD lesions by DNA photolyase.[447] The photolyase is bound to duplex DNA with the CPD lesion in a conformation (bent by 50 degrees) ready for blue light–mediated repair of CPD lesions. Abbreviations: CPD-DNA: duplex DNA with CPD lesion; FADH: flavin adenine dinucleotide; 8-HDF: light-harvesting antenna cofactor 8-hydroxy-5-deazaflavin. For details, see ref. [447] and text.

situation has changed very recently: The high-resolution crystal structure of a photolyase bound to CPD-like DNA lesion after *in-situ* repair was reported,[447] see Figure 5.47. The DNA photolyase is bound to duplex DNA that is bent by 50 degrees and comprises a synthetic CPD lesion. This CPD lesion is flipped into the active site and split there into two thymines by synchrotron radiation at 100 K. Although photolyases catalyze blue-light-driven CPD cleavage only above 200 K, this structure apparently mimics a structural substrate during light-driven DNA repair in which back-flipping of the thymines into duplex DNA has not yet taken place.

Both the electronic structure of the flavin cofactor and the protein–cofactor interactions play a central role in the repair process. Hence, precise information about both of them is indispensable for an understanding of its electron-transfer characteristics. Of particular interest are differences in the electronic structure of the flavin cofactor in the DNA photolyase in comparison to the (6–4) photolyase since the two enzymes, and their redox-active flavin cofactor, have different functional properties in their repair process. EPR and ENDOR have proven to be the methods of choice to extract information about the electronic structure of flavins and their interactions with the protein environment from the anisotropic components of the g- and hyperfine (A)-tensors.[20,21,25,28,454] Precise determination of the magnetic interaction tensors enables mapping of the spin-density distribution over the isoalloxazine ring[22] and may provide an understanding of how it is modulated by interactions with the protein environment. Furthermore, knowledge about the g- and A-tensors is essential if the identity and orientation of radicals and interacting radical pairs along the electron-transfer pathway are to be ascertained. In flavoproteins, blue-light-induced radical pairs involving a flavin and an oxidized amino acid have been observed, among others, in DNA photolyase,[455] in (6–4) photolyase[24] and in the cryptochromes.[456] The cryptochromes are not repair enzymes but blue-light photoreceptors that are involved in setting the circadian rhythm of plants and animals.

There is still an on-going debate about the enzyme–substrate docking configuration.[29] In CPD photolyase, FAD sits at the bottom of a cavity that leads to the protein surface. The opening of the cavity is flanked by positively charged amino-acid residues.[444] Though FAD is clearly the active site, it is not obvious which portion of the CPD, if any, sticks into the cavity, or if the dimer is in van der Waals contact with the FAD at the base of the hole. Model calculations at various levels of theory suggest FAD cofactor-to-dimer-distances between 3 Å and 15 Å.[457–459] For example, one model suggests the single-strand DNA to be in a conformation similar to that in a DNA duplex as formed with a complementary strand by Watson–Crick base pairing ("not flipped"), in another model the dimer is rotated from its usual position into the protein cavity ("flipped out"). Distances of 15 Å and 12 Å are predicted for the "not flipped" and "flipped out" configurations of the CPD lesion, respectively. Up to now, no computational study has decisively determined the cofactor-to-substrate distance. Experimental evidence was contributed from an ENDOR study on the substrate binding to CPD photolyase,[23] which is consistent with a large distance (≥ 6 Å) between the CPD lesion and the isoalloxazine ring of FAD. The ENDOR data show that the structure of the protein around the FAD-binding site does not change very much upon substrate binding. Small shifts of the isotropic proton hyperfine couplings within the isoalloxazine moiety of FAD can be understood in terms of the cofactor binding site becoming more nonpolar because of a displacement of water molecules upon CPD docking to the photolyase enzyme. This changed polarity of the matrix surrounding the FAD cofactor could be rationalized by DFT calculations employing an isodensity polarized continuum model.[23]

As was pointed out above, in addition to energy transfer and ET during the DNA repair reaction, another ET process, termed FAD cofactor photoactivation or photoreduction, is observed *in vitro*.[460] During protein purification under aerobic conditions, the FAD cofactor is reversibly oxidized to the catalytically inert neutral radical, FADH•, or the fully oxidized FAD^{ox}. Photoactivation is the light-initiated process, involving reversible formation of amino-acid radicals, to lower the redox state of FAD to the fully reduced, catalytically competent FADH$^-$. In CPD photolyase, the final electron donor is a conserved tryptophan (W), whereas in (6–4) photolyase this is a tyrosine (Y) residue.[24] Exogenous electron donors reduce W• or Y• in competition with charge recombination, thus stabilizing FADH$^-$ required for the DNA repair.

There is experimental evidence for distinct pathways of intraprotein ET in photoactivation in the two different classes, MTHF- or 8-HDF-type, of CPD photolyase. Preliminary evidence suggests that ET proceeds through a chain of tryptophan residues in *E. coli* photolyase (W382, W359, and W306),[444] whereas ET from a Y to the W• radical creates a transient Y• radical experimentally observed in *A. nidulans* photolyase.[460] Structural properties of the photolyase tyrosyl radical, such as spin densities, protonation, and β-CH$_2$ conformation, are similar to those of Y• radicals in PS II.[460]

In the following, some recent results of high-field EPR on photolyase systems will be summarized.

5.5.2 High-Field EPR and ENDOR Experiments

Besides enzymatic DNA repair, the second photoreaction, leading to photoactivation FAD (see above), is presently in the focus of the research on flavinenzymology.[29,461] In this photoreduction reaction, a chain of highly conserved tryptophan residues (W382, W359 and W306 in E. coli DNA photolyase) is utilized to lower the redox state of the FAD cofactor in a light-induced ET, starting from the catalytically inactive enzyme with FAD being either fully oxidized, FAD^{ox}, or one-electron reduced, FADH•.[29,461] W-306 serves as the terminal electron donor that is either re-reduced by backward ET from FADH• or by ET from exogeneous reductants to W-306 in the protein buffer.[460,462] In both photoreactions, DNA repair and flavin photoreduction, a semireduced flavin radical plays a crucial role. The photoreduction of the flavin chromophore can be studied on a nanosecond timescale using direct-detection transient EPR, TREPR, at various microwave frequencies and corresponding magnetic-field settings. The hyperfine couplings reflect the wavefunction of the unpaired electron, they are amenable by multifrequency ENDOR spectroscopy. Systematic mapping of the unpaired electron spin-density distribution in flavin radicals leads to knowledge of the electronic structure. This knowledge is crucial for an understanding of the reactivities of flavin cofactors embedded in different protein environments.

The relative orientation of the hyperfine- and g-tensors tells us something about the symmetry of the π-electron spin-density distribution over the conjugated molecule. Based on the results from ultrahigh-field EPR at 360 GHz/ 12.9 T,[25] see below, the orientation of the g-tensor of FADH• in *E. coli* DNA photolyase with respect to the frame of the flavin's dominant hyperfine-tensor of the proton H(5) of FADH• could be unambiguously established by pulse ENDOR at 95 GHz and 3.5 T,[28] taking advantage of the sufficiently high Zeeman magnetoselection to obtain single-crystal-like ENDOR data even from the disordered sample. This work was extended by an X- and W-band pulsed Davies-ENDOR study at 80 K to probe the N(5)–H bond of the isoalloxazine moiety of the flavin cofactor FADH• in DNA photolyase,[454] examining both the protonated and deuterated forms of this bond. By comparing the anisotropic hyperfine couplings measured for the protonated and deuterated samples, after scaling them by the magnetogyric ratio of the deuteron and the proton, subtle differences were revealed concerning the respective deuteron couplings as obtained by 95-GHz ENDOR on H → D buffer-exchanged samples. These differences are attributed to the different lengths of N(5)–H and N(5)–D bonds owing to the different masses of H and D. From the distance dependence of the dipolar hyperfine coupling it was estimated that the N(5)-D bond is about 2.5% shorter than the N(5)-H bond. It is believed that the strength of this hydrogen bond is crucial for noncovalent binding of the FAD to the protein. Subtle differences have been observed of the strength of this hydrogen bond in different types of photolyase.[27,454] The ENDOR method is highly sensitive to very small variations of the hydrogen-bond length, as was shown by comparing measured hyperfine couplings with those obtained by quantum-chemical calculations using DFT methods.[22,454]

While the hyperfine coupling parameters are *local* probes for the electron-spin density on individual nuclei in the flavin radical, the g-tensor represents a *global* probe for the electronic structure of the radical. The neutral flavin-adenin dinucleotid radical cofactor FADH• has an extremely small g-anisotropy, comparable to that of the primary-donor cation radical radical P•+ in bacterial photosynthesis (see Section 5.2.1). Hence, high-field cw EPR at 360 GHz/12.9 T was necessary to resolve the canonical g-tensor components in the powder-type EPR spectra.[25] In short, the main results of this study are as follows:

The measured g-tensor data of FADH• from *E. coli* DNA photolyase were the first to be extracted from a flavin radical spectrum in which the full rhombic symmetry of the g-tensor could be resolved. A fit of the spectrum, based on a spin Hamiltonian incorporating the electron Zeeman and electron–nuclear hyperfine interactions, yielded accurate principal values of the g-tensor that, indeed, show only a small anisotropy: $g_{XX} = 2.00431(5)$, $g_{YY} = 2.00360(5)$ and $g_{ZZ} = 2.00217(7)$, where X, Y, Z is the g-tensor axes system, and the experimental errors are given in parenthesis. The hyperfine splitting observed in the g_{YY} region could be assigned to an effective hyperfine-tensor component of the H(5) proton in the 7,8-dimethyl isoalloxazine moiety of FADH•. By combining the g-tensor results with additional information on the H(5) hyperfine coupling obtained by X-band Davies-ENDOR, the orientation of the g-tensor principal axes with respect to the H(5) hyperfine principal axes was deduced. Remaining ambiguities in the sign of the angle between the g-frame and the hyperfine-frame were tried to be solved from g-tensor calculations using DFT and semiempirical AM1-based methods. An unambiguous determination of the angle between the g-tensor principal axes and the molecular frame had to wait, however, for an additional, orientation-resolving W-band high-field ENDOR experiment to determine the principal axes of the H(5) proton hyperfine-tensor. This is discussed below.[25,30]

The g-tensor of the neutral radical form FADH$^{\bullet}$ of the flavin adenine dinucleotide cofactor of (6–4) photolyase from African clawed frogs *Xenopus (X.) laevis* have been determined by high-field/high-frequency EPR performed at 360 GHz/12.9 T. Due to the high spectral resolution the anisotropy of the g-tensor could be fully resolved in the frozen-solution cw EPR spectrum. By least-square-fittings of spectral simulations to experimental data, the principal values of the g-tensor have been established: $g_{XX} = 2.00433(5)$, $g_{YY} = 2.00368(5)$, $g_{ZZ} = 2.00218(7)$. A comparison between very high-field EPR data with proton and deuteron W-band ENDOR measurements of either the α-proton H(5)[25] or the methyl group α-protons H(8α)[28,454] yielded information concerning the orientation of the g-tensor with respect to the molecular frame (x, y, z). In other words: Both methods provide the rotation angle δ between the principal axes of the g-tensor and the molecular frame, as defined by the N(5)–N(10) axis (or the N(5)–H(5) bond), see Figure 5.48.

This data allowed a comparison to be made between the principal values of the g-tensors of the FADH$^{\bullet}$ cofactors of photolyases involved in the repair of two different DNA lesions: the cyclobutane pyrimidine dimer (CPD) and the (6–4) photoproduct. We see that g_{XX} and g_{ZZ} are similar in both enzymes, whereas the g_{YY} component is slightly larger in (6–4) photolyase. This shows the sensitivity of the g-tensor to subtle differences in the protein environment experienced by the flavin cofactor. Apparently, very subtle differences in the cofactor binding situation may lead to significant deviations in the g-tensor principal axes orientations in flavoproteins, while only moderately affecting their g-tensor principal values. This certainly underlines the need for calculations of g-tensor parameters with specific protein–cofactor interactions included.

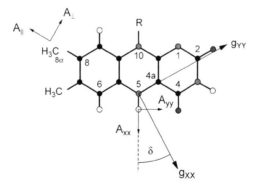

Figure 5.48 Molecular structure and IUPAC numbering scheme of the 7,8-dimethyl isoalloxazine moiety of neutral flavin semiquinones. Carbon, nitrogen, oxygen and hydrogen atoms are shown in black, light grey, dark grey, and white, respectively. R denotes the ribityl side chain of FAD. X, Y, and Z are the principal axes of the g-tensor. The principal values of the hyperfine-tensor $\tilde{A}$(H(5)) are A_{xx}, A_{yy}, and A_{zz}, and those of the hyperfine-tensor **A**(H(8α)) are $A_{\parallel}$ and $A_{\perp}$.

Quite surprisingly, the g-tensor principal axes in a flavin are not oriented as one would have expected for a 1,3-semibenzoquinone radical.[25] In FADH•, the large spin densities on N(5) and C(4a) apparently contribute to a significant reorientation of the g-tensor principal axes. Furthermore, different angles between the orientations of the X- and Y-axes of FADH• with respect to the N(5)–H(5) bond have been observed in DNA photolyase and (6–4) photolyase (see below). In order to understand the origin of the different g-tensor orientations, a comparison of the three-dimensional structures of both enzymes is required. At present, however, only that of the DNA photolyase enzyme from E. coli is available.[445] Fortunately, this lack of X-ray structure information can be compensated, at least partially, by taking advantage of the large magnetoselectivity of 360-GHz high-field spectra, even for an extremely small g-anisotropy and even for disordered frozen-solution samples like in these studies of flavin radicals.

As in the case of DNA photolyase, the 360-GHz EPR spectrum of FADH• in (6–4) photolyase shows a hyperfine splitting at the resonance position of g_{YY} (see Figure 5.49) that was attributed to the strongly coupled α-proton H(5), the assignment being based on EPR and ENDOR studies on deuterium-exchanged FADH• in DNA photolyase of E. coli.[25]

The observed difference between $A_{yy}(H(5))$ and the splitting at g_{YY}, as measured for FADH• in the two enzymes, DNA photolyase and (6–4) photolyase,

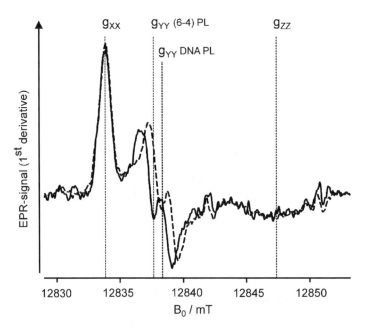

Figure 5.49 360-GHz EPR spectra of the neutral flavin semiquinone cofactors FADH• in the DNA photolyase (PL) of E. coli (dashed line; data taken from ref. [25]), and the (6–4) photolyase of X. laevis (drawn line). Both spectra have been recorded under the same experimental conditions.[30]

is likely to be due to matrix-induced differences in the rotation angle δ (see Figure 5.48) that determines the orientation of the X- and Y-axes of the g-tensor in the molecular plane. Assuming that Z is perpendicular to the molecular plane, then from fits to the 360-GHz cw-EPR spectrum of FADH$^\bullet$ bound to DNA photolyase[25] the value $|\delta|=(16\pm2)°$ was obtained for the angle between the Y- and y-axes, i.e. between the Y-axis of the g-tensor and the direction of $A_{yy}(H(5))$. Surprisingly, this value deviates significantly from the angle $\delta=-(29\pm2)°$ obtained for FADH$^\bullet$ bound to (6–4) photolyase.[30]

There could be several reasons for this discrepancy. For example, the π-system of the isoalloxazine moiety might deviate from planarity. This could be due to differences in the binding structure or the arrangement of the amino acids in the flavin-binding pocket. Such deviations from a planar structure would tilt the XY-plane relative to the principal axes systems (x, y, z) of the hyperfine-tensors of H(5) and H(8α). Although a nonplanar flavin radical cannot be entirely excluded, W-band ENDOR measurements[27,28] show that z is oriented almost parallel to $A\perp(H(8α))$ and perpendicular to $A_\|(H(8α))$. This implies a high degree of planarity, a result that is also supported by quantum-chemical calculations.[22] A promising method to remove remaining ambiguities in the relative orientation of the g- and hyperfine-tensors would be to perform orientation-selective ENDOR at 360 GHz/12.9 T, at which the corresponding proton Larmor frequency is 547 MHz. By exploiting the very high Zeeman resolution of the g-tensor components, it should be feasible to obtain single-crystal-like hyperfine information even for disordered radicals with very small g-anisotropies.[193]

Experiments to apply 360 GHz/12.9 T ENDOR to biologically functional organic radicals bound to proteins, including flavins, chlorophylls and quinones, are planned at FU Berlin.

5.6 Colicin A Bacterial Toxin

5.6.1 Introduction

The pore-forming protein colicin A is a member of a family of plasmid-encoded bacterial toxins that – different from most other toxins – are toxic to their own bacterial host, but not to other species.[463] Colicins are produced by and toxic to *Escherichia (E.) coli*,[463,464] but their plasmid bears an immunity protein that ensures protection of the organism from the colicin toxin. Seven members of the ion-channel forming colicins are currently identified and sequenced, denoted E1, A, B, K, N, Ia, Ib.[463,465,466] They are water-soluble proteins, mostly about 60–70 kD in size, and have varying homology among themselves.

Colicin A kills unprotected cells of attacked organisms by inserting specific portions of protein subdomains into the cytoplasmic membrane forming a voltage-gated ion channel. The open channel leads to electrical depolarization of the membrane and depletion of intracellular ion pools, which ultimately leads to cell death. The soluble channel-forming domain with its two hydrophobic α-helices buried inside the domain, has to undergo massive refolding to allow these helices to penetrate the nonpolar region of the membrane. Thus, the conformational changes preceding channel formation and escorting membrane insertion are crucial for colicin function. It is generally realized that understanding the mechanisms of protein entry into membranes and transmembrane channel formation on the molecular level is one of the fundamental aspects of membrane biology and toxicology (2003 Nobel Prize in Chemistry to R. MacKinnon for structural and mechanistic studies of ion channels in cell membranes). Such an understanding is not only of current biophysical interest, but also of biomedical significance, because insertion of protein domains into membranes and subsequent channel formation are common to many toxic proteins of bacterial pathogens, such as the diphtheria, tetanus and cholera toxins,[465–469] which can infect a large variety of organisms ranging from bacteria to humans.

The colicin toxins have to overcome the protecting barriers of the attacked cell and, to this end, colicins are composed of three functional globular domains: One domain regulates the target and binds to the receptor on the sensitive cell. The second is involved with translocation, co-opting the machinery of the target cell. The third is the "killing" domain and may produce a pore in the target cell membrane. Surprisingly, the wide colicin channel, in particular that of colicin A, is highly selective for protons over other cations and anions by many orders of magnitude. Currently, there is strong revived interest in understanding the mechanism behind the remarkable properties of the colicin channel.[470] The mechanisms of cell killing by colicins involve three distinct functional steps:[33,463] binding of the toxic protein to a receptor at the surface of the outer membrane, translocation across the outer membrane and periplasm, and insertion of specific helical segments of the toxin subdomain into the cytoplasmic membrane to form, after voltage activation, selective ion-conducting channels that destroy the membrane potential of the target cell. The three functional protein domains of colicin A are: the central receptor domain, R; the N-terminal translocation domain, T, to penetrate the outer membrane (aided by receptor proteins of the target cell) and to traverse the periplasm (aided by several membrane translocation proteins of the target), and the C-terminal channel-forming domain, C, to penetrate the inner membrane that protects the cytoplasm of the cell,[466] see Figure 5.50. These distinct functions of the colicins are reflected by their three-dimensional shape. The X-ray crystal structure of a complete (T-R-C) colicin Ia protein was recently determined to 2.3 Å resolution.[466,471] It reveals a harpoon-shaped molecule, 210 Å long, with the three functional domains well separated from each other. The long α-helices that link R with T and C enable the colicin molecule to span the periplasmic space and contact both the outer and inner membranes simultaneously during function,[471] thereby overcoming the protection barriers of the attacked cell.

The isolated C-domain retains its channel-forming ability in aqueous solutions of artificial membranes,[476,477] such as lipid vesicles. Hence, details of refolding processes of the C-domain can be studied by *in-vitro* experiments (see below). The X-ray crystal structure of the 204-residue (21 kD) channel-forming C-domain of

Applications of High-Field EPR on Selected Proteins and their Model Systems 325

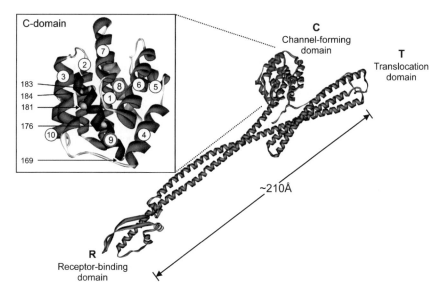

Figure 5.50 X-ray-crystallography-based ribbon structural model of a colicin bacterial toxin[458,463] with its three functional domains R, T, and C. The inset shows the X-ray structural model of the channel-forming C-domain of colicin A.[472–474] The encircled numbers designate the ten α-helices, the other numbers the sites on helix 9 that were successively marked by nitroxide spin labels. For details, see ref. [475] and text.

colicin A in its water-soluble conformation is available to 2.4 Å resolution[473] (see inset of Figure 5.50). Ten α-helices in a bundle, eight amphiphilic, two hydrophobic, are arranged in such a way that the amphiphilic helices surround the hydrophobic hairpin (helices 8 and 9) deeply buried in the center of the bundle. Thereby, the 8, 9 hairpin is shielded from contact with the water solvent.

In view of the difficulties encountered with determining X-ray structures of transient protein states in action, numerous spectroscopic techniques are being used to study colicins on the molecular level. Despite many years of work on colicins by solid-state NMR,[31,478–480] FT-IR,[481] FRET[472] and EPR,[32,33,482] the details of the refolding processes upon membrane association and channel formation are not yet known. Among the possible reasons for having failed so far in determining the three-dimensional structure of the membrane-associated state is the formation of a "molten-globule" membrane-insertion intermediate with poorly defined tertiary structure.[483] The molten globule intermediate in protein folding is discussed to be mainly characterized by: (1) conservation of the secondary structure of the native protein and its approximate volume; (2) loss of tertiary structure or poorly defined tertiary structure contacts; and (3) increased accessibility of the hydrophobic core to hydrophobic agents.[484] All in all, the channel-forming C-domain of colicin A is comparatively well characterized by now, and may serve as a paradigm system to answer an intriguing question of general interest in molecular biology and medicine: What is the mechanism to switch on the insertion of water-soluble pore-forming proteins into the nonpolar lipid environment of a membrane interior?

5.6.2 Models of Transmembrane Ion-Channel Formation

The C-domain of colicin A (and other members of the colicin family) can adopt two conformations, the water-soluble form and the transmembrane form. The transition between these conformations, which are energetically different from each other, requires massive refolding of the tertiary structure, probably with a molten-globule state as an intermediate. To date, two alternative models are being discussed to explain how the C-domain turns itself inside out to form the membrane-associated state with pore formation, the "umbrella" model[473] and the "penknife" model[472] (see Figure 5.51). Structurally, these models differ in the orientation of a helical hairpin formed by the hydrophobic helices C8 and C9 in the phospholipid bilayer. These models agree, however, in the view that most residues are located at the membrane-water interface. Mechanistically, they differ in the dynamics of the relatively slow (100 ms–s range) membrane-insertion step to be either spontaneous or voltage triggered. The docking process to the membrane surface initiates a slow refolding of the C-domain to bring the helices C8 and C9 to the outside of the protein complex. In the umbrella model the hydrophobic hairpin 8, 9 spontaneously traverses the

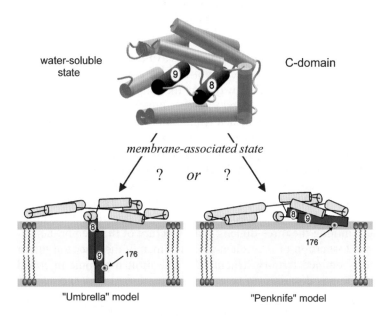

Figure 5.51 The "umbrella" model[476] and the "penknife" model[472] of the membrane-associated state of the channel-forming C-domain of colicin A. In the water-soluble state of the C-domain, the hydrophobic helices 8, 9 are deeply buried in the interior of the domain where a nonpolar microenvironment is energetically stabilized.

membrane, whereas in the penknife model the refolding leaves the 8, 9 hairpin close to the membrane surface, but a change of the electric transmembrane potential is required to trigger insertion of the 8, 9 hairpin into the membrane and, ultimately, to open the channel for ion flow.

For further reading about the biological background, structure and spectroscopy of colicin toxins see, for example, refs. [464,466,469,485,486].

5.6.3 95-GHz EPR Studies of Membrane Insertion

In the following, we review our work on colicin A transient states[34] applying high-field EPR at 95 GHz in conjunction with site-directed spin-labelling (SDSL) techniques using MTS as the nitroxide spin label (see Section 5.2.2.1). Owing to the high spectral resolution of 95-GHz EPR we could use both the A_{zz} hyperfine-tensor component of the ^{14}N nucleus and the g_{xx} tensor component as sensitive probes for the polarity and proticity of the microenvironment of the nitroxide side chain R1, and for its motional characteristics in three-dimensional space under the local constraints of the protein.[277,289,314] The toxic activity of the spin-labelled mutants was checked to be almost identical to that of the wild-type protein. In Figure 5.50, the five individual amino-acid residues of helix 9 of the C-domain of colicin A are indicated at residue positions 169, 176, 181, 183 and 184, which have been replaced by cysteines via exchange mutagenesis.[472,474] The single cysteines were spin labelled with the nitroxide label MTS providing a nitroxide side chain R1.[34] Our experimental strategy for detecting the refolding of the C-domain under membrane association was to compare the EPR spectra of colicin A in buffered aqueous solution under physiological conditions (pH 8) with those after adding lipids to the sample to form vesicles as artificial membranes.[34] To this end, small unilamellar vesicles were generated by sonification of DMPG (1,2-dimyristoyl-sn-glycero-3-phospho-rac-(1-glycerol)) in water.

To learn about the membrane-insertion mechanism, leading either to a transmembrane configuration (as predicted by the umbrella model) or to a configuration with the hydrophobic helix 9 staying close to the water/membrane interface (as predicted by the penknife model), the A_{zz} and g_{xx} tensor components were measured for the five mutants. Both tensor components have proven to be sensitive probes of the polarity profile of the nitroxide microenvironment along a helix structure.[277,288,289] Figure 5.52(a) shows, as a representative example, W-band spectra with and without lipid admixture of the colicin A mutant G176C. The spectra simulations clearly reveal shifts of the g_{xx} and A_{zz} components at the low-field and high-field regions, respectively. The signal-to-noise ratio is good enough to extract reliable shift values of the tensor components upon lipid addition.

In the water-soluble conformation (no lipid added) the A_{zz} and g_{xx} values reveal high polarity at both ends of helix 9 (positions 184, 169), and a lower polarity in the center (position 176). This finding is consistent with the X-ray crystal structure.[473] After addition of lipid a rather uniform, high-polarity character of the nitroxide microenvironment of all mutants results. The highest

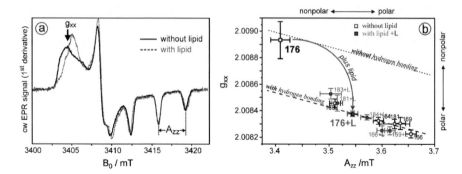

Figure 5.52 (a) W-band EPR spectra of the colicin A mutant G176C with the NO˙ spin-label at position 176. The aqueous-solution and membrane-associated (with lipid) states show different g_{xx} and A_{zz} values. Artificial membranes are formed by adding an aqueous suspension (pH 8) of DMPG phospholipid vesicles (25 mg/ml). (b) Polarity plot of g_{xx} vs. A_{zz} for the nitroxide spin-label positions in helix 9 of site-directed colicin A mutants. The measured tensor components with addition of lipid vesicles and without lipid addition are marked by full and open squares, respectively. The dashed lines define the limits between the nonhydrogen bonded (short dashes) and the fully hydrogen bonded (long dashes) cases. For details, see ref. [34] and text.

polarity change is experienced by the nitroxide side chain attached to position 176 in the central part of the helix. In comparison, the respective A_{zz} and g_{xx} data, determined for spin-label side chains attached to the outer surface of helix F of the transmembrane BR protein, shows a hydrophobic barrier[277] (see Section 5.2.2.2). Such a hydrophobic barrier does not exist for the membrane-associated configuration of helix 9 of colicin A, *i.e.* this helix does not span the membrane. Moreover, the changes of A_{zz} and g_{xx} upon addition of lipid reveal that all segments of helix 9 penetrate less than 1.0 nm into the artificial membrane.

In Figure 5.52(b) the W-band measurements of g_{xx} and A_{zz} are summarized. Both probes show that the central-region position 176 experiences a strong change in polarity of the microenvironment towards more polar character, whereas in the vicinity of the end-region positions 169, 184 only weak polarity changes occur upon adding lipids. This different behavior of polarity changes upon membrane association for position 176 and positions 169, 184 is especially evident from the plot g_{xx} vs. A_{zz} since such a representation is particularly informative with respect to local electric-field and H-bonding effects of the microenvironments of the nitroxide labels.[277] As is shown in Figure 5.52(b), addition of DMPG lipid results in a dramatic change of the microenvironment polarity in the vicinity of position 176 towards an increased polarity.

How can the W-band EPR results be rationalized in terms of the validity of either the umbrella or the penknife model of the membrane-associated configuration of the C-domain? In case the umbrella model is valid, helices 8 and 9 should penetrate the membrane spontaneously so that their central part, as probed by the nitroxide side chain 176 in helix 9, would be placed into the membrane's interior, *i.e.* in a highly nonpolar region. This means that in the umbrella model one would expect for spin label 176 no large changes of g_{xx} and A_{zz} upon adding lipid to the aqueous sample because in both states, water-soluble and membrane-associated, the microenvironment would remain nonpolar. When the penknife model is valid, however, helix 9 should remain for some time in a transient state close to the membrane surface – until a change of the electric membrane potential triggers helix insertion and pore formation in the membrane. In this situation, position 176 would experience a drastic change of the microenvironment from nonpolar to polar in the membrane-associated state.

This is exactly what we observe via the polarity probes g_{xx} and A_{zz} when adding vesicle-forming lipids to the aqueous sample. Hence, our data are not consistent with the umbrella model, but validate the penknife model for membrane association of the colicin A channel-forming domain. We conclude that multifrequency EPR in conjunction with SDSL shows that the penknife model of the molten globule intermediate state is an adequate description of the membrane-associated C-domain of colicin A. This conclusion is in agreement with earlier FRET experiments on colicin A double mutants with site-specifically attached fluorescence labels.[472] From these experiments it was concluded that the umbrella model has to be rejected because the postulated large distances between the hydrophobic hairpin and the remainder of the molecule were not observed (for more details, see ref. [34]).

On the other hand, solid-state NMR studies on the conformational changes of the colicin Ia channel-forming domain upon membrane binding[31,480,487] provided support for the umbrella model, although the data do not rule out the possibility that the umbrella and penknife structures exist in a dynamic equilibrium.[487] It is interesting to note that, among the various studies on colicin E1, the X-band SDSL/EPR results of Hubbell and coworkers[32,33,482] are in favor of the umbrella model and, thus, in contrast to our W-band SDSL/EPR results on colicin A.[34]

For colicin A, the refolding situation for channel formation with intermediate molten-globule state might be different from that of the homologous colicins E1 and Ia. The molten globule state, as originally proposed to be a compact denatured state with a native-like secondary structure but a largely flexible side-chain tertiary structure,[484] has been challenged.[488] The authors examined the intermediate conformational states of various proteins and observed that the amount of secondary structure and the extent of compactness are not necessarily close to those of the native structure, but vary significantly depending on the specific protein species. This suggests that there is a range of molten-globule states from relatively disordered to highly ordered conformational states.[489]

In principle, our results do not exclude the existence of yet another conformation of colicin A for the closed channel state, and a thermodynamic equilibrium between the two conformations, according to the umbrella and the penknife model, can be envisaged. The equilibrium may depend on the specific member of the colicin family, the composition of the artificial or natural membrane and the pH value. The influence of these parameters on the conformation of colicin A remains to be elucidated.

The extremely interesting, albeit very demanding extension of the *in-vitro* EPR experiments on the spin-labelled C-domain to *in-vivo* experiments would be a natural next step. Multifrequency SDSL/EPR and PELDOR studies of the transient states of the complete, physiologically intact protein complex of the whole cell in biological action are envisaged to provide more realistic insights of the specific mechanisms of toxic activity of the colicins than hitherto obtained from studying the C-domain in unilamellar vesicles alone.

We feel that multifrequency EPR spectroscopy in conjunction with SDSL mutagenesis techniques has now advanced to the level of spectral sensitivity and selectivity required to explore, even in such large protein complexes, details of long-range interactions between key protein domains and their modulation during biological action. In the last decade of protein research the strategy "from complexity to simplicity" was generally applied to first isolate the individual functional subunits and characterize them to atomic detail before mechanistic details of the complete system are tackled. Now, it seems that in the colicin case the strategy should be reversed to "from simplicity to complexity" in order to elucidate key aspects of the colicin *in-vivo* attack against the *E. coli* target organism. This attack involves all three domains, R, T, C, and their interactions with specific proteins located in the target cell's outer and inner membranes and the periplasm in between. These *in-vivo* experiments will be aggravated by the low concentration of spin-labelled colicin A proteins that, according to previous experience,[34] is difficult to increase biochemically. Spectroscopically, this challenging strategy of performing *in-vivo* experiments will require the highest performance of the dedicated EPR instrumentation to follow structural changes and conformational dynamics of large protein complexes in action over wide time scales (sub-µs to s).

References

1. *The Reaction Center of Photosynthetic Bacteria, Structure and Dynamics*, M. E. Michel-Beyerle, Ed., Springer, Berlin, 1996.
2. K. Gerwert, In: *Handbook of Vibrational Spectroscopy*, J. M. Chalmers, P. R. Griffiths, Eds., John Wiley, Chichester, 2002, pp. 1.
3. W. Mäntele, In: *The Photosynthetic Reaction Center*, J. Deisenhofer, J. R. Norris, Eds., Vol. **2**, Academic Press, New York, 1993, pp. 239.
4. A. Remy and K. Gerwert, *Nature Struct. Biol.*, 2003, **10**, 637.
5. C. Rödig, I. Chizhov, O. Weidlich and F. Siebert, *Biophys. J.*, 1999, **76**, 2687.
6. H. J. M. de Groot, *Curr. Opin. Struct. Biol.*, 2000, **10**, 593.
7. J. C. Lansing, J. G. Hu, M. Belenky, R. G. Griffin and J. Herzfeld, *Biochemistry*, 2003, **42**, 3586.
8. A. J. Mason, S. L. Grage, S. K. Straus, C. Glaubitz and A. Watts, *Biophys. J.*, 2004, **86**, 1610.

9. K. Möbius, *Chem. Soc. Rev.*, 2000, **29**, 129.
10. P. C. Riedi and G. M. Smith, In: *Electron Paramagnetic Resonance*, B. C. Gilbert, M. J. Davies, D. M. Murphy, Eds., Vol. **18**, Royal Society of Chemistry, Cambridge, 2002, pp. 254.
11. A. I. Smirnov, In: *Electron Paramagnetic Resonance*, B. C. Gilbert, M. J. Davies, D. M. Murphy, Eds., Vol. **18**, Royal Society of Chemistry, Cambridge, 2002, pp. 109.
12. H.-J. Steinhoff, *Front. Biosci.*, 2002, **7**, 97.
13. S. Weber, In: *Electron Paramagnetic Resonance*, N. M. Atherton, M. J. Davies, B. C. Gilbert, Eds., Vol. **17**, Royal Society of Chemistry, Cambridge, 2000, pp. 43.
14. K. Möbius, A. Savitsky, A. Schnegg, M. Plato and M. Fuchs, *Phys. Chem. Chem. Phys.*, 2005, **7**, 19.
15. W. Lubitz, In: *Electron Paramagnetic Resonance*, B. C. Gilbert, M. J. Davies, D. M. Murphy, Eds., Vol. **19**, Royal Society of Chemistry, Cambridge, 2004, pp. 174.
16. M. Plato, N. Krauss, P. Fromme and W. Lubitz, *Chem. Phys.*, 2003, **294**, 483.
17. S. Un, P. Dorlet and A. W. Rutherford, *Appl. Magn. Reson.*, 2001, **21**, 341.
18. K. Möbius and D. Goldfarb, In: *Biophysical Techniques in Photosynthesis*, T. J. Aartsma, J. Matysik, Eds., Vol. **II**, Springer, Dordrecht, 2008, pp. 267.
19. G. Richter, C. W. M. Kay, K. Struck, P. Sadewater, K. Möbius and S. Weber, In: *Flavins and Flavoproteins*, S. Ghisla, P. Kroneck, P. Macheroux, H. Sund, Eds., Rudolf Weber, Berlin, 1999, pp. 91.
20. S. Weber and R. Bittl, *Bull. Chem. Soc. Jpn.*, 2007, **80**, 2270.
21. C. W. M. Kay, R. Feicht, K. Schulz, P. Sadewater, A. Sancar, A. Bacher, K. Möbius, G. Richter and S. Weber, *Biochemistry*, 1999, **38**, 16740.
22. S. Weber, K. Möbius, G. Richter and C. W. M. Kay, *J. Am. Chem. Soc.*, 2001, **123**, 3790.
23. S. Weber, G. Richter, E. Schleicher, A. Bacher, K. Möbius and C. W. M. Kay, *Biophys. J.*, 2001, **81**, 1195.
24. S. Weber, C. W. M. Kay, H. Mögling, K. Möbius, K. Hitomi and T. Todo, *Proc. Natl. Acad. Sci. USA*, 2002, **99**, 1319.
25. M. R. Fuchs, E. Schleicher, A. Schnegg, C. W. M. Kay, J. T. Törring, R. Bittl, A. Bacher, G. Richter, K. Möbius and S. Weber, *J. Phys. Chem. B*, 2002, **106**, 8885.
26. C. W. M. Kay, H. Mögling, E. Schleicher, K. Hitomi, K. Möbius, T. Todo, A. Bacher, G. Richter and S. Weber, In: *Flavins and Flavoproteins 2002*, Rudolf Weber, Berlin, 2002, pp. 713.
27. C. W. M. Kay, E. Schleicher, K. Hitomi, T. Todo, R. Bittl and S. Weber, *Magn. Reson. Chem.*, 2005, **43**, S96.
28. C. W. M. Kay, R. Bittl, A. Bacher, G. Richter and S. Weber, *J. Am. Chem. Soc.*, 2005, **127**, 10780.
29. S. Weber, *Biochim. Biophys. Acta*, 2005, **1707**, 1.

30. A. Schnegg, C. W. M. Kay, E. Schleicher, K. Hitomi, T. Todo, K. Möbius and S. Weber, *Mol. Phys.*, 2006, **104**, 1627.
31. M. Hong, *Acc. Chem. Res.*, 2006, **39**, 176.
32. Y. K. Shin, C. Levinthal, F. Levinthal and W. L. Hubbell, *Science*, 1993, **259**, 960.
33. L. Salwinski and W. L. Hubbell, *Protein Sci.*, 1999, **8**, 562.
34. A. Savitsky, M. Kühn, D. Duche, K. Möbius and H. J. Steinhoff, *J. Phys. Chem. B*, 2004, **108**, 9541.
35. *Electron Transfer-From Isolated Molecules to Biomolecules, Part I*, J. Jortner, M. Bixon, Eds., John Wiley, New York, 1999.
36. J. Crystal and R. A. Friesner, *J. Phys. Chem. A*, 2000, **104**, 2362.
37. W. W. Parson and A. Warshel, *J. Phys. Chem. B*, 2004, **108**, 10474.
38. W. W. Parson, Z. T. Chu and A. Warshel, *Biochim. Biophys. Acta*, 1990, **1017**, 251.
39. A. Warshel and W. W. Parson, *Annu. Rev. Phys. Chem.*, 1991, **42**, 279.
40. M. R. Gunner, A. Nicholls and B. Honig, *J. Phys. Chem.*, 1996, **100**, 4277.
41. A. Warshel, M. Kato and A. V. Pisliakov, *J. Chem. Theory Comput.*, 2007, **3**, 2034.
42. W. W. Parson and A. Warshel, In: *Biophysical Techniques in Photosynthesis*, T. J. Aartsma, J. Matysik, Eds., Vol. II, Springer, Dordrecht, 2008, pp. 401.
43. M. Plato and C. J. Winscom, In: *The Photosynthetic Bacterial Reaction Center. J. Breton*, A. Verméglio, Eds., Plenum, New York, 1988, pp. 421.
44. M. E. Michel-Beyerle, M. Plato, J. Deisenhofer, H. Michel, M. Bixon and J. Jortner, *Biochim. Biophys. Acta*, 1988, **932**, 52.
45. M. Plato, K. Möbius, M. E. Michel-Beyerle, M. Bixon and J. Jortner, *J. Am. Chem. Soc.*, 1988, **110**, 7279.
46. P. O. J. Scherer and S. F. Fischer, *Chem. Phys.*, 1989, **131**, 115.
47. A. Warshel, S. Creighton and W. W. Parson, *J. Phys. Chem.*, 1988, **92**, 2696.
48. M. Bixon, J. Jortner and M. E. Michel-Beyerle, *Biochim. Biophys. Acta*, 1991, **1056**, 301.
49. P. O. J. Scherer, C. Scharnagl and S. F. Fischer, *Chem. Phys.*, 1995, **197**, 333.
50. L. Y. Zhang and R. A. Friesner, *Proc. Natl. Acad. Sci. USA*, 1998, **95**, 13603.
51. N. Ivashin, B. Kallebring, S. Larsson and O. Hansson, *J. Phys. Chem. B*, 1998, **102**, 5017.
52. D. Kolbasov and A. Scherz, *J. Phys. Chem. B*, 2000, **104**, 1802.
53. A. Petrenko and K. Redding, *Chem. Phys. Lett.*, 2004, **400**, 98.
54. R. A. Marcus, *J. Chem. Phys.*, 1956, **24**, 966.
55. R. A. Marcus, *J. Chem. Phys.*, 1956, **24**, 979.
56. C. J. F. Böttcher and P. Bordewijk, *Theory of Electric Polarization*, Elsevier, Amsterdam, 1978.

57. M. Bixon, J. Fajer, G. Feher, J. H. Freed, D. Gamliel, A. J. Hoff, H. Levanon, K. Möbius, R. Nechushtai, J. R. Norris, A. Scherz, J. L. Sessler and D. Stehlik, *Israel J. Chem.*, 1992, **32**, 369.
58. H. Levanon and K. Möbius, *Annu. Rev. Biophys. Biomol. Struct.*, 1997, **26**, 495.
59. W. B. Curry, M. D. Grabe, I. V. Kurnikov, S. S. Skourtis, D. N. Beratan, J. J. Regan, A. J. A. Aquino, P. Beroza and J. N. Onuchic, *J. Bioenerg. Biomembr.*, 1995, **27**, 285.
60. M. Bixon, J. Jortner and M. E. Michel-Beyerle, In: *Perspectives in Photosynthesis*, J. Jortner, B. Pullman, Eds., Kluwer Academic Publishers, Dordrecht, 1990, pp. 325.
61. M. Bixon and J. Jortner, In: *Electron Transfer-From Isolated Molecules to Biomolecules, Part I*, J. Jortner, M. Bixon, Eds., John Wiley, New York, 1999, pp. 35.
62. J. P. Lin, I. A. Balabin and D. N. Beratan, *Science*, 2005, **310**, 1311.
63. R. A. Marcus, *Discuss. Faraday Soc.*, 1960, **21**.
64. R. A. Marcus, *Annu. Rev. Phys. Chem.*, 1964, **15**, 155.
65. R. A. Marcus and N. Sutin, *Biochim. Biophys. Acta*, 1985, **811**, 265.
66. I. Rips and J. Jortner, *J. Chem. Phys.*, 1987, **87**, 2090.
67. I. Rips and J. Jortner, *Chem. Phys. Lett.*, 1987, **133**, 411.
68. P. Finckh, H. Heitele and M. E. Michel-Beyerle, *Chem. Phys.*, 1989, **138**, 1.
69. H. Heitele, M. E. Michel-Beyerle and P. Finckh, *Chem. Phys. Lett.*, 1987, **138**, 237.
70. H. Kurreck and M. Huber, *Angew. Chem. Int. Ed.*, 1995, **34**, 849.
71. F. Lendzian, J. Schlüpmann, J. von Gersdorff, K. Möbius and H. Kurreck, *Angew. Chem. Int. Ed.*, 1991, **30**, 1461.
72. F. Lendzian and B. von Maltzan, *Chem. Phys. Lett.*, 1991, **180**, 191.
73. K. Hasharoni and H. Levanon, *J. Phys. Chem.*, 1995, **99**, 4875.
74. G. Feher, *J. Chem. Soc., Perkin Trans.*, 1992, **2**, 1861.
75. A. J. Hoff and J. Deisenhofer, *Phys. Rep.*, 1997, **287**, 2.
76. C. C. Moser, C. C. Page, X. Chen and P. L. Dutton, *J. Biol. Inorg. Chem.*, 1997, **2**, 393.
77. L. Stryer, *Biochemistry*, 5th edn; W. H. Freeman Company, New York, 2001.
78. P. Jordan, P. Fromme, O. Klukas, H. T. Witt, W. Saenger and N. Krauß, *Nature*, 2001, **411**, 909.
79. A. Zouni, H. T. Witt, J. Kern, P. Fromme, N. Krauß, W. Saenger and P. Orth, *Nature*, 2001, **409**, 739.
80. J. Biesiadka, B. Loll, J. Kern, K. D. Irrgang and A. Zouni, *Phys. Chem. Chem. Phys.*, 2004, **6**, 4733.
81. J. Kern, B. Loll, C. Luneberg, D. DiFiore, J. Biesiadka, K. D. Irrgang and A. Zouni, *Biochim. Biophys. Acta*, 2005, **1706**, 147.
82. J. Kern, B. Loll, A. Zouni, W. Saenger, K. D. Irrgang and J. Biesiadka, *Photosynth. Res.*, 2005, **84**, 153.
83. B. Loll, J. Kern, W. Saenger, A. Zouni and J. Biesiadka, *Nature*, 2005, **438**, 1040.

84. J. Barber, *Biochem. Soc. Trans.*, 2006, **34**, 619.
85. N. Kamiya and J. R. Shen, *Proc. Natl. Acad. Sci. USA*, 2003, **100**, 98.
86. K. N. Ferreira, T. M. Iverson, K. Maghlaoui, J. Barber and S. Iwata, *Science*, 2004, **303**, 1831.
87. J. Barber, K. Ferreira, K. Maghlaoui and S. Iwata, *Phys. Chem. Chem. Phys.*, 2004, **6**, 4737.
88. W. Saenger, 2005, *personal communication*.
89. C. R. D. Lancaster and H. Michel, In: *Handbook of Metalloproteins*, K. Wieghardt, R. Huber, T. Poulos, A. Messerschmidt, Eds., Wiley, Chichester, 2001, pp. 119.
90. J. P. Allen and J. C. Williams, *FEBS Lett.*, 1998, **438**, 5.
91. J. Deisenhofer, O. Epp, K. Miki, R. Huber and H. Michel, *Nature*, 1985, **318**, 618.
92. J. P. Allen, G. Feher, T. O. Yeates, H. Komiya and D. C. Rees, *Proc. Natl. Acad. Sci. USA*, 1987, **84**, 5730.
93. M. H. B. Stowell, T. M. McPhillips, D. C. Rees, S. M. Soltis, E. Abresch and G. Feher, *Science*, 1997, **276**, 812.
94. A. Kuglstatter, P. Hellwig, G. Fritzsch, J. Wachtveitl, D. Oesterhelt, W. Mäntele and H. Michel, *FEBS Lett.*, 1999, **463**, 169.
95. Q. Xu, H. L. Axelrod, E. C. Abresch, M. L. Paddock, M. Y. Okamura and G. Feher, *Structure*, 2004, **12**, 703.
96. R. J. Cogdell, N. W. Isaacs, T. D. Howard, K. McLuskey, N. J. Fraser and S. M. Prince, *J. Bacteriol.*, 1999, **181**, 3869.
97. M. Z. Papiz, S. M. Prince, T. Howard, R. J. Cogdell and N. W. Isaacs, *J. Mol. Biol.*, 2003, **326**, 1523.
98. A. W. Roszak, T. D. Howard, J. Southall, A. T. Gardiner, C. J. Law, N. W. Isaacs and R. J. Cogdell, *Science*, 2003, **302**, 1969.
99. W. Kühlbrandt, *Nature*, 1995, **374**, 497.
100. G. McDermott, S. M. Prince, A. A. Freer, N. W. Isaacs, M. Z. Papiz, A. M. Hawthornthwaite-Lawless and R. J. Cogdell, *Protein Eng.*, 1995, **8**, 43.
101. A. Freer, S. Prince, K. Sauer, M. Papiz, A. Hawthornthwaite-Lawless, G. McDermott, R. Cogdell and N. W. Isaacs, *Structure*, 1996, **4**, 449.
102. R. J. Cogdell, P. K. Fyfe, S. J. Barrett, S. M. Prince, A. A. Freer, N. W. Isaacs, P. McGlynn and C. N. Hunter, *Photosynth. Res.*, 1996, **48**, 55.
103. X. C. Hu and K. Schulten, *Phys. Today*, 1997, **50**, 28.
104. X. C. Hu, A. Damjanovic, T. Ritz and K. Schulten, *Proc. Natl. Acad. Sci. USA*, 1998, **95**, 5935.
105. G. Gingras and R. Picorel, *Proc. Natl. Acad. Sci. USA*, 1990, **87**, 3405.
106. N. Srivatsan and J. R. Norris, *J. Phys. Chem. B*, 2001, **105**, 12391.
107. D. Kolbasoy, N. Srivatsan, N. Ponomarenko, M. Jager and J. R. Norris, *J. Phys. Chem. B*, 2003, **107**, 2386.
108. N. Srivatsan, S. Weber, D. Kolbasov and J. R. Norris, *J. Phys. Chem. B*, 2003, **107**, 2127.

109. N. Srivatsan, D. Kolbasov, N. Ponomarenko, S. Weber, A. E. Ostafin and J. R. Norris, *J. Phys. Chem. B*, 2003, **107**, 7867.
110. J. P. Zhang, H. Nagae, P. Qian, L. Limantara, R. Fujii, Y. Watanabe and Y. Koyama, *J. Phys. Chem. B*, 2001, **105**, 7312.
111. *The Photosynthetic Reaction Center*, Vols. **I** and **II**, J. Deisenhofer, J. R. Norris, Eds., Academic Press, San Diego, 1993.
112. C. C. Page, C. C. Moser, X. X. Chen and P. L. Dutton, *Nature*, 1999, **402**, 47.
113. D. Noy, C. C. Moser and P. L. Dutton, *Biochim. Biophys. Acta*, 2006, **1757**, 90.
114. M. Bixon and J. Jortner, *J. Chem. Phys.*, 1997, **107**, 5154.
115. S. S. Skourtis, J. N. Onuchic and D. N. Beratan, *Inorg. Chim. Acta*, 1996, **243**, 167.
116. I. Daizadeh, E. S. Medvedev and A. A. Stuchebrukhov, *Proc. Natl. Acad. Sci. USA*, 1997, **94**, 3703.
117. J. N. Onuchic, Z. Luthey-Schulten and P. G. Wolynes, *Annu. Rev. Phys. Chem.*, 1997, **48**, 545.
118. I. A. Balabin and J. N. Onuchic, *Science*, 2000, **290**, 114.
119. P. W. Fenimore, H. Frauenfelder, B. H. McMahon and F. G. Parak, *Proc. Natl. Acad. Sci. USA*, 2002, **99**, 16047.
120. A. V. Finkelstein and O. Ptitsyn, *Protein Physics: A Course for Lectures*, Academic Press, San Diego, 2002.
121. H. Frauenfelder, *Proc. Natl. Acad. Sci. USA*, 2002, **99**, 2479.
122. H. Frauenfelder, B. H. McMahon and P. W. Fenimore, *Proc. Natl. Acad. Sci. USA*, 2003, **100**, 8615.
123. J. M. Kriegl, F. K. Forster and G. U. Nienhaus, *Biophys. J.*, 2003, **85**, 1851.
124. J. M. Kriegl and G. U. Nienhaus, *Proc. Natl. Acad. Sci. USA*, 2004, **101**, 123.
125. W. L. Hubbell and C. Altenbach, *Curr. Opin. Struct. Biol.*, 1994, **4**, 566.
126. D. Stehlik and K. Möbius, *Annu. Rev. Phys. Chem.*, 1997, **48**, 745.
127. M. Fuhs and K. Möbius, In: *Lecture Notes in Physics*, C. Bertier, P. L. Levy, G. Martinez, Eds., Vol. **595**, Springer, Berlin, 2002, pp. 476.
128. K. Möbius, In: *Electron Spin Resonance*, N. M. Atherton, E. R. Davies, B. C. Gilbert, Eds., Vol. **14**, Royal Society of Chemistry, Cambridge, 1994, pp. 203.
129. J. B. Feix and C. S. Klug, In: *Electron Paramagnetic Resonance*, B. C. Gilbert, M. J. Davies, D. M. Murphy, Eds., Vol. **20**, Royal Society of Chemistry, Cambridge, 2006, pp. 50.
130. J. H. Freed, *Annu. Rev. Phys. Chem.*, 2000, **51**, 655.
131. P. P. Borbat and J. H. Freed, *EPR Newsletter*, 2007, **17**, 21.
132. T. Prisner, M. Rohrer and F. MacMillan, *Annu. Rev. Phys. Chem.*, 2001, **52**, 279.
133. O. Schiemann and T. F. Prisner, *Q. Rev. Biophys.*, 2007, **40**, 1.
134. W. Lubitz, *Pure Appl. Chem.*, 2003, **75**, 1021.

135. A. J. Chirino, E. J. Lous, M. Huber, J. P. Allen, C. C. Schenck, M. L. Paddock, G. Feher and D. C. Rees, *Biochemistry*, 1994, **33**, 4584.
136. J. P. Allen, G. Feher, T. O. Yeates, H. Komiya and D. C. Rees, *Proc. Natl. Acad. Sci. USA*, 1987, **84**, 6162.
137. W. Lubitz, *Phys. Chem. Chem. Phys.*, 2002, **4**, 5539.
138. A. J. Hoff, In: *Photosynthesis*, J. Amesz, Ed., Elsevier, Amsterdam, 1987, pp. 97.
139. G. Feher, *Photosynth. Res.*, 1998, **55**, 3.
140. J. R. Norris, R. A. Uphaus, H. L. Crespi and J. J. Katz, *Proc. Natl. Acad. Sci. USA*, 1971, **68**, 625.
141. G. Feher, A. J. Hoff, R. A. Isaacson and L. C. Ackerson, *Ann. N.Y. Acad. Sci.*, 1975, **244**, 239.
142. G. Feher, A. J. Hoff, R. A. Isaacson and J. D. McElroy, *Abstr. Biophys. Soc.*, 1973, **17**, 611a.
143. J. R. Norris, M. E. Druyan and J. J. Katz, *J. Am. Chem. Soc.*, 1973, **95**, 1680.
144. J. R. Norris, H. Scheer and J. J. Katz, *Ann. N.Y. Acad. Sci.*, 1975, **244**, 260.
145. F. Lendzian, M. Huber, R. A. Isaacson, B. Endeward, M. Plato, B. Bönigk, K. Möbius, W. Lubitz and G. Feher, *Biochim. Biophys. Acta*, 1993, **1183**, 139.
146. F. Lendzian, W. Lubitz, H. Scheer, C. Bubenzer and K. Möbius, *J. Am. Chem. Soc.*, 1981, **103**, 4635.
147. W. Lubitz, In: *Chlorophylls*, H. Scheer, Ed., CRC Press, Boca Raton, Florida, 1991, pp. 903.
148. M. Plato, K. Möbius and W. Lubitz, In: *Chlorophylls*, H. Scheer, Ed., CRC Press, Boca Raton, Florida, 1991, pp. 1015.
149. F. Lendzian, W. Lubitz, R. Steiner, E. Tränkle, M. Plato, H. Scheer and K. Möbius, *Chem. Phys. Lett.*, 1986, **126**, 290.
150. F. Lendzian, W. Lubitz, H. Scheer, A. J. Hoff, M. Plato, E. Tränkle and K. Möbius, *Chem. Phys. Lett.*, 1988, **148**, 377.
151. J. Rautter, F. Lendzian, W. Lubitz, S. Wang and J. P. Allen, *Biochemistry*, 1994, **33**, 12077.
152. M. Huber, E. J. Lous, R. A. Isaacson, G. Feher, D. Gaul and C. C. Schenck, In: *Reaction Centers of Photosynthetic Bacteria*, M. E. Michel-Beyerle, Ed., Springer, Berlin, 1990, pp. 219.
153. J. Rautter, C. Geßner, F. Lendzian, W. Lubitz, J. C. Williams, H. A. Murchison, S. Wang, N. W. Woodbury and J. P. Allen, In: *The Photosynthetic Bacterial Reaction Center II*, J. Breton, Ed., Plenum, New York, 1992, pp. 99.
154. H. Käss, J. Rautter, W. Zweygart, A. Struck, H. Scheer and W. Lubitz, *J. Phys. Chem.*, 1994, **98**, 354.
155. R. Klette, J. T. Törring, M. Plato, K. Möbius, B. Bönigk and W. Lubitz, *J. Phys. Chem.*, 1993, **97**, 2051.
156. J. P. Allen, G. Feher, T. O. Yeates, D. C. Rees, D. S. Eisenberg, J. Deisenhofer, H. Michel and R. Huber, *Biophys. J.*, 1986, **49**, A583.

157. D. E. Budil, S. S. Taremi, P. Gast, J. R. Norris and H. A. Frank, *Israel J. Chem.*, 1988, **28**, 59.
158. N. M. Atherton, *Principles of Electron Spin Resonance*, Ellis Horwood, New York, 1993.
159. T. F. Prisner, A. van der Est, R. Bittl, W. Lubitz, D. Stehlik and K. Möbius, *Chem. Phys.*, 1995, **194**, 361.
160. F. Müh, J. Rautter and W. Lubitz, *Biochemistry*, 1997, **36**, 4155.
161. F. Müh, C. Schulz, E. Schlodder, M. R. Jones, J. Rautter, M. Kuhn and W. Lubitz, *Photosynth. Res.*, 1998, **55**, 199.
162. M. Huber and J. T. Törring, *Chem. Phys.*, 1995, **194**, 379.
163. J. T. Törring, S. Un, M. Knüpling, M. Plato and K. Möbius, *J. Chem. Phys.*, 1997, **107**, 3905.
164. A. J. Stone, *Proc. R. Soc. London, Ser. A*, 1963, **271**, 424.
165. C. Schulz, F. Müh, A. Beyer, R. Jordan, E. Schlodder and W. Lubitz, In: *Photosynthesis: Mechanisms and Effects*, G. Garab, Ed., Vol. 2, Kluwer, Dordrecht, 1998, pp. 767.
166. W. Lubitz, F. Lendzian and R. Bittl, *Acc. Chem. Res.*, 2002, **35**, 313.
167. A. Ivancich, K. Artz, J. C. Williams, J. P. Allen and T.-A. Mattioli, *Biochemistry*, 1998, **37**, 11812.
168. M. R. Fuchs, A. Schnegg, M. Plato, C. Schulz, F. Müh, W. Lubitz and K. Möbius, *Chem. Phys.*, 2003, **294**, 371.
169. J. Rautter, F. Lendzian, C. Schulz, A. Fetsch, M. Kuhn, X. Lin, J. C. Williams, J. P. Allen and W. Lubitz, *Biochemistry*, 1995, **34**, 8130.
170. M. Huber, R. A. Isaacson, E. C. Abresh, D. Gaul, C. C. Schenck and G. Feher, *Biochim. Biophys. Acta*, 1996, **1273**, 108.
171. K. Artz, J. C. Williams, J. P. Allen, F. Lendzian, J. Rautter and W. Lubitz, *Proc. Natl. Acad. Sci. USA*, 1997, **94**, 13582.
172. F. Müh, M. Bibikova, F. Lendzian, D. Oesterhelt and W. Lubitz, In: *Photosynthesis: Mechanisms and Effects*, G. Garab, Ed., Vol. 2, Kluwer, Dordrecht, 1998, pp. 763.
173. F. Müh, F. Lendzian, M. Roy, J. C. Williams, J. P. Allen and W. Lubitz, *J. Phys. Chem. B*, 2002, **106**, 3226.
174. M. Huber, *Photosynth. Res.*, 1997, **52**, 1.
175. L. M. McDowell, D. Gaul, C. Kirmaier, D. Holten and C. C. Schenck, *Biochemistry*, 1991, **30**, 8315.
176. J. P. Allen, K. Artz, X. Lin, J. C. Williams, A. Ivancich, D. Albouny, T. A. Mattioli, A. Fetsch, M. Kuhn and W. Lubitz, *Biochemistry*, 1996, **35**, 6612.
177. O. Burghaus, M. Plato, M. Rohrer, K. Möbius, F. MacMillan and W. Lubitz, *J. Phys. Chem.*, 1993, **97**, 7639.
178. W. Wang, R. L. Belford, R. B. Clarkson, P. H. Davis, J. Forrer, M. J. Nilges, M. D. Timken, T. Walczak, M. C. Thurnauer, J. R. Norris, A. L. Morris and Y. Zhang, *Appl. Magn. Reson.*, 1994, **6**, 195.
179. M. Huber, J.T. Törring, M. Plato, U. Finck, W. Lubitz, R. Feick, C. C. Schenck, K. Möbius, In: *J. Sol. Energy Mater. Sol. Cells*, Vol. **38**, G. Calzaferri, Ed., 1995, pp. 119.

180. P. J. Bratt, E. Ringus, A. Hassan, H. van Tol, A.-L. Maniero, L.-C. Brunel, M. Rohrer, C. Bubenzer-Hange, H. Scheer and A. Angerhofer, *J. Phys. Chem. B*, 1999, **103**, 10973.
181. V. I. Gulin, S. A. Dikanov, Y. D. Tsvetkov, R. G. Evelo and A. J. Hoff, *Pure Appl. Chem.*, 1992, **64**, 903.
182. F. Müh, M. R. Jones and W. Lubitz, *Biospectroscopy*, 1999, **5**, 35.
183. M. Plato and K. Möbius, *Chem. Phys.*, 1995, **197**, 289.
184. G. Te Velde, F. M. Bickelhaupt, E. T. Baerends, C. F. Fonseca Guerra, S. J. A. van Gisbergen, J. G. Snijders and T. Ziegler, *J. Comp. Chem.*, 2001, **22**, 931.
185. G. Schreckenbach and T. Ziegler, *J. Phys. Chem. A*, 1997, **101**, 3388.
186. F. Neese and M. L. Munzarova, In: *The Quantum Chemical Calculation of NMR and EPR Properties*, M. Kaupp, M. Bühl, V. Malkin, Eds., Wiley-VCH, Weinheim, 2004.
187. G. Schreckenbach and T. Ziegler, *Theor. Chem. Acc.*, 1998, **99**, 71.
188. R. A. Isaacson, F. Lendzian, E. C. Abresch, W. Lubitz and G. Feher, *Biophys. J.*, 1995, **69**, 311.
189. M. Rohrer, M. Plato, F. MacMillan, Y. Grishin, W. Lubitz and K. Möbius, *J. Magn. Reson. A*, 1995, **116**, 59.
190. M. Flores, R. Isaacson, E. Abresch, R. Calvo, W. Lubitz and G. Feher, *Biophys. J.*, 2006, **90**, 3356.
191. M. Rohrer, F. MacMillan, T. F. Prisner, A. T. Gardiner, K. Möbius and W. Lubitz, *J. Phys. Chem. B*, 1998, **102**, 4648.
192. A. Carrington and A. D. McLachlan, *Introduction to Magnetic Resonance*, Harper and Row, New York, 1969.
193. A. Schnegg, A. A. Dubinskii, M. R. Fuchs, Y. A. Grishin, E. P. Kirilina, W. Lubitz, M. Plato, A. Savitsky and K. Möbius, *Appl. Magn. Reson.*, 2007, **31**, 59.
194. M. Rohrer, P. Gast, K. Möbius and T. F. Prisner, *Chem. Phys. Lett.*, 1996, **259**, 523.
195. T. F. Prisner, M. Rohrer and K. Möbius, *Appl. Magn. Reson.*, 1994, **7**, 167.
196. W. Lubitz and G. Feher, *Appl. Magn. Reson.*, 1999, **17**, 1.
197. M. Flores, R. Isaacson, E. Abresch, R. Calvo, W. Lubitz and G. Feher, *Biophys. J.*, 2007, **92**, 671.
198. M. L. Paddock, M. Flores, R. Isaacson, C. Chang, E. C. Abresch and M. Y. Okamura, *Biochemistry*, 2007, **46**, 8234.
199. S. Sinnecker, E. Reijerse, F. Neese and W. Lubitz, *J. Am. Chem. Soc.*, 2004, **126**, 3280.
200. H. Thomann, L. R. Dalton, L. A. Dalton, In: *Biological Magnetic Resonance*, Vol. **6**, L. J. Berliner, Ed., Plenum, New York, 1984, pp. 143.
201. S. Saxena and J. H. Freed, *J. Phys. Chem. A*, 1997, **101**, 7998.
202. E. P. Kirilina, T. F. Prisner, M. Bennati, B. Endeward, S. A. Dzuba, M. R. Fuchs, K. Möbius and A. Schnegg, *Magn. Reson. Chem.*, 2005, **43**, S119.

203. S. A. Dzuba, Y. D. Tsvetkov and A. G. Maryasov, *Chem. Phys. Lett.*, 1992, **188**, 217.
204. G. L. Millhauser and J. H. Freed, *J. Chem. Phys.*, 1984, **81**, 37.
205. D. Kleinfeld, M. Y. Okamura and G. Feher, *Biochemistry*, 1984, **23**, 5780.
206. F. Parak, E. N. Frolov, A. A. Kononenko, R. L. Mössbauer, V. I. Goldanskii and A. B. Rubin, *FEBS Lett.*, 1980, **117**, 368.
207. R. J. Debus, G. Feher and M. Y. Okamura, *Biochemistry*, 1986, **25**, 2276.
208. S. Weber, M. Fuhs, W. Hofbauer, W. Lubitz, K. Möbius **1998**, *unpublished results*.
209. U. Ermler, G. Fritzsch, S. K. Buchanan and H. Michel, *Structure*, 1994, **2**, 925.
210. J. Breton, C. Boullais, J.-R. Burie, E. Nabedryk and C. Mioskowski, *Biochemistry*, 1994, **33**, 14378.
211. W. B. S. van Liemt, G. J. Boender, P. Gast, A. J. Hoff, J. Lugtenburg and H. J. M. de Groot, *Biochemistry*, 1995, **34**, 10229.
212. R. Brudler, H. J. M. de Groot, W. B. S. van Liemt, W. F. Steggerda, R. Esmeijer, P. Gast, A. J. Hoff, J. Lugtenburg and K. Gerwert, *J. EMBO*, 1994, **13**, 5523.
213. M. Bosch, P. Gast, A. J. Hoff, A. P. Spoyalov and Y. D. Tsvetkov, *Chem. Phys. Lett.*, 1995, **239**, 306.
214. J. S. van de Brink, A. P. Spoyalov, P. Gast, W. B. S. van Liemt, J. Raap, J. Lugtenburg and A. J. Hoff, *FEBS Lett.*, 1994, **353**, 273.
215. R. A. Isaacson, E. C. Abresch, F. Lendzian, C. Boullais, M. L. Paddock, C. Mioskowski and W. G. F. Lubitz, In: *The Reaction Center of Photosynthetic Bacteria*, M. E. Michel-Beyerle, Ed., Springer, Berlin, 1995, pp. 353.
216. A. Savitsky, A. A. Dubinskii, M. Flores, W. Lubitz and K. Möbius, *J. Phys. Chem. B*, 2007, **111**, 6245.
217. A. Schnegg, M. Fuhs, M. Rohrer, W. Lubitz, T. F. Prisner and K. Möbius, *J. Phys. Chem. B*, 2002, **106**, 9454.
218. R. Calvo, E. C. Abresch, R. Bittl, G. Feher, W. Hofbauer, R. A. Isaacson, W. Lubitz, M. Y. Okamura and M. L. Paddock, *J. Am. Chem. Soc.*, 2000, **122**, 7327.
219. J. A. Nilsson, A. Lyubartsev, L. A. Eriksson and A. Laaksonen, *Mol. Phys.*, 2001, **99**, 1795.
220. T. N. Kropacheva, W. B. S. van Liemt, J. Raap, J. Lugtenburg and A. J. Hoff, *J. Phys. Chem.*, 1996, **100**, 10433.
221. G. Soda and T. Chiba, *J. Phys. Soc. Jpn.*, 1969, **26**, 249.
222. G. Soda and T. Chiba, *J. Chem. Phys.*, 1969, **50**, 439.
223. M. J. Hunt and A. L. Mackay, *J. Magn. Reson.*, 1974, **15**, 402.
224. M. J. Hunt and A. L. Mackay, *J. Magn. Reson.*, 1976, **22**, 295.
225. L. Mayas, M. Plato, C. J. Winscom and K. Möbius, *Mol. Phys.*, 1978, **36**, 753.
226. O. G. Poluektov, L. M. Utschig, M. C. Thurnauer and G. Kothe, *Appl. Magn. Reson.*, 2007, **31**, 123.

227. M. C. Thurnauer, O. G. Poluektov and G. Kothe, In: *Very High Frequency (VHF) ESR/EPR; Biological Magnetic Resonance*, O. Grinberg, L.J. Berliner, Eds., Vol. **22**, Kluwer/Plenum Publishers, New York, 2004, pp. 166.
228. A. D. Milov, A. B. Ponomarev and Y. D. Tsvetkov, *Chem. Phys. Lett.*, 1984, **110**, 67.
229. A. D. Milov, K. M. Salikhov and M. D. Shirov, *Fiz. Tverd. Tela*, 1981, **23**, 975.
230. L. V. Kulik, S. V. Paschenko and S. A. Dzuba, *J. Magn. Reson.*, 2002, **159**, 237.
231. V. P. Denysenkov, T. F. Prisner, J. Stubbe and M. Bennati, *Appl. Magn. Reson.*, 2005, **29**, 375.
232. V. P. Denysenkov, T. F. Prisner, J. Stubbe and M. Bennati, *Proc. Natl. Acad. Sci. USA*, 2006, **103**, 13386.
233. Y. Polyhach, A. Godt, C. Bauer and G. Jeschke, *J. Magn. Reson.*, 2007, **185**, 118.
234. A. Kamlowski, S. G. Zech, P. Fromme, R. Bittl, W. Lubitz, H. T. Witt and D. Stehlik, *J. Phys. Chem. B*, 1998, **102**, 8266.
235. D. Stehlik, C. H. Bock and J. Petersen, *J. Phys. Chem.*, 1989, **93**, 1612.
236. P. J. Hore, D. A. Hunter, C. D. McKie and A. J. Hoff, *Chem. Phys. Lett.*, 1987, **137**, 495.
237. G. L. Closs, M. D. E. Forbes and J. R. Norris, *J. Phys. Chem.*, 1987, **91**, 3592.
238. Y. Kandrashkin and A. van der Est, *Spectrochim. Acta A*, 2001, **57**, 1697.
239. R. Bittl and S. G. Zech, *J. Phys. Chem. B*, 1997, **101**, 1429.
240. L. V. Kulik, S. A. Dzuba, I. A. Grigoryev and Y. D. Tsvetkov, *Chem. Phys. Lett.*, 2001, **343**, 315.
241. G. M. Zhidomirov and K. M. Salikhov, *Sov. Phys. JETP*, 1969, **29**, 1037.
242. A. Schweiger and G. Jeschke, *Principles of Pulse Electron Paramagnetic Resonance*, Oxford University Press, Oxford, 2001.
243. M. C. Thurnauer and D. Meisel, *J. Am. Chem. Soc.*, 1983, **105**, 3729.
244. M. C. Thurnauer and J. R. Norris, *Chem. Phys. Lett.*, 1980, **76**, 557.
245. K. M. Salikhov, Y. Kandrashkin and A. K. Salikhov, *Appl. Magn. Reson.*, 1992, **3**, 199.
246. J. Tang, M. C. Thurnauer and J. R. Norris, *Chem. Phys. Lett.*, 1994, **219**, 283.
247. S. A. Dzuba, P. Gast and A. J. Hoff, *Chem. Phys. Lett.*, 1995, **236**, 595.
248. S. G. Zech, W. Lubitz and R. Bittl, *Ber. Bunsen-Ges. Phys. Chem.*, 1996, **100**, 2041.
249. S. G. Zech, R. Bittl, A. T. Gardiner and W. Lubitz, *Appl. Magn. Reson.*, 1997, **13**, 517.
250. I. V. Borovykh, S. A. Dzuba, I. I. Proskuryakov, P. Gast and A. J. Hoff, *Biochim. Biophys. Acta*, 1998, **1363**, 182.
251. M. Flores, A. Savitsky, E. Abresch, W. Lubitz and K. Möbius, *Photosynth. Res.*, 2007, **91**, 155.

252. V. P. Denysenkov, D. Biglino, W. Lubitz, T. F. Prisner and M. Bennati, *Angew. Chem. Int. Ed.*, 2008, **47**, 1224.
253. I. V. Borovykh, S. Ceola, P. Gajula, P. Gast, H. J. Steinhoff and M. Huber, *J. Magn. Reson.*, 2006, **180**, 178.
254. P. Gajula, I. V. Borovykh, C. Beier, T. Shkuropatova, P. Gast and H. J. Steinhoff, *Appl. Magn. Reson.*, 2007, **31**, 167.
255. J. R. Norris, D. E. Budil, P. Gast, C. H. Chang, O. Elkabbani and M. Schiffer, *Proc. Natl. Acad. Sci. USA*, 1989, **86**, 4335.
256. D. E. Budil and M. C. Thurnauer, *Biochim. Biophys. Acta*, 1991, **1057**, 1.
257. M. Huber, A. A. Doubinskii, C. W. M. Kay and K. Möbius, *Appl. Magn. Reson.*, 1997, **13**, 473.
258. F. Lendzian, R. Bittl and W. Lubitz, *Photosynth. Res.*, 1998, **55**, 189.
259. A. Marchanka, M. Paddock, W. Lubitz and M. van Gastel, *Biochemistry*, 2007, **46**, 14782.
260. S. V. Paschenko, P. Gast and A. J. Hoff, *Appl. Magn. Reson.*, 2001, **21**, 325.
261. A. Labahn and M. Huber, *Appl. Magn. Reson.*, 2001, **21**, 381.
262. S.V. Pachtchenko, PhD Thesis, Leiden University, **2002**.
263. R. H. Zeng, J. van Tol, A. Deal, H. A. Frank and D. E. Budil, *J. Phys. Chem. B*, 2003, **107**, 4624.
264. M. C. Thurnauer, *Rev. Chem. Intermed*, 1979, **3**, 197.
265. A. J. Hoff and I. I. Proskuryakov, *Chem. Phys. Lett.*, 1985, **115**, 303.
266. N. R. S. Reddy, S. V. Kolaczkowski and G. J. Small, *Science*, 1993, **260**, 68.
267. P. O. J. Scherer, S. F. Fischer, J. K. H. Hörber and M. E. Michel-Beyerle, In: *Antennas and Reaction Centers of Photosynthetic Bacteria*, M.E. Michel-Beyerle, Ed., Springer, Berlin, 1985.
268. A. Angerhofer, F. Bornhauser, V. Aust, G. Hartwich and H. Scheer, *Biochim. Biophys. Acta*, 1998, **1365**, 404.
269. V. Aust, A. Angerhofer, P. H. Parot, C. A. Violette and H. A. Frank, *Chem. Phys. Lett.*, 1990, **173**, 439.
270. C. A. Coulson and G. S. Rushbrooke, *Proc. Cambridge Philos. Soc.*, 1940, **36**, 193.
271. H. C. Longuet-Higgins, *J. Chem. Phys.*, 1950, **18**, 265.
272. K. Möbius, *Z. Naturforsch. A*, 1965, **A 20**, 1102.
273. I. Gutman, S. M. Patra and R. K. Mishra, *J. Serb. Chem. Soc.*, 1998, **63**, 25.
274. J. Fajer, I. Fujita, A. Forman, L. K. Hanson, G. W. Craig, D. A. Goff, L. A. Kehres and K. M. Smith, *J. Am. Chem. Soc.*, 1983, **105**, 3837.
275. M. Plato, F. Lendzian, W. Lubitz and K. Möbius, In: *Photosynthetic Bacterial Reaction Center II. Structure, Spectroscopy and Dynamics*, J. Breton, A. Vermeglio, Eds., Plenum Press, New York, 1992, pp. 109.
276. A. Angerhofer, In: *Chlorophylls*, H. Scheer, Ed., CRC Press, Boca Raton, Florida, 1991, pp. 945.

277. H.-J. Steinhoff, A. Savitsky, C. Wegener, M. Pfeiffer, M. Plato and K. Möbius, *Biochim. Biophys. Acta*, 2000, **1457**, 253.
278. H. Luecke, B. Schobert, H.-T. Richter, J.-P. Cartailler and J. K. Lanyi, *J. Mol. Biol.*, 1999, **291**, 899.
279. G. Váró and J. K. Lanyi, *Biochemistry*, 1991, **30**, 5008.
280. S. Druckmann, N. Friedmann, J. K. Lanyi, R. Needleman, M. Ottolenghi and M. Sheves, *Photochem. Photobiol.*, 1992, **56**, 1041.
281. B. Hessling, J. Herbst, R. Rammelsberg and K. Gerwert, *Biophys. J.*, 1997, **73**, 2071.
282. U. Haupts, J. Tittor and D. Oesterhelt, *Annu. Rev. Biophys. Biomol. Struct.*, 1999, **28**, 367.
283. S. Subramaniam and R. Henderson, *Nature*, 2000, **406**, 653.
284. C. Altenbach and W. L. Hubbell, In: *Foundations of Modern EPR*, G.R. Eaton, S. S. Eaton, K. M. Salikhov, Eds., World Scientific, Singapore, 1998, pp. 423.
285. W. L. Hubbell, H. S. McHaourab, C. Altenbach and M. A. Lietzow, *Structure*, 1996, **4**, 779.
286. W. L. Hubbell, A. Gross, R. Langen and M. A. Lietzow, *Curr. Opin. Struct. Biol.*, 1998, **8**, 649.
287. W. L. Hubbell, D. S. Cafiso and C. Altenbach, *Nature Struct. Biol.*, 2000, **7**, 735.
288. C. Wegener, A. Savitsky, M. Pfeiffer, K. Möbius and H.-J. Steinhoff, *Appl. Magn. Reson.*, 2001, **21**, 441.
289. M. Plato, H.-J. Steinhoff, C. Wegener, J. T. Törring, A. Savitsky and K. Möbius, *Mol. Phys.*, 2002, **100**, 3711.
290. H. J. Steinhoff, *Biol. Chem.*, 2004, **385**, 913.
291. I. V. Borovykh and H.-J. Steinhoff, In: *Biophysical Techniques in Photosynthesis*, Vol. II, T. J. Aartsma, J. Matysik, Eds., Springer, Dordrecht, 2008, pp. 345.
292. Y. Lou, M. T. Ge and J. R. Freed, *J. Phys. Chem. B*, 2001, **105**, 11053.
293. V. A. Livshits, D. Kurad and D. Marsh, *J. Magn. Reson.*, 2006, **180**, 63.
294. B. Dzikovski, K. Earle, S. Pachtchenko and J. Freed, *J. Magn. Reson.*, 2006, **179**, 273.
295. D. Kurad, G. Jeschke and D. Marsh, *Biophys. J.*, 2003, **85**, 1025.
296. R. Owenius, M. Engström, M. Lindgren and M. Huber, *J. Phys. Chem. A*, 2001, **105**, 10967.
297. M. G. Finiguerra, H. Blok, M. Ubbink and M. Huber, *J. Magn. Reson.*, 2006, **180**, 197.
298. T. I. Smirnova, T. G. Chadwick, M. A. Voinov, O. Poluektov, J. van Tol, A. Ozarowski, G. Schaaf, M. M. Ryan and V. A. Bankaitis, *Biophys. J.*, 2007, **92**, 3686.
299. O. H. Griffith, P. J. Dehlinge and S. P. Van, *J. Membr. Biol.*, 1974, **15**, 159.
300. T. Kawamura, S. Matsunam and T. Yonezawa, *Bull. Chem. Soc. Jpn.*, 1967, **40**, 1111.
301. C. Altenbach, D. A. Greenhalgh, H. G. Khorana and W. L. Hubbell, *Proc. Natl. Acad. Sci. USA*, 1994, **91**, 1667.

302. M. Pfeiffer, T. Rink, K. Gerwert, D. Oesterhelt and H.-J. Steinhoff, *J. Mol. Biol.*, 1999, **287**, 163.
303. M. Fuchs, PhD Thesis, Freie Universität Berlin, **1999**.
304. A. Schnegg, K. Möbius **2006**, *unpublished results.*
305. H. Brutlach, E. Bordignon, L. Urban, J. P. Klare, H. J. Reyher, M. Engelhard and H. J. Steinhoff, *Appl. Magn. Reson.*, 2006, **30**, 359.
306. J. T. Törring, PhD Thesis, Freie Universität Berlin, **1995**.
307. J. Ridley and M. C. Zerner, *Theor. Chim. Acta*, 1973, **32**, 111.
308. S. Un, M. Atta, M. Fontecave and A. W. Rutherford, *J. Am. Chem. Soc.*, 1995, **117**, 10713.
309. A. F. Gullá and D. E. Budil, *J. Phys.Chem. B*, 2001, **105**, 8056.
310. Z. Ding, A. F. Gullá and D. E. Budil, *J. Chem. Phys.*, 2001, **115**, 10685.
311. K. A. Earle, J. K. Moscicki, M. Ge, D. E. Budil and J. H. Freed, *Biophys. J.*, 1994, **66**, 1213.
312. M. A. Ondar, O. Y. Grinberg and Y. S. Lebedev, *Sov. J. Chem. Phys.*, 1985, **3**, 781.
313. D. Marsh, D. Kurad and V. A. Livshits, *Chem. Phys. Lipids*, 2002, **116**, 93.
314. N. Radzwill, K. Gerwert and H.-J. Steinhoff, *Biophys. J.*, 2001, **80**, 2856.
315. S. Subramaniam, I. Lindahl, P. Bullough, A. R. Faruqi, J. Tittor, D. Oesterhelt, L. Brown, J. Lanyi and R. Henderson, *J. Mol. Biol.*, 1999, **287**, 145.
316. J. Tittor, S. Paula, S. Subramaniam, J. Heberle, R. Henderson and D. Oesterhelt, *J. Mol. Biol.*, 2002, **319**, 555.
317. P. P. Borbat, A. J. Costa-Filho, K. A. Earle, J. K. Moscicki and J. H. Freed, *Science*, 2001, **291**, 266.
318. Z. C. Liang and J. H. Freed, *J. Phys. Chem. B*, 1999, **103**, 6384.
319. J. H. Freed, In: *Very High Frequency (VHF) ESR/EPR, Biological Magnetic Resonance*, O. Grinberg, L. J. Berliner, Eds., Vol. **22**, Kluwer/Plenum Publishers, New York, 2004, pp. 19.
320. K. A. Earle, B. Dzikovski, W. Hofbauer, J. K. Moscicki and J. H. Freed, *Magn. Reson. Chem.*, 2005, **43**, S256.
321. O. Klukas, W. D. Schubert, P. Jordan, N. Krauss, P. Fromme, H. T. Witt and W. Saenger, *J. Biol. Chem.*, 1999, **274**, 7361.
322. G. Füchsle, R. Bittl, A. van der Est, W. Lubitz and D. Stehlik, *Biochim. Biophys. Acta*, 1993, **1142**, 23.
323. R. Bittl and S. G. Zech, *Biochim. Biophys. Acta*, 2001, **1507**, 194.
324. C. Teutloff, W. Hofbauer, S. G. Zech, M. Stein, R. Bittl and W. Lubitz, *Appl. Magn. Reson.*, 2001, **21**, 363.
325. A. van der Est, T. Prisner, R. Bittl, P. Fromme, W. Lubitz, K. Möbius and D. Stehlik, *J. Phys. Chem. B*, 1997, **101**, 1437.
326. O. G. Poluektov, L. M. Utschig, S. L. Schlesselman, K. V. Lakshmi, G. W. Brudvig, G. Kothe and M. C. Thurnauer, *J. Phys. Chem. B*, 2002, **106**, 8911.
327. B. Zybailov, A. van der Est, S. G. Zech, C. Teutloff, T. W. Johnson, G. Z. Shen, R. Bittl, D. Stehlik, P. R. Chitnis and J. H. Golbeck, *J. Biol. Chem.*, 2000, **275**, 8531.

328. W. Xu, P. Chitnis, A. Valieva, A. van der Est, Y. N. Pushkar, M. Krzystyniak, C. Teutloff, S. G. Zech, R. Bittl, D. Stehlik, B. Zybailov, G. Z. Shen and J. H. Golbeck, *J. Biol. Chem.*, 2003, **278**, 27864.
329. M. Fuhs, A. Schnegg, T. Prisner, I. Köhne, J. Hanley, A. W. Rutherford and K. Möbius, *Biochim. Biophys. Acta*, 2002, **1556**, 81.
330. A. Savitsky, B. V. Trubitsin, K. Möbius, A. Y. Semenov and A. N. Tikhonov, *Appl. Magn. Reson.*, 2007, **31**, 221.
331. J. Raymond and R. E. Blankenship, *Biochim. Biophys. Acta*, 2004, **1655**, 133.
332. W. Hofbauer, A. Zouni, R. Bittl, J. Kern, P. Orth, F. Lendzian, P. Fromme, H. T. Witt and W. Lubitz, *Proc. Natl. Acad. Sci. USA*, 2001, **98**, 6623.
333. P. Dorlet, L. Xiong, R. T. Sayre and S. Un, *J. Biol. Chem.*, 2001, **276**, 22313.
334. L. Kulik, B. Epel, J. Messinger and W. Lubitz, *Photosynth. Res.*, 2005, **84**, 347.
335. L. V. Kulik, B. Epel, W. Lubitz and J. Messinger, *J. Am. Chem. Soc.*, 2005, **127**, 2392.
336. B. Kok, B. Forbush and M. McGloin, *Photochem. Photobiol.*, 1970, **11**, 457.
337. R. D. Britt, K. Sauer and V. K. Yachandra, *Biochim. Biophys. Acta*, 2001, **1503**, 2.
338. V. J. DeRose, I. Mukerji, M. J. Latimer, V. K. Yachandra, K. Sauer and M. P. Klein, *J. Am. Chem. Soc.*, 1994, **116**, 5239.
339. V. K. Yachandra, K. Sauer and M. P. Klein, *Chem. Rev.*, 1996, **96**, 2927.
340. P. Glatzel, J. Yano, U. Bergmann, H. Visser, J. H. Robblee, W. W. Gu, F. M. F. de Groot, S. P. Cramer and V. K. Yachandra, *J. Phys. Chem. Solids*, 2005, **66**, 2163.
341. P. Glatzel, U. Bergmann, J. Yano, H. Visser, J. H. Robblee, W. W. Gu, F. M. F. de Groot, G. Christou, V. L. Pecoraro, S. P. Cramer and V. K. Yachandra, *J. Am. Chem. Soc.*, 2004, **126**, 9946.
342. V. K. Yachandra, *Philos. Trans. R. Soc. London, Ser. B*, 2002, **357**, 1347.
343. W. Junge, V. K. Yachandra, C. Dismukes and L. Hammarstrom, *Philos. Trans. R. Soc. London, Ser. B*, 2002, **357**, 1357.
344. J. Messinger, J. H. Robblee, U. Bergmann, C. Fernandez, P. Glatzel, H. Visser, R. M. Cinco, K. L. McFarlane, E. Bellacchio, S. A. Pizarro, S. P. Cramer, K. Sauer, M. P. Klein and V. K. Yachandra, *J. Am. Chem. Soc.*, 2001, **123**, 7804.
345. M. Haumann, C. Müller, P. Liebisch, L. Iuzzolino, J. Dittmer, M. Grabolle, T. Neisius, W. Meyer-Klaucke and H. Dau, *Biochemistry*, 2005, **44**, 1894.
346. M. Haumann, P. Liebisch, C. Müller, M. Barra, M. Grabolle and H. Dau, *Science*, 2005, **310**, 1019.

347. R. D. Britt, K. A. Campbell, J. M. Peloquin, M. L. Gilchrist, C. P. Aznar, M. M. Dicus, J. Robblee and J. Messinger, *Biochim. Biophys. Acta*, 2004, **1655**, 158.
348. J. M. Peloquin and R. D. Britt, *Biochim. Biophys. Acta*, 2001, **1503**, 96.
349. J. M. Peloquin, K. A. Campbell, D. W. Randall, M. A. Evanchik, V. L. Pecoraro, W. H. Armstrong and R. D. Britt, *J. Am. Chem. Soc.*, 2000, **122**, 10926.
350. L. V. Kulik, W. Lubitz and J. Messinger, *Biochemistry*, 2005, **44**, 9368.
351. J. Yano, J. Kern, K. D. Irrgang, M. J. Latimer, U. Bergmann, P. Glatzel, Y. Pushkar, J. Biesiadka, B. Loll, K. Sauer, J. Messinger, A. Zouni and V. K. Yachandra, *Proc. Natl. Acad. Sci. USA*, 2005, **102**, 12047.
352. C. Policar, M. Knüpling, Y. M. Frapart and S. Un, *J. Phys. Chem. B*, 1998, **102**, 10391.
353. J. Kurreck, D. Niethammer and H. Kurreck, *Chem. unserer Zeit*, 1999, **33**, 72.
354. D. Gust, T. A. Moore and A. L. Moore, *Acc. Chem. Res.*, 2001, **34**, 40.
355. D. M. Guldi, *Pure Appl. Chem.*, 2003, **75**, 1069.
356. T. Galili, A. Regev, A. Berg, H. Levanon, D. I. Schuster, K. Möbius and A. Savitsky, *J. Phys. Chem. A*, 2005, **109**, 8451.
357. E. T. Chernick, Q. X. Mi, A. M. Vega, J. V. Lockard, M. A. Ratner and M. R. Wasielewski, *J. Phys. Chem. B*, 2007, **111**, 6728.
358. Q. Mi, E. A. Weiss, M. A. Ratner and M. R. Wasielewski, *Appl. Magn. Reson.*, 2007, **31**, 253.
359. R. F. Kelley, M. J. Tauber and M. R. Wasielewski, *Angew. Chem. Int. Ed.*, 2006, **45**, 7979.
360. M. R. Wasielewski, *J. Org. Chem.*, 2006, **71**, 5051.
361. Q. X. Mi, E. T. Chernick, D. W. McCamant, E. A. Weiss, M. A. Ratner and M. R. Wasielewski, *J. Phys. Chem. A*, 2006, **110**, 7323.
362. M. Jakob, A. Berg, E. Stavitski, E. T. Chernick, E. A. Weiss, M. R. Wasielewski and H. Levanon, *Chem. Phys.*, 2006, **324**, 63.
363. E. T. Chernick, Q. X. Mi, R. F. Kelley, E. A. Weiss, B. A. Jones, T. J. Marks, M. A. Ratner and M. R. Wasielewski, *J. Am. Chem. Soc.*, 2006, **128**, 4356.
364. L. Sinks, M. J. Fuller, W. H. Liu, M. J. Ahrens and M. R. Wasielewski, *Chem. Phys.*, 2005, **319**, 226.
365. E. A. Weiss, L. E. Sinks, A. S. Lukas, E. T. Chernick, M. A. Ratner and M. R. Wasielewski, *J. Phys. Chem. B*, 2004, **108**, 10309.
366. S. Shaakov, T. Galili, E. Stavitski, H. Levanon, A. Lukas and M. R. Wasielewski, *J. Am. Chem. Soc.*, 2003, **125**, 6563.
367. K. Hasharoni, H. Levanon, J. von Gersdorff, H. Kurreck and K. Möbius, *J. Chem. Phys.*, 1993, **98**, 2916.
368. A. Berg, Z. Shuali, M. Asano-Someda, H. Levanon, M. Fuhs and K. Möbius, *J. Am. Chem. Soc.*, 1999, **121**, 7433.
369. E. A. Weiss, M. A. Ratner and M. R. Wasielewski, *J. Phys. Chem. A*, 2003, **107**, 3639.

370. M. Volk, T. Häberle, R. Feick, A. Ogrodnik and M. E. Michel-Beyerle, *J. Phys. Chem.*, 1993, **97**, 9831.
371. M. Di Valentin, A. Bisol, G. Agostini, M. Fuhs, P. A. Liddell, A. L. Moore, T. A. Moore, D. Gust and D. Carbonera, *J. Am. Chem. Soc.*, 2004, **126**, 17074.
372. T. Yago, Y. Kobori, K. Akiyama and S. Tero-Kubota, *J. Phys. Chem. B*, 2002, **106**, 10074.
373. Y. E. Kandrashkin, M. S. Asano and A. van der Est, *J. Phys. Chem. A*, 2006, **110**, 9607.
374. Y. E. Kandrashkin, M. S. Asano and A. van der Est, *J. Phys. Chem. A*, 2006, **110**, 9617.
375. P. W. Anderson, *Phys. Rev.*, 1959, **115**, 2.
376. M. Di Valentin, A. Bisol, G. Agostini and D. Carbonera, *J. Chem. Inform. Model.*, 2005, **45**, 1580.
377. H. Fröhlich, *Theory of Dielectrics*, Oxford University Press, Oxford, 1990.
378. A. J. Martin, G. Meier and A. Saupe, *Symp. Faraday Soc.*, 1971, **5**, 119.
379. M. Fuhs, G. Elger, A. Osintsev, A. Popov, H. Kurreck and K. Möbius, *Mol. Phys.*, 2000, **98**, 1025.
380. M. Asano-Someda, H. Levanon, J. L. Sessler and R. Z. Wang, *Mol. Phys.*, 1998, **95**, 935.
381. G. P. Wiederrecht, W. A. Svec, M. R. Wasielewski, T. Galili and H. Levanon, *J. Am. Chem. Soc.*, 2000, **122**, 9715.
382. H. Levanon and K. Hasharoni, *Prog. React. Kinet.*, 1995, **20**, 309.
383. G. L. Closs and M. D. E. Forbes, *J. Phys. Chem.*, 1991, **95**, 1924.
384. N. I. Avdievich and M. D. E. Forbes, *J. Phys. Chem.*, 1995, **99**, 9660.
385. M. D. E. Forbes, N. I. Avdievich, J. D. Ball and G. R. Schulz, *J. Phys. Chem.*, 1996, **100**, 13887.
386. J. Schlüpmann, F. Lendzian, M. Plato and K. Möbius, *J. Chem. Soc. Faraday Trans.*, 1993, **89**, 2853.
387. G. Elger, H. Kurreck, A. Wiehe, E. Johnen, M. Fuhs, T. Prisner and J. Vrieze, *Acta Chem. Scand.*, 1997, **51**, 593.
388. J. Fajer, K. M. Barkigia, D. Melamed, R. M. Sweet, H. Kurreck, J. von Gersdorff, M. Plato, H. C. Rohland, G. Elger and K. Möbius, *J. Phys. Chem.*, 1996, **100**, 14236.
389. K. M. Salikhov, J. Schlüpmann, M. Plato and K. Möbius, *Chem. Phys.*, 1997, **215**, 23.
390. G. Elger, M. Fuhs, P. Müller, J. von Gersdorff, A. Wiehe, H. Kurreck and K. Möbius, *Mol. Phys.*, 1998, **95**, 1309.
391. F. J. J. de Kanter and R. Kaptein, *J. Am. Chem. Soc.*, 1982, **104**, 4759.
392. F. J. J. de Kanter, J. A. den Hollander, A. H. Huizer and R. Kaptein, *Mol. Phys.*, 1977, **34**, 857.
393. K. A. Earle, J. K. Moscicki, A. Polimeno and J. H. Freed, *J. Chem. Phys.*, 1997, **106**, 9996.
394. D. S. Leniart, H. D. Connor and J. H. Freed, *J. Chem. Phys.*, 1975, **63**, 165.

395. J. S. Hwang, R. P. Mason, L. P. Hwang and J. H. Freed, *J. Phys. Chem.*, 1975, **79**, 489.
396. H. Dieks, Ph.D. Thesis, Freie Universität Berlin, **1996**.
397. A. Wiehe, M. O. Senge, A. Schäfer, M. Speck, S. Tannert, H. Kurreck and B. Röder, *Tetrahedron*, 2001, **57**, 10089.
398. K. Hasharoni, H. Levanon, S. R. Greenfield, D. J. Gosztola, W. A. Svec and M. R. Wasielewski, *J. Am. Chem. Soc.*, 1996, **118**, 10228.
399. K. Hasharoni, H. Levanon, S. R. Greenfield, D. J. Gosztola, W. A. Svec and M. R. Wasielewski, *J. Am. Chem. Soc.*, 1995, **117**, 8055.
400. H. Kurreck, S. Aguirre, H. Dieks, J. Gaschmann, J. von Gersdorff, H. Newman, H. Schubert, M. Speck, T. Stabingis, J. Sobek, P. Tian and A. Wiehe, *Radiat. Phys. Chem.*, 1995, **45**, 853.
401. P. L. Dutton, M. Seibert and J. S. Leigh, *Biochem. Biophys. Res. Commun.*, 1972, **46**, 406.
402. A. Regev, R. Nechushtai, H. Levanon and J. P. Thornber, *J. Phys. Chem.*, 1989, **93**, 2421.
403. H. Levanon and J. R. Norris, *Chem. Rev.*, 1978, **78**, 185.
404. M. C. Thurnauer, J. J. Katz and J. R. Norris, *Proc. Natl. Acad. Sci. USA*, 1975, **72**, 3270.
405. Y. Kobori, S. Yamauchi, K. Akiyama, S. Tero-Kubota, H. Imahori, S. Fukuzumi and J. R. Norris, *Proc. Natl. Acad. Sci. USA*, 2005, **102**, 10017.
406. H. Murai, S. Tero-Kubota and S. Yamauchi, In: *Electron Paramagnetic Resonance*, B. C. Gilbert, M. J. Davies, K. A. McLauchlan, Eds., Vol. 17, Royal Society of Chemistry, Cambridge, 2000, pp. 130.
407. K. Ishii, J. Fujisawa, Y. Ohba and S. Yamauchi, *J. Am. Chem. Soc.*, 1996, **118**, 13079.
408. J. Fujisawa, K. Ishii, Y. Ohba, S. Yamauchi, M. Fuhs and K. Möbius, *J. Phys. Chem. A*, 1997, **101**, 5869.
409. K. Ishii, J. Fujisawa, A. Adachi, S. Yamauchi and N. Kobayashi, *J. Am. Chem. Soc.*, 1998, **120**, 3152.
410. N. Mizuochi, Y. Ohba and S. Yamauchi, *J. Chem. Phys.*, 1999, **111**, 3479.
411. J. Fujisawa, K. Ishii, Y. Ohba, S. Yamauchi, M. Fuhs and K. Möbius, *J. Phys. Chem. A*, 1999, **103**, 213.
412. J. Fujisawa, Y. Iwasaki, Y. Ohba, S. Yamauchi, N. Koga, S. Karasawa, M. Fuhs, K. Möbius and S. Weber, *Appl. Magn. Reson.*, 2001, **21**, 483.
413. S. Yamauchi, Y. Iwasaki, H. Yashiro, H. Murai, S. Weber, K. Möbius, T. Berthold and G. Kothe, In: *Advanced EPR Applied to Bioscience*, A. Kawamori, Ed., Kwansei Gakuin Univ., Japan, 2002, pp. 29.
414. V. F. Tarasov, I. S. M. Saiful, Y. Iwasaki, Y. Ohba, A. Savitsky, K. Möbius and S. Yamauchi, *Appl. Magn. Reson.*, 2006, **30**, 619.
415. C. Corvaja, M. Maggini, M. Prato, G. Scorrano and M. Venzin, *J. Am. Chem. Soc.*, 1995, **117**, 8857.
416. N. Mizuochi, Y. Ohba and S. Yamauchi, *J. Phys. Chem. A*, 1997, **101**, 5966.
417. F. Conti, C. Corvaja, A. Toffoletti, N. Mizuochi, Y. Ohba, S. Yamauchi and M. Maggini, *J. Phys. Chem. A.*, 2000, **104**, 4962.

418. E. Sartori, A. Toffoletti, C. Corvaja and L. Garlaschelli, *J. Phys. Chem. A*, 2001, **105**, 10776.
419. M. Carano, C. Corvaja, L. Garlaschelli, M. Maggini, M. Marcaccio, F. Paolucci, D. Pasini, P. P. Righetti, E. Sartori and A. Toffoletti, *Eur. J. Org. Chem.*, 2003, **374**.
420. Y. Ohba, M. Nishimura, N. Mizuochi and S. Yamauchi, *Appl. Magn. Reson.*, 2004, **26**, 117.
421. S. Yamauchi, *Bull. Chem. Soc. Jpn.*, 2004, **77**, 1255.
422. S. Yamauchi, Y. Iwasaki, Y. Ohba, B. Z. S. Awen and A. Ouchi, *Chem. Phys. Lett.*, 2005, **411**, 203.
423. M. Bortolus, M. Prato, J. van Tol and A. L. Maniero, *Chem. Phys. Lett.*, 2004, **398**, 228.
424. L. Franco, M. Mazzoni, C. Corvaja, V. P. Gubskaya, L. S. Berezhnaya and I. A. Nuretdinov, *Chem. Comm.*, 2005, **2128**.
425. A. Bencini and D. Gatteschi, *EPR of Exchange Coupled Systems*, Springer, Berlin, 1990, p. 55.
426. P. A. Liddell, G. Kodis, L. de la Garza, J. L. Bahr, A. L. Moore, T. A. Moore and D. Gust, *Helv. Chim. Acta*, 2001, **84**, 2765.
427. D. M. Guldi, *Chem. Soc. Rev.*, 2002, **31**, 22.
428. B. Chance and M. Nishimura, *Proc. Natl. Acad. Sci. USA*, 1960, **46**, 19.
429. J. J. Regan and J. N. Onuchic, In: *The Reaction Centers of Photosynthetic Bacteria*, M. E. Michel-Beyerle, Ed., Springer, Berlin, 1996, pp. 117.
430. A. Berman, E. S. Izraeli, H. Levanon, B. Wang and J. L. Sessler, *J. Am. Chem. Soc.*, 1995, **117**, 8252.
431. J. P. Sumida, P. A. Liddell, S. Lin, A. N. Macpherson, G. R. Seely, A. L. Moore, T. A. Moore and D. Gust, *J. Phys. Chem. A*, 1998, **102**, 5512.
432. M. J. Therien, M. Selman, H. B. Gray, I. J. Chang and J. R. Winkler, *J. Am. Chem. Soc.*, 1990, **112**, 2420.
433. R. A. Marcus, *Pure Appl. Chem.*, 1997, **69**, 13.
434. S. S. Skourtis, D. N. Beratan, In: *Electron Transfer-From Isolated Molecules to Biomolecules, Part I*, J. Jortner, M. Bixon, Eds., John Wiley, New York, 1999, pp. 377.
435. P. J. F. Derege, S. A. Williams and M. J. Therien, *Science*, 1995, **269**, 1409.
436. J. L. Sessler, B. Wang and A. Harriman, *J. Am. Chem. Soc.*, 1993, **115**, 10418.
437. P. J. Hore, In: *Advanced EPR, Applications in Biology and Biochemistry*, A. J. Hoff, Ed., Elsevier, Amsterdam, 1989, pp. 405.
438. R. H. Felton, In: *The Porphyrins*, D. Dolphin, Ed., Vol. **5**, Academic Press, New York, 1978, pp. 53.
439. Landolt-Börnstein, *Numerical Data and Functional Relationships in Science and Technology*, New Series, Group II, Part d1, K.-H. Hellwege, Ed., Vol. **9**, Springer, Berlin, 1980, pp. 619.
440. A. Sancar, *Chem. Rev.*, 2003, **103**, 2203.
441. L. O. Essen and T. Klar, *Cell. Mol. Life Sci.*, 2006, **63**, 1266.
442. T. Todo, *Mutat. Res.-DNA Repair*, 1999, **434**, 89.

443. T. Carell, L. T. Burgdorf, L. M. Kundu and M. Cichon, *Curr. Opin. Chem. Biol.*, 2001, **5**, 491.
444. H. W. Park, S. T. Kim, A. Sancar and J. Deisenhofer, *Science*, 1995, **268**, 1866.
445. T. Tamada, K. Kitadokoro, Y. Higuchi, K. Inaka, A. Yasui, P. E. deRuiter, A. P. M. Eker and K. Miki, *Nature Struct. Biol.*, 1997, **4**, 887.
446. H. Komori, R. Masui, S. Kuramitsu, S. Yokoyama, T. Shibata, Y. Inoue and K. Miki, *Proc. Natl. Acad. Sci. USA*, 2001, **98**, 13560.
447. A. Mees, T. Klar, P. Gnau, U. Hennecke, A. P. M. Eker, T. Carell and L. O. Essen, *Science*, 2004, **306**, 1789.
448. K. Hitomi, S. T. Kim, S. Iwai, N. Harima, E. Otoshi, M. Ikenaga and T. Todo, *J. Biol. Chem.*, 1997, **272**, 32591.
449. K. Hitomi, H. Nakamura, S. T. Kim, T. Mizukoshi, T. Ishikawa, S. Iwai and T. Todo, *J. Biol. Chem.*, 2001, **276**, 10103.
450. G. B. Sancar, M. S. Jorns, G. Payne, D. J. Fluke, C. S. Rupert and A. Sancar, *J. Biol. Chem.*, 1987, **262**, 492.
451. P. F. Heelis and A. Sancar, *Biochemistry*, 1986, **25**, 8163.
452. M. S. Jorns, E. T. Baldwin, G. B. Sancar and A. Sancar, *J. Biol. Chem.*, 1987, **262**, 486.
453. G. Payne, P. F. Heelis, B. R. Rohrs and A. Sancar, *Biochemistry*, 1987, **26**, 7121.
454. S. Weber, C. W. M. Kay, A. Bacher, G. Richter and R. Bittl, *Chem. Phys. Chem.*, 2005, **6**, 292.
455. Y. M. Gindt, E. Vollenbroek, K. Westphal, H. Sackett, A. Sancar and G. T. Babcock, *Biochemistry*, 1999, **38**, 3857.
456. B. Giovani, M. Byrdin, M. Ahmad and K. Brettel, *Nature Struct. Biol.*, 2003, **10**, 489.
457. D. B. Sanders and O. Wiest, *J. Am. Chem. Soc.*, 1999, **121**, 5127.
458. J. Hahn, M. E. Michel-Beyerle and N. Rösch, *J. Phys. Chem. B*, 1999, **103**, 2001.
459. J. Antony, D. M. Medvedev and A. A. Stuchebrukhov, *J. Am. Chem. Soc.*, 2000, **122**, 1057.
460. C. Aubert, M. H. Vos, P. Mathis, A. P. M. Eker and K. Brettel, *Nature*, 2000, **405**, 586.
461. R. Bittl and S. Weber, *Biochim. Biophys. Acta*, 2005, **1707**, 117.
462. Y. F. Li, P. F. Heelis and A. Sancar, *Biochemistry*, 1991, **30**, 6322.
463. J. H. Lakey, F. G. van der Goot and F. Pattus, *Toxicology*, 1994, **87**, 85.
464. W. A. Cramer, J. B. Heymann, S. L. Schendel, B. N. Deriy, F. S. Cohen, P. A. Elkins and C. V. Stauffacher, *Annu. Rev. Biophys. Biomol. Struct.*, 1995, **24**, 611.
465. R. M. Stroud, *Curr. Opin. Struct. Biol.*, 1995, **5**, 514.
466. R. M. Stroud, K. Reiling, M. Wiener and D. Freymann, *Curr. Opin. Struct. Biol.*, 1998, **8**, 525.
467. K. J. Oh, H. Zhan, C. Cui, K. Hideg, R. J. Collier and W. L. Hubbell, *Science*, 1996, **273**, 810.

468. P. D. Huynh, C. Cui, H. Zhan, K. J. Oh, R. J. Collier and A. Finkelstein, *J. Gen. Physiol.*, 1997, **110**, 229.
469. D. B. Lacy and R. C. Stevens, *Curr. Opin. Struct. Biol.*, 1998, **8**, 778.
470. S. L. Slatin, A. Finkelstein and P. K. Kienker, *Biochemistry*, 2008, **47**, 1778.
471. M. Wiener, D. Freymann, P. Ghosht and R. M. Stroud, *Nature*, 1997, **385**, 461.
472. J. H. Lakey, D. Duché, J.-M. González-Manas, D. Baty and F. Pattus, *J. Mol. Biol.*, 1993, **230**, 1055.
473. M. W. Parker, J. P. M. Postma, F. Pattus, A. D. Tucker and D. Tsernoglou, *J. Mol. Biol.*, 1992, **224**, 639.
474. D. Duché, M. W. Parker, J.-M. González Manas, F. Pattus and D. Baty, *J. Biol. Chem.*, 1994, **269**, 6332.
475. K. Möbius, A. Savitsky, C. Wegener, M. Rato, M. Fuchs, A. Schnegg, A. A. Dubinskii, Y. A. Grishin, I. A. Grigor'ev, M. Kuhn, D. Duche, H. Zimmermann and H. J. Steinhoff, *Magn. Reson. Chem.*, 2005, **43**, S4.
476. J. R. Dankert, Y. Uratani, C. Grabau, W. A. Cramer and M. Hermodson, *J. Biol. Chem.*, 1982, **257**, 3857.
477. A. Nardi, S. L. Slatin, D. Baty and D. Duché, *J. Mol. Biol.*, 2001, **307**, 1293.
478. Y. Kim, K. Valentine, S. J. Opella, S. L. Schendel and W. A. Cramer, *Protein Sci.*, 1998, **7**, 342.
479. S. Lambotte, P. Jasperse and B. Bechinger, *Biochemistry*, 1998, **37**, 16.
480. D. Huster, Y. L. Yao, K. Jakes and M. Hong, *Biochim. Biophys. Acta*, 2002, **1561**, 159.
481. A. Menikh, M. T. Saleh, J. Gariepy and J. M. Boggs, *Biochemistry*, 1997, **36**, 15865.
482. A. P. Todd, J. Cong, F. Levinthal, C. Levinthal and W. L. Hubbell, *Proteins*, 1989, **6**, 294.
483. F. G. van der Goot, J. M. Gonzalez-Manas, J. H. Lakey and F. Pattus, *Nature*, 1991, **354**, 408.
484. O. B. Ptitsyn, In: *Protein Folding*, T. E. Creighton, Ed., W. H. Freeman, New York, 1992, pp. 243.
485. E. Cascales, S. K. Buchanan, D. Duche, C. Kleanthous, R. Lloubes, K. Postle, M. Riley, S. Slatin and D. Cavard, *Microbiol. Mol. Biol. Rev.*, 2007, **71**, 158.
486. J. H. Lakey and S. L. Slatin, *Curr. Top. Microbiol. Immunol.*, 2001, **257**, 131.
487. D. Huster, X. L. Yao and M. Hong, *J. Am. Chem. Soc.*, 2002, **124**, 874.
488. I. Nishii, M. Kataoka and Y. Goto, *J. Mol. Biol.*, 1995, **250**, 223.
489. Y. Hagihara, M. Oobatake and Y. Goto, *Protein Sci.*, 1994, **3**, 1418.

CHAPTER 6
Conclusions and Perspectives

In molecular biology, particularly in the emerging field of proteomics, attempts are being made to understand, on the basis of structure and dynamics data, the dominant factors that control the specificity and efficiency of electron- and ion-transfer processes in proteins. In parallel with high-resolution X-ray crystallography, spectroscopy in all its facets is being used to elucidate the hidden structure–dynamics–function relations in their multiparameter complexity.

Among the spectroscopic techniques, modern cw and pulse EPR, operated at high magnetic fields and microwave frequencies and extended by multiple-resonance capabilities, is playing an important role in this endeavor, particularly in view of the fact that no single-crystal protein preparations are needed to obtain detailed structural information at atomic resolution. During the last decade, the combined efforts of biologists, chemists and physicists in developing high-field EPR techniques and applying them to functional proteins demonstrated that this type of spectroscopy is particularly powerful for characterizing the structure and dynamics of transient states of proteins in action on biologically relevant time scales. The information obtained is unique in its specificity and is complementary to that of protein crystallography, solid-state NMR and laser spectroscopy.

The book describes how high-field EPR methodology, in conjunction with site-specific isotope or spin-labelling mutation strategies, and with support of modern quantum-chemical computation methods for data interpretation, is capable of providing new insights into the electronic structure of cofactors participating in biological transfer processes. Specifically, the theoretical and instrumental background of cw and pulse high-field EPR and its multiple-resonance extensions ENDOR, TRIPLE and PELDOR as well as high-field RIDME and ESEEM are discussed. Multifrequency EPR applications to representative examples from photochemical reactions in liquid phase or frozen solution are described as well as paradigmatic protein complexes and their

---

High-Field EPR Spectroscopy on Proteins and their Model Systems: Characterization of Transient Paramagnetic States
By Klaus Möbius and Anton Savitsky
© 2009 Klaus Möbius and Anton Savitsky
Published by the Royal Society of Chemistry, www.rsc.org

biomimetic model systems. The transient paramagnetic states detected after photoinitiation of the reaction often show characteristic electron spin-polarization (CIDEP) effects, which not only contain valuable information about structure, dynamics and reaction pathways of the short-lived intermediates, but can also be exploited for signal enhancement in EPR and ENDOR experiments. The power of multifrequency time-resolved EPR (TREPR) spectroscopy is demonstrated by combining the results from X-band and W-band TREPR to unravel the complex electron-transfer and electron-spin dynamics of donor–acceptor biomimetic model systems. As examples, covalently linked porphyrin–quinone dyad and triad systems as well as Watson–Crick base-paired porphyrin-quinone and porphyrin–dinitrobenzene donor–acceptor systems are discussed.

Concerning applications of high-field EPR to topical proteins, the book focuses on reviewing recent 95- and 360-GHz studies (EPR, ENDOR, TRIPLE, ESEEM, RIDME, PELDOR) at FU Berlin highlighting the following systems: *(i)* Light-induced electron-transfer intermediates in wild-type and mutant reaction-center (RC) proteins from photosynthetic organisms. *(ii)* Light-induced proton-transfer intermediates of site-specifically nitroxide spin-labelled mutants of bacteriorhodopsin. *(iii)* Light-generated flavin radical intermediates in DNA photolyase enzymes, which utilize long-wavelength light to repair UV-induced DNA damages. *(iv)* Intermediate states of refolding protein domains of site-specifically nitroxide spin-labelled mutants of the colicin A bacterial toxin, which initiates the formation of lethal transmembrane ion channels.

When we now ask for an answer to the question – what do we learn from high-field EPR spectroscopy on proteins – some conclusions can be made by summarizing the results of the studies elaborated on in several chapters of the book. Although we have restricted our overview largely to work performed at our laboratory at FU Berlin, our concluding statements on specific advantages of high-field EPR pertinent to biomolecular research are equally based on the work of esteemed colleagues in the field:

1. Many organic cofactors in proteins have only small g-anisotropies and, therefore, require much higher magnetic fields than available in X-band EPR to resolve the canonical g-tensor orientations in their powder spectra ("Zeeman magnetoselection"). Thereby, even on disordered samples, orientation-selective hydrogen bonding and polar interactions in the protein-binding sites can be traced. This is important information complementary to what is available from high-resolution X-ray diffraction of protein crystals.

2. In photochemical reactions and electron-transfer processes of organic systems often several radical species are generated as transient intermediates with overlapping EPR spectra. To distinguish them by the notoriously small differences in their g-factors and hyperfine interactions, high Zeeman fields are required and, hence, high-frequency EPR becomes the method of choice.

3. Often, high-purity protein samples can be prepared only in minute quantities, for example RCs of site-directed mutants or with isotopically labelled cofactors. The problem of small concentration of paramagnetic states

Conclusions and Perspectives

is even the rule for single crystals of membrane proteins, which often are tiny in all dimensions. Accordingly, to study them by EPR, very high detection sensitivity is needed, which often can be accomplished only by dedicated high-field/high-frequency spectrometers.

4. The enhanced low-temperature Boltzmann electron-spin polarization at high Zeeman fields allows extraction of the absolute sign of the zero-field splitting parameter, D, of a two-spin system like a biradical or triplet state. At high fields, considerable thermal spin polarization can be achieved already well above helium temperature provided the sample temperature becomes comparable with the Zeeman temperature, $T_Z = g\,\mu_B \cdot B_0/k_B$. At $T < T_Z$, the Boltzmann distribution leads to increased populations of the low-energy levels, resulting in asymmetric lineshapes from which the absolute sign of D can be directly read off. For $g = 2$ systems, pronounced thermal spin polarization, as a means to determine the absolute sign of D in high-spin systems, requires sample temperatures comparable with $T_Z \approx 0.4\,\text{K}$ (at 9.5 GHz), $T_Z \approx 4\,\text{K}$ (at 95 GHz), $T_Z \approx 6.5\,\text{K}$ (at 140 GHz) and $T_Z \approx 15.5\,\text{K}$ (at 360 GHz).

5. High-field/high-frequency cw EPR generally provides, by lineshape analysis, shorter time windows down into the ps range for studying correlation times and fluctuating local fields over a wide temperature range. They are associated with characteristic dynamic processes, such as protein motion and refolding or cofactor libration and reorientation in the binding site.

6. On the other hand, pulsed high-field/high-frequency EPR, for instance in the form of two-dimensional field-swept ESE spectroscopy, gives real-time access to specific cofactor/protein slow motions on the ns time scale. Even their motional anisotropy can be traced, which is generated by anisotropic interactions, e.g., hydrogen bonding within the binding site, and leads to temperature-dependent anisotropic relaxation. This relaxation contribution is due to librational motion of a cofactor about specific molecular axes. They are determined by the network of specific H-bonds of different strength to the amino-acid residues in the different directions. Thus, pulsed high-field ESE measurements of the anisotropic T_2 relaxation dynamics of frozen-solution samples are decisive in answering the questions whether a cofactor is hydrogen bonded or not, and what is the likely binding partner in the protein pocket (assuming that its X-ray structure is known).

7. ENDOR at high magnetic fields takes advantage of additional orientation selection of molecular subensembles in powder or frozen-solution samples (double orientation selection by the Zeeman field and the ENDOR frequency). Thereby, even in the case of small g-anisotropies, ENDOR can provide single-crystal-like information about hyperfine interactions, including anisotropic hydrogen bonding to the protein.

8. By properly adjusting the Zeeman field in multifrequency pulsed EPR experiments, strongly and weakly hyperfine-coupled nuclei in the protein system can be differentiated by their ENDOR and ESEEM spectra. The different sensitivity of ENDOR and ESEEM for strongly and weakly coupled nuclei can be exploited, for example, to differentiate between remote and coordinated nitrogens in histidines of metallo-proteins.

9. In metallo-protein high-spin systems, such as the Mn cluster of the oxygen-evolving complex in the photosynthetic reaction center PS II, the analysis of the EPR spectrum can be drastically simplified at high Zeeman fields owing to suppression of second-order effects. EPR transitions of certain high-spin metallo-proteins with large zero-field splittings cannot be observed at all at X-band frequencies, but become accessible at the higher quantum energies of mm or sub-mm microwave fields.

10. The combination of cw and pulse high-field EPR spectroscopy with site-directed spin-labelling (SDSL) techniques employing nitroxide radicals has turned out to be particularly powerful in revealing subtle changes of the polarity and proticity profiles along segments of proteins or their micro-environment of the plasma or the membrane. This information can be obtained by orientation-selective high-field EPR on frozen-solution samples, thereby resolving the principal components of the nitroxide Zeeman and hyperfine-tensors of the spin labels attached to specific molecular sites. The g_{xx} and A_{zz} components of the g- and nitrogen hyperfine-tensors are particularly sensitive probes for the polarity and proticity of the microenvironment of the spin label. In addition to the g_{xx} and A_{zz} components, the ^{14}N ($I=1$) quadrupole-interaction tensor of the nitroxide spin label can be exploited for probing effects of the microenvironment of functional protein sites. To this end, the ^{14}N quadrupole-tensor components P_{xx}, P_{yy} of a nitroxide spin label in disordered frozen solution can be measured with high accuracy by 95-GHz high-field ESEEM.[1] It turned out that the P_{yy} component is especially sensitive to the proticity and polarity of the nitroxide environment in H-bonding and non-H-bonding situations. When using P_{yy} as a testing probe of the environment, its ruggedness towards temperature changes represents an important advantage over the g_{xx} and A_{zz} parameters usually employed for probing matrix effects. Thus, beyond measurements of g_{xx} and A_{zz} of spin-labelled protein sites in disordered solids, high-field ESEEM studies of ^{14}N quadrupole interactions open a new avenue to reliably probe subtle environmental effects on the electronic structure. This is a significant step forward on the way to differentiating between effects from matrix polarity and hydrogen-bond formation. Thereby, our understanding of the relation between structure, dynamics and function will be improved considerably. This is especially true with respect to the fine tuning of electronic properties of cofactor reaction partners by means of weak interactions with their protein environment, such as hydrogen bonding to specific amino-acid residues.

11. Pulsed high-field EPR spectroscopy at 95 GHz in combination with its extensions to PELDOR and RIDME offers powerful tools for obtaining, beyond information on the electronic structure of the redox partners, also information on the three-dimensional structure of radical-pair systems with large interspin distances (up to about 8 nm) even in disordered frozen solutions. In photosynthetic electron-transfer proteins, for example, distance and relative orientation of functional cofactors within the protein domains and their conformational changes during the photocycle determine the selectivity and efficiency of the biological processes. By the novel approach of orientation-resolving

electron–electron dipolar spectroscopy, such as 95-GHz high-field PELDOR, not only interspin distances but the full 3D-structure of laser-flash induced transient charge-separated radical pairs $P_{865}^{\bullet+}Q_A^{\bullet-}$ in frozen-solutions reaction centers (RCs) from the photosynthetic bacterium *Rb. sphaeroides* could be solved.[2]

High-field RIDME and PELDOR can also be used to determine distance and relative orientation of two weakly coupled nitroxide spin labels attached to specific positions in site-specific double mutants. In metallo-proteins one could introduce a single nitroxide spin label at a specific site to weakly couple with the paramagnetic metal-ion center. The problem with conventional nitroxide spin labels for precise structure determination is the orientational flexibility of their tether with which they are linked to the protein backbone. Although still allowing for rather reliable distance measurements, this flexibility leads to a broad distribution of orientations. There appears to exist two general options for a solution of the current problem of nonrigidity of the available nitroxide spin labels when used to construct 3D-structure models: *(i)* To develop synthetic strategies in molecular engineering to incorporate rigid nitroxide headgroups into the protein in the form of artificial nitroxyl amino acids without substantial perturbation of the wild-type structure. This will allow for a precise determination of distances and relative orientations of the spin-carrying protein moieties by high-field PELDOR or related variants of electron–electron dipolar spectroscopy. On this basis, a relatively small number of double mutants will suffice for providing enough well-defined structural constraints to construct a 3D molecular model. A larger number of double mutants will be needed if only distance constraints can be used, as are provided by electron–electron dipolar spectroscopy at low microwave frequencies, *e.g.*, from the Pake-pattern-type spectra at X-band or even S-band. *(ii)* To use, as the other option, a large number of double mutants with conventional, *i.e.* nonrigid, nitroxide spin labels attached at numerous specific sites for weak interelectron dipolar couplings. By executing spin triangulation via numerous high-field PELDOR-type experiments, a sufficient number of both distance and orientation constraints will be obtained, compensating for the broad orientation distributions of nonrigid spin labels.

12. In conclusion, it may fairly be said that cw and pulse multifrequency EPR up to high magnetic fields has matured over the last years to add substantially to the capabilities of "classical" spectroscopic and diffraction techniques for determining structure–dynamics–function relations of biosystems, since transient intermediates can be observed in real time in their working states at biologically relevant time scales. The role of high-field EPR in biology, chemistry and physics is growing rapidly. This growth has been recognized in recent years by an increasing number of national and international high-field EPR research programs and specialized symposia. Accordingly, the scientific literature on high-field EPR is also rapidly growing, and only a small sector could be covered in this book.

Now, after these 12 conclusion theses, what do we see as perspectives?

We started the book with a delineation of the historical roots of EPR and NMR, both starting around 1945 within ± 1 year, both originally employing cw radiofrequency fields, but soon separating from each other, NMR banking exclusively on pulse-irradiation schemes, EPR staying for many years in the cw domain and needing almost a decade longer before the first electron-spin echoes were detected. And even now, almost 65 years after Zavoisky's discovery of EPR, there still exist many good reasons to develop and apply EPR both in the cw and pulse mode of operation. On the other hand, the magnetic resonance community is currently observing a converging process of EPR and NMR in terms of applying higher and higher Zeeman fields with corresponding resonance frequencies. The turning point from divergence to convergence can be assigned to the early 1980s when several research groups completed their development of prototype spectrometers, both in the area of pulse EPR[3] and high-field EPR,[4] and turned to a variety of impressive applications. Particularly in structural biology one can appreciate how high-field extensions of NMR and EPR methods can provide detailed information on structure and dynamics of complex protein systems, and the trend to higher and higher magnetic fields appears to be unbroken. In NMR the magic GHz limit has been passed already,[5] and in EPR the THz domain has been reached (without yet hitting the g-strain limit of the Zeeman field beyond which there will be no further resolution improvement).[6] Naturally, this puts extreme demands on the magnet systems generating the correspondingly high fields that probably can be met only in large-scale magnet facilities.

The current state and future trends in molecular biology and biophysics are predominantly characterized by two subject areas, "understanding biological function at the atomic level" and "extending molecular characterization from nanostructures to whole cells" (no wonder that these topical subject areas are also the main themes of the 2008 Meeting of the German Biophysical Society in Berlin). We claim that high magnetic fields in magnetic resonance spectroscopy are the very means to stay in the forefront of research in the biosciences. Finally, we want to substantiate this personal view by briefly reviewing four examples from the magnetic-resonance literature that clearly demonstrate how modern EPR and NMR meet again, to the benefit of them both, after fifty years of alienation.

Fourier-Transform High-Field EPR with Broadband Stochastic Microwave Excitation[7,8]

Most EPR experiments are performed by measuring either the absorption of cw irradiated monochromatic microwaves or the amplitude of a spin echo after a train of resonant microwave pulses. Especially for high-frequency EPR, with its notoriously low available excitation power, the spectral bandwidths of the pulses usually are small compared to the EPR spectrum of the sample. Therefore, cw and pulse high-field EPR spectra are normally recorded while sweeping or stepping the external magnetic field. There are only very few laboratories that have already developed[9–11] or are currently in the process of developing[12] high-field, high-pulse-power EPR spectrometers that produce, at

W-band, kW pulses to generate $\pi/2$ pulses as short as a few ns. Such short $\pi/2$ pulses are necessary to increase the excitation bandwidth and allow full excitation of many paramagnetic systems including nitroxide spin labels. Provided that the deadtime of the spectrometers can be reduced to a few ns, FID detection strategies may be implemented that would allow 2D methods to be used in high-field EPR spectroscopy that are common in NMR or X-band EPR, including Fourier-transform (FT) spectroscopy. It appears that high-field FT-EPR with coherent $\pi/2$ pulses short enough for broadband excitation of, say, spin-labelled proteins, is still not at hand for most laboratories.

On the other hand, it is known from NMR or X-band EPR that FT-spectroscopy, in which the total spin system is excited by one radiofrequency or microwave pulse, and the system response is detected in the time domain, offers several distinct advantages. Among them is the feasibility of measuring correlations between two spins. Two-dimensional correlation spectroscopy (COSY) allows, for instance, resonances belonging to the same molecule from those belonging to different molecules to be distinguished.[13] Another important point is that FT-spectroscopy brings the so-called multiplex advantage into NMR or EPR, enabling a shortening in the data-scanning time by orders of magnitude, and enabling a concomitant increase in the signal-to-noise ratio achievable per unit time.

Clearly, for good sensitivity in pulse EPR one has to turn the magnetization by $\pi/2$ with one pulse. The bandwidth of this coherent excitation has the approximate magnitude of the reciprocal pulse length that depends on the available microwave power. As was pointed out above, most high-frequency microwave sources with good noise characteristics do not supply enough power for such a coherent broadband excitation.

To overcome these difficulties an alternative approach was tried[7,8] by using a broadband incoherent continuous microwave "noise" irradiation with a spectral FWHM bandwidth of about 250 MHz. This "stochastic resonance" approach, in which the computed crosscorrelation function of the input noise and the output noise is equivalent to the FID of pulse FT-spectroscopy, was introduced to NMR by R.R. Ernst[14] and R. Kaiser.[15] They showed theoretically and experimentally that the sensitivity for measuring one-dimensional spectra equals that of coherent pulse FT-spectroscopy. Quite a number of publications report on applications of stochastic spin excitation in NMR correlation spectroscopy, see for example.[16–18] In contrast, there are only a few applications of EPR correlation spectroscopy with stochastic excitation, for example at X-band[19] and at an electron Larmor frequency of 300 MHz,[20] in addition to the stochastic-excitation experiment in W-band reviewed here.[8] For optimum sensitivity in stochastic resonance, one has to apply the same average power as in pulse FT-spectroscopy. Consequently, for stochastic resonance with continuous microwave irradiation, one needs only peak powers that are reduced by a factor of τ_p/T_1 compared to pulse spectroscopy, where τ_p is the pulse length and T_1 is the spin-lattice relaxation time. This reduction factor, τ_p/T_1, contains the ratio of the duty cycles of pulse and stochastic spectroscopy and may be as small as 10^{-2} to 10^{-3}. Thus, the optimum microwave excitation field strength for stochastic FT-EPR depends on the excitation bandwidth and on T_1.[14] It is this requirement of only relatively low excitation power that makes stochastic resonance particularly interesting for high-field EPR.

Compared to the previously realized stochastic NMR at 280 MHz, with spectral bandwidths up to 200 kHz, stochastic W-band FT-EPR operates at 95 GHz, and typical samples require to cover a bandwidth of about 250 MHz. Thus, the frequency spread to be covered is three orders of magnitude larger than that in NMR. This large bandwidth requirement poses severe experimental problems. To cope with these problems, a novel stochastic W-band microwave bridge and bimodal induction-mode transmission resonator had to be built. The broadband microwave excitation was generated by using pseudostochastic maximum-length binary sequences by which the 95-GHz carrier is phase-modulated. The bimodal probehead was realized as a Fabry–Perot transmission resonator operating in the TEM_{008} mode with the detection arm rotated by 90° with respect to the excitation arm in order to decouple the input noise from the signal-encoded output noise. The resonator is overcoupled to decrease the quality factor Q to 200 and, thereby, provide the required high bandwidth. The first measurements with this instrumentation on a test nitroxide-nitronyle radical showed the soundness of the chosen excitation scheme. All the expected five ^{14}N hyperfine lines from the two equivalent nitrogens in the test radical could be resolved over a spectral width of 90 MHz (3 mT) at one fixed value of the Zeeman field B_0.

On the other hand, the sensitivity of the stochastic excitation experiment in its present version, employing a Fabry–Perot resonator, was smaller than that in cw EPR, employing a cylindrical cavity as the probehead. Fabry–Perot resonators have a smaller microwave power-to-field conversion factor than cylindrical cavities have. This feature and the small output power of the used 95-GHz microwave source are responsible for the fact that the optimum microwave excitation field strength at the given relaxation time T_1 of about 1 μs is not reached. In ref. 8 concrete steps in instrumental developments are discussed that are necessary to realize for future applications of stochastic W-band FT-EPR to stable organic radicals or transient radical pairs in biological processes.

Photochemically Induced Dynamic Nuclear Polarization (Photo-CIDNP) in Magic-Angle-Spinning (MAS) NMR[21-23]

As was discussed in Section 2.2.1.2, in photochemical reactions one often observes spin-polarization effects of the reaction partners that manifest themselves in non-Boltzmann magnetization with resonance lines in enhanced absorption or emission. In an EPR spectrum, this phenomenon is observed by the non-Boltzmann behavior of the lines of the paramagnetic reaction partners and is called CIDEP (chemically induced dynamic electron polarization). In NMR an analogous spin-polarization phenomenon (CIDNP: chemically

Conclusions and Perspectives

induced dynamic nuclear polarization) can be observed by the non-Boltzmann behavior of the lines of diamagnetic reaction partners, which interact with paramagnetic intermediates of the photoreaction. In solid-state photoreactions, photo-CIDNP polarization effects have been observed in various light-induced electron-transfer reactions (for a historical account, see ref. 23), particularly in photosynthetic reaction centers (RCs), by magic-angle-spinning (MAS) NMR. The origin of photo-CIDNP in solution is described by the radical-pair mechanism (RPM), see Section 2.2.1.2). In the solid state or frozen solutions, RPM cannot explain the observed photo-CIDNP effects because the necessary chemical branching does not occur in frozen RCs. However, the photo-CIDNP effect in solids can be explained by a combination of several new polarization mechanisms,[23] which transfer the electron-spin polarization obtained in the initial singlet-state radical pair of the primary donor and acceptor cofactors to their nuclei via hyperfine interaction. Therefore, as an alternative and a complement to EPR, photo-CIDNP-enhanced MAS NMR appears to be a powerful spectroscopic method to study the electronic structure of both the ground state and the radical-pair state of the primary donor–acceptor system. The electronic ground-state properties are deduced from the chemical shifts, while the radical-pair properties are affecting the NMR intensities. Most remarkably, very large nuclear polarizations up to a factor of 10^5 above Boltzmann equilibrium have been observed in photo-CIDNP ^{13}C MAS NMR experiments, which opens new avenues for applying solid-state NMR to photochemical processes.

Recently, photo-CIDNP was observed in quinone-depleted frozen photosynthetic RCs from the purple bacterium *Rb. sphaeroides* wild type (WT) by ^{13}C solid-state NMR at three different magnetic fields.[21] All light-induced signals turned out to be emissive at all three fields. At 4.7 T (200 MHz proton frequency), the strongest enhancement of NMR signals is observed, which is more than a factor of 10^5 above the Boltzmann polarization. At higher fields, the enhancement factor decreases. At 17.6 T, the enhancement factor is about 60. The field dependence of the enhancement is the same for all nuclei. The observed field dependence agrees with simulations that assume two competing mechanisms of polarization transfer from electrons to nuclei, namely three-spin mixing (TSM) and differential decay (DD). These simulations indicate that the ratio of the average electron spin densities on the L- and M-half of the special-pair primary donor bacteriochlorophylls is 3:2 in favor of the L-BChl during the radical cation state. The good agreement of the simulations with the experiments raises the expectation that photo-CIDNP ^{13}C MAS NMR can be applied to investigate the electronic structure of other natural and artificial solid-state reaction centers in the near future.

Somewhat earlier, the same authors[22] had studied, by photo-CIDNP ^{13}C MAS NMR, the photochemically induced nuclear spin polarization in intact cells of the carotenoid-less strain R26 of *Rb. sphaeroides*. The ^{13}C solid-state NMR experiments were performed at three different magnetic fields (4.7, 9.4, and 17.6 T). In this carotenoid-less mutant the signals of the donor appear in enhanced absorption (positive) and of the acceptor in emission (negative).

This observation is in contrast to the photo-CIDNP data of RCs from wild-type *Rb. sphaeroides* reported previously[21] in which all signals appear in emission.

This difference in spin polarization is explained by considering an additional triplet-involving three-spin mixing (TSM) polarization mechanism occurring in RCs from R26, in which the transient triplet state of the primary donor is not quenched by a nearby carotenoid that, however, is present in the wild-type RC. This is an exciting observation because it allows differentiation between different mutants with and without triplet-quenching cofactors by solid-state NMR methods. In other words: The long-lived triplet state of the donor allows for spectral editing by the different enhancement mechanisms.

The overall shape of the NMR spectra remains independent of the strength of the magnetic field. The strongest enhancement is observed at 4.7 T (200 MHz proton frequency), enabling the observation of photo-CIDNP enhanced NMR signals from RC cofactors at nanomolar concentration of the donor cofactor in entire bacterial cells, striving for molecular and atomic resolution. This would allow the detection of subtle changes in the electronic structure of the donor cofactor during the photocycle. Furthermore, the photo-CIDNP MAS NMR method could be used to study the effects of detergents and other chemicals, *e.g.*, cryoprotectants, commonly used for RC preparations, on the electronic structure of photosynthetic cofactors.

Dynamic Nuclear Polarization at High Magnetic Fields[24,25]

With the same goal – to enhance the NMR signal intensity by introducing interactions of the nuclear spins with unpaired electron spins – a different approach is being made in numerous laboratories by using stable radicals as mediator. This approach is generally termed DNP-enhanced NMR, and a prominent representative in this endeavor is the research group of R. G. Griffin at the Francis Bitter Magnet Laboratory, Massachusetts Institute of Technology. In the following, the very recent work of this group is briefly introduced.[24,25] DNP (dynamic nuclear polarization) is a method that permits NMR signal intensities of solids and liquids to be enhanced significantly. Therefore, it is potentially an important tool in structural and mechanistic studies of biologically relevant molecules. During a DNP experiment, the large spin polarization of an unpaired electron located either outside or inside the molecular system under study, is transferred to the nuclei of interest in the NMR spectrum by microwave irradiation of the sample. The maximum theoretical NMR enhancement achievable is given by the gyromagnetic ratios of the electron and the respective nucleus, being approximately 660 for protons. In the early 1950s, the DNP phenomenon was demonstrated experimentally, and intensively investigated in the following four decades, primarily at low magnetic fields. The review article of Maly *et al.*[25] focuses on recent developments in the field of DNP with a special emphasis on work done at high magnetic fields (≥ 5 T), *i.e.* in the regime where contemporary NMR experiments are being performed. The classical cw DNP mechanisms for explaining the various polarization features comprise the Overhauser effect, the solid effect, the cross effect, and the effect of thermal mixing. However, coherent polarization-transfer mechanisms are potentially more efficient at high fields than classical polarization schemes. Naturally, the implementation of DNP capability at high magnetic fields has required the development of new NMR probeheads and microwave instrumentation in the mm-band frequency range. The review article gives an overview of DNP applications to biological systems in the solid and liquid state, and possible areas for future developments of DNP-enhanced NMR are outlined.

The second reviewed article from the Griffin group[24] describes high-frequency/high-field DNP studies using nitroxide biradicals as polarization-transfer mediators. This work demonstrates once more the intimate links between modern NMR and EPR, in particular with regard to optimizing the sensitivity and resolution performance.

To date, the cross effect (CE) and thermal mixing (TM) mechanisms have consistently provided the largest enhancements in DNP experiments performed at high magnetic fields. Both mechanisms involve a three-spin electron–electron–nucleus process. Its efficiency depends primarily on the electron–electron interactions, *i.e.* the interelectron distance R and the correct EPR-frequency separation that matches the nuclear Larmor frequency, which is mandatory for polarization transfer: $|\omega_{e2}-\omega_{e1}| = \omega_n$. It is anticipated that biradicals, for example two 2,2,6,6-tetramethyl-piperidine-1-oxyl (TEMPO) radicals tethered with a molecular linker, can, in principle, constrain both the distance and relative g-tensor orientation between the two unpaired electrons in the TEMPOs, allowing these two spectral parameters to be optimized for the CE and TM. To verify this hypothesis, several biradicals (bis-TEMPO tethered by n ethylene glycol units, abbreviated as BtnE) were synthesized that show an increasing DNP enhancement with a decreasing tether length. Specifically, at 90 K and 5 T the enhancement factor grew from ≈ 40 observed with 10 mM monomeric TEMPO, where the average $R \approx 5.6$ nm, corresponding to an electron–electron dipolar coupling constant $\omega_d/2\pi = 0.3$ MHz, to ≈ 175 with 5 mM BT2E (10 mM electrons) which has $R \approx 1.3$ nm with $\omega_d/2\pi = 24$ MHz. In addition, these DNP enhancements were compared with those from other nitroxide biradicals having shorter and more rigid tethers. The interelectron distances and relative g-tensor orientations of all of these biradicals were characterized via an analysis of their cw EPR lineshapes at 9 and 140 GHz. The results show that the largest DNP enhancements are observed with BT2E and biradicals that have shorter tethers and for which the two TEMPO moieties are oriented in such a way to satisfy the matching condition for the CE polarization-transfer mechanism.

2-GHz Proton NMR in Pulsed Magnets[5]

Here, we will briefly review an article by J. Haase and coworkers (Leibniz Institute for Solid State and Materials Research (IFW), Dresden) on their recent achievement to perform proton NMR at 2 GHz in a 50-T pulsed field magnet. Moreover, the paper gives an impressive account of the current trend in NMR towards higher static magnetic fields, B_0. In NMR this is driven not only by the higher signal-to-noise and chemical-shift resolution attainable, as in biological application areas, but also by the desire to study materials with markedly field-dependent properties that may only be observable at very high B_0 values. Examples include field-induced phase transitions, strongly correlated electron systems, materials with anisotropic magnetic susceptibility.

Commercial 22.3 T (950 MHz) high-resolution superconducting magnets have become available in recent years. The National High Magnetic Field Laboratory (NHMFL) at Tallahassee has reported the design of an ultrawide bore (105 mm diameter) 900-MHz magnet.[26] With currently available superconducting materials, still higher fields cannot be created in persistent mode. Electromagnets with coils made from resistive rather than superconducting wire do not have such constraints, and they are being employed for high-field NMR.[27] However, the high ohmic losses make cooling and stabilization of these magnets difficult and expensive. For example, NMR spectra at 1.06 GHz using a 25.0-T electromagnet have been reported.[28] The mandatory cooling increases the size of the main magnet coil that, for a given field strength, results in strongly increased field energy required (and concomitant losses to be cooled off). This is why resistive magnets rarely exceed 33 T. In the unbroken quest for still higher static B_0 fields hybrid magnets have recently been built in several national and international high magnetic field laboratories that can reach 45 T, despite the enormous amounts of electric energy and cooling water that these magnets consume. In the hybrid magnets, a resistive magnet (33 T) is placed inside a large nonpersistent superconducting magnet. Given the cost for construction and operation a further increase in field strength with similar systems seems unlikely.[5]

An alternative way to create very high magnetic fields is to use small coils and short field pulses so that cooling of the coil during the pulse can be avoided.[29] The small volume of the coil then makes high fields possible even at relatively low energies. Pulsed magnets have found wide applications over the last few years;[30] peak field strengths of up to 80 T over periods of some 15 ms can be achieved.

Until recently, such magnets had not attracted attention from the NMR community (and even less so from the EPR community), because the magnetic field fluctuates rapidly both in time and space. Several attempts to record NMR signals had failed before the first spectra of ^{63}Cu NMR at fields up to 33 T (370 MHz Larmor frequency) could be observed by using a pulsed field magnet.[31,32] Shortly afterwards, ^{2}H NMR signals were reported at fields up to 58 T (380 MHz).[33,34] The next challenge was the construction of a data-acquisition system for obtaining ^{1}H NMR data above 2 GHz, which represents the first NMR experiment[5] at these S-band microwave frequencies.

The pulsed magnet used for these experiments consists of a 50-T, 340-kJ, short-pulse coil with 24 mm warm bore. The coil has a length and diameter of about 20 cm. It is energized by a 1-MJ capacitor bank (for comparison, a state-of-the-art 800-MHz cryomagnet has a stored energy > 10 MJ). The capacitor bank can be charged to the full energy of 1 MJ at 10 kV in about 40 s. Into the main coil bore a cryostat is introduced for adjusting the sample temperature between 1.8 and 330 K. Inside the cryostat and center field region an insert carries sensors for measuring the temperature and magnetic field, thereby limiting the available space for the NMR probe to a diameter of ≈ 7 mm.

The home-built NMR probe is based on a small solenoid of copper wire wrapped around a quartz sample capillary (OD = 0.7 mm, ID = 0.53 mm) filled with a 3:1 $H_2O:D_2O$ aqueous solution containing paramagnetic $GdBr_3$ in such a concentration that the ^1H spin–lattice relaxation time T_1 is about 200 μs. The probe coil is part of a capacitively coupled parallel circuit with unloaded quality factor $Q \approx 35$ after impedance matching to 50 Ω at a frequency of 2.0 GHz, corresponding to a bandwidth of 57 MHz. The small size of the probe coil and sample is advantageous for minimizing the spatial inhomogeneity of the magnetic field over the sample.

Applying an appropriate train of 2-GHz pulses, a free-induction-decay (FID) signal was observed with a signal-to-noise of about 55 (at 8 MHz bandwidth). Within the current experimental setup several factors need further optimization, for example the optimum sample position in the main coil bore for minimum field inhomogeneity, the reduction of possible contributions of eddy currents to line broadening. In the long run, after having demonstrated that NMR data can be collected in pulsed field magnets, the design of the magnet coil itself may need to be optimized for NMR performance, for example a longer magnet coil would result in higher B_0 homogeneity.

The authors conclude that the very high B_0 field may help to further improve sensitivity and resolution in specific areas of material research but, in general, NMR in pulsed high fields is unlikely to compete with that in static B_0 fields with larger homogeneity. However, the relatively low cost of such pulsed systems, as compared to their static high-resolution counterparts, may become an important factor in adopting this new capability of high-field NMR for widespread applications.

Epilogue

At the end of this book on high-field EPR, some concluding remarks remain, perhaps, to be said. We have seen that what is a high magnetic field in EPR spectroscopy, for example 14 T in 360-GHz EPR, is meanwhile standard in NMR spectroscopy (600 MHz). But this fact should not lead the EPR community to develop an inferiority complex. Throughout the book we have, we hope, shown that multifrequency EPR at correspondingly high magnetic fields has solved, and will continue to solve very relevant problems in chemistry as well as in molecular and structural biology, that cannot be solved by diffraction methods or other spectroscopic techniques. This holds, in particular, with regard to the characterization of transient intermediates of chemical or biological processes at the molecular and atomic level. Cooperation between scientists from biology, chemistry and physics is one vital prerequisite for the spectacular results obtained from multifrequency EPR at higher and higher magnetic fields during the last decades. Dissemination of these results to a wider scientific and public audience is another prerequisite, essential for receiving the attention and recognition modern EPR spectroscopy deserves, and vital for getting the necessary support. With both prerequisites in mind we have written this book.

References

1. A. Savitsky, A. A. Dubinskii, M. Plato, Y. A. Grishin, H. Zimmermann, K. Möbius, *J. Phys. Chem. B*, 2008, **112**, 9079.
2. A. Savitsky, A. A. Dubinskii, M. Flores, W. Lubitz and K. Möbius, *J. Phys. Chem. B*, 2007, **111**, 6245.
3. A. Schweiger, *Angew. Chem. Int. Ed.*, 1991, **30**, 265.
4. Y. S. Lebedev, In: *Modern Pulsed and Continuous-Wave Electron Spin Resonance*, L. Kevan, M. K. Bowman, Eds., Wiley, New York, 1990, pp. 365.
5. J. Haase, M. B. Kozlov, A. G. Webb, B. Buchner, H. Eschrig, K. H. Müller and H. Siegel, *Solid State Nucl. Magn. Reson.*, 2005, **27**, 206.
6. H. J. Schneider-Muntau, B. L. Brandt, L. C. Brunel, T. A. Cross, A. S. Edison, A. G. Marshall and A. P. Reyes, *Phys. B*, 2004, **346**, 643.
7. M. Fuhs, PhD Thesis, Freie Universtät Berlin, 1999.
8. M. Fuhs, T. Prisner and K. Möbius, *J. Magn. Reson.*, 2001, **149**, 67.
9. K. A. Earle, B. Dzikovski, W. Hofbauer, J. K. Moscicki and J. H. Freed, *Magn. Reson. Chem.*, 2005, **43**, S256.
10. W. Hofbauer, K. A. Earle, C. R. Dunnam, J. K. Moscicki and J. H. Freed, *Rev. Sci. Instrum.*, 2004, **75**, 1194.
11. J. H. Freed, *EPR Newsletter*, 2007, **16**, 10.
12. G.M. Smith: 41st Annual International Meeting of the ESR Group of the Royal Society of Chemistry, 2008, London.
13. R. R. Ernst, G. Bodenhausen and A. Wokaun, *Principles of Nuclear Magnetic Resonance in One and Two Dimensions*, Clarendon Press, Oxford, 1987.
14. R. R. Ernst, *J. Magn. Reson.*, 1970, **3**, 10.
15. R. Kaiser, *J. Magn. Reson.*, 1970, **3**, 28.
16. B. Blümich and D. Ziessow, *J. Chem. Phys.*, 1983, **78**, 1059.
17. B. Blümich, *Prog. Nucl. Magn. Reson. Spec.*, 1987, **19**, 331.
18. D. K. Yang, J. E. Atkins, C. C. Lester and D. B. Zax, *Mol. Phys.*, 1998, **95**, 747.
19. T. Prisner and K. P. Dinse, *J. Magn. Reson.*, 1989, **84**, 296.
20. R. H. Pursley, J. Kakareka, G. Salem, N. Devasahayam, S. Subramanian, R. G. Tschudin, M. C. Krishna and T. J. Pohida, *J. Magn. Reson.*, 2003, **162**, 35.
21. S. Prakash Alia, P. Gast, H. J. M. de Groot, G. Jeschke and J. Matysik, *J. Am. Chem. Soc.*, 2005, **127**, 14290.

22. S. Prakash Alia, P. Gast, H. J. M. de Groot, J. Matysik and G. Jeschke, *J. Am. Chem. Soc.*, 2006, **128**, 12794.
23. E. Daviso, G. Jeschke and J. Matysik, In: *Biophysical Techniques in Photosynthesis*, Vol. II, T. J. Aartsma, J. Matysik Eds., Springer, Dordrecht, 2008, pp. 385.
24. K. N. Hu, C. Song, H. H. Yu, T. M. Swager and R. G. Griffin, *J. Chem. Phys.*, 2008, **128**, 052302.
25. T. Maly, G. T. Debelouchina, V. S. Bajaj, K. N. Hu, C. G. Joo, M. L. Mak-Jurkauskas, J. R. Sirigiri, P. C. A. van de Wel, J. Herzfeld, R. J. Temkin and R. G. Griffin, *J. Chem. Phys.*, 2008, **128**, 052211.
26. W. D. Markiewicz, I. R. Dixon and T. A. Painter, *Adv. Cryo. Eng. Mater.*, 2004, **50B**, 921.
27. F. Herlach, *Phys. B*, 2002, **319**, 321.
28. Y. Y. Lin, S. Ahn, N. Murali, W. Brey, C. R. Bowers and W. S. Warren, *Phys. Rev. Lett.*, 2000, **85**, 3732.
29. D. Shoenberg, *Nature*, 1952, **170**, 569.
30. G. S. Boebinger, A. H. Lacerda, H. J. Schneider-Muntau and N. Sullivan, *Phys. B*, 2001, **294**, 512.
31. J. Haase, F. Steglich, D. Eckert, D. Siegel, H. Eschrig and K. H. Müller, *Solid State Nucl. Magn. Reson.*, 2003, **23**, 263.
32. H. Krug, M. Doerr, D. Eckert, H. Eschrig, F. Fischer, P. Fulde, R. Groessinger, A. Handstein, F. Herlach, D. Hinz, R. Kratz, M. Loewenhaupt, K. H. Müller, F. Pobell, L. Schultz and H. Siegel, *Phys. B*, 2001, **294**, 605.
33. J. Haase, D. Eckert, H. Siegel, H. Eschrig, K. H. Müller, A. Simon and F. Steglich, *Phys. B*, 2004, **346**, 514.
34. J. Haase, *Appl. Magn. Reson.*, 2004, **27**, 297.

Subject Index

Page references in *italic* indicate figures. Page reference in **bold** indicate tables.

6-4 photolyase 322–323

ab-initio molecular modeling 197
ACERT (National Biomedical Center for Advanced ESR Technology) 2
adenosine triphosphate (ATP) 211
 synthesis in bacteria 272
Adrian quasiadabiatic model 57
α-protons, hyperfine splitting 41
aluminium, ENDOR 42
Anacystis nidulans 316
anisotropy, Zeeman interactions, spin-1/2 entities 26
antenna systems, photosynthetic 210
applications 31
 ENDOR 78–79
archaea 216–217, 272
ATP (adenosine triphosphate) 211, 272
azurin 201

B880 unit 214
bacteria
 photosynthetic 213–217
 photosynthetic reaction centers, EPR experiments 217–219
bacteriochlorophyll 214
 oxidation 219
bacteriorhodopsin (BR) 80, 216–217
 conformational changes 281–283
 multifrequency EPR experiments 272–283

polarity gradient 275–281
proton-transfer channel, hydrophobic barrier 275–281
spin labelling 274–275
X-ray structural model *282*
bandwidth, heterodyne diode mixer systems 142
base-paired complexes 308–314
basis sets, density functional theory (DFT) 203
Bell Telephone Laboratories 133
p-benzoquinone 63
Berlin,Free University, instruments 160–172
β-protons, hyperfine splitting 41
Bloch, M. 9
Blume, R.J 10
bolometers 140
boron, ENDOR 42
BR *see* bacteriorhodopsin
Bruker Biospin 15

C-domain, colicin A 326–327
cavities 146–148, *146*
 FU Berlin 95GHz instrument 166–169, *168*, **169**
 liquid-phase ENDOR 42
CCRP (spin-correlated coupled radical pair mechanism) 58–60
 dipolar coupling frequencies in frozen solution 252–254

Subject Index

in photosynthetic bacterial reaction
   centers 250–262
quinone-acceptor 265–267
PELDOR spectra *267*
spin Hamiltonians 59
Cederquist, A.L. 11
cell killing 324
charge-recombination rate constants,
   quinone-acceptor CCRPs 265–266,
   266
Chelyabinsk 31
chemically induced dynamic electron
   polarization *see* CIDEP
Chernobyl disaster 31, 32
Chlamydomas reinhardtdii 288
chlorine, ENDOR 42
chlorophyll ion radicals 29
   ENDOR 52–53
chloroplast 212
cholera toxin 324
CIDEP (chemically induced dynamic
   electron polarization) effects
   multiplet-effect polarization 62–63
   net-effect polarization 62, 63
   photochemistry
      radical pair mechanism 55–60
      reaction intermediates 65–67
      triplet mechanism 54–55
   spin-correlated radical pair
      mechanism (CCRP) 58–60
colicin A 207, 323–330
   C-domain 326–327
   cell-killing mechanisms 324
   transmembrane ion-channel
      formation 326–327
computational chemistry
   density functional theory (DFT)
      software 202–203
   state of art 195–197
concanavalin A 78
conformational changes,
   bacteriorhodopsin 282–283
continuous-wave EPR, instruments
   125–126
cooling systems 164
CPD lesions 314–319, *317*

cross effect (CE) 361
cryostat, FU Berlin 95Ghz instrument
   163–164
cyclic voltammetry 292
cyclobutane pyrimidine dimer (CPD)
   lesions 314–319, *317*

DEER *see* PELDOR
delocalization 269
density functional theory (DFT) 40
   state of art 195–197
   basis sets 203
   computer programs 202–203
   disadvantages 197–198
   g-tensor calculations 199, 200–201
   for bacterial photosynthetic
      reaction centers 233–234
   hyperfine couplings 201–202
   origins 197
deuterated samples, high-field
   condition 29
deuterium 77
Deutsche Forschungsgemeinschaft
   (DFG) 1–2
DFT *see* density functional theory
dibenzylketone 63
1,2-dimyristoyl-*sn*-glycero-3-phospho-
   rac-(1-glycerol) (DMPG) 327
dinitrobenzene-porphyrin complexes
   308–314
Dinse, K.P. 12
dipolar vector, CCRPs *260*
diptheria toxin 324
distance measurements 13
DMPG (1,2-dimyristoyl-*sn*-glycero-
   3-phospho-*rac*-(1-glycerol)) 327
DNA photolyases 314–319, *317*
   high-field EPR and ENDOR
      experiments 318–323
donor pigments 210
donor-acceptor model system ABC
   303–304
dosimetry, tooth-enamel 31
double electron-electron resonance
   (DEER) *see* PELDOR
double ENDOR 12

double quantum coherence (DQC) 13
doublet states
  ENDOR resolution *37*
  photosystem I 284–287
  photosystem II 288–290
  see also spin-1/2 entities
dynamic nuclear polarization (DNP) 360–361

EIKA (extended interaction klystron amplifier) 145
Einstein-Stokes-Debye relation, porphyrin-quinone dyads and triads 300–301
electron
  magnetic moment 8
  spin, spin-1/2 23–32
electron acceptors 210
  quinones *see* quinones
electron spin-echo envelope modulation *see* ESEEM
electron transfer chain (ETC), oxygenic photosynthesis 287
electron transfer (ET) 3, 207–209
  in biomimetic model systems, overview 290–294
  in DNA repair processes 318–319
  in photosynthesis 210, 211–213
    nonoxygenic 213–216
    oxygenic 287
electron-electron dipolar interaction and electron-nuclear dipolar interaction 94–95
  Hamiltonian 94
electron-electron dipolar spectroscopy
  overview 93–97
  PELDOR 97–100
  RIDME 100–101
  see also PELDOR
electron-nuclear dipole (END)
  interactions 25–26, 44
  Hamiltonian 25
electron-spin polarization 353
electron-spin-echo (ESE), spin-correlated radical pairs 248–250

ENDOR (electron-nuclear double resonance) 11
  advantages 353
  applications 78–79
  CIDEP-enhanced 61–64
  compared to EPR 11–12
  DNA repair photolyases 319–323
  double 12
  FU Berlin 95GHz instrument 169–170
  FU Berlin 360-GHz instrument 178–179, 179–180
  hyperfine splitting 74
  liquid crystals 50–51
  liquid-state 37–41
    spectral lineshape changes, due to field coherence effects 42–46
    overview 35–36
  photosynthetic intermediates, X-band 219–223
  pulse sequences 74
  resolution
    doublet states *37*
    vs. EPR 36
    vs. TRIPLE 47
  sensitivity enhancement using TRIPLE 75–76
  signal intensity 42–44, 74–75
  solid-state
    pulsed
      for systems with quadrupole interaction 76–79
      for systems without quadrupole interaction 73–76
    spectral analysis 77
    spectral resolution, photosynthetic intermediates 222
energy absorption, near EPR resonance 137
EPR
  95GHz, colicin A membrane-insertion mechanisms 327–330
  compared to ENDOR 11–12
  DNA repair photolyases 319–323
  FU Berlin 95GHz instrument 169–170
  high-field advantages 353
  history 9–17

Subject Index

multifrequency, on doublet states in photosystem II 288–290
photosynthetic intermediates, X-band 219–223
pulse vs. continuous-wave 8–9
pulsed 129–130, 353
  cavities 168
  deadtime 130
  reaction centers
    mutant, donor cations 223–234
    spectral resolution 222
    triplet states, in photosynthetic bacterial reaction centers 268–272
Escherichia coli 79, 316, 323
  DNA photolyase, g-tensor 320
ESEEM (electron spin-echo envelope modulation) 12, 15–16, 79–93, 80
  polarity effects 92–93
  spin labelling 80–81
ET *see* electron transfer
ETC (electron transfer chain) 287
Europe, high-field EPR groups *3*
extended interaction klystron amplifier (EIKA) 145

F-band, definition *7*
Fabry-Perot resonator 134, *146*, 148–152
  field amplitude plots *150*
  Fourier transform EPR 358
  FU Berlin 95 GHz instrument 164–166, *167*
  modification for PS I experiments 286
FAD *see* flavin-adenine dinucleotide
FAD chromophore 314–323
FADH 320
  6-4 photolyase 322–323
  g-tensor principal axes 322
  Xenopus laevis 321
far-infrared (FIR) lasers 145
Fermi contact interaction 38–39
  modeling 201
flavin radicals, nonresonant probehead 155
flavin-adenine dinucleotide, *see also* FADH
flavin-adenine dinucleotide (FAD)
  chromophore 314–323, 315–316
  cofactor photoactivation 318–319

fluorescence energy transfer (FRET) 94
Fourier transform (FT) spectroscopy 165–166
  high-field EPR 356–358
  NMR 9–10
Franck-Condon factor 207–208, 218–219, 228
free induction decay (FID) 129–130, 183–184
high-field proton NMR 363
FT *see* Fourier transform
FU Berlin, EPR instrumentation 160–172

g-tensor 29–30
  bacterial RCs
    mutant 232–234
    orientation 224, *225*
    triplet states 270–271
  bacteriorhodopsin, hyperfine dependence and H-bonding 279–280
  density functional theory (DFT) 199, 200–201
  FADH 320
  modeling 200–201
  photosystem I 285
  quinones *243*
  transient radical pairs 248–250
  triplet states 270–271
general TRIPLE 47–50

half-integer spin systems 34
halobacteria 216–217, 272
Halobacterium salinarium 216–217
Hamiltonian
  CCRP (spin-correlated coupled radical pair) 59
  electron-electron dipolar interaction 94–95
  organic radicals and low-spin metal-transition ions 23–32
  organic radicals and low-spin transition-metal ions 23–32
  spin-1/2 entities, quadrupole interaction 24
  triplet states and high-spin metal ions 23–32, 33–35
harmonic multipliers 143–144

HEMT (hot-electron mobility
  transistors) 141
heterodyne systems, phase noise 142
high-field conditions 29
  photosystem II 289
  quinone-acceptor spin-correlated
    radical pairs 251–252
high-field EPR
  advantages 3–7
  development 13–17
histidine His(M202) 229
HOMO (highest unoccupied molecular
  orbital), triplet states 269
hot-electron mobility transistors
  (HEMT) 141
Hückel π-electron theory 39–40
Hyde, J.S. 11
hydrogen, magnetic moment ratio 8
hydrogen bonding
  effect on g-tensor vs hyperfine tensor
    dependence, bacteriorhodopsin
    279–280
  and quinones 247–248
hyperfine coupling
  constants, and molecular structure
    39–41
  EPR and ENDOR, photosynthesis
    222
  measurement 38–39
  modeling with DFT 201–202
  nitroxide radicals 89
  spin-1/2 entities 24–25
  tensor, DNA photolyases 319–320
  TRIPLE 76
hyperfine spectroscopy *see* individual
  spectroscopic methods (ENDOR,
  TRIPLE, ESEEM)
hyperfine splitting
  6–4 photolyase 322–323
  β-protons 41
  ENDOR 74

Institute of Chemical Physics, Moscow
  134
instrumentation
  general requirements 124
  bolometers 140

commercial development 135
diode detectors 141
history 131–133
  first-generation instruments 133–134
  second-generation instruments
    134–135
laboratory-built instruments **136**
magnets 158–160
multipurpose
  FU Berlin
    95-GHz 160–172
    360-GHz 172–186
  preamplifiers 141–142

Japan, EPR research facilities 2

Karplus-Fraenkel relation 40–41
klystrons 144

Larmor frequency 36–37
  CCRP 59–60
  quinones, photosynthetic bacteria 242
lasers, as microwave source 145
Lebedev, Yakovs 13–14, 134
LGR (loop-gap resonator) 152–153
light-harvesting (LH1 and LH2)
  complexes 214, *215*
Liouville formalism 57
liquid crystals
  and biomimetic electron transfer
    rates 293–294
  ENDOR 50–51
liquid-phase ENDOR 11
local oscillator (LO) 142
loop-gap resonators (LGR) 152–153
lowest unoccupied molecular orbital
  (LOMO), triplet states 269

McConnell equation 39–40, 40
Mg-tetraphenylporphyrin 307–308
magnet systems 158–160
  FU Berlin 95 GHz instrument 163–164
  FU Berlin 360 GHz instrument 178
  small-coil 362–363
magnetic dipole, modeling 201
magnetic field strength, and quadrupole
  interaction measurement, proteins 81

Subject Index

magnetic moment 8
magnetic spin quantum numbers,
  spin-1/2 radicals 26
Maki, A.S. 11
manganese ions, as reference sample 34
Mn_4uO_xCa complex 289
Marcus electron-transfer theory 207–209,
  293
metmyoglobin 33
microwave bridge
  FU Berlin 95GHz instrument 162–163,
    162
  FU Berlin 360-GHz instrument
    174–176
microwave detectors 140
microwave excitation, stochastic
  broadband 357–358
microwave frequency
  and orientational selectivity 4, *6*
  and sensitivity 4, 5–6
  and spectral resolution 4
  see also high-field conditions
microwave sources 143–145
  FU Berlin 360-GHz instrument
    174–176, 182–186
  Orotron 15, 145
microwave transmission lines 156–158
  quasioptical 172–174
Mims, W.B. 11, 12
Mock, J.B. 133
molecular dynamics (MD) 195
  quinones 241–247
molecular modeling
  and hyperfine structure 39–41
  see also density functional theory
molten-globule membrane insertion,
  colicin A 325
monolithic microwave integration
  circuits (MMIC) 141
MTS (1-oxyl-2,2,2,5,5,-
  tetramethylpyrroline-3-methyl)
  spin label 80

National Biomedical EPR Center 2
National High Magnetic Field
  Laboratory (NHMFL) 2, 362

nitrogen 76
  magnetic moment ratio 8
nitroxide radicals 355
  B1, PELDOR spectra *108*
  dipolar hyperfine coupling, and
    temperature 89
  ESEEM
    effect of solvent 90–91
    quadrupole interactions 82–89
    decay modulations 84–85
  magnetic interactions **83**, 107–108
  PELDOR 106–112
  quadrupole interaction 81
  RIDME 102–106
  spin labelling of bacteriorhodopsin
    274–275
  structural determination, RIDME
    and PELDOR 110–111
NMR
  compared with EPR 7–9
  pulsed-magnet proton 361–363
noise
  cw EPR 125
  phase noise 142
  and premaplifiers 141–142
  solid-state microwave sources 143
  TREPR 127–128
nonresonant systems 154–155
nuclear accidents 31–32
nuclear magnetic resonance (NMR)
  spectroscopy 7–9, 361–363

orientational selectivity
  and field frequency 4
  PELDOR 111
Orotron 15, 145
  pulsed 182–186
oxygen, ENDOR 42
oxygen-evolving complex (OEC) 289–290

Pake pattern 251, *253*
PELDOR 10, 13, 16, 355
  95GHz, spin-correlated radical pairs
    in bacterial photosynthetic
    reaction centers 250–262
  CCRP, quinone-acceptor 266–267

deadtime 99
development 96–97
electron-electron dipolar interaction 94
field-jump 170–172
FU Berlin 95GHz instrument 170–172
orientational selectivity 99–100, 111
oxygen-evolving complex (OEC) 289
pulse schemes 97–99, *97*
  on CCRPs generated by pulsed laser *258*
  nitroxide biradicals *107*
  SSE 99
spin-labelling requirements 112
temperature 75
perelenyl-ion 186
photo-CIDNP 358–360
photochemistry
  cavities 168–169
  definition 3
  reaction intermediates 61, 64–73
    DMPA (2,2-dimethoxy-1,2-diphenylethan-1-one) 65–67
    TMDPO(2,4,6,-trimethylbenzoyl-diphenyl-phosphineoxide) 67–73
  transient EPR and ENDOR 53–54
photosynthesis 210–213
  definition 3
  bacterial, quinone acceptors 222–223
  nonoxygenic, electron transfer 213–216
  oxygenic 283–290
    net reaction 210–211
    photosystems I & II 210–211
    proton transfer 216–217
photosynthetic reaction centers *see* reaction centers
photosystem I 284–287
photosystem II 288–290
pigments, photosynthetic 210
porphyrin-quinone complexes
  base-paired 309, *310*
  dyads and triads 294–308
    base-paired 308–314
    'bulky' 298–300
    molecular dynamics 301–308
    nonbulky 297

porphyrinoid ions 52–53
power loss, due to microwave transmission 156–157
photosynthesis, nonoxygenic, electron transfer 213–216
preamplifiers 141–142
probeheads *see* resonators
proteins
  bacterial photosynthetic reaction centers *see* reaction centers
  light-harvesting 214–215
  metalated 201
proton-transfer channel, bacteriorhodopsin 275–281
protonated samples, high-field condition 29
protons 41
publication rates *2*
pulse ELDOR *see* PELDOR
pulse EPR 129–130, 353
  advantages over TREPR 129–130
  cavities 168
  deadtime 130
  development 10–11
  Fourier transform spectroscopy 357
  and pulse NMR 8
  sensitivity 138
pulse high-field electron dipolar spectroscopy (PELDOR) 14, 16
3-pulse electron-electron double resonance (PELDOR) 13
pulse electron-electron double resonance (PELDOR) 10
Purcell, E.M. 9
purple bacteria 213–217
3-pyridylphenyl ketone anion radical 47, *48*

quadrupole interaction
  ENDOR in liquid crystals 50–51
  nitroxide radicals 82–89
  nitroxide spin labels 81
  solid-state pulse ENDOR for systems exhibiting 76–79
  spin-1/2 entities, Hamiltonians 24
  and Zeeman field value 81

Subject Index

quasioptical microwave propagation 172–174
quinones
  covalently linked porphyrin diads and triads 294–308
  EPR and ENDOR experiments 234–248
  in hydrogen-bond networks 247–248
  in spin-correlated radical pairs
    ESE-detected EPR experiments 248–250
    RIDME and PELDOR experiments 250–268
    spectra for individual radicals 251–252, *253*
  ubiquinone-10 241

radiation dosimetry 31–32
radical pair mechanism (RPM) 55–60
radical pairs *see* CCRP
radicals
  nitroxide *see* nitroxide radicals
  organic, spin Hamiltonians 23–32
  spin-1/2 26–28
  TEMPO (2,2,6,6,-tetramethyl-piperidine-1-oxyl) 361
  tyrosil 288
radiolysis, reaction intermediates 61
reaction centers 217–219
  CCRP mechanism 250–262
  EPR and ENDOR experiments 233–234
    quinones 234–248
    X-band 219–223
  primary donor cations 223–228
  g-tensor calculations 232–234
  mutant RCs 228–231
  triplet states 268–272
reaction intermediates 61–64
reactivity, RPM CIDEP radical pairs 56
resolution
  bio-organic samples in frozen solution 28
  ENDOR, doublet-state systems *37*
  ENDOR vs. EPR 36
  ENDOR vs. TRIPLE 47

resonators 145–155
  dielectric 153–154
  Fabry-Perot 148–152
  FU Berlin 360-GHz instrument 178–179
  single-mode cavities 146–148
  see also nonresonant systems
Rhodobacter sphaeroides 29, 213, 221, 222, 223, 269, 359–360
  comparison of radical pairs with PS I 284–286
  mutant reaction centers 228–231
  transient radical pairs 248–250
Rhodopseudomonas acidophila 214
Rhodopseudomonas palustris 214
Rhodopseudomonas viridis 213–214, 269
rhodopsins 217
RIDME (relaxation-induced dipolar modulation enhancement) 13
  95GHz, spin-correlated radical pairs in bacterial photosynthetic reaction centers 250–264
  advantages over PELDOR 101
  pulse sequence, nitroxide biradicals *103*
  spin-labelling requirements 112
rotation patterns 27–28

Salikhov, K.M. 10
sample holders
  nonresonant 154–155
  resonant *see* resonators
samples, biological, mounting in thin films 165
Schreckenbach-Ziegler (SZ) method 196–197
SDSL (site-directed spin labeling) 80, 274–275
Sellafield 31
sensitivity
  absolute and relative 30
  EPR 125, 137–138
  and mw transmission losses 156
  and resonator choice 148, *149*
silicon, detection via ENDOR 42

single crystals, spin-1/2 radicals 27–28
singlet-triplet spin polarization mechanism 300
singlet-triplet (ST) mixing 61–64
singly occupied molecular orbital (SOMO), and g-tensor 227–228
site-directed spin-labeling (SDSL) 80, 274–275
software, density functional theory (DFT) 202–203
solid-state microwave sources 143
solvents
  biomimetic long-range electron transfer 293–294
  interaction with bacteriorhodopsin 278–281
special TRIPLE 12, 46–47
  ENDOR sensitivity enhancement 75–76
spectral analysis, ENDOR 77
spectral resolution, and mw frequency 4
spin energy levels, spin-1/2 radicals 26–27, *27*
spin Hamiltonian *see* Hamiltonian
spin interaction energies, spin-1/2 entities 24
spin labeling
  bacteriorhodopsin (BR) 274–275
  ESEEM 80–81, 274–275
spin labels
  MTS 80
  nitroxide *see* nitroxide radicals
  TOAC 111–112
spin polarization 299–300
  and mw frequency 4–5
spin relaxation 299–300
  organic radicals 44
  porphyrin-quinone dyads and triads 300–301
spin-1/2 entities
  ESEEM modulation-depth factor 86–87
  spin Hamiltonians 23–32
spin-1 entities, zero-field parameter by thermal polarization of triplet levels 35
spin-orbit coupling 299

spin-rotational (SR) interactions 44
stochastic Liouville equation (SLE) 73
stochastic resonance 357
subharmonic mixer 175–176
sulfur, detection via ENDOR 42
superconducting magnets 159–160
symmetry, single-crystal bacterial photosynthetic reaction centers 224–227, *225*
Synechococcus lividus, photosystem I 249–250
SZ method 196–197

temperature
  PELDOR 75
  and quadrupole splitting, nitroxde radicals 89
TEMPO (2,2,6,6,-tetramethyl-piperidine-1-oxyl), radicals 361
tetanus toxin 324
thermal mixing (TM) 361
Thermosynechococcus elongatus 212–213
Thermus thermophilus 316
thylakoid phospholipid membrane *212*
thymine 314
TOAC spin label 111–112
Tokaimura 31–32
tooth enamel 31, 32
transfer lines 156–158
transient radicals 12–13, 61, 248–250
transition levels, liquid-phase ENDOR *38*, 42–43
transition topologies 47–49, *49*
transmission lines 156–158
TREPR (time-resolved transient EPR) 12–13, 126–129
  compared to pulse EPR 129–130
  electron transfer in biomimetic model systems 292
  exchange interaction 292
  high-field 15
  microwave detectors 140
  photosystem I 286
  porphyrin-quinone dyads and triads 294–308, *297*, 306–308
  pyrino-fullerenes 305–308
  ringing time 127

sensitivity 127–128
time resolution 126–127
waveguides 181–182
tribochemistry 3
TRIPLE 36–37, 75–76
  general 47–50
  special 12, 46–47
triplet states 299
  in bacterial RCs 268–272
  bacterial RCs, g-tensor 270–271
  electron delocalization 269
  singlet-triplet (ST) mixing 61–64
  spin Hamiltonians 33–35
  zero-field splitting 202
tryptophan residues 319
tyrosil radicals 288

unidirectionality 223
USA, research groups 2

vacuum permeability 25
varactor-diode harmonic multipliers 143–144

W-band, definition 7
Watson-Crick base-paired complexes 309, *310*
waveguides 156–158
  materials 176

quasioptical 172–174, 176
transient EPR bridge 181
whispering-gallery mode (WGM) resonators 154–155
Windscale fire 31

X-band, definition 7
X-band EPR and ENDOR, photosynthetic bacterial reaction centers 220–228
X-ray crystallography
  comparison to PELDOR results 261, **262**
  photosynthetic proteins 212–213
X-ray spectroscopy, resolution, photosystem I 284–287

Zavoisky, E.K. 9, *10*
Zeeman field
  magnet systems 158–160
  magnitude 29
  and sensitivity 30–31
  and spectral resolution 4–5, *5*
  and transient EPR signal detection 13
  see also high-field conditions
zero-field splitting (ZFS) 33–35
  modeling 202
zero-order regular approximation (ZORA) 196, 200